The Chemistry of Inorganic Homo- and Heterocycles

Volume 1

The Chemistry of Inorganic Homo- and Heterocycles

Volume 1

IONEL HAIDUC

Babes-Bolyai University, Cluj-Napoca, Romania

D. B. SOWERBY

University of Nottingham, United Kingdom

1987

ACADEMIC PRESS
Harcourt Brace Jovanovich, Publishers
London Orlando San Diego New York Austin
Boston Sydney Tokyo Toronto

CHEMISTRY

0.2641367

ACADEMIC PRESS INC. (LONDON) LTD.
24/28 Oval Road,
London NW1

United States Edition published by
ACADEMIC PRESS INC.
Orlando, Florida 32887

British Library Cataloguing in Publication Data

The Chemistry of inorganic homo- and
  heterocycles.
1. Cyclic compounds   2. Chemistry, Inorganic
I. Haiduc, Ionel   II. Sowerby, D. B.
546   QD197

ISBN 0-12-655775-6 v.1
ISBN 0-12-655776-4 v.2

Typset by Bath Typesetting Ltd.
Printed in Great Britain at T. J. Press, (Padstow) Ltd.

Contributors

Professor Christopher W. Allen Department of Chemistry, University of Vermont, Burlington, Vermont 05405, USA.

Professor M. Baudler Institut für Anorganische Chemie, Universität Köln, D-5000 Köln, Federal Republic of Germany.

Professor M. Cesari Eniricerche, San Donato Mil., 20097 Milan, Italy.

Dr T. Chivers Department of Chemistry, University of Calgary, Calgary, Alberta T2N 1N4, Canada.

Dr S. Cucinella Enichem Elastomeri, San Donato Mil., 20097 Milan, Italy.

Dr V. Chvalovský Institute of Chemical Process Fundamentals, Czechoslovak Academy of Science, Praha 6, Czechoslovakia.

Professor Martin Dräger Institut für Anorganische Chemie, Universität Mainz, D-6600 Mainz, Federal Republic of Germany.

Dr A. Durif Laboratoire de Cristallographie. CNRS-166 X, 38042 Grenoble Cedex, France.

Professor G. Fritz Institut für Anorganische Chemie, Universität Fridericiana Karlsruhe, Kaiserstrasse 12, D-7500 Karlsruhe, Federal Republic of Germany.

Dr K. Glinka Institut für Anorganische Chemie, Universität Köln, D-5000 Köln, Federal Republic of Germany.

Professor Ionel Haiduc Department of Chemistry, Babes-Bolyai University, 3400 Cluj-Napoca, Romania.

Dr J. Härer Institut für Anorganische Chemie, Universität Fridericiana Karlsruhe, Kaiserstrasse 12, D-7500 Karlsruhe, Federal Republic of Germany.

Dr P. G. Harrison Department of Chemistry, University of Nottingham, Nottingham NG7 2RD, UK.

Dr Karl Hassler Institut für Anorganische Chemie, Technische Universität, A-8010 Graz, Austria.

Professor Edwin Hengge Institut für Anorganische Chemie, Technische Universität, A-8010 Graz, Austria.

Dr Rodney Keat Department of Chemistry, University of Glasgow, Glasgow G12 8QQ, UK. Present address: Chemdata Ltd., Helston Cornwall TR13 0NQ, UK.

Dr U. Klingebiel Institut für Anorganische Chemie, Universität Göttingen, D-3400 Göttingen, Federal Republic of Germany.

Dr A. Maringgele Institut für Anorganische Chemie, Universität Göttingen, D-3400 Göttingen, Federal Republic of Germany.

Dr B. Mathiasch Institut für Anorganische Chemie, Universität Mainz, D-6500 Mainz, Federal Republic of Germany.

Professor Alfred Schmidpeter Institut für Anorganische Chemie, Universität München, D-8000 München, Federal Republic of Germany.

Professor W. Siebert Institut für Anorganische Chemie, Universität Heidelberg, D-6900 Heidelberg, Federal Republic of Germany.

Dr D. B. Sowerby Department of Chemistry, University of Nottingham, Nottingham NG7 2RD, UK.

Professor E.-M. Strauss Institut für Anorganische Chemie, Technische Universität, D-1000 Berlin, Germany.

Professor Ralf Streudel Institut für Anorganische Chemie, Technische Universität, D-1000 Berlin, Germany.

Professor M. Veith Anorganische Chemie, Universität des Saarlandes, D-6600 Saarbrücken 11, Federal Republic of Germany.

Preface

The chemistry of inorganic homo- and heterocycles is becoming a self-consistent branch of modern inorganic chemistry, and international symposia on inorganic ring systems have already taken place (IRIS Symposia: Besançon 1975, Madrid 1977, Göttingen 1978, Graz 1981, Paris 1985), thus recognizing the legitimacy of this science.

Although inorganic ring structures were correctly suggested for several compounds as early as the second half of the 19th century (for example for metaphosphates and cyclophosphazenes), it was only after diffraction methods for structure determination became available, that cyclic structures were confirmed for elements other than carbon. This was then followed by intensive research in the field.

The classical period of inorganic ring chemistry was dealt with in 1970 in a comprehensive two-volume monograph "The Chemistry of Inorganic Ring Systems" (by Ionel Haiduc), which covered the literature published up to 1969. More recent literature, published since 1979/1980 is reviewed in a series of annual surveys, initiated by the same author. Thus the review covering the year 1980 was published in *Reviews in Inorganic Chemistry* **5** (1983) 7–121, and will be followed by further annual reviews, now in preparation. This left a gap of about ten years (1969–1979) to be covered by a comprehensive monograph, which would collect the information published during a decade of research and progress. This book is intended to fill this gap, and also to bring in some more recent results. Because of the rapid development in the chemistry of inorganic ring systems and the accumulation of a large body of information, most of it of great significance, the task of reviewing the literature of a whole decade could not be performed by a single person.

The present book was written by a team of internationally recognized experts in the field, who have actively contributed to its development. Their first-hand knowledge and expertise in the field was invaluable in preparing an authoritative survey of inorganic homo- and heterocyclic chemistry. Some short chapters were written by the editors themselves.

A difficult task was that of finding acceptable nomenclature. Since no

IUPAC rules for naming inorganic rings are available, the editors suggested to all contributors that they use the same nomenclature as in Haiduc's 1970 monograph. Some, however, preferred other naming procedures, and the editors felt that they should not impose their own preferences. The nomenclature is briefly explained in the Introduction.

Readers will be aware of the substantial editorial difficulties in producing a work of this complexity, not least of which are those concerned with meeting deadlines and operating within a given overall length. As a consequence of the latter, most articles were subject to editorial revision.

We are grateful to the many contributors to this book for their participation in this endeavour, for their patience and understanding of the somewhat difficult circumstances and long delay in assembling such a large work with so many participants. We acknowledge assistance from the British Council and the Romanian Ministry of Education and are indebted to our families for their understanding.

Ionel Haiduc
D. Bryan Sowerby

Contents of Volume I

Contributors   v

Preface   vii

Contents of Volume II   xvii

Introduction   1

IONEL HAIDUC and D. B. SOWERBY

1   What is an inorganic ring?   1
2   Classification of inorganic rings   5
3   Nomenclature of inorganic rings   9
    Bibliography   13
    References   14

1   Boron Homocycles (Cyclopolyboranes)   15

IONEL HAIDUC

    References   16

2   Boron–Nitrogen Heterocycles   17

W. MARINGGELE

2.1   General remarks   17
2.2   Nomenclature   17
2.3   Borazines $(RBNR')_3$ 19
      2.3.1   Unsubstituted borazine $(HBNH)_3$ 19
      2.3.2   Inorganic derivatives of borazine   19
      2.3.3   Functional derivatives of borazine   21
      2.3.4   1,3,5-Triorgano-2,4,6-triorganoborazines   25
      2.3.5   Miscellaneous borazine derivatives   27
      2.3.6   Polycyclic systems containing one borazine ring   28

2.3.7   Macrocyclic boron-nitrogen ring systems   30
2.3.8   Reactions of the borazines   32
2.3.9   Metal derivatives of borazines   37
2.3.10 Spectroscopic and other physical properties   39
2.3.11 Potential commercial uses   49
2.4 Heterocycles containing three-coordinate boron atoms   50
2.4.1   Borazocines $(RBNR')_4$ 50
2.4.2   1,3,2,4-Diazadiboretidines $(RBNR')_2$ 51
2.4.3 Δ^2-Tetrazaboroline   53
2.4.4   1,2,4,5-Tetraza-3,6-diborine   55
2.4.5   1,2,4-Triaza-3,5-diborolidine   57
2.4.6   Boron–nitrogen–silicon heterocycles   60
2.4.7   Boron–nitrogen–phosphorus heterocycles   62
2.4.8   Boron–nitrogen–arsenic heterocycles   66
2.4.9   Boron–nitrogen–oxygen heterocycles   67
2.4.10 Boron–nitrogen–oxygen–silicon heterocycles   68
2.4.11 Boron–nitrogen–sulphur heterocycles   69
2.4.12 Boron–nitrogen–selenium heterocycles   72
2.4.13 Fused-ring systems containing boron, nitrogen and a   72
        heteroatom
2.5 Boron–nitrogen coordination heterocycles. Systems containing
     four-coordinate boron   74
     2.5.1   Four-membered P_2N_2 rings   74
     2.5.2   Six-membered B_3N_3 rings (trimeric aminoboranes)   80
     2.5.3   Dimeric and tetrameric hydrazinoboranes   81
     2.5.4   Pyrazaboles   83
     2.5.5   Other ring systems, containing only boron and nitrogen 84
     2.5.6   Ring systems with boron, oxygen and nitrogen   85
     2.5.7   Boron–nitrogen–phosphorus rings   87
     2.5.8   Boron–nitrogen–sulphur ring systems   87
     2.5.9   Boron–nitrogen–aluminium ring systems   88
     Bibliography   89
     References   90

3 Boron–Phosphorus and Boron–Arsenic Heterocycles   103

D. B. SOWERBY

3.1 Introduction   103
3.2 Cyclodiphosphinoborines $(R_2BPR'_2)_2$ and $(RBPR')_2$ 104
3.3 Cyclotriphosphinoborines $(R_2BPR'_2)_3$ 105
3.4 Cyclotetra- and condensed phosphinoborines   107
3.5 Cycloarsinoborines   107
     References   108

4 Boron–Oxygen Heterocycles   109

IONEL HAIDUC

4.1 Introduction   109
4.2 Cyclodiboroxanes   110
4.3 Cyclodiboratrioxanes   110
4.4 Cyclotriboroxanes   111
   4.4.1 Inorganic derivatives   112
   4.4.2 Cyclic borates   114
   4.4.3 Organic derivatives $(RBO)_3$ 122
   4.4.4 Functional derivatives   124
   4.4.5 Physical properties and structure   124
   4.4.6 Chemical properties   126
   4.4.7 Uses   128
4.5 Cyclodiboratetroxanes   129
   Bibliography   130
   References   131

5 Boron–Sulphur and Boron–Selenium Heterocycles   143

W. SIEBERT

5.1 Introduction   143
5.2 Rings with three-coordinate boron   144
   5.2.1 The B_8S_{16} molecule   144
   5.2.2 1,3,2,4-Dithiadiboretanes $(RBS)_2$ 145
   5.2.3 1,3,5,2,4,6-Trithiatriborinanes $(RBS)_3$ 146
   5.2.4 1,2,4,3,5-Trithiadiborolanes   149
   5.2.5 Tetrathiadiborinanes   153
   5.2.6 1,3,2,4,5-Dithiatriborolane   155
   5.2.7 1,4,2,3,5,6-Dithiatetraborinane   156
   5.2.8 Thiasilaborolanes   157
5.3 Rings with four-coordinate boron   157
   5.3.1 Intermolecular donar–acceptor compounds   157
   5.3.2 Methylthiodiborane and alkylthioboranes   158
   5.3.3 Alkylthiodihalogenoboranes and alkylthio(halogeno)-
      alkylboranes   159
5.4 Metal thioborates   160
5.5 Boron–selenium rings with three-coordinate boron   161
   5.5.1 1,3,5,2,4,6-Triselenatriborinane   161
   5.5.2 1,2,4,3,5-Triselenadiborolane   161
   5.5.3 1,3,4,2,5-Selenadiazadiborolidine and 1,2,4,3,5-
      Diselenaazadiborolidine   162
5.6 Boron–selenium rings with four-coordinate boron   163
   5.6.1 Dimeric and trimeric alkylselenodihalogenoboranes
      $(RSeBX_2)_{2,3}$ 163

5.6.2 μ-Methylselenadiborane 163
Bibliography 163
References 164

6 Aluminium–Nitrogen Rings and Cages 167
M. CESARI and S. CUCINELLA

6.1 Introduction 167
6.2 Synthesis 167
  6.2.1 Aluminium–nitrogen ring compounds 167
  6.2.2 Aluminium–nitrogen cage compounds 171
6.3 Structure and bonding 178
6.4 Spectra 183
6.5 Chemical properties and reactions 184
  6.5.1 Chemical stability 184
  6.5.2 Reactions 184
6.6 Uses 185
Bibliography 187
References 187

7 Silicon Homocycles (Cyclopolysilanes) and Related Heterocycles 191
EDWIN HENGGE and KARL HASSLER

7.1 Homocyclic silanes 191
  7.1.1 Cyclotrisilanes 191
  7.1.2 Cyclotetrasilanes 192
  7.1.3 Cyclopentasilanes 197
  7.1.4 Cyclohexasilanes 202
  7.1.5 Cyclosilanes $(SiR_2)_n$ with $n > 6$ 206
  7.1.6 Polycyclic systems 206
  7.1.7 Radical anions and cations of cyclosilanes 208
7.2 Heterocyclic silanes 209
  7.2.1 Heterocyclic silanes containing Group III elements 209
  7.2.2 Heterocyclic silanes containing Group IV elements 209
  7.2.3 Heterocyclic silanes containing Group V elements 211
  7.2.4 Heterocyclic silanes containing Group VI elements 213
Bibliography 216
References 216

8 Silicon–Nitrogen Heterocycles 221
U. KLINGEBIEL

8.1 Introduction 221
8.2 Cyclosilazanes 224
  8.2.1 Cyclodisilazanes 224
  8.2.2 Cyclotri- and cyclotetrasilazanes 228
  8.2.3 Polycyclic silazanes 230

8.2.4   Structure and bonding   233
8.2.5   Spectra   236
8.2.6   Other physical properties   236
8.2.7   Chemical properties and reactions   237
8.2.8   Uses   244
8.3   Silicon–nitrogen heterocycles with N–N bonds   245
    8.3.1   Cyclosiladiazanes   245
    8.3.2   Cyclodisilatriazanes   245
    8.3.3   Cyclodisilatetrazanes   246
    8.3.4   Cyclotrisilatetrazanes   247
    8.3.5   Cyclotetrasilatetrazanes   247
8.4   Silicon–nitrogen heterocycles with an Si–Si bond   247
    8.4.1   Cyclotrisiladiazanes   247
    8.4.2   Cyclotetrasilatriazanes   248
    8.4.3   Cyclopentasilatetrazanes   249
8.5   Heterocyclosilazanes containing Group IV elements   249
    8.5.1   Germanium-containing heterocyclosilazanes   249
    8.5.2   Tin-containing heterocyclosilazanes   251
    8.5.3   Lead-containing heterocyclosilazanes   252
8.6   Heterocyclosilazanes containing Group V elements   252
    8.6.1   Phosphorus-containing heterocyclosilazanes   253
    8.6.2   Arsenic- and antimony-containing heterocyclosilazanes   255
8.7   Cyclosilazoxanes   257
    8.7.1   Cyclodisilazoxanes   257
    8.7.2   Cyclotrisiladiazoxanes   257
    8.7.3   Cyclotrisilazadioxanes   259
    8.7.4   Cyclotetrasilazoxanes   259
    8.7.5   Cyclopentasiladiazatrioxanes   261
    8.7.6   Cyclohexasiladiazatetroxanes   261
    8.7.7   Rearrangements of cyclosilazoxanes to rings of other sizes 261
    8.7.8   Cyclosilazoxanes with N–N ring units   263
    8.7.9   Cyclosilazoxanes with Si–Si ring units   264
    8.7.10 Cyclosilazoxanes with N–O ring units   264
    8.7.11 Uses   264
8.8   Silicon–nitrogen–sulphur heterocycles   264
    8.8.1   Cyclodisilathia(VI)triazanes and related systems   265
    8.8.2   Cyclosilathia(II)azanes   265
    8.8.3   Cyclosilthiazanes with Si–Si ring units   266
8.9   Heterocyclosilazoxanes   266
    8.9.1   Boron-containing cyclosilazoxanes   268
    8.9.2   Germanium-containing cyclosilazoxanes   268
    8.9.3   Tin-containing cyclosilazoxanes   268
    8.9.4   Phosphorus- and arsenic-containing cyclosilazoxanes   268
    8.9.5   Sulphur-containing cyclosilazoxanes   269
    Bibliography   269
    References   269

9 Silicon–Phosphorus Heterocycles   277

G. FRITZ and J. HÄRER

9.1 Introduction   277
9.2 Cyclophosphapolysilanes   278
9.3 Cyclosilaphosphanes   278
9.4 Silaphosphanes with cage structures   282
9.5 Chemical behaviour of cyclosilaphosphanes   283
     Bibliography   285
     References   286

10 Silicon–Oxygen Heterocycles (Cyclosiloxanes)   287

V. CHVALOVSKÝ

10.1 Introduction   287
10.2 Preparation   289
      10.2.1 Hydrolysis   289
      10.2.2 Heterofunctional condensation   290
      10.2.3 Depolymerization   290
10.3 Chemical properties and reactions   291
      10.3.1 Polymerization   291
      10.3.2 Reactions on substituents of organocyclosiloxanes   293
10.4 Silsesquioxanes   293
10.5 Heterocyclosiloxanes   294
      Bibliography   320
      References   320

11 Silicon–Sulphur Heterocycles   349

IONEL HAIDUC

11.1 Introduction   349
11.2 Cyclosilathianes   350
      11.2.1 Preparation   350
      11.2.2 Structure   352
      11.2.3 Chemical reactions   353
      11.2.4 Uses   354
11.3 Cyclo-1,2,4-trisila-3,5-dithianes   355
11.4 Cycloborasilathianes   355
11.5 Cyclosilathioxanes   356
      Bibliography   358
      References   358

12　Germanium Homocycles (Cyclopolygermanes) and Related Heterocycles　　361

IONEL HAIDUC and MARTIN DRÄGER

12.1　Cyclopolygermanes　362
12.2　Heterocyclogermanes　364
　　　Bibliography　364
　　　References　365

13　Germanium-Containing Heterocycles　　367

IONEL HAIDUC

13.1　Introduction　367
13.2　Germanium–nitrogen heterocycles　368
13.3　Germanium–phosphorus heterocycles　370
13.4　Germanium–oxygen heterocycles　371
13.5　Germanium–sulphur heterocycles　372
13.6　Germanium–selenium heterocycles　373
13.7　Germanium–tellurium heterocycles　374
　　　Bibliography　374
　　　References　374

14　Cyclostannanes　　377

P. G. HARRISON

14.1　Syntheses　377
14.2　Structural data　378
14.3　Spectra　380
14.4　Chemical reactions　380
　　　References　381

15　Tin–Nitrogen and Tin–Phosphorus Heterocycles　　383

M. VEITH

15.1　Cyclostannazanes　383
　15.1.1　Preparation and structure　383
　15.1.2　Reactions of cyclostannazanes　386
15.2　Cyclic silastannazanes and other heterocyclostannazanes　388
15.3　Polycyclic Sn–N compounds and related molecules　391
15.4　Cyclostannaphosphanes　395
　15.4.1　Cyclodistannadiphosphanes and cyclotristannatriphosphanes　395
　15.4.2　Phosphorus-rich and tin-rich phosphorus-tin cycles
　　　　　and polycycles　397
　　　Bibliography　399
　　　References　399

16 Tin–Oxygen, Tin–Sulphur, Tin–Selenium and Tin–Tellurium Heterocycles 401

B. MATHIASCH

16.1 Introduction 401
16.2 Cyclostannoxanes $(R_2SnO)_n$, $n = 2$ or 3 402
  16.2.1 Preparation and structure 402
  16.2.2 Spectra 404
16.3 Tin–sulphur rings 405
  16.3.1 Cyclostannatetetrasulphane R_2SnS_4 405
  16.3.2 Cyclostannathianes $(R_2SnS)_n$, $n = 2$ or 3 406
  16.3.3 Polycyclic stannathianes $(RSn)_4S_6$ 408
  16.3.4 Cyclotri(tetra)stannadithianes $(R_2Sn)_nS_2$, $n = 3, 4$ 409
  16.3.5 Cyclostannadithiadiazenes $Me_2SnS_2N_2$ 410
16.4 Tin–selenium rings 410
  16.4.1 Cyclostannaselenanes $(R_2RnSe)_n$, $n = 2$ or 3 410
  16.4.2 Polycyclic stannaselenane $(MeSn)_4Se_6$ 411
  16.4.3 Cyclotri(tetra)stannadiselenanes $(R_2Sn)_nSe_2$ 412
16.5 Tin–tellurium rings 413
  16.5.1 Cyclostannatelluranes $(R_2SnTe)_n$ $n = 2$ or 3 413
  16.5.2 Cyclotri(tetra)stannaditelluranes $(R_2Sn)_nTe_2$, $n = 3$ or 4 414
  References 415

Subject Index I

Contents of Volume II

17 Nitrogen Homocycles   417

IONEL HAIDUC

17.1 Introduction   417
17.2 Three-membered rings (cyclotriazanes, triaziridines)   417
17.3 Four-membered rings   418
17.4 Five-membered rings (pentazoles)   419
17.5 Six-membered rings   419
      References   420

18 Cyclophosphanes and Related Heterocycles   423

M. BAUDLER and K. GLINKA

18.1 Neutral monocyclic phosphanes   423
   18.1.1 General   423
   18.1.2 Syntheses   425
   18.1.3 Miscellaneous formation reactions   427
   18.1.4 Spectra, structure and bonding   429
   18.1.5 Chemical properties   431
18.2 Monocyclic phosphorus anions   435
18.3 Polycyclic phosphorus homocycles   437
   18.3.1 General   437
   18.3.2 Syntheses   438
   18.3.3 Spectra, structure and bonding   442
   18.3.4 Chemical properties and reactions   446
18.4 Heterocyclic phosphorus monocycles   447
   18.4.1 General   447

18.4.2   Syntheses   447
18.4.3   Spectra, structure and bonding   454
18.4.4   Chemical properties and reactions   459
         Bibliography   460
         References   461

**19   Phosphorus(III)–Nitrogen Heterocycles and
       Heterocyclophosph(III)azanes**   467

RODNEY KEAT

19.1   General   467
19.2   Preparation methods   468
       19.2.1   Reactions of phosphorus(III)halides with amines or amine
                hydrochlorides   468
       19.2.2   Reactions of phosphorus(III)halides with silylamines 472
       19.2.3   Transamination and related reactions   474
       19.2.4   Other methods of cyclophosph(III)azane synthesis   476
       19.2.5   Cyclophosph(III)azane-related compounds   476
       19.2.6   Addition of unsaturated species to phosph(III)azenes 478
19.3   Structure and bonding   478
       19.3.1   Structural studies of cyclodiphosph(III)azanes   478
       19.3.2   Other cyclophosph(III)azanes   481
       19.3.3   Bonding in cyclophosph(III)azanes   482
19.4   Spectra   482
       19.4.1   Nuclear magnetic resonance spectra   482
       19.4.2 ^{31}P NMR spectroscopy   482
       19.4.3 ^{1}H and ^{13}C NMR spectroscopy   483
       19.4.4   NMR of other nuclei   483
       19.4.5   Other spectroscopic methods   484
19.5   Chemical properties and reactions   484
       19.5.1   Thermal, hydrolytic and oxidative stability   484
       19.5.2   Substitution reactions   485
       19.5.3   Oxidative addition   488
       19.5.4   Complex formation   492
       19.5.5   Rearrangement to rings of other sizes   493
       19.5.6   Other chemical properties   493
         Bibliography   495
         References   496

20   Cyclophosphazenes and Heterocyclophosphazenes   501

CHRISTOPHER W. ALLEN

20.1   Introduction   501

20.2   Cyclophosphazenes   502
   20.2.1   Preparation methods   502
   20.2.2   Diffraction studies   507
   20.2.3   Spectroscopy   508
   20.2.4   Other physical methods   515
   20.2.5   Electronic structure   521
   20.2.6   Chemical properties and reactions   523
   20.2.7   Applications   576
20.3   Heterocyclophosphazenes   577
   Bibliography   587
   References   589

21   Cyclophosph(v)azanes with Five- and Six-Coordinate Phosphorus   617

ALFRED SCHMIDPETER

21.1   Introduction   617
21.2   The phosphazene–cyclodi(phosphazane) alternative   618
21.3   Monocyclic diazadiphosphetidines   620
   21.3.1   Preparation   620
   21.3.2   Reactions   623
   21.3.3   Structure and spectra   627
21.4   Spirocyclic diazadiphosphetidines   629
   21.4.1   Preparation   629
   21.4.2   Structure and spectra   635
21.5   Condensed diazadiphosphetidines   636
   21.5.1   Preparation   636
   21.5.2   Reactions   650
   21.5.3   Structure and spectra   650
21.6   Thiadoazaphosphetidines   651
   Bibliography   652
   References   653

22   Phosphorus–Oxygen Heterocycles   659

A. DURIF

22.1   Introduction   659
22.2   Cyclotriphosphates   660
   22.2.1   Preparation   660
   22.2.2   Crystal chemistry   663
22.3   Cyclotetraphosphates   667
   22.3.1   Preparation   667
   22.3.2   Crystal chemistry   668
22.4   Cyclopentaphosphates   671
22.5   Cyclohexaphosphates   671
   22.5.1   Preparation   671

22.5.2 Crystal chemistry 672
22.6 Cyclooctaphosphates 673
22.6.1 Preparation 673
22.6.2 Crystal chemistry 673
22.7 Cyclodecaphosphates 674
22.8 Cyclododecaphosphates 675
22.9 Cyclophosphate anions in other types of compounds 676
22.9.1 Phosphate-tellurates 676
22.9.2 Cyclophosphate esters 676
References 677

23 Phosphorus–Sulphur and Phosphorus–Selenium Rings and Cages 681

D. B. SOWERBY

23.1 Introduction 681
23.2 Cyclodiphosph(v)adithianes and related ring compounds 682
23.2.1 Preparation and structure 682
23.2.2 Reactions 684
23.3 Other phosphorus–sulphur ring compounds 686
23.4 Phosphorus–sulphur cage compounds 688
23.4.1 P_4S_2 688
23.4.2 P_4S_3 688
23.4.3 P_4S_4 690
23.4.4 P_4S_5 691
23.4.5 P_4S_7 691
23.4.6 P_4S_9 692
23.4.7 P_4S_{10} 693
23.4.8 P_4S_2 694
23.5 Phosphorus–selenium cage compounds 695
23.6 Mixed phosphorus–arsenic sulphides 696
Bibliography 697
References 697

24 Cycloarsanes 701

IONEL HAIDUC and D. B. SOWERBY

24.1 Introduction 701
24.2 Monocyclic compounds 702
24.2.1 Preparations 702
24.2.2 Physical properties and structure 702
24.2.3 Chemical properties and reactions 703
24.2.4 Metal complexes 704
24.3 Polycyclic and cage compounds 707
24.3.1 Inorganic systems 707

24.3.2   Organic-inorganic systems   708
Bibliography   709
References   709

25   Arsenic–Nitrogen, –Oxygen, –Sulphur and –Selenium Heterocycles   713

D. B. SOWERBY

25.1   Introduction   713
25.2   Arsenic–nitrogen rings   714
    25.2.1   Arsenic(III) compounds   714
    25.2.2   Arsenic(V) compounds   716
25.3   Arsenic–oxygen rings   718
    25.3.1   Arsenic(III) compounds   718
    25.3.2   Arsenic(V) compounds   719
    25.3.3   Cage compounds   720
25.4   Arsenic–sulphur rings   721
    25.4.1   Monocyclic compounds   721
    25.4.2   Cage compounds   722
25.5   Arsenic–selenium rings   724
25.6   Other arsenic-containing rings   725
    Bibliography   727
    References   727

26   Antimony and Bismuth Homocycles and Heterocycles   729

D. B. SOWERBY

26.1   Antimony homocycles   729
26.2   Antimony heterocycles   731
26.3   Bismuth compounds   733
    References   735

27   Sulphur Homocycles   737

RALF STEUDEL

27.1   Introduction   737
27.2   Neutral rings S_n 738
    27.2.1   Preparation of the crystalline homocyclic sulphur allotropes   738
    27.2.2   Structures   741
    27.2.3   Molecular spectra   745
    27.2.4   Thermodynamic properties   746
    27.2.5   Chromatographic separation   748
    27.2.6   Occurrence of S_n molecules in mixtures   748
    27.2.7   Reactions of S_n molecules   750
27.3   Homocyclic sulphur cations   751
    27.3.1   Monocations   752

27.3.2 Dications 752
27.4 Homocyclic sulphur oxides 756
27.4.1 Preparation 756
27.4.2 Structure and bonding 757
27.4.3 Spectra 758
27.4.4 Chemical properties and reactions 760
27.5 Homocyclic sulphur-halide cations 760
27.5.1 The S_7I^+ ion 760
27.5.2 The $S_{14}I_3^{3+}$ ion 761
References 762

28 Selenium Homocycles and Sulphur–Selenium Heterocycles 769

RALF STEUDEL and EVA-MARIA STRAUSS

28.1 General 769
28.2 Neutral selenium-ring molecules 770
28.2.1 Cyclopentaselenium Se_5 770
28.2.2 Cyclohexaselenium Se_6 770
28.2.3 Cycloheptaselenium Se_7 771
28.2.4 Cyclooctaselenium Se_8 772
28.2.5 Other ring molecules 774
28.2.6 Thermodynamic properties 775
28.3 Homocyclic selenium cations 777
28.3.1 Mass spectra of selenium vapour 777
28.3.2 Dications Se_n^{2+} in solids and solutions 778
28.4 Cyclic selenium sulphides 782
28.4.1 Preparation 782
28.4.2 Crystal and molecular structures 785
28.4.3 Vibrational spectra 786
28.4.4 High-pressure liquid chromatography 788
28.4.5 Thermodynamic properties 788
28.4.6 Reactions 788
References 789

29 Sulphur–Nitrogen Heterocycles 793

T. CHIVERS

29.1 Introduction 793
29.2 Unsaturated sulphur–nitrogen heterocycles containing two-coordinate sulphur, and other binary sulphur–nitrogen molecules and ions 794
29.2.1 Tetrasulphur tetranitride (cyclotetrathiazene) S_4N_4 794
29.2.2 Disulphur dinitride (cyclodithiazene) S_2N_2 801
29.2.3 Tetrasulphur dinitride S_4N_2 803

29.2.4   Pentasulphur hexanitride S_5N_6 805
29.2.5   The thiodithiazyl cation $S_3N_2^+$ (and the dimer $S_6N_4^{2+}$)   806
29.2.6   The thiotrithiazyl cation $S_4N_3^+$ 807
29.2.7   The cyclotetrathiazyl dication $S_4N_4^{2+}$ 809
29.2.8   The tetrasulphur pentanitride (+1) cation $S_4N_5^+$ 810
29.2.9   The cyclopentathiazyl cation $S_5N_5^+$ 811
29.2.10 The trisulphur trinitride (cyclotrithiazene) anion $S_3N_3^-$ 812
29.2.11 The tetrasulphur pentanitride (−1) anion $S_4N_5^-$ 814
29.2.12 Heterocyclothiazenes   815
29.2.13 Metallocyclothiazenes   820
29.3 Unsaturated sulphur–nitrogen heterocycles containing three-
      coordinate sulphur   822
29.3.1   Thiodithiazyl dichloride $[S_3N_2Cl^+]Cl^-$ 822
29.3.2   Thiodithiazyl oxide S_3N_2O 823
29.3.3   Other thiodithiazyl derivatives S_3N_2X 824
29.3.4   Monosubstituted cyclotrithiazyl derivatives S_3N_3X 825
29.3.5   Cyclotrithiazyl halides $S_3N_3X_3$ 826
29.3.6   1,5-Disubstituted derivatives of cyclotetrathiazene $S_4N_4X_2$ 829
29.3.7   Tetrasubstituted derivatives of cyclotetrathiazene $S_4N_4X_4$ 831
29.3.8   Heterocyclothiazenes $(Ph_2PN)_x(NSX)_y$ 831
29.4 Unsaturated sulphur–nitrogen heterocycles containing four-
      coordinate sulphur   834
29.4.1   Sulphanuric halides (trioxocyclotrithiazenes) and their
         derivatives $[NS(O)X]_3$ 834
29.4.2   Mixed sulphanuric-phosphazene rings $[NS(O)X]_x(NPCl_2)_y$ 838
29.4.3   Mixed sulphanuric-thiazyl rings   841
29.4.4   Sulphur-nitrogen heterocycles containing the NSO_2 group   842
29.4.5   Cyclotetrathiazene tetroxide $S_4N_4O_4$ 845
29.4.6   The tetrasulphur pentanitride oxide anion $S_4N_5O^-$ 845
29.5 Saturated sulphur–nitrogen heterocycles containing two-
      coordinate sulphur   846
29.5.1   Cyclic sulphur imides (cycloazasulphanes) based on the
         S_8 ring, $S_{8-x}(NH)_x$ (x = 1−3)   847
29.5.2   Tetrasulphur tetraimide (cyclotetrathiazane) $S_4N_4H_4$ 852
29.5.3   Cyclic sulphur hydrazides   854
29.5.4   Metal complexes of the $S_2N_2H^-$ ligand   854
         Bibliography   857
         References   858

30   Sulphur–Oxygen, Selenium–Oxygen and Selenium–Nitrogen
     Heterocycles   871

IONEL HAIDUC

30.1 Sulphur–oxygen compounds   871

30.2   Selenium–oxygen compounds   873
30.3   Selenium–nitrogen compounds   874
       Bibliography   876
       References   877

Subject Index   I

Introduction

IONEL HAIDUC

Babes–Bolyai University, Cluj-Napoca, Romania

D. B. SOWERBY

University of Nottingham, England

1 What is an inorganic ring?

Cyclic structures, consisting of closed chains of carbon atoms, are the basis of a vast area of organic chemistry. Partial replacement of carbon by so-called heteroatoms leads to an extensive group of cyclic systems, called heterocycles. These facts are known and understood even by beginners in the study of organic chemistry. The ability of other elements, primarily nonmetals, to reproduce the structural variety of organic chemistry with atoms other than carbon is a much less known or understood subject. Inorganic (carbon-free) rings were described in the literature many years ago (see B 1 for a brief historic outline) and representatives are now known for every nonmetal (except the noble gases) and for many metals. However, the development of inorganic ring chemistry occurred in the shade of more spectacular achievements (for example coordination and organometallic compounds) and for many years went almost unnoticed by the scientific community at large. Unlike organic rings, inorganic homo- and heterocycles were studied mostly in relation to the chemistry of specific elements (for example sulphur, phosphorus, silicon, boron); some excellent monographs based upon this approach were published.[B 2–B 6] An integrated view and treatment of inorganic ring chemistry as a system, within a self-consistent discipline, has been attempted only in the last twenty years.[B 1, B 7–B 10] In addition, the recognition of inorganic ring chemistry as a discipline is underscored by the regular organization of international symposia on this topic, but so far only one inorganic chemistry textbook has devoted a

THE CHEMISTRY OF INORGANIC HOMO-
AND HETEROCYCLES VOL 1 ISBN 0-12-655775-6

special chapter to "Inorganic Chains, Rings, Cages and Clusters".[B 11] Some review articles, based on this integrated view, have also been published recently.[B 12 – B 16]

In defining the area of inorganic ring chemistry, some (perhaps expected) difficulties are encountered, related to the diffuse borderline with organic chemistry and with other areas of inorganic chemistry. These borders are not deep gaps, but rather connecting bridges, which make any decision somewhat arbitrary or at least highly subjective.

If in a cyclic structure carbon is successively replaced by a heteroatom, full replacement obviously produces an inorganic ring. Thus, starting from cyclopentadiene we end up with pentazole (ignoring position isomers):

$$C_5 \qquad C_4N \qquad C_3N_2 \qquad C_2N_3$$

$$CN_4 \qquad N_5$$

The question is: which is the last organic ring in this series? Organic chemists tend to regard even pentazole (N_5) as an organic ring ("heteroatom-only ring"), while the inorganic chemist will perhaps consider tetrazole (CN_4) as an inorganic ring with carbon as heteroatom. In this book, as in the previous one,[B 1] we include only carbon-free rings, and thus the borderline between organic and inorganic rings is conventionally fixed by the presence or absence of carbon in the ring. This decision was determined to some extent by a pragmatic reason: even if only one-carbon heterocycles were included, then a very large number of additional rings would have to be treated—and there is simply not enough space available to do this. Furthermore, the synthesis of even one-carbon heterocycles starts with typical organic compounds and uses organic-type reactions. In conclusion, *only those rings that contain no carbon atom (carbon-free rings) are considered as inorganic rings.*

In addition to typical cyclic compounds of nonmetals, such as borazines, cyclosilanes, cyclosiloxanes or cyclosilazanes, cyclopolyphosphines and cyclophosphazenes, which are inorganic analogues of organic cyclic hydrocarbons and heterocycles, in inorganic chemistry we often find closed formations of atoms (i.e. rings) in compounds that do not have formal organic analogues. Should these be recognized as inorganic rings? The matter arises especially in boron chemistry and in transition-metal chemistry. Thus boranes do contain cyclic arrangements of atoms, either as molecular skeletons or as molecular fragments:

$$B_2H_6 \qquad\qquad B_4H_{10}$$

In polyhedral boranes B_3 triangular faces are frequently present. The formation of polynuclear borane structures (as well as those of carboranes and their derivatives) is the result of multicentre electron-deficient-bond interactions; these compounds constitute a special area of inorganic chemistry in its own right, which cannot be squeezed in the chemistry of inorganic rings. We shall therefore consider that boranes, carboranes and related compounds should not be included in this book or in any review of inorganic ring systems. The same applies to transition-metal clusters, where the triangular faces M_3, square M_4 fragments or pentagonal M_5 fragments, etc, should not be regarded as rings in a polycyclic structure. The cages are stable as a whole owing to collective electronic interactions. We suggest that cluster chemistry should be regarded as a self-contained area of inorganic chemistry.

Some transition-metal compounds, however, are situated on the borderline between ring chemistry and coordination chemistry. Trinuclear species, usually containing a metal–metal bond and a nonmetal bridge, as illustrated by the following examples, can be regarded as inorganic rings as well as coordination or organometallic compounds:

$$(CO)_4Fe\!-\!Fe(CO)_4 \qquad (CO)_3Co\!-\!Co(CO)_3 \qquad (C_5H_5)(CO)_2Mn\!-\!Co(CO)_3$$

An extremely large number of dinuclear doubly bridged compounds are known; the bridges can be halogens, oxygen, sulphur, phosphorus and other nonmetals:

$$\begin{bmatrix} Cl & Cl & Cl \\ & Cu & Cu \\ Cl & Cl & Cl \end{bmatrix}^{2-} \qquad L_xCu \overset{\overset{\displaystyle R}{O}}{\underset{\underset{\displaystyle R}{O}}{}} CuL_x \qquad L_xFe \overset{\overset{\displaystyle R}{S}}{\underset{\underset{\displaystyle R}{S}}{}} FeL_x \qquad L_xFe \overset{\overset{\displaystyle R_2}{P}}{\underset{\underset{\displaystyle R_2}{P}}{}} FeL_x$$

Usually, these are simply treated as binuclear coordination compounds. But it can also be argued that they should be regarded as inorganic rings. Five-membered, six-membered and larger rings, whose formation involves coordination bonds, can also be cited:

$$(CO)_4Mn \overset{\overset{\displaystyle R_2}{As}}{\underset{\underset{\displaystyle R_2As-AsR_2}{}}{}} Mn(CO)_4 \qquad \overset{Cl}{\underset{L}{>}}Pt\overset{Cl}{\underset{N-N}{<}}Pt\overset{Cl}{\underset{L}{<}} \qquad (CO)_2Co \overset{\overset{\displaystyle R}{N}}{\underset{\underset{OC \quad CO}{F_2P \quad PF_2}}{}} Co(CO)_2$$

$$LPd \overset{\overset{\displaystyle R_2}{P=S}}{\underset{\underset{\displaystyle R_2}{S=P}}{}} PdL \qquad LNi \overset{N-N}{\underset{N-N}{}} NiL \qquad R_2Al \overset{\overset{\displaystyle R \quad R}{P}}{\underset{\underset{\displaystyle R \quad R}{P}}{}} AlR_2$$

Such cyclic coordination compounds are again borderline cases, which can equally rightfully be considered as belonging either to inorganic ring chemistry or to coordination chemistry.

Another large borderline family includes the carbon-free chelate rings, illustrated by four-, five- and six-membered ring species:

$$M \overset{\overset{\displaystyle R}{N}}{\underset{\underset{\displaystyle R}{N}}{}} N \qquad M \overset{\overset{\displaystyle F_2}{P}}{\underset{\underset{\displaystyle F_2}{P}}{}} NR \qquad M \overset{\overset{\displaystyle R}{N=N}}{\underset{\underset{\displaystyle R}{N}}{}} N \qquad M \overset{\overset{\displaystyle R_2}{P}}{\underset{\underset{\displaystyle R_2}{P}}{}} \overset{PR}{\underset{PR}{}}$$

$$M \overset{O-P}{\underset{O=P}{}} N \qquad M \overset{\overset{\displaystyle R_2}{O=P}}{\underset{\underset{\displaystyle R_2}{O=P}}{}} NR \qquad M \overset{N-N}{\underset{N-N}{}} BH_2$$

In addition, compounds with (sometimes multiple) metal–metal bonds, bridged by a trinuclear nonmetal fragment, as in the following examples, cannot be ignored as inorganic ring compounds:

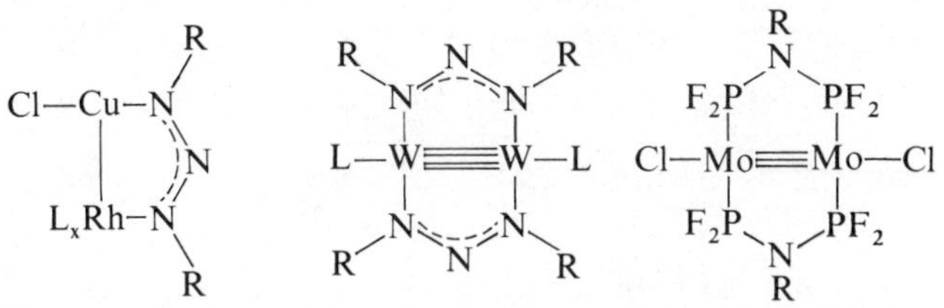

The borderline examples cited demonstrate the infinite variety of possibilities of building cyclic arrangements of atoms other than carbon. As a result, the task of clearly defining the borders of inorganic ring chemistry is extremely complex.

In this book the editors have chosen to include only noncontroversial types of inorganic rings: first of all, nonmetal derivatives, which can be regarded as analogues of organic ring systems (in the language of organic chemistry, "heteroatom-only rings"). Derivatives of main-group metals are also included, but transition-metal derivatives, whose cyclic compounds involve participation of coordination bonds, are not treated in this book. These could be the subject of a special monograph, dealing with carbon-free metallocycles, and underscoring the overlap of inorganic ring chemistry and coordination chemistry. This editorial decision may be regarded as arbitrary, but space limitations were one of the important factors in this choice.

In conclusion, the question asked in the heading of this section can be answered as follows: *an inorganic ring is a finite polynuclear system of atoms other than carbon, arranged to form a closed (planar or close to planar) structure, made up of identical atoms (homocycle) or different atoms (heterocycle).* In the present book this definition is applied only to main-group elements, leaving open the possibility of extending it to transition-metal atoms as well. Rings can be combined to form tridimensional structures of finite size, i.e. cages (molecular compounds) or of infinite size (polymeric or macromolecular compounds). Cages are distinguished from clusters by the nature of the chemical bonds involved: in cages a skeleton is formed through localized two-centre two-electron bonds, while clusters are formed via delocalized multicentre multielectron collective interactions.

2 Classification of inorganic rings

Inorganic rings can be classified according to their composition and structure, in the following types:[B 1] (*a*) *homocycles*, i.e. rings containing identical atoms; well known for silicon, phosphorus, sulphur and other nonmetals:

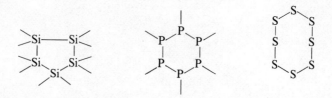

(*b*) *heterocycles formed by insertion of a heteroatom* into a parent homocyclic system (maintaining some bonds between identical atoms):

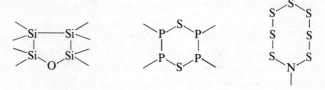

(*c*) *heterocycles formed by regular alternation* of two different elements (called "pseudo-heterocycles" because of the presence of identical repeating units or atom pairs):

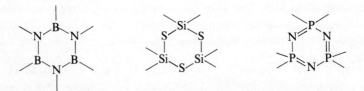

(*d*) *mixed-type heterocycles*, in which the regular alternation is disturbed by introducing a third element; these heterocycles are formed from different units or atom pairs):

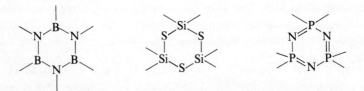

The types (*a*)–(*d*) are known mainly for non-metallic elements and main group metals.

Polycyclic systems are often formed by connection of monocyclic units via bridges (*bridged rings*), one common atom (*spirocyclic systems*) or several (two or more) common atoms (*fused rings*):

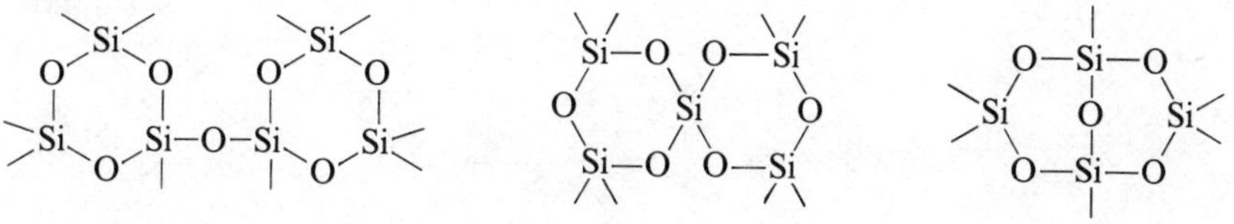

A particular case of polycyclic systems includes *cage structures*, consisting of several rings connected in a finite, three-dimensional molecular skeleton:

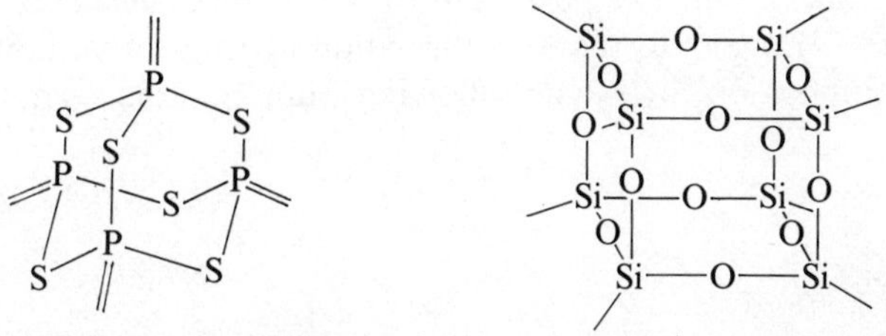

Polycyclic high-polymeric structures can be formed from cyclic units, connected in one of the three modes mentioned, leading to either mono-, bi- or tridimensional infinite chains, sheets and networks respectively.

The various types of inorganic rings are interrelated. Thus the transition from homocyclic systems to heterocycles formed by regular alternation formally occurs via successive insertion of heteroatoms:

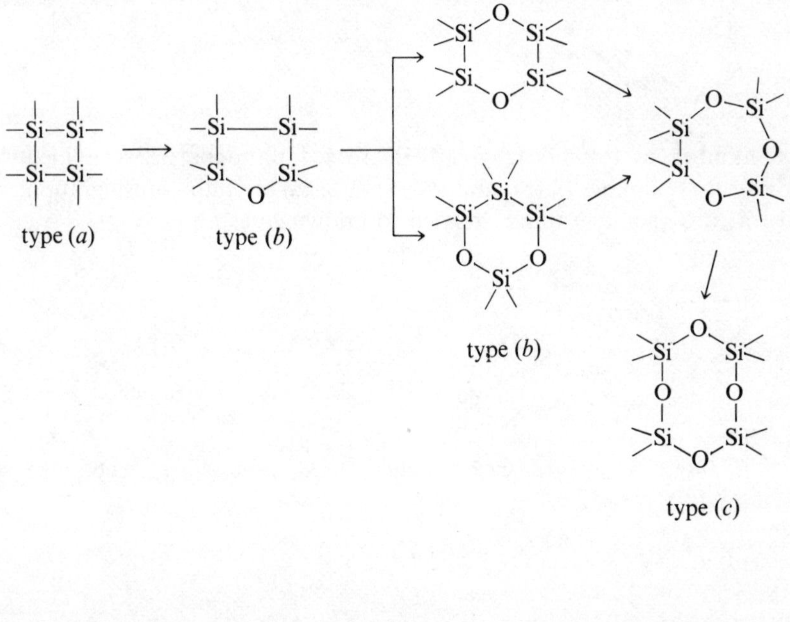

type (*a*)   type (*b*)

type (*b*)

type (*c*)

A pair of symmetric rings formed by regular alternation (type (c)) is related to another pair through a series of irregular (unsymmetric) heterocycles (type (d)):

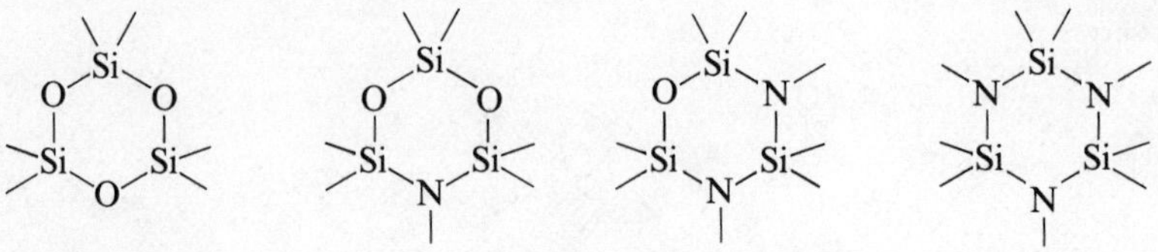

type (c)   type (d)   type (c)

Additional structural diversity is achieved in some cases (in particular, the Si–N pair, but in principle possible for other pairs of elements, for example P–N or As–N) owing to the easy formation of rings of variable size. Thus for the composition Si_4N_4 the following isomeric atom arrangements are possible:

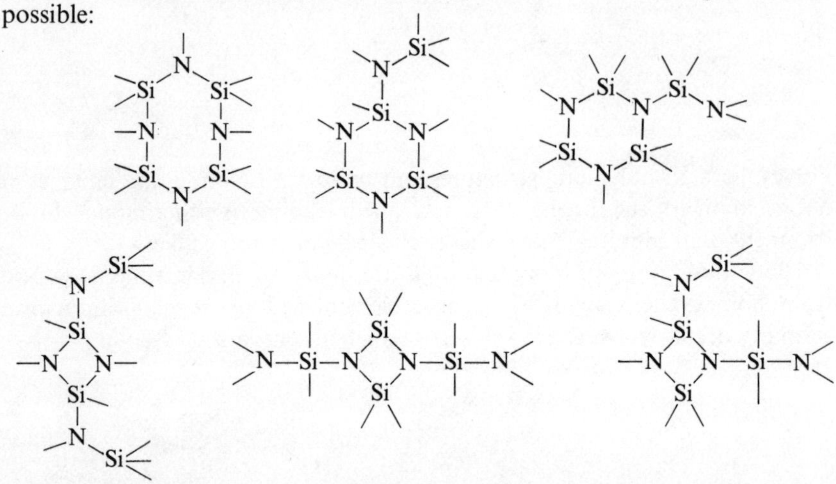

The number of isomers is greatly increased if bonds between identical atoms, for example Si–Si and N–N, are taken into consideration, as illustrated by some examples related to known rings:

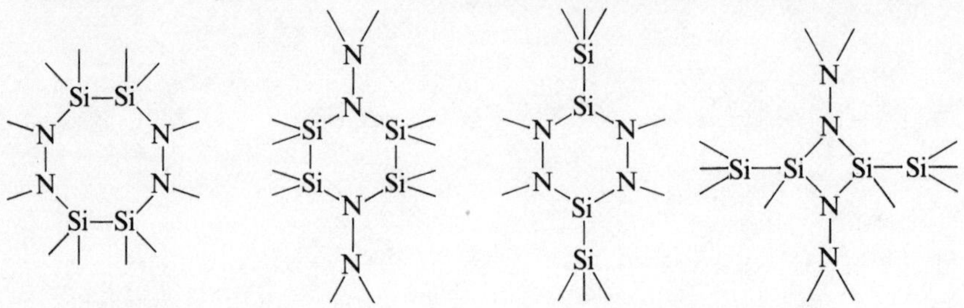

3 Nomenclature of inorganic rings

The nomenclature of inorganic rings is still a matter of controversy and no generally accepted rules are available. The present IUPAC rules for naming inorganic compounds[1] pay little attention to inorganic ring systems. The IUPAC Commission for Nomenclature in Inorganic Chemistry presently has the matter under consideration and a working group was formed some years ago.[2] The nomenclature of inorganic rings was also discussed at the IRIS Symposia in Madrid (1977) and in a special session at Göttingen (1978). The current practice in naming inorganic rings was reviewed recently and proposals for a unified nomenclature of inorganic ring systems were made.[B 12] The reader is referred to this article for a more detailed presentation of the subject.

At present there are several procedures for naming inorganic rings, and these have certain common features which may serve as a basis for unification of the nomenclature. These will be briefly presented, to aid the reader in deciphering the names used in this book.

The nomenclature of inorganic rings cannot ignore the rules developed for organic rings, and in many cases the procedure has simply been transferred from the latter to the former. There are two main ways in naming *organic* rings: (*a*) the Hantzsch–Widman system and (*b*) the replacement nomenclature.

The Hantzsch–Widman system uses specific suffixes to indicate the ring size:

for 4-membered rings	-etene, -etane
for 5-membered rings	-ole, -olene, -olane
for 6-membered rings	-in, -ine, -inane
for 7-membered rings	-epane, -epine
for 8-membered rings	-ocin, -ocine, -ocane

The application of this nomenclature to inorganic rings results in names such as triazatriborine for B_3N_3, tetrazatetraborocine for B_4N_4, pentazole for N_5, triazatriphosphorine for N_3P_3, which are encountered in the literature.

In the replacement nomenclature a heterocycle is considered to result by replacing carbon atoms with other elements (for example oxygen, nitrogen, sulphur, silicon). This procedure has mainly been used for silicon-containing rings (oxa–aza–thia convention)[3] and has resulted in names like diazatetrasilacyclohexane for Si_4N_2 and triazadisilaboracyclohexane for Si_2BN_3. Abbreviated forms of replacement names have been used[3] for silicon-containing rings based on a regular alternation of two elements, for example cyclotrisiloxane for Si_3O_3, cyclotrisilazane for Si_3N_3, and cyclodisilthiane for Si_2S_2, indicating simply the number of repeating units in the ring. The extension of this nomenclature to phosphorus-containing rings produces names like cyclotriphosphazene for P_3N_3 and cyclotetraphosphazene for P_4N_4. The "repeating-unit" system of nomenclature was extended by Allcock[B 9] to all alternant inorganic rings. In this system a ring name indicates the number of repeating units, preceded by the prefix "cyclo". For rings not containing a regular alternation, an additive-type nomenclature is more appropriate, and permits the naming of any inorganic ring, regardless of its degree of symmetry or complexity, by using the prefix "cyclo" and then simply joining the name roots of ring elements and adding the termination "-ane" (for saturated rings) or "-ene" (for unsaturated rings). The replacement nomenclature and additive nomenclature are closely related, as shown by the following example:

replacement name : dioxaazatrisilacyclohexane

additive name : cyclotrisilazadioxane

The additive names are preferable, since they make the nomenclature of inorganic rings independent of the names of cyclic hydrocarbons.

Two important literature sources of inorganic ring chemistry are Chemical Abstracts and the multivolume Gmelin Handbook of Inorganic Chemistry. Chemical Abstracts uses in some cases the Hantzsch–Widman system, borrowed from organic chemistry (for example P_3 triphosphirane, P_4 tetraphosphetane, P_5 pentaphospholane) and the replacement nomenclature with full names (for example B_2SiN_3 triazasiladiboracyclohexane, B_2PN_3 triazaphosphadiborine, N_3P_3 triazatriphosphorine) or abbreviated forms (for example Si_3O_3 cyclotrisiloxane, Si_4N_4 cyclotetrasilazane, Si_2S_2 cyclodisilthiane). For some boron-containing rings trivial names, derived from the systematic ones, are currently used by Chemical Abstracts (for example borazine for B_3N_3 and boroxin for B_3O_3).

The Gmelin Handbook uses a rational system of nomenclature, unfortunately applied so far only to boron–nitrogen heterocycles. In this system the prefix "cyclo" is used in all names, numerical prefixes indicate the number of each kind of atoms in a ring and the nitrogen atom is numbered first. This results in names like cyclo-1,3-diaza-2,4,5-triboran for B_3N_2, cyclo-1,3,5-triaza-2,4,6-triboran for B_3N_3 and cyclo-1,2,4,5-tetraaza-3,6-diboran for B_2N_4:

$$B_3N_2 \qquad\qquad B_3N_3 \qquad\qquad B_2N_4$$

The "repeating-unit nomenclature" has been used in some monographs dealing with inorganic rings.[B 1, B 9] Without being formulated as a set of rules, the following principles were used in the nomenclature of Allcock[B 9]: (a) a ring system was indicated by the prefix "cyclo-"; (b) the ring size was indicated by the prefixes "di-", "tri-," "tetra-", "penta-", "hexa-", showing the number of repeating units; (c) the termination "-ane" was used for saturated rings and the termination "-ene" for unsaturated rings; (d) the order of root names of elements was that of increasing electronegativity.

The rules suggested in the monograph by Haiduc[B 1] were further developed in a set[B 12] which also incorporated a proposal made by Wannagat[4] for indicating the ring size in square brackets. These rules are briefly listed here:

Rule 1   A cyclic structure is indicated by the use of the prefix "cyclo" as a part of the ring name.

Rule 2   The termination "-ane" is used for saturated rings and "-ene" for naming formally unsaturated rings (i.e. rings that must be written with formal double bonds).
Examples: cyclophosphazenes, cyclosiloxanes (as class names)

Rule 3   The prefix "cyclo" is followed by the roots of element names and prefixes indicating the number and nature of the atoms involved.

   3.1   The prefixes "di-," "tri-", "tetra-", "penta-", "hexa-", "hepta-", "octa-", etc. derived from Greek or Roman numerals are used to indicate the number of each kind of atoms in a ring.
   Examples: B_2N_4 cyclodiboratetrazane
   Si_4N_2 cyclotetrasiladiazane
   The rings based on a regular alternation can be named by indicating the number of repeating units in the ring (preferably with parentheses).
   Examples: Si_3N_3 cyclotri(silazane)

B_3O_3 cyclotri(boroxane)
P_3N_3 cyclotri(phosphazene)

3.2 The ring size is indicated by a numeral in square brackets.
Examples: Si_3N_3 cyclo[6]trisilazane
          B_2S_3 cyclo[5]diboratrithiane

3.3 The following roots can be used for naming the basis elements of a ring:

B bor(a)   Si sil(a)   P phosph(a)   S thia
Al alum(a)   Ge germ(a)   As ars(a)   Se sel(a)
Ga gall(a)   Sn stann(a)   Sb stib(a)   Te tellura(a)
In ind(a)   Pb plumb(a)   Bi bismut(a)

The electronegative heteroatoms present in the ring are indicated at the end, with the following terminations:

O   -oxane
N   -azane (for saturated rings)
     -azene (for unsaturated rings; nitrogen atom bearing no exocyclic substituent, like in S_4N_4 or S_2N_2)
S(II) -thiane
Se(II) -selane

Rule 4 The position of each element in a ring, when necessary (for example in unsymmetrical rings) is indicated with the aid of numeric locants. The numbering starts with the least-electronegative element.
Examples: cyclo[6]-1,3-dibora-5-sila-4-aza-2,6-dioxane
           cyclo[6]-1-bora-3-sila-5-phospha-2,6-diaza-4-oxane

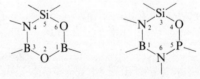

Numeric locants are also used to indicate the position of formal double bonds.

Rule 5 In the names of compounds derived from a parent ring, the substituents are indicated before the prefix cyclo; their position can be shown by numeric locants or by using the symbol of the element bearing the substituent.
Example:

$$Et—N_4^{Me} \overset{\overset{Me}{\underset{|}{Si}}}{\underset{|}{{}_5 \quad {}_6}} N—Et$$
$$Ph—B_3^{} {}_2 \overset{|}{\underset{O}{B}}—Ph$$

1,3-diphenyl-5,5-dimethyl-4,6-diethyl-
cyclo[6]-1,3-dibora-5-sila-4,6-diaza-2-oxane

These rules can serve for naming any inorganic heterocycle, regardless of its complexity of structure and composition. Examples of using the rules for general and particular cases of inorganic rings and their derivatives were presented in some detail in the article cited.[B 12]

Bibliography

B 1   I. Haiduc, *The Chemistry of Inorganic Ring Systems* (in two volumes) (Wiley-Interscience, London, 1970).

B 2   M. Goehring, *Ergebnisse und Probleme der Chemie der Schwefelverbindungen* (Akademie Verlag, Berlin, 1957).

B 3   S. Pantel and M. Becke-Goehring, *Sechs- und achtgliedrige Ringsysteme in der Phosphor-Stickstoff Chemie* (Springer, Berlin, 1969).

B 4   H. R. Allcock, *Phosphorus–Nitrogen Compounds. Cyclic, Linear and High Polymeric Systems* (Academic Press, New York, 1972).

B 5   H. G. Heal, *The Inorganic Heterocyclic Chemistry of Sulfur, Nitrogen and Phosphorus* (Academic Press, London, 1980).

B 6   K. Niedenzu, *Boron–Nitrogen Compounds*, (Academic Press, London, 1965).

B 7   I. Haiduc, *Introducere în chimia ciclurilor anorganice [Introduction to the chemistry of inorganic rings]* (Editura Academiei, Bucuresti, 1960—in Romanian). Translated into Polish as *Wstep do chemii nieorganicznych zwiazkow pierscienyowych* (PWN, Warszawa, 1964).

B 8   H. Garcia-Fernandez, *Les hétérocycles en chimie minérale.* (Hermann, Paris, 1964); *Quimica heterocíclica inorgánica* (Editorial Alhambra, Madrid, 1973).

B 9   H. R. Allcock, *Heteroatom Ring Systems and Polymers* (Academic Press, New York, 1967).

B 10   D. A. Armitage, *Inorganic Rings and Cages* (E. Arnold, London, 1972).

B 11   J. E. Huheey, *Inorganic Chemistry. Principles of Structure and Reactivity*, 3rd edn, (Harper & Row, New York, 1983).

B 12   I. Haiduc, *Rev. Inorg. Chem.* **2** (1980) 219 (a survey of the nomenclature of inorganic ring systems together with proposals for a unified nomenclature).

B 13   I. Haiduc, *Rev. Inorg. Chem.* **4** (1982) 179 (interconversion reactions of inorganic heterocycles).

B 14   H. W. Roesky and M. Witt, *Rev. Inorg. Chem.* **4** (1982) 45 (small inorganic rings).

B 15   E. Hengge, *Chem. Labor. Betr.* **33** (1982) 307 (recent results in inorganic ring chemistry).

B 16   I. Haiduc, *Rev. Inorg. Chem.* **5** (1983) 7 (inorganic cyclic compounds: an annual review covering the year 1980).

References

1   IUPAC, *Nomenclature of Inorganic Chemistry*, 2nd edn, (Butterworth, London, 1970).
2   *IUPAC Information Bulletin* No. 54 (1977) 39.
3   E. J. Crane, *Chem. Engng News* **30** (1962) 4513.
4   U. Wannagat, *Primer Simposio Internacional de Quimica Heterociclica Inorganica (Segunda Reunion), Madrid, 22–26 Junio 1977*, p. 405.

Boron Homocycles (Cyclopolyboranes)

IONEL HAIDUC

Babes–Bolyai University, Cluj-Napoca, Romania

Boron hydrides are polynuclear compounds, with cage structures formed through polycentre polyelectronic bonds. Lewis electronic structures can be written for cyclotetra-, -penta- or hexaboranes (1–3):

| (1) | (2) | (3) |

but these are largely hypothetical.

The best-known derivative of tetraborane(4) is B_4Cl_4, but this is a tetrahedral molecule, with polycentre bonds, rather than a monocyclic system (cyclotetraborane derivative).[1]

The possible existence of a cyclotetraborane (1) was inferred from a theoretical study of the stereochemical nonrigidity and isomerization of hypothetical $B_4H_4{}^{2-4}$ and its derivative B_4F_4.[3,4] The degenerate rearrangement of these two tetrahedral species passes through the square-planar monocyclic structure:

$$T_d \longrightarrow D_{4h} \longrightarrow T_d$$

THE CHEMISTRY OF INORGANIC HOMO-
AND HETEROCYCLES VOL 1 ISBN 0-12-655775-6

The molecular-orbital correlation diagram for the $T_d \rightarrow D_{4h}$ transformation was calculated and revealed that the rearrangement is accompanied by an orbital crossing of the HOMO–LUMO type, which generates a large barrier for the process. This barrier is 85 kcal/mol for B_4H_4 and lower for B_4F_4; for both molecules the square homocyclic structure is said to be a possible stable intermediate along the reaction pathway.[3,4]

When two electrons are added to the tetraboron system, to obtain the $B_4H_4^{2-}$ anion, MNDO calculations[5] predict that a nonplanar D_{2d} structure will be more stable than a square-planar D_{4h} structure by 3.4 kcal/mol.

The reaction of B_4Cl_4 with t-BuLi gives the compound $B_4Bu_4^t$. It is not clear whether in this compound the B_4 tetrahedron is preserved or a square cyclic structure (of type 1) is formed. The preparation of this compound has the merit of demonstrating that the B_4 framework is stable in the absence of backbonding ligands,[6] but in the absence of convincing physical evidence the structure of this compound cannot be ascertained with any degree of confidence. The same tetra-t-butyltetraborane(4) $B_4Bu_4^t$, was formed by reductive dimerization of di-t-butyldichlorodiborane(4), $B_2Cl_2Bu_2^t$; the radical anion from $B_4Bu_4^t$ was prepared by Na/K alloy reduction of $B_2Cl_2Bu_2^t$ and identified by ESR[7].

Cyclopentaboranes (2) have not been mentioned in the literature.

The first fully confirmed cycloborane was the hexakis(dimethylamino) derivative of (3), $B_6(NMe_2)_6$, which slowly crystallized, in low yield, from the distillation residues of the Na/K alloy dehalogenation of Me_2N—BCl_2.[8] The main reaction products were $B_3(NMe_2)_5$ and $B_4(NMe_2)_6$. The cyclic compound was characterized by NMR spectroscopy (^{11}B singlet $\delta = 65$ ppm; ^1H singlet $\delta = 2.87$ ppm; ^{13}C singlet $\delta = 46.65$ ppm). A single-crystal X-ray diffraction study revealed a cyclic structure, in which the B_6 homocycle adopts a chair conformation, with B–B distances (related in pairs by symmetry) of 1.704, 1.778 and 1.676 Å, the mean value being 1.70 Å.[8] It can be speculated that there is some electron backdonation from the NMe_2 substituents, with the effect of stabilizing the ring.

References

1   I. Haiduc, *The Chemistry of Inorganic Ring Systems* (Wiley-Interscience, London, 1970), p. 50.

2   D. A. Kleier, J. Bicerano and W. N. Lipscomb, *Inorg. Chem.* **19** (1980) 216.

3   M. L. McKee and W. N. Lipscomb, *Inorg. Chem.*, **20** (1981) 4148.

4   W. N. Lipscomb, in *Boron Chemistry*—4, edited by R. W. Parry and G. Kodama (Pergamon Press, Oxford, 1979), p. 1.

5   M. J. S. Dewar and M. L. McKee, *Inorg. Chem.* **17** (1978) 1569.

6   T. Davan and J. A. Morrison, *J. Chem. Soc. Chem. Commun.* **1981**, 250.

7   H. Klusik and A. Berndt, *J. Organomet. Chem.* **234** (1982) C 17

8   H. Nöth and H. Pommerening, *Angew. Chem.* **92** (1980) 481.

2

Boron–Nitrogen Heterocycles

W. MARINGGELE

Universität Göttingen, Federal Republic of Germany

2.1 General remarks

In this chapter the literature on boron–nitrogen heterocycles from 1969 to 1981 is reviewed. For the most intensively investigated ring system, borazine, as well as for other boron–nitrogen heterocycles,[B 1 – B 6] new syntheses have been established. Thermochemical[B 10] and spectroscopic data have been published.[B 7 – B 9, B 11 – B 19] Especially important in this field are nuclear magnetic resonance,[B 8, B 11 – B 14] photoelectron spectroscopy,[B 15 – B 17] mass spectrometry,[B 19] and infrared and Raman spectroscopic investigations.[B 7] Spectroscopic data for different ring systems have been correlated and have led to better understanding of bonding characteristics. On the other hand, a number of new ring systems have been synthesized that, in addition to boron and nitrogen, contain elements like silicon, aluminium, phosphorus, oxygen, sulphur, selenium and arsenic.

2.2 Nomenclature

Essentially, three systems of nomenclature for monocyclic systems are used in the literature: (*a*) the extended Hantzsch–Widman system; (*b*) the "a" nomenclature (replacement nomenclature); and (*c*) for rings with repeating units the prefix "cyclo", followed by a numerical affix "di-," "tri-," "tetra-," the "a"-term and the suffix "-ane" or "-ene" that specifies the state of hydrogenation. This third system is also applied to rings that do not contain repeating units, in which case the numerical affix is omitted. For fused heterocyclic systems the fusion nomenclature is used.[B 20 – B 26]

THE CHEMISTRY OF INORGANIC HOMO-
AND HETEROCYCLES VOL I ISBN 0-12-655775-6

The nomenclature used in this chapter is in agreement with the Ring Index of Chemical Abstracts and with the IUPAC rules.[B 24]

For 4–8-membered monocyclic systems that contain boron and nitrogen atoms in an sp^2 hybridized state and additional O, S, Se, P, As heteroatoms the extended Hantzsch–Widman system is used. The suffixes "-etidine," "-olidine," "-oline," "-ine" and "-ocine" are used for 4-, 5-, 6- and 8-membered ring systems. For compounds containing silicon in addition to B and N and for B–N coordination heterocycles, the "a" nomenclature will be used. The seniority list of "a" terms, in decreasing order of priority, is: oxa, thia, selena, aza, phospha, arsa, sila, stanna, bora, alumina, and the "a" terms for uncharged atoms have priority over the ionic "a" terms: oxonia, thionia, azonia and borata. In addition, some trivial names are accepted: borazine B_3N_3, silaborazines BSi_2N_3 and B_2SiN_3, and borazocine B_4N_4. The following examples illustrate the use of this nomenclature:

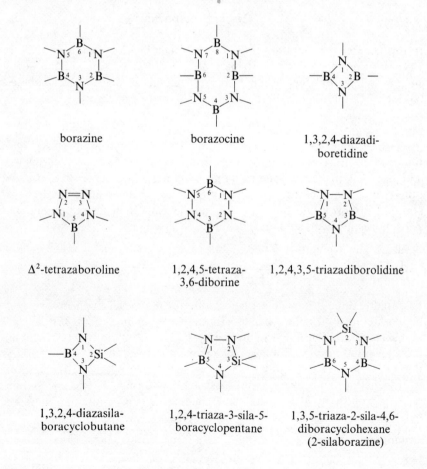

borazine

borazocine

1,3,2,4-diazadi-
boretidine

Δ^2-tetrazaboroline

1,2,4,5-tetraza-
3,6-diborine

1,2,4,3,5-triazadiborolidine

1,3,2,4-diazasila-
boracyclobutane

1,2,4-triaza-3-sila-5-
boracyclopentane

1,3,5-triaza-2-sila-4,6-
diboracyclohexane
(2-silaborazine)

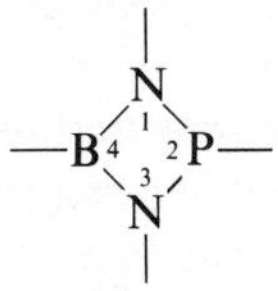

1,3,2,4-diazaphospha-
boretidine

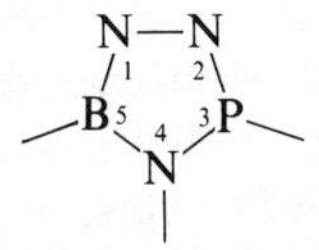

1,2,4-triaza-
3-phospha-5-borolidine

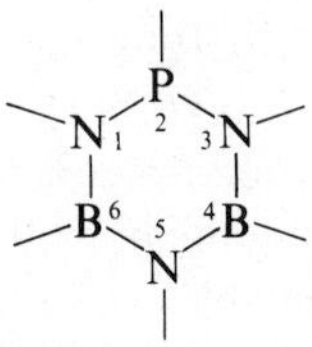

1,3,5-triaza-2-
phospha-4,6-
diborine

2.3   Borazines $(RBNR')_3$

2.3.1   Unsubstituted borazine $(HBNH)_3$

Borazine has been prepared by the reaction of ammonium chloride and lithium or sodium tetrahydridoborate. The yields of borazine as a function of the molar ratios of the reagents and the reaction temperature have been investigated and thermo-gravimetric measurements made.[1–3]

When $H_3B.NH_3$ is dissolved in a suitable glycol ether (b.p. 100 °C above that of borazine) and rapidly heated to 160 °C, borazine is produced in 69% yield.[4]

The relatively strong base ammonia reacts with B_5H_9 during heating at 160–180 °C to form borazine.[5,6] The reaction is believed to occur via BH_3 split-off (with formation of up to 0.25 B_2H_6 per B_5H_9 consumed), through $B_4H_6.NH_3$, H_2BNH_2, $H_2NB_2H_5$ and $(H_2BNH_2)_3$ intermediates.

Diborane reacts rapidly and quantitatively with 2-dimethylamino-borazine in a heptane solution or neat to form a simple adduct, $(BH_3.Me_2N)H_2B_3N_3H_3$, which decomposes in diethyl ether at room temperature to form $(HBNH)_3$, μ-$Me_2NB_2H_5$, H_2BNMe_2 and a nonvolatile residue.[7] Excess of diborane is consumed by the adduct to form $(HBNH)_3$ and μ-$Me_2NB_2H_5$ quantitatively.

2.3.2   Inorganic derivatives of borazine

Inorganic derivatives of borazine, including halogeno-, amino-, alkoxy- and pseudohalogeno- derivatives have been reported in the recent literature.

2,4,6-Trichloroborazine $(ClBNH)_3$ has been synthesized by a convenient procedure[8] that is a modification of the reaction of boron trichloride with ammonium chloride. First, the acetonitrile–boron-trichloride adduct

MeCN.BCl$_3$ is prepared in chlorobenzene, ammonium chloride is added and the mixture is refluxed until evolution of hydrogen chloride ceases. The formation process of 2,4,6-trichloroborazine from boron trichloride and ammonium chloride has been studied and a mechanism has been suggested[9] that involves triaminoborane B(NH$_2$)$_3$ as an intermediate.

If a pyridine complex of (ClBNH)$_3$ is reacted with LiBH$_4$ in diethyl ether, 2-chloroborazine ClH$_2$B$_3$N$_3$H$_3$ is obtained along with the C$_5$H$_5$N.BH$_3$ adduct.[10]

2,4-Dichloroborazine Cl$_2$HB$_3$N$_3$H$_3$ was obtained in good yield by the reaction of Cl$_2$(Me$_2$N)B$_3$N$_3$H$_3$ with diborane.[11] If gaseous BF$_3$ is reacted with Cl$_2$(Me$_2$N)B$_3$N$_3$H$_3$, the compound Cl$_2$FB$_3$N$_3$H$_3$ (2-fluoro-4,6-dichloroborazine) can be isolated in 20% yield.[11] The low yield is attributed to the disproportionation of the unsymmetrical product to give (ClBNH)$_3$ and ClF$_2$B$_3$N$_3$H$_3$.[11]

The reaction of boron trichloride and nitrogen trichloride, in which hexachloroborazine (ClBNCl)$_3$ is formed, has been studied in detail. It has been found that under special reaction conditions tetrachloroaminoborane Cl$_2$BNCl$_2$ can be isolated.[12,12a]

Photochemical preparations of B-monochloro- and B-monobromoborazine have been performed. Gas-phase mixtures of borazine and CCl$_4$, CHCl$_3$, CH$_3$Cl, CH$_3$Br or HBr were irradiated using a medium-pressure mercury lamp.[13]

1,3,5-Trichloro-2,4,6-trifluoroborazine (FBNCl)$_3$ was obtained by the reaction of ClN(SiMe$_3$)$_2$ with Et$_2$O.BF$_3$. The partially N-silylated borazines F$_3$B$_3$N$_3$Cl$_2$SiR$_3$ and F$_3$B$_3$N$_3$Cl(SiR$_3$)$_2$ (R = Me) are formed as by-products (identified by mass spectrometry).[13]

On reacting BF$_3$ with BrN(SiMe$_3$)$_2$ in CCl$_4$, 1-bromo-3,5-bis(trimethylsilyl)-2,4,6-trifluoroborazine F$_3$B$_3$N$_3$(SiMe$_3$)$_2$Br is obtained in 16% yield, the symmetrical compound F$_3$B$_3$N$_3$(SiMe$_3$)$_3$ being a by-product.[14]

2,4,6-Trifluoroborazine (FBNH)$_3$ was obtained from the reaction of 2,4,6-trichloroborazine with NaF in acetonitrile.[15]

By thermal decomposition of bis(trimethylsilyl)aminodifluoroborane F$_2$B—N(SiMe$_3$)$_2$ on heating at 275 °C for several days, the fluoro derivative [FBN(SiMe$_3$)]$_3$ was obtained. Another efficient method for preparing this compound is the reaction of Me$_3$SiNH$_2$.BF$_3$ with EtPr$^i_2$N.BF$_3$ in ether, [16] or by reacting 1,3,5-tris(trimethylsilyl)-2,4,6-trichloroborazine with NaF in refluxing acetonitrile.[16]

2,4,6-Tribromo-1,3,5-tris(dibromoboryl)borazine [BrBN(BBr$_2$)]$_3$ resulted from the reaction of N-chlorovanadiumnitride trichloride Cl$_3$VNCl with boron tribromide.[17] A large excess of BBr$_3$ favours formation of the borazine derivative.

The preparation of 1,3,5-tri(arylsulphonyl)-2,4,6-trihalogenoborazines

$(XBNSO_2Ar)_3$ in high yields from $ArSO_2NX_2$ or $ArSO_2N = SbPh_3$ has been claimed.[18]

Mercaptoborazines $(MeS — BNH)_3$ react exothermically at room temperature with trialkoxystibines to give the corresponding 2,4,6-tris(alkoxy)borazine $(RO—BNH)_3$ (e.g. $R = n\text{-butyl}$).[19] The driving force of the reaction has been attributed to the high affinity of boron for oxygen and of antimony for sulphur.

A compound with methoxy groups attached to nitrogen, $(FBNOMe)_3$, has been obtained by reaction of disilylated methoxyhydroxylamine $MeON(SiMe_3)_2$, either with BF_3 or its diethyl etherate.[20]

Unsymmetrical derivatives 2,4-dichloro-6-dimethylaminoborazine $(Me_2N)Cl_2B_3N_3H_3$ and 2-chloro-4,6-bis(dimethylamino)borazine $(Me_2N)_2\text{-}ClB_3N_3H_3$ have been prepared from $(ClBNH)_3$ and dimethylamine; the latter compound is unstable at room temperature.[11] It can also be obtained by substituent redistribution between $(Me_2N\cdot BNH)_3$ and BCl_3.[11] For the synthesis of unsymmetrically substituted pseudohalogeno derivatives, the reaction of borazine with a silver(I) salt has been studied:[21]

$$(HBNH)_3 + AgX \longrightarrow XH_2B_3N_3H_3 + Ag + \tfrac{1}{2} H_2 \qquad (1)$$
$$X = CN, NCO, SO_3Me, MeCOO$$

The reactions of Ag(I) salts with borazine are very complex. Another route for the synthesis of B-monosubstituted pseudohalogenoborazines $XH_2B_3N_3H_3$ is the reaction of 2-chloroborazine with a silver(I) salt AgX ($X = CN, NCO$).[21] The reactions of $HCl_2B_3N_3H_3$ with silver(I) salts in acetonitrile result in the isolation of $H(Cl)(CN)B_3N_3H_3$, $H(Cl)(OCN)B_3N_3H_3$ and $H(OCN)_2B_3N_3H_3$.[22]

2.3.3 Functional derivatives of borazine

2,4,6-Trihydroborazines $(HBNR)_3$. The reaction of aromatic amines with triethylamineborane $Et_3N.BH_3$ has been investigated, and some borazine derivatives $(HBNR)_3$, with $R = Ph$, $o\text{-ClC}_6H_4$, $o\text{-CF}_3C_6H_4$, C_6F_5, $2,6\text{-Cl}_2C_6H_3$, 2-Cl-6-FC_6H_3, $2,6\text{-Me}_2C_6H_3$, have been obtained by heating the reactants to 140–180 °C.[23]

The procedure for the synthesis of $H_3B_3N_3H_2Me$ from $NaBH_4$, NH_4Cl and $MeNH_2.HCl$ in diglyme has been modified.[24]

When the tris(mono-n-alkylamine)pentaborane(9) complexes $B_5H_9.3NH_2R$ are heated to 150–180 °C they decompose to give hydrogen and 1,3,5-trialkylborazines $(HBNR)_3$.[25,26] A mechanism for this reaction has been suggested,[27,27a] involving intermediate formation of monoalkylaminoborane $(RHNBH_2)_n$.

The reaction of monoalkylamines with phenylborate $B(OPh)_3$, aluminium powder and hydrogen, under pressure and heating, gives borazines $(HBNR)_3$ in high yield (82–93%) (with R = Et, Pr^i, Bu^t).[28]

Bis(mercaptoboranes), $HB(SR)_2$, R = Et, Bu^t, and bis(alkoxy)boranes have been reacted with primary amines to yield borazines $(HBNR)_3$.[29]

Anilino-dihydroborane, $PhNHBH_2$, on heating to 120 °C eliminates hydrogen and forms 1,3,5-triphenylborazine, $(HBNPh)_3$.[30]

6-Hydrogen- and 4,6-dihydrogenborazines. On treatment of $H_3B_3N_3H_2Me$ with $HgCl_2$, a mixture of two isomers is obtained, which can be separated by fractionation:[24]

$$\text{para isomer} \qquad \text{ortho isomer} \tag{2}$$

By-products of this reaction are the dichloro isomers of l-methylborazine, which, however, could not be obtained in pure state.

The dimethylamino derivatives of l-methylborazine are readily prepared from the corresponding chloro derivatives by reaction with dimethylamine in n-pentane at −78 °C.[24]

1,3,5-Triorgano-2,4,6-trihalogenoborazines (XBNR)₃. The adduct of heptamethyldisilazane with BF_3, $(Me_3Si)_2NMe.BF_3$, decomposes on heating to eliminate trimethylfluorosilane with formation of the borazine derivative $(FBNMe)_3$.[31] In the synthesis of the corresponding chloro- and bromo-derivatives, the compounds X_2B—NMe—$SiMe_3$ (X = Cl and Br) can be isolated after elimination of Me_3SiX from $(Me_3Si)_2NMe.BX_3$; they decompose readily at room temperature to form borazines $(XBNR)_3$.[31]

The conversion of 1,3,5,7-tetramethyl-2,4,6,8-tetrafluoroborazocine $(FBNMe)_4$ into the corresponding borazine derivative $(FBNMe)_3$ (equation 3), occurring during sublimation, has been reported.[32] This demonstrates the higher stability of the six-membered ring.

(Trimethylsilyl)alkylamino difluoroboranes F_2B—$N(SiMe_3)(NHR)$, which have been obtained by reacting $Et_2O.BF_3$ with Me_3SiNHR in the presence of $EtPr^i_2N$ (dehydrofluorination), undergo thermal decomposition to give the corresponding fluoroborazines $(FBNR)_3$ with R = Me, Et, Pr^n, Pr^i and Bu^n.[33]

$$
3 \quad
\begin{array}{c}
\text{F} \\
\text{R—N}\overset{\text{B}}{\diagup}\diagdown\text{N—R} \\
| \quad\quad | \\
\text{FB} \quad\quad \text{BF} \\
| \quad\quad | \\
\text{R—N}\diagdown\underset{\text{B}}{\diagup}\text{N—R} \\
\text{F}
\end{array}
\quad\xrightarrow{\Delta}\quad 4 \quad
\begin{array}{c}
\text{F} \\
\text{R—N}\overset{\text{B}}{\diagup}\diagdown\text{N—R} \\
| \quad\quad | \\
\text{FB}\diagdown\quad\diagup\text{BF} \\
\text{N} \\
| \\
\text{R}
\end{array}
\tag{3}
$$

Another method for preparing 2,4,6-trifluoroborazine derivatives is the reaction of 2,4,6-tris(alkylamino)borazines $(RNH \cdot BNR)_3$ with BF_3 ($R = Me$, C_6F_5, p-ClC_6H_4). The intermediate formation of an unstable aminodifluoroborane $RNHBF_2$ was suggested; this decomposes to the fluoroborazine $(FBNR)_3$ and the adduct $BF_3.RNH_2$. [34,34a]

Primary amine–boron trifluoride adducts and ammonium tetrafluoroborates can be dehydrofluorinated using ethyldiisopropylamine–boron trifluoride complex as an HF-trapping agent:[35]

$$3\ R'NH_2.BF_3 + 6\ R_3N.BF_3 \longrightarrow (FBNR')_3 + 6\ R_3NH^+BF_4^- \tag{4}$$

$$3\ R'NH_3^+BF_4^- + 9\ R_3N.BF_3 \longrightarrow (FBNR')_3 + 9\ R_3NH^+BF_4^- \tag{5}$$

$$R = Me,\ Et,\ Pr^i,\ Bu^t,\ Ph,\ etc.$$

Modified procedures for the synthesis of $(ClBNMe)_3$ from BCl_3 and $MeNH_2.HCl$ [36] and by cleavage of the exocyclic B–N bond of tris-(amino)borazines[37,38] have been reported. Some 2,4,6-trichloroborazines $(ClBNR)_3$, with $R = o$-MeC_6H_4, o-BrC_6H_4, have been prepared by the reaction of amines with boron trichloride, via intermediate $RNH_2.BCl_3$.[39–43]

1,3,5-Trimethyl-2-butyl-4,6-difluoroborazine, 1,3,5-trimethyl-2,4-dibutyl-6-fluoroborazine, 2,4-dibutyl-6-fluoroborazine and 1,2,3,4,5-pentamethyl-6-fluoroborazine have been obtained by refluxing the corresponding chloro derivatives with NaF in acetonitrile.[42a]

The silylaminophosphine Ph_2P—NMe—$SiMe_3$ reacts with BCl_3 to give $(ClBNMe)_3$, Ph_2PCl and Me_3SiCl; a mechanism has been postulated for this reaction.[43]

Several 1,3,5-triorgano-2,4,6-trichloroborazines $(ClBNR)_3$, with $R = \alpha$-naphthyl, o-MeC_6H_4, Ph, o-ClC_6H_4, α-$C_{10}H_6Cl$, and α-$C_{10}H_6Br$ have been obtained by reaction of the BCl_3 adducts of the corresponding aromatic amines with triethylamine.[43a, 43b]

The thermolysis of silylaminohalogenoboranes Cl_2B—$NR(SiMe_3)$ at 500–600 °C yields 2,4,6-trichloroborazines $(ClBNR)_3$ ($R = Me$, $SiMe_3$) after elimination of Me_3SiCl.[44] Another 2,4,6-trichloroborazine $(ClBNEt)_3$ has been obtained by decomposition of $EtN(BClNMe_2)_2$ on attempted distillation. The second product was $ClB(NMe_2)_2$.[45]

Substituted aminoborazines. The formation of tris(methylamino)borazine $(MeNH \cdot BNMe)_3$ from the reaction of BCl_3 with methylamine has been investigated in some detail, and several reaction steps involving $B(NHMe)_3$, $ClB(NHMe)_2$ and $ClB(NHMe)_2 \cdot MeNH_2$ have been proposed.[46] Depending on the amount of methylamine used, either tris(methylamino)borane $B(NHMe)_3$ or the tris(methylamino)borazine can be isolated.

By thermal decomposition of t-butyldimethylsilylaminoborane Bu^tMe_2 $SiN(R)$—$B(NH_2)(NMe_2)$ (R = H, Me) even on standing at 25 °C, dimethylamine is eliminated, with formation of 2,4,6-tris(amino)borazines $[Bu^tMe_2$ $SiN(R) \cdot BNH]_3$.[47,48] A 2,4,6-tris(pentafluorophenylamino)borazine $(C_6F_5-$ $NH \cdot BNC_6F_5)_3$ was prepared from pentafluoroaniline and the BCl_3 adduct of triethylamine, by dehydrochlorination with triethylamine.[34] 1,3,5-Triphenyl-2,4,6-tris(anilino)borazine[35] has also been obtained by a dehydrofluorination procedure:

$$3\ R'NH_2 \cdot BF_3 + 3\ R''NH_2 + 9\ R_3N \cdot BF_3 \longrightarrow$$
$$(R'NHBNR'')_3 + 9\ R_3NH^+BF_4^- \qquad (6)$$

Lithiated hexamethyldisilazane $(Me_3Si)_2NLi$ reacts with 2-chloro-1,3,4,5,6-pentamethylborazine $ClMe_2B_3N_3Me_3$ to give 2-hexamethyldisilazanyl-1,3,4,5,6-pentamethylborazine $(Me_3Si)_2N(Me_2)B_3N_3Me_3$.[49] Similarly, trimethyl-2,4,6-trichloroborazine $(ClBNMe)_3$ reacts with lithiohexamethyldisilazane (mole ratio 1 : 3) to give 1,3,5-trimethyl-2,4,6-tris[bis-(trimethylsilyl)amino]borazine $[(Me_3Si)_2N \cdot BNMe]_3$.[50]

On treatment of silylated carbonic-acid amides and thioamides R'—$C(:X)$—$NR(SiMe_3)$ with 2-chloropentamethylborazine, several amido- and thioamidoborazine derivatives $R'C(X)NR.Me_2B_3N_3Me_3$ are obtained (X = O, S; R = Me, CF_3 or Ph; R' = Me, Ph, H, 2,6-$Me_2C_6H_3$ or 2,4-$Me_2C_6H_3$).[49,49a]

It has been shown that 2-chloro-1,3,4,5,6-pentamethylborazine reacts with silylated amidines, $RC(:NR')$—$NR''(SiMe_3)$, with elimination of Me_3SiCl, at 150–160 °C, to give the corresponding 2-amidino-1,3,4,5,6-pentamethylborazine derivatives; compounds with R = H, CF_3; R' and R'' = Ph, 2-FC_6H_4, 3-$CF_3C_6H_4$ and 2,4,6-$Me_3C_6H_2$ were thus obtained.[51]

Borazines with exocyclic O-substituents. Hydroxoborazines $(HO \cdot BNC_6H_4-Me_2)_3$ and $(HO \cdot BNC_6H_3Me_2\text{-}2,6)_3$ have been obtained by the reaction of the corresponding chloroborazine with water. The two compounds are hydrolytically stable owing to the presence of the methyl groups in *ortho* positions.[40]

By reaction of 2,4,6-trihalogenoborazines $(XBNR)_3$ with (*N,O*)-bis(trimethylsilyl)acetamide Me—$C(OSiMe_3)(:NSiMe_3)$, some trimethylsiloxyborazines $(Me_3SiO \cdot BNR)_3$ (R = H, Me, Ph; X = F or Cl) have been obtained.[50]

2,4,6-Tris-n-butoxyborazine $(BuOBNH)_3$ has been obtained by the reaction of 2,4,6-tris(methylthio)borazine with tri-n-butoxystibine.[19]

2.3.4   1,3,5-Triorgano-2,4,6-triorganoborazines

Several hexaorganoborazines have been obtained by ring-closure reactions[51] of bis(trimethylsilyl)methylamino diboryl-amines Me—N[BR—N—Me(SiMe$_3$)]$_2$ with dihalogeno-organoboranes $R'BX_2$, with elimination of trimethylchlorosilane; the reaction gave borazines of the type $R_2R'B_3N_3Me_3$, with R = Me, Ph and R' = Me, Ph.

α-Naphthylamine reacts with boron trichloride to give 1,3,5-tris(α-naphthyl)-2,4,6-trichloroborazine, which can be converted to 1,3,5-tris(α-naphthyl)-2,4,6-trimethylborazine in a reaction with MeMgI.[53] Similarly, methylation of 2,4,6-trichloroborazines was used for the synthesis of 1,3,5-triorgano-2,4,6-trimethylborazines, $Me_3B_3N_3RR'R''$, where the RR'R'' sets are various combinations of α-$C_{10}H_7$, Ph, 2-ClC_6H_4, 2-MeC_6H_4 and other aromatic substituents.[43a,43b]

Tris(carbimino)borazines   (RN=CH·BNR')$_3$ (R = cyclo-C_6H_{11}, Ph; R' = Me or Ph) have been obtained by hydroboration of isonitriles RN≡C with borazines (HBNR')$_3$ at 160 °C in a sealed tube.[54]

1,3,5-Trimethyl-2-butylborazine and 1,3,5-trimethyl-2,4-dibutylborazine react with isonitriles in the mole ratio 1 : 2 and 1 : 1, to yield the unsymmetrically substituted carbiminoborazines, $Bu(RN=CH)_2B_3N_3Me_3$ and $Bu_2(RN=CH)B_3N_3Me_3$ (R = Ph, cyclo-C_6H_{11}).[54]

(7)

$$R = C_6H_{13}$$

1,3,5-Tris(4-pyridyl)-2,4,6-triphenylborazine $(PhBN·C_5H_4N)_3$ has been obtained by refluxing a slurry of 4-aminopyridine with bis(dimethylamino)-phenylborane.[55,56]

If a sample of (2-pyridylamino)diphenylborane is slowly heated under vacuum to 180 °C (30 min) a nonvolatile residue of 1,3,5-tris(2-pyridyl)-2,4,6-triphenylborazine is obtained.[57]

1,3,5-Trimethyl-2,4,6-tris(methylcyclopentadienyl)borazine has been prepared by addition of a solution of 1,3,5-trimethyl-2,4,6-trichloroborazine in benzene to a suspension of potassium methylcyclopentadienyl in benzene.[58] Similarly, 1,3,5-trimethyl-2,4,6-tris(cyclopentadienyl)borazine has been obtained.[58]

The reaction of 1,3,5-trimethyl-1,3,5-tris(azonia)-2,4,6-tris(borata)cyclohexane with l-hexene has been studied,[59] and by pyrolytic dehydrogenation the intermediate 1,3,5-trimethyl-2,4,6-trihexyl-1,3,5-triazonia-2,4,6-triboratacyclohexane is converted to 2,4,6-trimethyl-1,3,5-trihexylborazine (see equation 7).

In an attempt to synthesize the S=P—N—B skeleton, starting from $Ph_2P(S)Cl$ and the silylaminoboranes $Me_3SiNHBEt_2$ and $Me_3Si-NMe-BEt_2$ (at 100 °C), besides a small amount of Me_3SiCl and most of the starting materials, 2,4,6-triethylborazine and 1,3,5-trimethyl-2,4,6-triethylborazine have been identified. In contrast with this, $Me_2P(S)Cl$ reacts with $Me_3Si-NMe-BEt_2$ with the formation of 1,3,5-trimethyl-2,4,6-triethylborazine $(EtBNMe)_3$ in 70% yield,[60] as a result of the decomposition of $Me_2P(S)-NMe-BEt_2$ formed as an intermediate.

N-{Bis(dimethylamino)thiophosphoryl}methylamino dimethylborane $(Me_2N)_2P(S)-NMe-BMe_2$ decomposes upon storage for a prolonged time and reacts with bromodimethylborane to yield hexamethylborazine $(MeBNMe)_3$ and $(Me_2N)_2P(S)Me$ or $(Me_2N)_2P(S)Br$ respectively. If an unsymmetrically substituted diborylamine, for example $Ph_2B-NMe-BMe_2$, is heated to 60 °C, formation of a borazine $(PhBNMe)_3$ takes place.[61] Another asymmetrically substituted diborylamine by thermolysis gives some hexamethylborazine (R = Me):[62]

$$\begin{array}{c} R \\ | \\ N \\ \diagup \; \diagdown \\ \left| \begin{array}{c} \\ \end{array} \right. \quad B-N-BR_2 \\ \diagdown \; \diagup \quad | \\ N \quad R \\ | \\ R \end{array} \xrightarrow[20\,h]{140\,°C} \tfrac{1}{3}(RBNR)_3 + \begin{array}{c} R \\ | \\ N \\ \diagup \; \diagdown \\ \left| \begin{array}{c} \\ \end{array} \right. \quad B-R \\ \diagdown \; \diagup \\ N \\ | \\ R \end{array} \qquad (8)$$

In the attempted synthesis of Me_2B-NEt-$SnMe_3$ from Me_2B-NEtLi and Me_3SnCl, 1,3,5-triethyl-2,4,6-trimethylborazine has been identified, besides BMe_3 and $(Me_3Sn)_2NEt$; this is an indication of the stability of the borazine ring compared to noncyclic B–N compounds.[63]

N-Lithiomethylaminodimethylborane $Me_2BNLiMe$ reacts with trimethylborane and methylaminodimethylborane to yield hexamethylborazine $(MeBNMe)_3$ and $Li[BMe_4]$ and $Li[Me_3B\text{-}NHMe]$ respectively.[64]

1,3,5-Triethyl-2,4,6-tris(o-tolyl)borazine has been prepared by the reaction of o-tolylmagnesium bromide with 1,3,5-triethyl-2,4,6-trichloroborazine. In this case atropisomerism is observed. The *cis* isomer was isolated in a pure state by crystallization from a benzene–hexane mixture (m.p. 198.5–199.5 °C).[1] *Trans*-enriched samples (m.p. 193.5–195.5°C) have been obtained by column chromatography.[65]

1,3,5-Tris(o-tolyl)-2,4,6-triethylborazine has been prepared by reaction of ethylmagnesium bromide with 1,3,5-tris(o-tolyl)-2,4,6-trichloroborazine.[65]

1,3,5-Tris(buten-3-yl)-2,4,6-triphenylborazine and 1,3,5-tris(buten-3-yl)-2,4,6-trimethylborazine have been prepared from buten-3-yl-amine by reaction with dichlorophenylborane and dibromomethylborane respectively.[79]

By thermolysis of silylaminohalogenoboranes ClRB—NR′(SiMe$_3$) at 500–600 °C, depending on the nature of R and R′, either 1,3-diaza-2,4-diboretidines (RBNR′)$_2$ or borazines (RBNR′)$_3$ are formed.[44] Borazines are obtained with R = Ph, R′ = Me, Bu^t and 4-MeC$_6$H$_4$ and with R = 2,4,6-Me$_3$C$_6$H$_2$ and R′ = Me.

2.3.5 Miscellaneous borazine derivatives

1,3,5-Trimethyl-2,4,6-tris(methyl-d_3)borazine was prepared by treating the organomagnesium compound obtained from 1,3,5-trimethyl-2,4,6-trichloroborazine with methyl-d_3 iodide.[66]

1,3,5-Tris-(ferrocenyl)borazine has been prepared by a convenient procedure, from C$_5$H$_5$FeC$_5$H$_4$BCl$_2$ and ammonia, in toluene or hexane, in 80–90% yield.[67,67a]

1-Trimethylsilyl-2,3,4,5,6-pentamethylborazine Me$_3$B$_3$N$_3$Me$_2$(SiMe$_3$) has been synthesized in 47% yield by the ring closure of Me$_3$Si—N(BMeBr)$_2$ with Me$_2$B-NMeLi; pentamethyldiborylamine MeN(BMe$_2$)$_2$ is formed as a by-product.[68]

When bis(9-borabicyclo[3.3.1]nonane-9-yl)amine is lithiated with t-BuLi and treated with chlorodiethylborane, several parallel reaction paths are followed, one leading to a borazine with N-(9-borabicyclo[3.3.1]nonan-9-yl) substituents.[69]

The cyclotrisilazane ring system has been attached to a borazine ring by reacting lithiated hexamethylcyclotrisilazane with 2-chloro-pentamethylborazine (R = Me):[70]

(9)

Hexamethylcyclotrisilazane $(Me_2SiNH)_3$ reacts with dichlorophenyl borane $PhBCl_2$ and gives 2,4,6-triphenylborazine $(PhBNH)_3$, 2-chloro-1-(dimethylphenylsilyl)-4,6-diphenylborazine, $ClPh_2B_3N_3H_2(SiMe_2Ph)$ and a B_2SiN_3 ring derivative.[71]

Table 2.1 presents some reactions in which the borazine ring system has been formed as a by-product. Such investigations have proved useful for the interpretation of reaction mechanisms.

Table 2.1 Borazine derivatives as by-products in the synthesis of other compounds.

Reactants	Borazine derivative	Reference
$Me-N(SiMe_3)_2 + MeBBr_2$	$(MeBNMe)_3$	72
$Me-N(SiMe_3)_2 + PhBCl_2$	$(MeBNPh)_3$	72
$Me-N(SiMe_2Br)_2 + MeBBr_2$	$(MeBNMe)_3$	72
$(Me_2Si-NMe)_3 + BBr_3$	$(BrBNMe)_3$	72
$MeB(NMeLi)_2 + Ph_2SiCl_2$	$(MeBNMe)_3$	73
$LiNMe-BMe_2 + SnCl_2$	$(MeBNMe)_3$	74
$MeB(SMe)_2 + MeNH-NHMe$	$(MeBNMe)_3$	75
$Me_2B_2S_3 + NH_3$	$(MeBNH)_3$	76
$Me_2B_2S_3 + MeNH_2$	$(MeBNMe)_3$	76
$Me_2B-NH-BMe_2 + LiMe$	$(MeBNH)_3$	76a
$Me-N(SiMe_3)_2 + Ph_2BCl$	$(PhBNMe)_3$	31

Reactions between aminoborazine and B_2H_6 produced white polymeric substances which evolved hydrogen at room temperature. After the evolution of hydrogen ceased, needlelike crystals sublimed at 50°. These crystals proved to be diborazinylamine:[77,78]

$$
\begin{array}{c}
\text{H} \quad\quad \text{H} \quad\quad \text{H} \\
\text{N} \quad\quad \text{N} \quad\quad \text{N} \\
HB \diagup \quad\backslash B \diagup \quad\backslash B \diagup \quad\backslash BH \\
| \quad\quad | \quad\quad | \quad\quad | \\
HN \quad\quad NH \quad HN \quad\quad NH \\
\quad\backslash B \diagup \quad\quad\quad \backslash B \diagup \\
\quad\quad \text{H} \quad\quad\quad\quad \text{H}
\end{array}
$$

2.3.6 Polycyclic systems containing one borazine ring

This type of compound contains the borazine ring incorporated in a fused ring system, formed mainly of organic rings, as in (2)-(8).

Reaction of buten-3-ylamine with diborane or with trimethylamine–borane yielded tris(tetramethylene)borazine (2) in 7.4% and 65% yield respectively.[79]

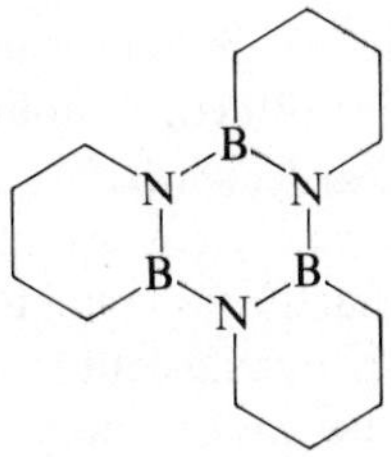

(2)

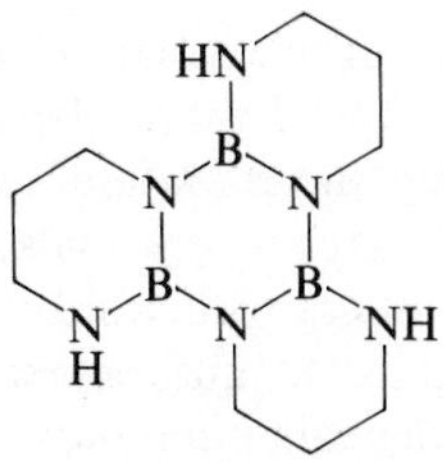

(3)

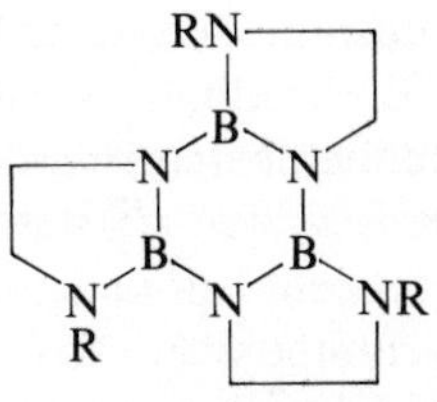

(4)

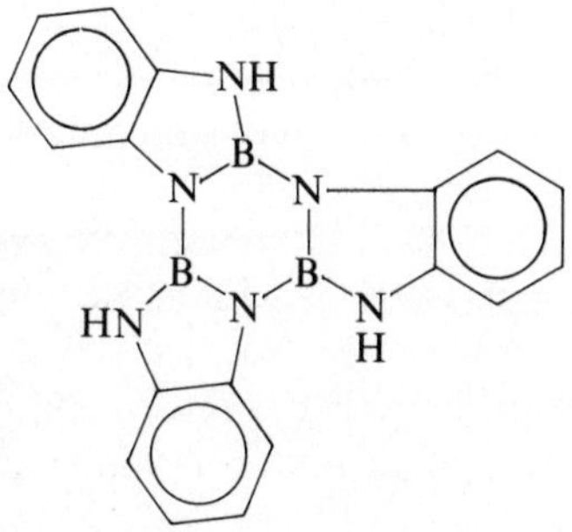

(5)

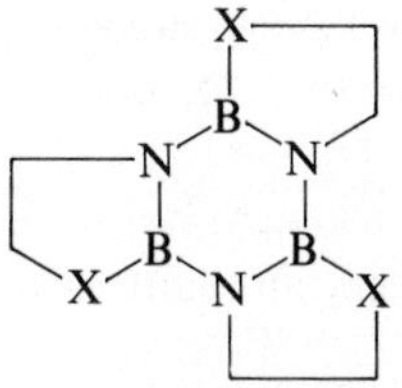

(6a) X = S
(6b) X = O

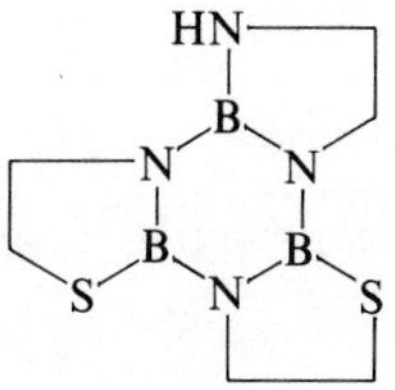

(7)

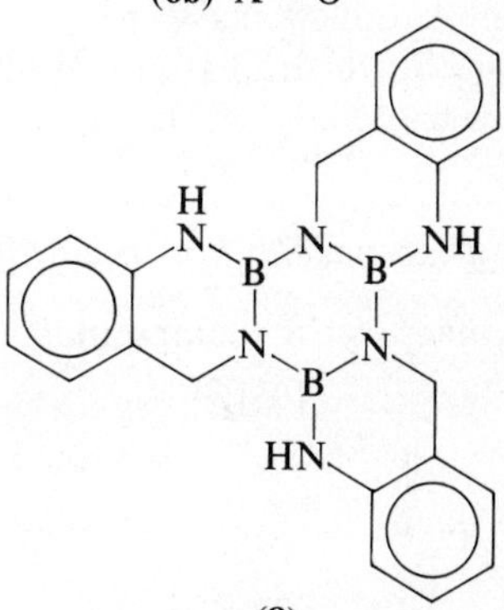

(8)

In the reaction of tris(dimethylamino)borane $B(NMe_2)_3$ with 1,3-propylenediamine $H_2N(CH_2)_3NH_2$, the primary product 2-dimethylamino-1,3,2-diazaboracyclohexane, on attempted distillation, eliminated dimethylamine and formed a polycyclic borazine (**3**):[80]

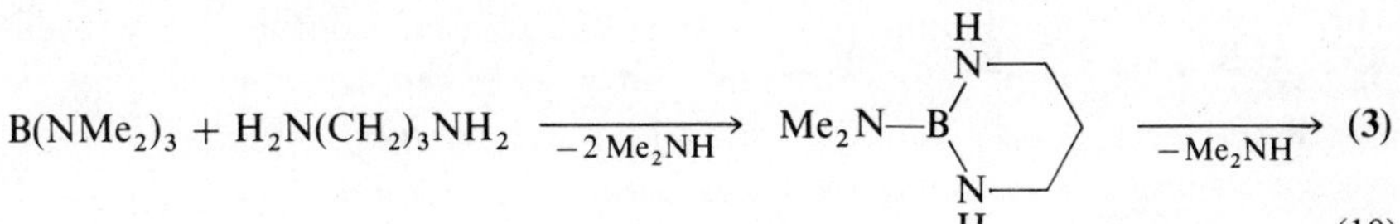

$$B(NMe_2)_3 + H_2N(CH_2)_3NH_2 \xrightarrow[-2\,Me_2NH]{} Me_2N{-}B \qquad \xrightarrow[-Me_2NH]{} \textbf{(3)}$$

(10)

By using N-monosubstituted ethylenediamine derivatives $H_2N(CH_2)_2NHR$ in the reaction with $B(NMe_2)_3$, a polycyclic system is formed, containing five-membered rings fused to the B_3N_3 heterocycles, (4) R = Me.[80] It has been shown[81,82] that 1,3,2-diazaboracycloalkanes with reactive groups at the boron atom tend to undergo intermolecular condensation reactions, unless the nitrogen is protected by an organic exocyclic substituent.

The reaction of o-phenylenediamine and alkylborate gave a polycyclic system (5), which is stable against boiling water and boiling acetic acid. It is destroyed by hydrochloric acid.[83] The same compound has been obtained by the reaction of o-phenylenediamine with $B(NMe_2)_3$.[82a]

Other polycyclic systems (6) are obtained in high yields by the reaction of a tris(thioborane), $B(SPr)_3$ with hydroxy- and mercaptoaminoethanes H_2N—CH_2CH_2—XH (X = O, S).[84-86]

An unsymmetrical borazine derivative (7) has been obtained by refluxing tris(ethylenethio)borazine (6a) X = S, with ethylenediamine in toluene for one week.[87]

It has been found that o-aminobenzonitrile is able to react with diborane at 85 °C, and, depending on the molar ratios and reaction conditions, different products have been obtained. One of them is the polycyclic borazine (8). In this case the molar ratio o-aminobenzonitrile : diborane = 2 : 1 was used.[88]

2.3.7 Macrocyclic boron–nitrogen ring systems

Some macrocyclic compounds, (9)-(12), incorporating several borazine rings, have been prepared:

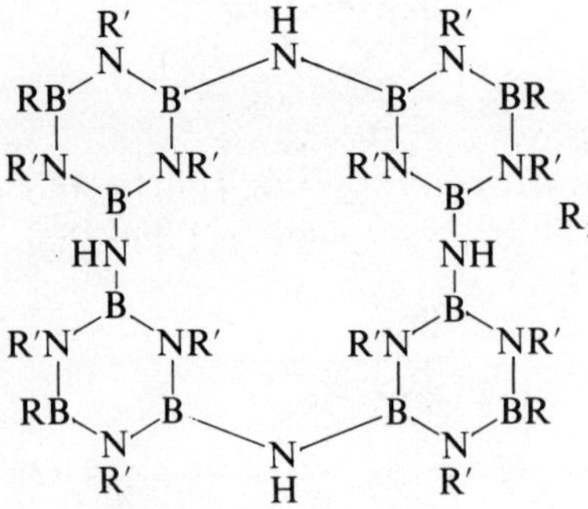

(9)

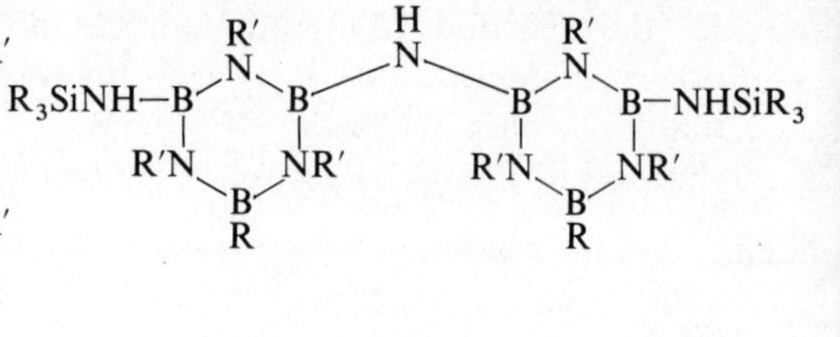

(10)

(11)

(12a) X = Y = O
(12b) X = O, Y = NH

By reaction of 1,2,3,5-tetraorgano-4,6-dichloroborazines $Cl_2R'B_3N_3R_3$ with hexamethyldisilazane, the first $(BN)_{12}$ ring system (**9**) has been made (R = Me, Et, Pr^i; R′ = Me).[89,90] The use of highly purified 1,2,3,5-tetraorgano-4,6-dichloroborazine is essential for such reactions. In case of R′ = Me two by-products of the reaction were isolated: a small amount of 1,2,3,5-tetramethyl-4,6-bis(trimethylsilylamino)borazine and bis(1,3,4,5-tetramethyl-6-trimethylsilylamino-2-borazinyl)amine (**10**) (R = R′ = Me) in 25% yield.[89,90]

By the reaction of 2,4-dichloro-1,3,5,6-tetramethylborazine with N,N-dimethylformamide and dimethylamine six borazine molecules have been linked together by O-bridges and a ring system with 36 atoms in the ring, (11), was formed (R = Me). The mass spectrum of this compound shows the molecular peak as base peak.[91] Studies on molecular models have shown that this molecule must have a crown-like configuration with a threefold symmetry axis.

Macrocyclic systems have also been prepared by reactions of 2,4-difunctional borazine derivatives with m-difunctional aromatic compounds. By this method systems have been formed that contain an alternating sequence of 2,4-borazines and aromatic 1,3-dihydroxy-, 1,3-diamino- or 1-amino-3-hydroxy- compounds. The structure of these compounds is based on a skeleton illustrated by formula (12) (X = O, Y = O or NH, R = Me). They have been obtained in yields of 5–48% by interaction of 2,4-dichloro-1,3,5,6-tetramethylborazine with 1,3-$C_6H_4(ONa)_2$, 1,3-$(NaO)_2$-5-Me-C_6H_3, 1,3-$C_{10}H_6(ONa)_2$ and 1-NH_2-3-NaO-C_6H_4 respectively in refluxing benzene.[92,92a]

2.3.8 Reactions of the borazines

Some of the borazine reactions have already been mentioned in previous sections dealing with the synthesis of functional derivatives or of polycyclic systems.

Photochemical reactions. Photochemical reactions of borazine with HCl, CH_3Cl, $CHCl_3$, CCl_4 or HSO_3Cl in the gas phase yield 2-chloroborazine. Irradiation of a borazine–CH_3Br mixture produces 2-bromoborazine. The 4,6-dideuterated derivatives have been formed by irradiating 2,4,6-trideutero-borazine with methyl halides. Mixtures of borazine and halogenating agents were subjected to 1849 Å radiation. The reaction mechanism has been discussed.[93]

From preparative-scale photolysis experiments with borazine–ammonia mixtures, 2-aminoborazine has been isolated. A small amount of the B–N analogue of naphthalene has also been obtained. This molecule results by elimination of hydrogen from the vibrationally excited species $B_3N_3H_6^*$. The resulting intermediate reacts with another borazine molecule and forms the fused naphthalene-like system.[94]

The photochemical exchange of borazine with deuterium has been studied under 1849 Å radiation. The quantum yield for the production of 2-deutero-borazine increased with increase in D_2 pressure and was greatly affected by the addition of He, N_2 or H_2. The excited-state mechanism, which was discussed as well as a radical mechanism, has been shown to be consistent with the experimental data.[95] Exchange of hydrogen and deuterium reaches an equilibrium with the products $DH_2B_3N_3H_3$, $D_2HB_3N_3H_3$, $D_3B_3N_3H_3$, HD and H_2.[96]

Photolysis of gaseous borazine at 1849 Å in the absence of another reactant produced naphthalene- and biphenyl-like bicyclic boron–nitrogen ring systems (13) and (14):

(13)   (14)

The formation of (13) was interpreted as an attack of $H_2\dot{B}_3\dot{N}_3H_2$ (formed by decomposion of the excited species $H_3B_3N_3H_3{}^* \rightarrow H_2\dot{B}_3\dot{N}_3H_2$) on a borazine molecule in a Diels–Alder-type reaction with borazine as the "diene".[97,98]

2-Methylaminoborazine and 2-dimethylaminoborazine have been prepared by photochemical reactions of borazine with methylamine and dimethylamine.[98]

Mixtures of borazine and hexafluoroacetone have been photolysed (mole ratio 4:1, total pressure 5 mm, block radiation between 2200 and 2700 Å). By this process $H_2[(CF_3)_2HCO]B_3N_3H_3$ and $H_2[(CF_3)_3CO]B_3N_3H_3$ have been prepared. A reaction mechanism has been proposed in which a hydrogen atom is abstracted almost exclusively from the boron site, followed by radical recombination to 2-(2H-hexafluoro-2-propoxy)borazine.[99]

The photolysis of neat N-methylborazine produced H_2 and CH_4 in addition to several monocyclic and bicyclic borazine derivatives (B–N analogues of naphthalene). 1-Methylborazine photolysed with CH_3OH and with $HNMe_2$ gave 1-methyl-2-methoxyborazine and 1-methyl-2-dimethylaminoborazine respectively. The photolysis of 1,3,5-trimethylborazine and ammonia gave 1,2-bis(3′,5′-dimethylborazinyl)ethane and bis(N-trimethylborazinyl)amine as the main products. In addition, 1-methyl-2-aminoborazine (ortho isomer), 1,3-dimethyl-2-aminoborazine and 1,3,5-trimethyl-2-aminoborazine have been discussed. It was assumed that the process involves the formation of a longer-lived excited state $H_3B_3N_3H_2Me^*$, which may undergo collision reactions with ammonia molecules to form the cited product with elimination of H_2. Competing with this process a free-radical mechanism has been proposed.[100]

The mercury-sensitized photolysis of $(HBNMe)_3$ with hydrogen, using 2537 Å radiation, has produced 1,2-di(3′,5′-dimethylborazinyl)ethane (15) (R = Me):[101]

$$\underset{R}{\overset{R\diagdown}{\underset{N-B}{\underset{N-B}{\overset{N}{\bigcirc}}}}}\underset{H}{\overset{H}{N}}-CH_2-CH_2-\underset{H}{\overset{H}{N}}\underset{R}{\overset{B-N\diagup R}{\underset{B-N\diagdown}{\overset{N}{\bigcirc}}}}$$

(15)

Again, a radical mechanism has been proposed. Analogous diborazinyl-ethane derivatives have been obtained by irradiation of 1-methylborazine and 1,3,5-trimethylborazine with H_2. The major products of photolysis of $(CD_3)_3B_3N_3H_3$ with D_2 was $D(CD_3)_2B_3N_3H_3$.[102]

Chemical reactions. A method has been described for the synthesis of 2,4,6-trideuteroborazines by direct deuteration of a borazine derivative with a deuterium–HBF_2 mixture, which were allowed to react thermally for 16 h. This method has been found suitable for preparation of $D_3B_3N_3Me_3$ and $D_3B_3N_3Et_3$.[103]

Reactions of 1,3,5-trimethylborazine $(HBNMe)_3$ with hydrogen halides have been investigated. At room temperature three molecules of hydrogen halide add to a molecule of borazine to form $(HBNMe)_3.3HX$ (X = Cl, Br). Pyrolysis studies have shown that hydrogen–halogen exchange occurs to yield 2-halogeno-, 2,4-dihalogeno- and 2,4,6-trihalogenoborazine derivatives. In the reaction of $(HBNMe)_3$ with bromine, 1,3,5-trimethyl-2,4-dibromoborazine has been obtained.[104]

Exchange reactions of borazines with heavy-metal halides have been investigated.[105–107] *N*-Trimethylborazine will react with metal halides to yield partially halogenated derivatives of 1,3,5-trimethylborazine. For example, 1,3,5-trimethylborazine upon treatment with tin(IV) halides produces the corresponding 1,3,5-trimethyl-2,4-dihalogenoborazines (X = Cl, Br). In the reaction of the same $(HBNMe)_3$ with Hg_2Cl_2 and Hg_2Br_2, 1,3,5-trimethyl-2-halogenoborazines have been prepared. From $(HBNMe)_3$ and TiF_4 the trifluoro derivative $(FBNMe)_3$ was obtained.[105]

A series of *B*-monosubstituted amine–boranes and borazines have been compared for their chemical reactivity. Mercury(II) halides have been found to be more reactive towards $H_3B.NMe_3$ than to $(HBNH)_3$. This was attributed to π-bonding in the borazine ring system.[106]

1-Methylborazine $H_3B_3N_3H_2Me$ has been reacted with a variety of halogenating agents, such as $HgCl_2$, $SnCl_4$, BCl_3 and BBr_3.[107]

Reactions of hexaorganoborazines with Grignard reagents led to unsymmetrically substituted borazine derivatives.[108] Thus hexamethylborazine, $(MeBNMe)_3$ with PhMgBr gave $Me_2PhB_3N_3Me_3$, and $(PhBNMe)_3$ with MeMgBr gave $Ph_2MeB_3N_3Me_3$.

Another way for the synthesis of unsymmetrically substituted borazine derivatives starts from 2,4,6-trihalogenoborazines, which are reacted with Grignard reagents. Thus from $(ClBNMe)_3$ and RMgBr (R = Me, Ph), $RCl_2B_3N_3Me_3$ and $R_2ClB_3N_3Me_3$ were obtained. These reacted with a second Grignard reagent R′MgBr (R′ = Me, Ph) to form $RR′ClB_3N_3Me_3$ and $R_2R′B_3N_3Me_3$.[108]

Two borazine molecules have been linked together through a B–B bond by heating $ClMe_2B_3N_3Me_3$ with potassium sand in heptane. Probably owing to some alkali-metal oxide present on the metal surface, small amounts of $B,B′$-bis(pentamethylborazinyl) oxide have also been found.[109]

Position isomers of some tetrasubstituted borazines have been observed on treating $H_3B_3N_3R_2R′$ (R = Me, Et; R′ = cyclo-C_6H_{11}) with methyl or ethyl Grignard reagents, 1,3,5,2,4-pentaorganoborazine being formed to a small extent (R″ = Me, Et):

$$\text{(11)}$$

Steric effects have been found to be important in determining the reactivity of 1,3,5-trialkylborazines towards Grignard reagents. For this reason, position 2 has been alkylated rather than position 4, and therefore 1,2,3,5-tetraorganoborazines are predominant. The isomers have been separated by gas chromatography.[110]

In the reaction of MeMgBr with 2,4,6-tris(n-butylmercapto)borazine $(BuS·BNH)_3$, 2-methyl-4,6-di(butylmercapto)borazine $Me(BuS)_2B_3N_3H_3$ and 2,4-dimethyl-6-butylmercapto-borazine $Me_2(BuS)B_3N_3H_3$ have been obtained. The latter compound reacts with ethylenediamine to give $N,N′$-di(4,6-dimethylborazin-2-yl)ethylenediamine, after elimination of BuSH.[111]

A transhalogenation reaction has been studied: 2-methyl-4,6-di(n-butyl-mercapto)-1,3,5-trimethylborazine $(BuS)_2MeB_3N_3Me_3$ reacted with mercury(II) chloride to yield 2,4-dichloroborazine $Cl_2MeB_3N_3Me_3$, and this was converted to the corresponding 2,4-difluoro derivative $F_2MeB_3N_3Me_3$ by treatment with NaF.[111]

From the reaction of B-chloroborazine derivatives $Cl_nMe_{3-n}B_3N_3Me_3$ with stoichiometric amounts of Ph_3GeK or GeH_3K in tetrahydrofuran or monoglyme, the corresponding germyl- and triphenylgermylborazines $(R_3Ge)_nMe_{3-n}B_3N_3Me_3$ (n = 1, 2, 3; R = H or Ph) have been obtained in high yield.[112]

It has been reported that 2,4,6-trichloroborazine reacts with glacial acetic acid (equation 12) to tetrakis(acetato)bis(diaminoborine):[113,114]

$$2(ClBNH)_3 + 12 \ AcOH \longrightarrow 3 \ [(AcO)_2BNH_2]_2 + 6 \ HCl \qquad (12)$$

The reaction of $(ClBNH)_3$ with silver acetate in benzene results in the formation of 2,4,6-triacetatoborazine $(AcOBNH)_3$, and 1,3,5-trimethyl-2,4,6-triacetatoborazine has been prepared similarly from $(ClBNMe)_3$ with silver acetate.[114]. By reaction of 2,4,6-trichloroborazine with appropriate haloacetic acids, ring contraction occurs and various acetatobis(aminobori-nes) $[(CXYZCOO)_2BNH_2]_2$ (X, Y, Z = H, F, Cl, Br in various combinations) have been synthesized.[115]

Transamination reactions with aliphatic diamines have been carried out with 2,4-bis(dialkylamino)borazine derivatives to form polycondensates. Starting materials were 2,4-di(butylamino)-1,3,5,6-tetramethylborazine, 2,4-dianilino-1,3,5,6-tetramethylborazine and 2-anilino-1,3,4,5,6-penta-methylborazine; the following diamines have been used: hexamethylene-diamine, piperazine, benzidine. Polycondensates also resulted from the reactions with diisocyanates.[116]

Hexamethylborazine reacts with N,N'-dimethylhydrazine to form the permethylated 1,2,4-triaza-3,5-diborolidine (R = Me):[117]

$$2\,(RBNR)_3 + 3 \ RNH—NHR \xrightarrow{80\,°C} \quad R—B\underset{\underset{R}{N}}{\overset{RN—NR}{\diagdown}}B—R + 3 \ RNH_2 \qquad (13)$$

Table 2.2 Hydrolysis of borazines.

Compound	Half-life
$(HBNPh)_3$	4.3 min
$(HBN·C_6H_4Me\text{-}2)_3$	2.41 h
$(HBN·C_6H_4Cl\text{-}2)_3$	10.4 min
$H_2MeB_3N_3Ph_3$	30 s
$MeH_2B_3N_3Me_3$	30 s
$(HO)_2HB_3N_3(C_6H_3Me_2\text{-}2,6)_3$	no hydrolysis after 1 month
$(HO)_2HB_3N_3(C_6H_3Et_2\text{-}2,6)_3$	no hydrolysis after 1 month
$(HO)_2HB_3N_3(C_6H_3Cl_2\text{-}2,6)_3$	215 days
$PhH_2B_3N_3Ph_3$	30 s
$(PhBNPh)_3$	1.37 h
$MePh_2B_3N_3Ph_3$	23 min

The hydrolysis of 25 borazine derivatives has been studied at 25 °C in dilute 90% THF–10% H_2O solutions. Some borazines, with sterically large substituents on nitrogen, are not hydrolysed even after prolonged contact time. Depending on the nature of substituents attached to boron and nitrogen, the half-life (time to hydrolyse 50% of all B–N bonds in the molecule) has been found to vary from seconds to hours. Selected examples are given in Table 2.2.[118]

Substituted aminoborazines were found to react with diborane and deuterodiborane, to produce the Lewis adducts $H_2(H_3B.Me_2N)B_3N_3H_3$ and $(D_3B.Me_2N)H_2B_3N_3H_3$.[7]

2.3.9 Metal derivatives of borazines

Two types of metal derivatives are known: in the first case the borazine ring is incorporated in complexes as a π-system,[B 31] in the second metal-substituted σ-derivatives can be formed. For the latter case, metal atoms can be substituted on two different sites of the borazine ring:[B 27]

Hexamethylborazine $(MeBNMe)_3$ and 1,3,5-trimethyl-2,4,6-triethylborazine $(EtBNMe)_3$ react with tris(acetonitrile)chromium tricarbonyl $(MeCN)_3Cr(CO)_3$ to form π-complexes of the type $(RBNR')_3Cr(CO)_3$ (16) (R = Me, Et, R' = Et):[119]

(16)

Hexaalkylborazine- and hexaalkylbenzene-chromium tricarbonyls differ only slightly in their properties. The bonding of the metal atom to the borazine ring is similar to that of benzene. At present a relatively large number of tri- and hexaalkylborazinechromium tricarbonyl complexes are known,[119-125] and a similar complex of triphenylborazine has also been reported.[126] In the latter the metal is attached to the B_3N_3 ring, rather than to a phenyl group. Carbonyl stretching frequencies of these complexes have been found at 1930–1950 cm^{-1} and 1830–1850 cm^{-1}.[120,125] Their ^{1}H NMR spectra and the solvent dependence have been discussed.[119-123,126] The crystal and molecular structure of hexaethylborazinechromium tricarbonyl $(EtBNEt)_3Cr(CO)_3$ has been analysed by X-ray diffraction.[127,128] The borazine ring is slightly nonplanar in this case; the three boron atoms and the three nitrogen atoms are located in two different planes, so that borazine adopts a chair conformation in this complex. This indicates that complexes between transition metals and borazine rings are intermediate between pure π-complexes, like dibenzenechromium, and donor complexes of σ-type.

Kinetic and thermochemical studies have been carried out with complexes of the type $(RBNR')_3Cr(CO)_3$ in reactions with phosphines and phosphites; the latter phosphorus(III) reagents displace the borazine from the complex:[129,130]

$$(RBNR')_3Cr(CO)_3 + 3\,P(OR)_3 \longrightarrow Cr(CO)_3[P(OR)_3]_3 + (RBNR')_3$$
(14)

$$R,R' = Me,\ Et,\ Pr^n,\ Pr^i$$

The results indicate that steric factors are most important for the rate of ring ligand displacement. For reactions of $(MeBNMe)_3Cr(CO)_3$ with tetraphenylarsonium salts, tetrahydrofuran was employed as solvent:[131]

$$(MeBNMe)_3Cr(CO)_3 + 3\,[AsPh_4][ECl_3] \longrightarrow$$
$$(MeBNMe)_3 + [AsPh_4][Cr(CO)_3(ECl_3)_3]$$ (15)
$$E = Ge,\ Sn$$

The thermochemistry of chromium, molybdenum and tungsten complexes $(MeBNMe)_3M(CO)_3$ (M = Cr, Mo, W) has been investigated.[132]

A manganese–borazine compound, $[Mn(CO)_5]_3B_3N_3H_3$, formed in a reaction of trichloroborazine $(ClBNH)_3$ with $NaMn(CO)_5$ by nucleophilic displacement of chlorine, contains a stable boron–manganese σ-bond.[133] The chemistry of some other σ-bonded metal carbonyl derivatives of borazine have been investigated. The spectral absorptions of 2-pentamethylborazinylmanganese tetracarbonyl(triphenylphosphine) $(Ph_3P)(CO)_4Mn$—$Me_2B_3N_3Me_3$, and 2-pentamethylborazinylmanganese pentacarbonyl $(CO)_5MnMe_2B_3N_3Me_3$ were correlated with the substituent change. The existence of 2,2'-bis(1,3,5-trimethylborazinyl)iron tetracarbonyl, $(H_2B_3N_3Me_3)_2Fe(CO)_4$ and 2-(4,6-dichloroborazinyl)dicarbonyl(π-cyclopentadienyl)iron $(Cl_2B_3N_3H_3)Fe(CO)_2C_5H_5$ have been postulated on the basis of spectroscopic evidence.[134]

2.3.10   Spectroscopic and other physical properties

During the last few years nuclear magnetic resonance spectra (^{1}H, ^{19}F, ^{14}N, ^{11}B, ^{13}C, ^{17}O), photoelectron (PE), infrared (IR), Raman (R) and mass spectra (MS) of many borazines have been investigated. The ^{11}B NMR data for a large number of borazine derivatives have been summarized.[B 11]

NMR spectra. Deuterium NMR of several boron deuterides has been studied in the solid state, and deuteron quadrupole coupling constants have been obtained from the powder spectra. The value for borazine has been determined to be 116 ± 2 kHz. The deuterium quadrupole coupling constant has been correlated with the B–H stretching frequency.[135]

Proton NMR spectra have been reported for ^{15}N-enriched borazine derivatives: borazine, 1-methylborazine, 1,3-dimethylborazine and 1-methyl-2-aminoborazines.[136] Substituent effects on chemical shifts and coupling constants for borazine derivatives have been compared with those of analogous benzene derivatives.

The use of ^{15}N-enriched borazine in proton NMR studies results in an easily observed doublet for different N–H environments, owing to the coupling of the proton with ^{15}N $(I = \frac{1}{2})$.

Variable-temperature proton NMR spectra of borazine and borazine-^{10}B have been analysed for chemical shifts, coupling constants and line widths. The line-width data suggest that the broadening is due to a combination of quadrupole relaxation resulting from the high-spin nuclei present and long-range spin coupling. The chemical shifts and coupling constants for borazine are[137]

$$^{11}\text{B–H} \quad \delta = 4.46 \pm 0.02 \text{ ppm} \ (-40 \text{ to } +80\,^\circ\text{C})$$

$$^{14}\text{N–H} \quad \delta = 5.47 \text{ ppm} \ (-40\,^\circ\text{C}) \ \delta = 5.40 \ (60\,^\circ\text{C})$$

$$\left. \begin{array}{l} J(^{11}\text{B–H}) = 138.4 \pm 0.5 \text{ Hz} \\ J(^{14}\text{N–H}) = 55.1 \pm 0.2 \text{ Hz} \end{array} \right\} \text{ in the range } -20 \text{ to } +80\,^\circ\text{C}$$

The nuclear quadrupole coupling constant e^2qQ/h for ^{11}B has been estimated to be 3.6 MHz. This leads to an estimate of 67% double-bond character in borazine.[137]

The effect of substituents Me, NMe_2, OMe, F, Cl, Br on the ^{1}H and ^{11}B NMR spectra has been investigated for a series of 2-monosubstituted borazines $XH_2B_3N_3H_3$. The ^{1}H and ^{11}B chemical-shift data are consistent with the hypothesis that the π-electrons of borazine are delocalized and the substituents interact with this system by means of a resonance effect to alter the π-electron density at the *ortho* and *para* positions.[138]

From ^{1}H and ^{11}B NMR spectra it has been concluded that in

$H_3B_3N_3H_2Me$, $H_3B_3N_3HMe_2$, $H_2MeB_3N_3H_3$ and $HMe_2B_3N_3H_3$ the methyl group is acting as an electron-releasing substituent.[139]

The proton NMR spectra of several substituted borazine derivatives have been determined in noninteracting (CCl_4) and interacting (C_6H_6 or C_6D_6) solvents. The spectral differences have been attributed to the association of anisotropic benzene molecules with molecular dipoles of the borazines.[140-142a] The $\delta^{11}B$ value of $(HBNMe)_3$ has been determined to be 32.2 ppm.[143]

The ^{17}O chemical shifts of boron–oxygen compounds containing three-coordinate boron atoms have been interpreted in terms of B–O(p–p)π interaction. The $\delta^{17}O$ value for $(MeOBNH)_3$ is 15.0 ppm (relative to external H_2O).[144]

Spin–spin coupling constants over five, six and seven bonds between protons in different methyl groups have been reported for hexamethylborazine, 1-methylborazine and 2-methylborazine.[145]

The proton NMR spectra of ^{15}N-labelled borazine, 1-methylborazine, 1,3-dimethylborazine and 1-methyl-2-aminoborazine have been analysed.[146] The ^{11}B and ^1H NMR data have been recorded for a large number of borazine derivatives.[147,148]

The ^{13}C NMR chemical shifts for 2,4,6-triphenylborazine (in THF) have been measured:[149] δ(ortho) = 127.9 ppm, J(CH) = 158 Hz; δ(meta) = 132.3 ppm, J(CH) = 157 Hz; δ(para) = 129.7 ppm, J(CH) = 160 Hz.

A comparative study of borazine and six methylborazines has been conducted by NMR. It became evident that the chemical shift of N–CH_3 protons hardly varies between mono-, di- and tri-N-methylborazines, whereas this parameter undergoes a strong change when one passes from 1,3,5-trimethylborazine to hexamethylborazine.[150]

The ^1H NMR spectra of hexamethylborazine, 2-ethylpentamethylborazine, B,B'-decamethyldiborazinyl, B-phenylpentamethylborazine and decamethylbiphenyl have been studied in an attempt to elucidate the electronic effects in the borazine ring system. For B,B'-decamethyldiborazinyl and 2-pentamethylphenylpentamethylborazine, the methyl-proton chemical shifts have been determined.[151]

The $\delta^{14}N$ and $\delta^{11}B$ chemical shifts of borazines have been interpreted in terms of delocalized p_z electron pairs of the nitrogen atoms in the ring system. Calculated π-charge densities (by using the CNDO/2 method)[152] at the N-atoms show a linear correlation with $\delta^{14}N$. These data were tabulated for a large number of borazine derivatives.[152]

Chemical shifts $\delta^{13}C$ of borazines $(RBNR')_3$ have been explained by the action of a γ-effect exerted by the substituents on the deshielding of $^{13}C(BC)$ and $^{13}C(NC)$. Analogies have been found with similar benzene derivatives. Heteronuclear ^{13}C, ^1H, ^{11}B triple-resonance experiments have permitted the observation of sharp $^{13}C(BC)$ resonance signals.[153]

The ^{1}H NMR data of many mono- and disubstituted borazine derivatives have been published in the literature.[10,11,14a,21,22,24,107] It was supposed that there is an additivity effect per substituent on the ^{1}H chemical shift of the NH protons, in which the *ortho* and *para* NH protons are magnetically equivalent.[22] The solvent dependence of ^{1}H NMR chemical shifts has been investigated.[14a]

Photoelectron spectra. Photoelectron (PE) spectra have been intensively investigated in the last few years. The borazine molecule has been compared with the isosteric benzene molecule. In benzene the bands with lowest ionization energy have been attributed to the five highest molecular orbitals. Compared with benzene, in borazine the symmetry has been lowered from D_{6h} to D_{3h}. According to symmetry correlation tables the following symmetry types are obtained:

$$D_{6h} \longrightarrow D_{3h}$$
$$\pi: \quad e_{1u} \text{ and } e_{2u} \longrightarrow e''$$
$$b_{2g} \text{ and } a_{2u} \longrightarrow a_2''$$
$$\sigma: \quad e_{2g} \text{ and } e_{1u} \longrightarrow e'$$

In a qualitative MO model this means that the bonding π-orbitals $1e''$ and $1a_2''$ are lowered by interaction with the antibonding orbitals $2e''$ and $2a_2''$. The occupied σ-orbital is not changed essentially owing to 3 bonding σ- and 4 antibonding σ-orbitals of the same symmetry type e'.[154] The PE spectra of $(MeBNH)_3$, $(HBNMe)_3$, $(MeBNMe)_3$, $(FBNH)_3$, $(FBNMe)_3$, $(ClBNH)_3$ and $(ClBNMe)_3$ have been reported.[154a] With the halogeno substituents $X = F$, Cl, the highest occupied σ-level is lowered; in the π-system the mesomeric effect of X is acting against the inductive effect.[155] The first band in the PE spectra of 2,4,6-trifluoroborazine at 10.66 eV and of 2,4,6-trichloroborazine at 10.55 eV has been assigned to π-ionization. A 584 Å photoelectron spectrum of borazine has been published,[156] and the ionization potentials compared with data estimated by an INDO calculation are given in Table 2.3.

Table 2.3 Experimental and theoretical ionization potentials (in eV) of borazine.[156]

Orbital symmetry	Calculation		Experimental vertical IP
	CNDO/2	INDO	
e'	10.2	9.8	10.09 ± 0.02
e''	11.0	10.4	11.42 ± 0.04
a_1'	14.6	14.0	12.82 ± 0.04
e'	14.9	14.8	13.98 ± 0.02
a_2'	16.7	16.6	14.70 ± 0.05
a_2''	18.3	17.7	17.10 ± 0.10

The He(I) PE spectrum and the measured vertical ionization potentials of 2,4,6-trimethylborazine and 1,3,5-trimethylborazine have been reported. It has been shown that the highest occupied orbital is of π-type. Methylation of borazine at 2,4,6 (boron) sites of borazine leaves the energy separation of the first three orbital sets almost unchanged, but the methyl groups in 1,3,5 positions (N sites) increase the separation between the two sets by at least 0.6 eV (compared with borazine). Similar effects to those of 2,4,6-trimethylation are observed upon 1,3,5-trifluorination.[157]

The extent of the π-bonding in borazine, as measured by the energy separation of the π-type orbitals, has been determined to be c.85% of that in benzene. By fluorination of borazine, σ-orbitals are stabilized to a greater extent than π-orbitals.[158]

The photoelectron spectra and the gas phase far-UV spectra of borazines $(XBNY)_3$ (X = H, Me, F, Cl; Y = H, Me) have been reported. Modified CNDO/CI calculations with new parameters for boron and fluorine give satisfactory predictions, not only of orbital sequences and of the ionization energies determined by PE spectroscopy, but also of substituent effects on the UV excitation energies.[159]

It has been stated again that the highest MOs of borazine are π_3, π_2 rather than σ, and these are followed by σ and π_1. The PE spectra of borazine and 2,4,6-trifluoroborazine have been nicely complemented by ab initio calculations. The ab initio borazine calculations are not in quantitative agreement with the observed vertical ionization potentials, but do concur with the π_3, π_2, σ, π_1 ordering dictated by the perfluoro effect.[160]

The electronic structures of benzene and borazine have been described by a straightforward extension of Frost's floating spherical Gaussian orbital (FSGO) model. The results of variational calculations using three-centre two-electron π-orbitals have been discussed mainly in connection with evaluations of the molar, diamagnetic and paramagnetic susceptibility tensors for the two molecules.[161]

Two proposed models of borazine have been examined by ab initio calculations using Gaussian-type atomic orbitals. It has been found that the planar D_{3h} model is energetically preferred to the twisted-boat C_2 model. The electron-population analysis reveals that nitrogen and boron atoms in borazine possess a negative and positive charge respectively. The trimerization energy of boronimide HBNH has been calculated to be -193.8 kcal/mol.[162] The results of a Mulliken population analysis[163] for planar D_{3h} borazine are shown in Table 2.4. This reveals that the nitrogen atoms possess large negative charges ($-0.735e$) arising from a large σ-electron drift ($1.144e$) from the boron atoms ($0.422e$) and from hydrogen ($0.299e$) attached to nitrogen. Superimposed on this is a substantial π-electron transfer from nitrogen to boron ($0.409e$). This leads to an overall positive charge ($0.473e$) at the boron atoms, with a σ-contribution of $0.884e$.[163]

Table 2.4 Electronic population for the planar D_{3h} model of borazine

Nitrogen	Boron	Hydrogen
σ 6.144	σ 4.118	(N) 0.701
π 1.592	π 0.409	(B) 1.037
Total 7.735	Total 4.527	

Ionization potentials of borazine have been calculated by the MNDO method, which gives better results than the MINDO/3 method. The proton affinity of (HBNH)$_3$ has been calculated to be 184.3 kcal/mol; the experimental value has been found as 203 kcal/mol.[164] *Ab initio* calculations give for the proton affinity a value of 203.4 kcal/mol, and the most energetically favourable structure of the borazinium ion is one in which very little structural change occurs relative to the neutral borazine, with the exception of the geometry about the protonated nitrogen atom. Charge distributions and bond lengths have been used to explain bonding changes upon protonation.[165]

A CNDO/2 method has been applied to borazine and the first two ionization potentials have been calculated exactly.[166] Calculations by a modified CNDO/2 method have indicated that benzene–borazine and borazine–borazine complexes in which the molecules are symmetrically disposed in parallel planes can exist in the ground state. The stabilization energies have been calculated to be in the range 2–5 kcal/mol for benzene–borazine and 5–18 kcal/mol for borazine–borazine, with interplanar separations of nearly 3 Å in both cases.[167]

The bonding in borazine and symmetric trisubstituted derivatives (FBNH)$_3$, (HBNF)$_3$, (MeBNH)$_3$ and (HBNMe)$_3$ has been discussed by means of all electron *ab initio* SCF–MO calculations. *B*-Trifluoroborazine has been predicted to be 12.9 eV more stable than *N*-trifluoroborazine, and *B*-trimethylborazine 2.8 eV more stable than the *N*-trimethyl derivative. The calculated wave functions have been used to interpret the low-energy photoelectron spectra of these molecules. π-Donation into the borazine ring is markedly reduced on 1,3,5-trisubstitution relative to 2,4,6-trisubstitution. The calculated sequence for the highest occupied valence MO's is πσπ for all derivatives except 2,4,6-trifluoroborazine (FBNH)$_3$, where the sequence is ππσ.[168]

The results of the Mulliken population analysis for (HBNH)$_3$, (HBNF)$_3$, (FBNH)$_3$ and F$_3$B$_3$N$_3$H$_2$Me have been discussed in terms of the effect of fluorine bonded to boron and/or nitrogen. The correspondence between experiment and calculation for the orbital energies has also been discussed and the net charges from the Mulliken population analysis have been calculated.[169] All valence-electron calculations of the series 2,4,6-trichloro-, 1,3,5-trichloro- and hexachloroborazine have been reported.[170]

The molecular structures of borazine and boroxine have been compared within the framework of an *ab initio* double-zeta-type calculation.[171] Vertical excitation energies for borazine have been calculated. Low-lying Rydberg states of borazine have also been included.[172]

Ab initio SCF MO and CI calculations for two different Gaussian basis sets have been carried out for benzene and borazine in order to investigate the effect of increasing the flexibility in the representation of their respective π-systems. In both treatments $^1E_{2g}$ has been predicted to be at lower level than $^1E_{1u}$.[173] The charge density ($q_\mu^{\sigma+\pi}$, q_μ^π) and the bond order ($p_{\mu\nu}^{\sigma+\pi}$, $p_{\mu\nu}^\pi$) of hexamethylborazine have been evaluated.[173,173a]

On the basis of the MO–LCAO theory, the total charges on the B and N atoms, the bond orders and the bond lengths have been calculated for linear polyborazines.[174,175] The same has been done for the diborazine compounds (17) and (18):[176,177]

(17)   (18)

Mass spectra. The mass spectra of *B*-substituted borazine derivatives and of *N*-trimethylborazine have been recorded and analysed in terms of the relative contribution of the parent ion and certain fragment-ion intensities. The relative contributions of the P and P−H species (P = parent ion) to the overlapping peak have been calculated from a set of simultaneous equations using a matrix-inversion technique.[178]

The mass spectra of the mixed *B*-methyl- and *B*-phenylborazines indicate the relative peak intensity of the P−CH₃ fragment decreases from hexamethylborazine to *N*-trimethyl-*B*-triphenylborazine. From this result, it has been suggested that the methyl lost in the fragmentation comes from the boron site; an *N*-methyl group is markedly less labile than a *B*-methyl group. From this, it has been concluded that a positive charge is more readily stabilized at the boron site. The loss of a *B*-substituent can be stabilized by an increase in electron donation from the adjacent nitrogen atom. The resistance to loss of H atoms from the nitrogen sites is supported by a small relative contribution of the P−H species. Molecular ion species are stabilized by halogen substitution in the order F ∼ Cl < Br < I.[178] The ions that result by loss of two or three H atoms from the parent ion also have a contribution.[179]

The mass spectra of a series of 1,3,5-trialkylborazines (HBNR)₃ (R = Et, Prⁿ, Prⁱ, Buⁿ, Buˢ, Buⁱ, Buᵗ, CH₂CD₃, CD₂CH₃) and (DBNR)₃

(R = CD_2CH_3, CD_2CD_3) have been reported and discussed. The dominant feature of the spectra is a group of ions which arise through cleavage of an alkyl radical from the α-carbon atom of an N-substituent in the molecular ion. These fragment species are isoelectronic with the benzyl or tropylium ions observed in the mass spectra of alkylbenzenes.[180]

From ion kinetic energy spectra of a series of 1,3,5-trimethylborazines it was suggested the possible formation of cyclopentadienyl-like ions, by metastable decomposition of the tropylium-like ions:[181]

R = Me

The mass spectra of C-deuterated borazine derivatives, $(RBNC_6D_5)_3$ and $(C_6D_5BNR')_3$ have been investigated. The mass spectra contain strong peaks of doubly charged ions, which are stabilized by exocyclic ring closure.[227]

Chemical-ionization mass spectra of borazine in methane, ethane, n-butane, ammonia, hydrogen, neon and krypton have been obtained.[182-186]

Crystal and molecular structure. The B–N distances in several borazines have been determined and are listed in Table 2.5.

Table 2.5 The B–N bond length in borazines.

Compound	B–N distance (Å)	Reference
$(HBNH)_3$	1.436	187
$(HBNMe)_3$	1.42	187
$(ClBNH)_3$	1.41	187
$(FBNH)_3$	1.426	187
$H_2(NH_2)B_3N_3H_3$	1.415^a	187
$[(Me_2N)BNH]_3$	1.433	188
$(EtBNH)_3$	1.423	189
$(MeBNH)_3$	1.39	189

^{a} Ring B–N bond.

The molecular structure of 2-aminoborazine in the gas phase has been studied by electron diffraction. Three models have been considered: a planar ring, a borazine ring in a chair conformation and a borazine ring in a boat conformation. For all models the amine nitrogen was assumed to be coplanar with the NBN group to which it is attached. The three models fitted the data equally well. The bond distances were determined to be: B–N(ring) 1.418 ± 0.004 Å, B–N(amine) 1.498 ± 0.008 Å, B–H 1.25 ± 0.03 Å, N–H(ring) 1.033 ± 0.02 Å, N–H(amine) 1.062 ± 0.03 Å.[189]

2,4,6-Tris(dimethylamino)borazine $[(Me_2N)BNH]_3$ (space group $P2_1/a$, $Z = 4$) contains a ring that exhibits only small deviations from a planar structure of D_{3h} symmetry. The endocyclic and exocyclic B–N bonds are equal within the standard deviation (1.43 Å). The ring angles are different at the boron atoms (c. 117°) and at the nitrogen atoms (c. 123°).[190]

The crystal structure of hexaphenylborazine $(PhBNPh)_3$ (orthorhombic, space group $Pna2_1$, $Z = 4$) has been determined as a part of a study on sterically overcrowded borazines. It has been found that hexaphenylborazine, like the corresponding benzene derivative, exhibits a propeller conformation with the following torsion angles of the attached carbon rings: 66.8°, 69.5°, 60.7°, 71.4°, 62.6° and 64.5°.[191]

1,3,5-Triphenyl-2,4,6-trichloroborazine $(ClBNPh)_3$ (space group $C2/c$, $Z = 4$) exhibits molecular symmetry C_2. The conformation of the ring is planar and the atoms linked directly to the ring are in the same plane. Steric interaction between the phenyl groups results in a tilting angle of 77° (ring C_{11}–C_{14}) and 87° (ring C_{21}–C_{26}) with respect to the plane of the borazine ring.[192]

The structure of borazine $(HBNH)_3$ has been reinvestigated by electron diffraction. The bond distances thus determined are B–N 1.4355 ± 0.0021 Å, B–H 1.258 ± 0.14 Å and N–H 1.050 ± 0.012 Å. The ring angles are NBN 117.7° ± 1.2° and BNB 121.1° ± 1.2°.[193]

The geometrical parameters (B–N, B–R, N–R', NBN, BNB) of $(HBNH)_3$, $(FBNH)_3$, $(ClBNH)_3$, $(MeBNH)_3$, $(Me_2N·BNH)_3$, $(NH_2)B_3N_3H_5$, $(HBNMe)_3$, $(EtBNEt)_3$, $(ClBNCl)_3$ have been summarized.[194]

Hexachloroborazine $(ClBNCl)_3$[195–197] is built up from planar layers of $(ClBNCl)_3$ molecules lying perpendicular to the trigonal c-axis and can be derived from a cubic closest packing of Cl atoms by substituting the cube corner atoms by $(BN)_3$ rings. The shape and dimensions of the hexachloroborazine is almost identical with those of hexachlorobenzene, which crystallizes in a similar structure.[195]

Aluminium bromide forms a 1 : 1 adduct with hexamethylborazine.[198] The coordination takes place at a nitrogen atom, therefore the borazine ring loses its planarity. The ring conformation is influenced by weak intramolecular and intermolecular B–Br interactions.

The bond lengths and angles are $Al–N_1$ 2.011 Å, $N_1–B_2$ 1.517Å, $N_3–B_2$ 1.401 Å, $N_3–B_4$ 1.490 Å, $N_5–B_4$ 1.413 Å, $N_5–B_6$ 1.387 Å and $N_1–B_6$ 1.543 Å; $B_4N_5B_6$ 124.0°, $N_1B_2N_3$ 118.2°, $N_3B_4N_5$ 117.3°, $N_5B_6N_1$ 114.1°, AlN_1B_2 106.0°, AlN_1B_6 104.6°, $B_6N_1B_2$ 112.8° and $B_2N_3B_4$ 120.4°.

The B–N bond distances for tris(dimethylamino)borazine are B–N(endocyclic) 1.433 Å and B–N(exocyclic) 1.429 Å.[198a]

Nuclear quadrupole resonance. The $(ClBNH)_3$ molecule has *m*-point symmetry, and, since only two ^{35}Cl signals are observed in the NQR spectrum, the mirror plane must be perpendicular to the plane of the molecule, passing through one of the chlorine atoms, in perfect agreement with the X-ray crystal-structure analysis.[199]

The quadrupole coupling constants of borazine[200] and other NQR parameters of borazine, trichloroborazine and hexamethylborazine[197] have been studied. The ^{35}Cl NQR frequencies of 2,4,6-trichloroborazine are 19.936 and 20.258 MHz at 77 K.[197a]

Magnetooptic studies. The mean magnetic susceptibility of some borazines $(XBNY)_3$ with X = H, R, Cl and Y = H, R, have been measured; in all cases the borazine ring exhibits an increase in diamagnetism that has been taken as a proof of its aromatic character.[201]

The Faraday effect of D_{3h} substituted borazines leads to the conclusion that the large increase observed in their magnetic rotation can be attributed to electronic delocalization along the cycle.[202] The magnetooptical properties of borazines have been used for determining the extent of electronic delocalization in this ring. The Faraday measurements confirmed the existence of a ring current. Energy levels and bond structures of borazines and boroxines have been compared.[203]

UV spectra. The UV spectra of benzene and borazine have been compared. In borazine the excited states have $^1A_2'$, $^1A_1'$, and $^1E'$ symmetries.[204]

In the vapour phase and matrix-isolated borazine absorption spectrum in the 2000–1500 Å region three electronic transitions have been observed. The strongest absorption, with a maximum at 1650 Å has been assigned to the allowed $^1E'-^1A_1'$ transition. The two weaker absorptions, with forbidden origins at 1975 and 1889 Å have been assigned to the forbidden transitions $^1A_2'-^1A_1'$ and $^1A_1'-^1A_1'$ respectively.[205]

The absorption spectrum of borazine vapour, at pressures between 0.1 and 5 mmHg, has been recorded in the region 2015–1800 Å. It has been concluded that the $^1A_1'$ state is the lowest singlet excited state of borazine, and the spectrum has been compared with that of benzene.[206] Other UV spectra investigated are those of 2,4,6-triphenylborazine and 2,4,6-tris(p-chlorophenyl)borazine.[207]

An attempt was made to locate experimentally the triplet state of borazine. In a mixture of benzene and borazine vapour excited at 2580 Å, borazine had no effect on the first excited singlet state of benzene, but did strongly quench the triplet state.[208].

Singlet excitation energies and oscillator strengths for $\pi-\pi^*$ transitions in borazine have been calculated using a Pariser–Parr–Pople model.[209] The zero-field splitting parameters of C_6H_6 and C_6D_6 in a borazine crystal host have been measured.[210,211] A Mulliken-type study has shown that the transition $^1A_2'-^1A_1'$ is symmetry-forbidden, but becomes allowed through coupling with a vibration of e' species.[212] Energy transfer processes in 1,3,5-trimethylborazine have been investigated.[213]

Ion cyclotron resonance trapped-electron spectra of borazine and 1,3,5-trimethylborazine have been reported. Triplet states of these molecules have been located at 7.9 eV and 6.6 eV respectively, in good agreement with expectations from optical spectra and theoretical calculations.[214]

The electronic spectra and structures of halogenoborazines,[215,154a] alkylborazines[154a,215] and alkylaminoborazines[216] have been discussed.

Raman spectroscopy, The Raman spectrum of borazine has been analysed in detail.[217,218] The centrifugal distortion constants D_J for borazine have been determined.[219]

IR spectra. The vibrational spectra of 1,3,5-trimethylborazine,[220] trihalogenoborazines (Cl, Br derivatives),[221,221a,221b,222] and several ^{10}B, D- and ^{15}N-labelled borazines[223–226], as well as of some C_6D_5 derivatives[227] have been investigated.

The vibrational frequencies of borazine and its symmetrically deuterated derivatives have been used in order to calculate a modified valence force field; nine in-plane force constants have been adjusted to fit the 42 observed frequencies of A_1' and E' species, and five out-of-plane symmetry force constants were refined by using the 12 observed A_2'' frequencies.[228,229]

Valence force constants of the borazine ring have been evaluated[230–236] and a normal coordinate analysis has been carried out for borazine and borazine-d_6 using Urey–Bradley nonbonded interaction force constants.[238,239] The vapour-phase infrared spectra of a large number of borazines have been recorded and assigned.[237,240] Isotopically labelled species,[241,242] B-monosubstituted borazines $XH_2B_3N_3H_3$ (X = NMe_2, OMe, OAc, F, Cl, Br, NH_2, NCO, SCN, CN)[243] and trimethylborazines $(RBNR')_3$ (R = Cl, D, H, Me, CD_3, Ph, C_6D_5; R' = CD_3, Me, C_6D_5)[244] have been analysed in detail. It was concluded that hydrogen bonding is absent or exceedingly weak in borazines[243] and the characteristic B–N stretch is coupled with the symmetric N–CH_3 deformation.[244]

Miscellaneous physical properties. The dipole moment of 2-aminoborazine has been determined by the Guggenheim method (μ = 1.21 $\pm$ 0.7 D).[78]

The fluorescence spectra have been described for several aromatic borazine derivatives.[245]

The heats of hydrolysis of $(BrBNMe)_3$ and $(BrBNH)_3$ have been reported to be -84.4 ± 0.5 and -120.9 ± 0.5 kcal/mol and their heats of formation -246.5 and -215.9 kcal/mol respectively. The boron–nitrogen bond energies in these two compounds are 113.2 and 105.5 kcal/mol respectively.[246]

2.3.11 Potential commercial uses

The properties of borazines have been investigated in relation with the production of potentially heat-resistant polymers.[247–260]

The poly(borazinylene oxides) (19), prepared from 1,3,5-triarylborazines in boiling DMF were crosslinked if the reaction was carried out longer than 48 h, and had increased hydrolytic stability if *ortho* substituents were present:[251]

(19)

Reaction of 1,3,5-trimethyl-1,3,5-trazonia-2,4,6-triboratacyclohexane with 1,3,5-triphenyl-2,4,6-triallylborazine at 100–300 °C for 20 h gave an insoluble polymer containing units of alternating N-phenyl and N-methyl borazine rings.[253] The synthesis of several polyborazines has been reported.[254–259] Thus polymers were obtained by heating $(H_2N \cdot BNPh)_3$ at

250 °C for 4 h,[255] by reaction of some borazines with tetramethyldisilox-ane,[257] by copolymerization of borazines with bis(acrylamides) and other bis-amides,[258] or by reaction of borazines with diisocyanates.[259] Copo-lymers of pentamethyl-6-vinyl-borazine with styrene, and of 2,4-dimethyl-6-isopropenylborazine with 4-vinylpyridine[260] have been described.

The condensation product of borazines with p-dioxophenylene (20) has been recommended for use as electrostatographic latent and toner image supporting materials.[261,261a]

(20)

Many borazine derivatives have been tested for their ability to impart flame-retardant properties to various materials,[262-264] for stabilization of polyolefins;[265] borazines have been recommended as counters for thermal neutrons,[266,267] for use as propellants,[268,268a] and as regulators of the detonation properties of $C(NO_2)_4$.[269]

The properties of BN films, produced by decomposition of borazine in high-frequency plasma,[270] by thermal decomposition of 2,4,6-trichlorobor-azine vapour at 1000–1600 °C [271-273] or by decomposition of $(ClBNCl)_3$ at 900 °C,[274] have been investigated. Other studies of such BN films obtained from borazines have been reported.[275-279]

1,3,5-Trimethyl-2,4,6-triphenylborazine and 1,3,5-tris(isobutyl)-2,4,6-tris(isobutylamino)borazine have been tested for potential use as insecti-cides.[280]

Studies of heteroaromatic B–N compounds have illustrated chemothera-peutical possibilities.[B 29]

2.4 Heterocycles containing three-coordinate boron atoms

2.4.1 Borazocines (RBNR')₄

Eight-membered boron–nitrogen heterocycles (21) are relatively rare. They usually form in reactions similar to those used for the synthesis of bora-zines, if sterically bulky groups are present in the starting materials.

R
|
B
R′—N⟋ ⟍N—R′
| |
R—B   B—R
| |
R′—N⟍   ⟋N—R′
B
|
R

(21)

By reaction of methylammonium fluoroborate with aluminium dust, besides 1,3,5-trimethyl-2,4,6-trifluoroborazine, the tetramer, 1,3,5,7-tetramethyl-2,4,6,8-tetrafluoroborazocine, (21) R = F, R′ = Me, has been obtained:[32]

$$MeNH_3^+ \ BF_4^- + Al \longrightarrow \tfrac{1}{3}(FBNMe)_3 + AlF_3 + \tfrac{3}{2}H_2 \qquad (16)$$

$$MeNH_3^+ BF_4^- + Al \xrightarrow{280\,°C} \tfrac{1}{4}(FBNMe)_4 + AlF_3 + \tfrac{3}{2}H_2 \qquad (17)$$

Both compounds were also formed in the reaction of methylamine–trifluoroborane adduct $BF_3.MeNH_2$ heated with aluminium dust.[32]

In reactions of sterically hindered amines with boron trihalides, in the presence of triethylamine, borazine formation can be suppressed and 1,3,5,7-tetra(t-butyl)-2,4,6,8-tetrachloroborazocine (21) R = Cl, R′ = Bu^t, can be isolated from Bu^{t}NH$_2$ and BCl$_3$. By heating t-butyltrimethylsilyl-aminodichloroborane Cl$_2$B—NBu^t(SiMe$_3$) at 150 °C, a smooth decomposition proceeds, with elimination of Me$_3$SiCl and formation of (ClBNBu^t)$_4$. It is interesting to note that this borazocine is highly resistant to hydrolysis.[39,40,47,281]

In the reaction of (F$_2$BNHBu^t)$_2$ with LiN(SiMe$_3$)$_2$, besides (FBNBu^t)$_3$, the eight-membered ring compound (FBNBu^t)$_4$ has been isolated. If (FBNBu^t)$_4$ is heated to 300 °C the trimer is formed by ring contraction. The reaction of (FBNBu^t)$_4$ with MeLi leads to an unsymmetrically substituted borazocine derivative, MeF$_3$B$_4$N$_4$Bu$_4^t$.[282a]

Recently, high-resolution mass-spectral studies have facilitated the identification of the molecular ions of (XBNR)$_4$, which have not been observed previously in some cases.[282b]

2.4.2 1,3,2,4-Diazadiboretidines (RBNR′)$_2$

By thermolysis of silylamino–halogenoboranes at 600 °C in an apparatus where the reaction products can be quenched by liquid nitrogen, a series of 1,3,2,4-diazadiboretidines (22) (R = C$_6$F$_5$, 2,4,6-Me$_3$C$_6$H$_2$, R′ = Bu^t, SiMe$_3$) has been obtained:[44]

$$\underset{\underset{Cl}{\overset{R}{\diagdown}}{\overset{}{B}}{-}\underset{SiMe_3}{\overset{R'}{\diagup}}{\overset{}{N}} \quad \xrightarrow[-Me_3SiCl]{600\,°C} \quad \underset{R'}{\overset{R}{\diagdown}}\overset{B \; N}{\underset{N \; B}{\square}}\underset{R}{\overset{R'}{\diagup}}$$

(18)

(22)

The molecular structure of 1,3-di-t-butyl-2,4-bis(pentafluorophenyl)-1,3,2,4-diazadiboretidine $(C_6F_5BNBu^t)_2$ has been confirmed by X-ray diffraction. The bond distances for the ring are B–N 1.436 Å and B'–N 1.426 Å; the bond angles are NBN' 95.70° and BNB' 84.30°.[44] The compound $(C_6F_5BNBu^t)_2$ reacts with ketones (in enol form) and aldehydes, with opening of the ring and with insertion of RCHO units (to form a six-membered ring).[44]

The crystal and molecular structure of hexakis(trimethylsilyl)-2,4-diamino-1,3,2,4-diazadiboretidine (22), R = $N(SiMe_3)_2$, R' = $SiMe_3$, has also been determined by X-ray diffraction.[282] The four-membered ring is planar and the substituents at the N atoms are also arranged in the plane. The Si atoms attached to the exocyclic N atoms are arranged almost perpendicular to the plane of the ring. The bond lengths are B–N 1.45 Å, Si–N 1.75 Å and Si–C 1.87 Å. The exocyclic B–N bond is relatively short, which is surprising with regard to the conformation of the exocyclic substituent.[282]

Another synthesis of 1,3,2,4-diazadiboretidines starts from diazastannetidine $(Me_2SnNBu^t)_2$, which reacts under mild conditions with RBX_2 and BX_3 to form diazadiboretidines (22), R = Me, Cl, R' = Bu^t, which are very sensitive towards hydrolysis and oxidation.[310]

Lithiation of bis(t-butylamino)phenylborane $PhB(NHBu^t)_2$ with MeLi, followed by addition of $PhBCl_2$, leads to (22), R = Ph, R' = Bu^t.[14a] By heating bis(t-butyltrimethylsilyl)fluoroborane $FB(NBu^tSiMe_3)_2$ to 185 °C, the four-membered ring (22), R = NBu^tSiMe_3, R' = Bu^t, is formed. The ^{1}H NMR spectrum of the compound suggests cis–trans isomerism.[310a]

A 1,3,2,4-diazadiboretidine (22), R = NMe_2, R' = cyclohexenyl, has been reported from the reaction of bis(dimethylamino)boron azide $Me_2B{-}N_3$ with cyclohexene on heating under elimination of nitrogen and dimethylamine. The photolytic decomposition (UV irradiation) of bis (diethylamino)boron azide $(Et_2N)_2B{-}N_3$ in benzene, cyclohexane or cyclohexene yields (22), R = NEt_2, R' = NEt_2, in addition to $B(NEt_2)_3$ and nitrogen.[283] Tetrakis(diethylamino)diazadiboretidine (22), R = R' = NEt_2, which contains exocyclic N–N bonds, is very easily oxidized by oxygen; with methanolic potassium hydroxide it is cleaved with formation of diethylamine and 1,1-diethylhydrazine.[283]

Ab initio STO–NG calculations have been performed for the B_2N_2 ring of diazadiboretidine using an optimum B–N distance of 1.47 Å and an optimum NBN angle of 93°. The $(HBNH)_2$ ring has been calculated to be about 74.3 kcal/mol more stable than the two monomeric species. The highest occupied MO is localized on the nitrogen $2p_\pi$ orbitals and has a rather low stabilization (predicted ionization potential 8.1 eV, compared with 11.1 eV for $H_2B—NH_2$). The vertical excitation energy to the lowest triplet state T_1 is also relatively low at 119 kcal/mol (compared with 157 for $H_2B—NH_2$) since the transition involved is from a nonbonding N to a nonbonding B π-orbital.[284] The B–N stretching modes in $(HBNH)_2$ are about 300 cm^{-1} higher than those in $(H_2B—NH_2)_2$; therefore it has been suggested that the B–N bonds in $(HBNH)_2$ are stronger than those in $(H_2B—NH_2)_2$. The calculated MNDO geometry and assignments for calculated N–H stretching, B–H stretching, B–H bending, N–H bending vibrations have been made.[285]

2.4.3 Δ^2-Tetrazaboroline

B-Halogen substituted Δ^2-tetrazaborolines (**23**) have been prepared by heating methylamine–trichloroborane $Cl_3B.MeNH_2$ with methyl azide in the presence of a tertiary amine (R = Cl, R' = Me):

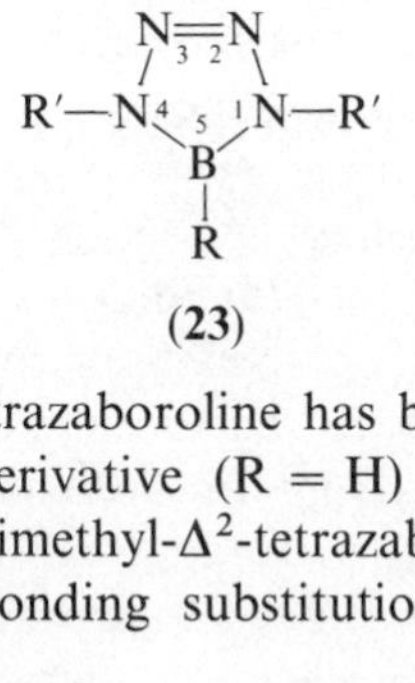

(**23**)

5-Bromo-1,4-dimethyl-Δ^2-tetrazaboroline has been prepared in good yield by reaction of the B–H derivative (R = H) with *N*-bromosuccinimide. Treatment of 5-chloro-1,4-dimethyl-Δ^2-tetrazaboroline with silver pseudohalides yielded the corresponding substitution products (**23**), R = CN, SCN, SeCN, R' = Me.[286]

A diboron compound (**24**) has been obtained by treating 5-chloro-1,4-dimethyl-Δ^2-tetrazaboroline with sodium amalgam:[286]

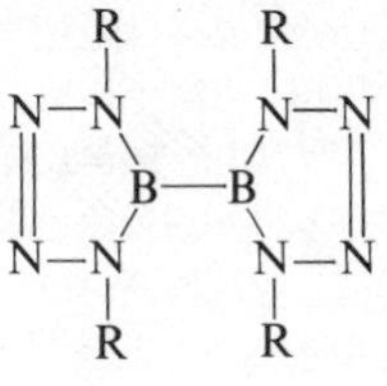

(**24**)

The molecular structure of 1,4-dimethyl-Δ^2-tetrazaboroline has been investigated. The bond lengths and angles are B—N 1.413 $\pm$ 0.010 Å, N—N 1.375 $\pm$ 0.005 Å, N=N 1.291 $\pm$ 0.006 Å, C—H 1.096 $\pm$ 0.010 Å, NBN 101.8 $\pm$ 0.6° and CNN 115.8 $\pm$ 0.3°. According to these values, the presence of extensive delocalization of the π-electrons in the ring has been suggested. The B–N bond length is shorter than in borazines.[187]

The calculated values for geometrical variables of 1,4-dimethyl-Δ^2-tetrazaboroline (23), R = H, R′ = Me, have been compared with the experimentally determined data (in parentheses)[164] B—N 1.445 Å (1.413 Å), B—H 1.158 Å (1.195 Å), N—N 1.357 Å (1.376 Å), N=N 1.255 Å (1.291 Å), NBN 100.3° (101.8°).

1,4-Di-p-chlorophenyl-Δ^2-tetrazaboroline (23), R = H, R′ = p-ClC$_6$H$_4$, has been prepared by the reaction of diborane with an ethereal solution of p-chloroaniline, followed by addition of p-chlorophenyl azide.[287] Syntheses of unsymmetrically substituted Δ^2-tetrazaborolines have been attempted by using different amine and azide substituents. However, only mixtures of the symmetrically substituted derivatives have been obtained:

$$ 2\ RNH_2.BH_3 + 2\ R'N_3 \longrightarrow HBN_4R_2 + HBN_4R_2' + 4\ H_2 \qquad (19) $$

This randomization of substituents appears to be a general feature in preparations of unsymmetrically substituted Δ^2-tetrazaborolines.[287]

Self-consistent-field calculations have been performed on some Δ^2-tetrazaborolines using both *ab initio* and semi-empirical (CNDO) approaches. A Mulliken population analysis of Δ^2-tetrazaborolines indicates that the boron is positively charged ($+0.475$), and this originates from a σ-electron flow (0.950) to N$_1$ and N$_4$, opposed by a much smaller π-electron drift (0.494) from these nitrogens to the boron. Nitrogen atoms 1 and 4 are highly negatively charged, owing to electron movement from both the hydrogens and the boron atom. In contrast, nitrogen atoms 2 and 3 are virtually electrically neutral. The π-electron system consists of a partially delocalized six-electron arrangement, in which the electrons are largely concentrated in the N$_2$—N$_3$ and the two B—N bonds. The highest filled orbital is of π-type and is equally distributed over the atoms (ionization potential calculated 11.9 eV, observed 10.4 eV). The first virtual orbital is also of π-symmetry, but contains no contribution from the boron. The second virtual orbital (also of π-type) is largely concentrated on boron. It is the first example of a planar boron compound whose first π-type virtual orbital is not concentrated on the boron atom. Therefore boron has no acceptor properties in Δ^2-tetrazaboroline.[288]

The heats of atomization of Δ^2-tetrazaborolines have been calculated by using the IOC-ω technique.[289]

1,4,5-Trimethyl-Δ^2-tetrazaboroline (**23**), R = R' = Me, (L) has been reacted with $SnCl_4$, $SbCl_5$ and BCl_3 to give the complexes $SnCl_4 2L$, $SbCl_5.L$ and $BCl_3.L$.[290] With $TiCl_4$ several types of complexes have been obtained.[290,291] The vibrational spectra of $SnCl_4$ and $SbCl_5$ adducts[292] and a charge-transfer complex of 1,4,5-trimethyl-Δ^2-tetrazaboroline and 1,3,5-trinitrobenzene have been studied.[293]

2.4.4 1,2,4,5-Tetraza-3,6-diborine

Tetrazadiborines are six-membered rings, containing two N–N bonds (**25**):

(**25**)

3,6-Diphenyl-1,2,4,5-tetraza-3,6-diborine (**25**), R = Ph, R' = R'' = H, is formed in the reaction of bis(dimethylamino)phenylborane $PhB(NMe_2)_2$ with hydrazine in n-hexane.[294] The same compound (but with a different melting point reported) is said to result from the reaction of triphenylboroxine, $(PhBO)_3$ with hydrazine, or from phenylboronic acid and hydrazine.[295]

Several 1,2,4,5-tetraza-3,6-diborines (**25**), R = H, Me, Ph; R' and R'' = H, Me, Ph, have been obtained in high yields by hydrazinolysis of bis(aminoboranes) or by reaction of hydrazines with $PhBCl_2$ in the presence of triethylamine. Derivatives with different substituents on the boron atoms have been obtained by using a mixture of $PhB(NMe_2)_2$ and $MeB(NMe_2)_2$. Dimethylamino derivatives of boron are used owing to the high volatility of the dimethylamine formed. From dipole moment measurements and ^1H NMR spectra, a planar structure of the B_2N_4 ring has been ruled out, and a boat conformation for the ring has been suggested.[296]

Organo derivatives of tetrazadiborines (**25**), R = Me, Et, Ph; R' = R'' = Me, Et, Ph, can also be prepared by reaction of bis(mercapto)-boranes, $RB(SMe)_2$ or 1,3,4,2,5-trithiadiborolanes $R_2B_2S_3$ with symmetrically substituted hydrazines.[297,298]

The condensation of aminoboranes $B(NMe_2)_3$ and organothioboranes $B(SMe)_3$ with N,N'-dimethylhydrazine MeHN—NHMe, has been proved to be a convenient synthesis for B,B'-functional derivatives (**25**), R = NMe_2, SMe, R' = R'' = Me.[299] Substitution reactions of MeS groups in 3,6-bis(methylthio)-1,2,4,5-tetramethyl-1,2,4,5,3,6-tetrazadiborine (**25**), R = SMe, R' = R'' = Me, in reactions with dimethylamine, methanol and phosphorus(III) chloride, to give the corresponding (25), with $R = NMe_2$, OMe and Cl respectively, have been carried out.[299]

The partial transamination product of 1,2,4,5-tetramethyl-3,6-diphenyl-tetrazadiborine $(PhBNMeNMe)_2$ was obtained by a ring-contraction reaction with cyclohexylamine; one mole of dimethylhydrazine is eliminated and the resulting compound is a five-membered ring (26), $R = Ph$, $R' = Me$, $R'' = cyclo$-C_6H_{11}:[299a]

$$(20)$$

$$(26)$$

Polymers containing B_2N_4 rings (as $Ph_2B_2N_4H_2$ units) connected through p-biphenylene or $-(CH_2)_6-$ bridges, have been synthesized from $PhB(NMe_2)_2$ and $4\text{-}H_2N\text{—}NH\text{—}C_6H_4\text{—}C_6H_4\text{—}NHNH_2\text{-}4'$ and $H_2N\text{—}NH(CH_2)_6NHNH_2$ respectively.[299b]

The reaction of 3,6-diphenyl-1,2,4,5-tetraza-3,6-diborine, (25) $R = Ph$, $R' = R'' = H$, with two moles of methylisocyanate leads to a bis-addition product (25), $R = Ph$, $R' = H$, $R'' = CONHMe$.[299c]

With acetic anhydride (containing some acetic acid), ring opening of (25) $(R = Bu^t, Ph; R' = R'' = H)$ occurs; pure acetic anhydride gave a diacetylated product (25), $R = Ph$, $R' = H$, $R'' = Ac$, in good yield.[299c]

The dimethylamino group attached to boron in (25), $R = NMe_2$ $R' = R'' = Me$, has been substituted with halogens on treatment with boron halides (BCl_3, BBr_3).[299]

Hexamethyltetrazadiborine (25), $R = R' = R'' = Me$, reacts with methylamine quantitatively to yield the 1,2,4,3,5-triazadiborolidine (26) as a result of ring contraction.[117]

The 1,2,4,5-tetraza-3,6-diborine ring system forms radical cations (27), $R = NMe_2$, $R' = R'' = Me$, under the action of anhydrous $AlCl_3$ in dichloromethane:

$$(27)$$

The twisted-boat conformation of the initial B_2N_4 ring is transformed to a planar ring system, which is isoelectronic with benzene radical anions. The solutions containing these radical cations are blue. Their ESR spectra have been investigated.[300] The radical cations are stabilized by formation of a trigonal planar system.[301,302]

1,2,4,5-Tetraza-3,6-diborine derivatives (25) (R = Me, Ph, Cl, MeS, MeO, NMe_2; R' = R'' = Me) undergo ring contraction on heating, eliminating Me—N=N—Me, to form five-membered ring compounds (26).[303]

The crystal and molecular structure of 3,5-bis(dimethylamino)-1,2,4,5-tetramethyl-1,2,4,5,3,6-tetrazadiborine (25), R = NMe_2, R' = R'' = Me, has been investigated by X-ray diffraction. The B_2N_4 ring is nonplanar, as expected for an 8π electron heterocycle. The bond lengths are B–N 1.43 Å, N–N 1.43 Å and C–N 1.44 Å. All B–N bonds are of equal length. MO calculations of various 1,2,4,5-tetraza-3,6-diborine derivatives have shown that a twisted ring (D_2 symmetry) is the most stable conformation.[188]

The photoelectron spectra of some B_2N_4 derivatives (25), R = Me, Cl, OMe, SMe, NMe_2; R' = R'' = Me, have been reported. Orbital energies have been calculated by the CNDO/2 method. Evidence for strong π-interactions has been deduced by comparison with the spectra of diaminoboranes. EHT calculations have shown that the twisted form (D_2) has an energetic advantage of ~ 7 kcal/mol compared with the planar system (D_{2h}). Chair and boat conformations are much less favoured energetically than a twist and a planar system.[304]

A ^{13}C NMR analysis of (25), R = Ph, R' = R'' = Me, has been carried out.[304a]

2.4.5   1,2,4-Triaza-3,5-diborolidine

Some five-membered ring derivatives, 1,2,4-triaza-3,5-diborolidines (26), have already been mentioned in the previous section. Other derivatives have been synthesized by reactions of organobis(methylthio)boranes $RB(SMe)_2$ with a hydrazine derivative in a molar ratio 2 : 1, followed by addition of ammonia or a primary amine (R = Me, Ph; R' = Me, Ph; R'' = H, Me, NMe_2, cyclo-C_6H_{11}):[75]

$$2\,RB(SMe)_2 + R'NH\!-\!NHR' \xrightarrow{\ Et_2O\ } RB \overset{\displaystyle \overset{R'}{N}\!-\!\overset{R'}{N}}{\underset{\displaystyle SMe\quad SMe}{\diagup \qquad \diagdown}} BR \xrightarrow{\ R''NH_2\ } (26) \qquad (21)$$

Hexamethylborazine (MeBNMe)$_3$ reacts quantitatively with N,N'-dimethylhydrazine to form the 1,2,4-triaza-3,5-diborolidine systems (**26**), R = R' = R" = Me.[117] Unsymmetrically substituted 1,2,4-triaza-3,5-diborolidine derivatives (**26a**), R = Me, R' = H, R" = H, Me, R''' = Me, resulted from the interaction of methylbis(methylthio)borane MeB(SMe)$_2$ with monosubstituted hydrazine MeNHNH$_2$:[305]

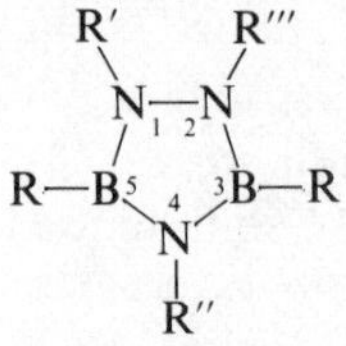

(26a)

The 3,5-bis(methylthio) derivative (**26**), R = SMe, R' = R" = Me, has proved to be a suitable starting compound for the preparation of a series of B-functional derivatives:[306]

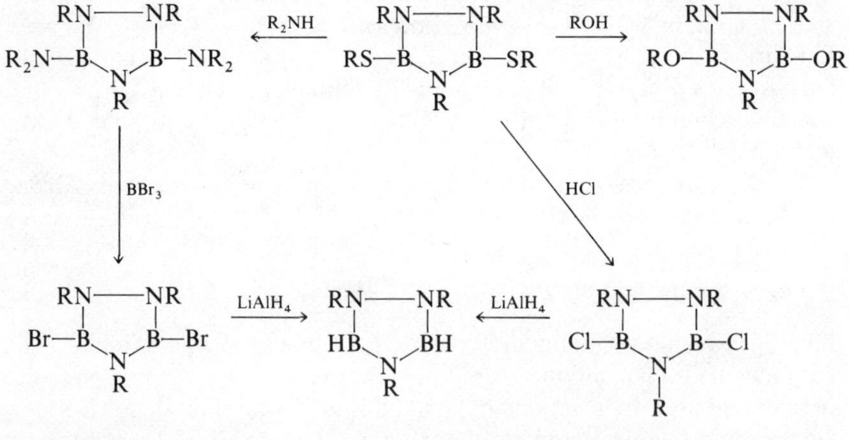

R = Me

N-Functional derivatives of 1,2,4-triaza-3,5-diborolidines (**26**) with R = R' = Me, R" = SiMe$_3$, SnMe$_3$, B(NMe$_2$)$_2$, P(NMe$_2$)$_2$ and PMe$_2$ were obtained by lithiation of 4H-triazadiborolidines, followed by treatment with a variety of halides. Similar treatment of 1H-triazadiborolidines gives analogous unsymmetrically substituted derivatives.[306]

The PE spectra of 1,2,4-triaza-3,5-diborolidines have been discussed, CNDO/S charge densities, and bond orders have been evaluated and a linear correlation between the ^{11}B chemical shifts and the CNDO/S π-charge densities has been established.[307]

3,5-Unsymmetrically substituted 1,2,4-triaza-3,5-diborolidine derivatives have been prepared from a cyclic triazaborasilidine by treatment with boron functional compounds (R = Me, X = Cl, Br, SMe):[308]

$$
\begin{array}{c}
\text{RB}\underset{\underset{R}{N}}{\overset{\overset{R}{N}-\overset{R}{N}}{\diagdown}}\text{SiR}_2 + BX_3 \longrightarrow \text{RB}\underset{\underset{R}{N}}{\overset{\overset{R}{N}-\overset{R}{N}}{\diagdown}}\text{BX} + R_2SiX_2
\end{array}
\qquad (22)
$$

(26b)

The lability of the B–X bond in (26b), X = SMe, has been used for the preparation of other unsymmetrically substituted derivatives; by treatment with LiAlH$_4$, SbF$_3$, HNMe$_2$, MeOH, Bu^{t}OH and PhBCl$_2$, the compounds (26b) with X = H, F, NMe$_2$, OMe, OBu^t and Cl have been obtained.[308] A bicyclic compound, containing two five-membered B$_2$N$_3$ rings bridged by an NMe—NMe group, has also been prepared by treating (26b), R = Me, X = SMe, with N,N'-dimethylhydrazine.[308]

One-electron oxidation of 4-dimethylamino-1,2,3,5-tetramethyltriazadiborolidine produces a six-membered cyclic radical cation (blue colour), following a ring expansion (R = Me):

$$
R\text{—B}\underset{\underset{R}{N}}{\overset{\overset{R}{N}-\overset{R}{N}}{\diagdown}}\text{B—R} \xrightarrow[-e^-]{AlCl_3 \text{ in } CH_2Cl_2} R\text{—B}\underset{N-N}{\overset{N-N}{(\cdot +)}}\text{B—R}
\qquad (23)
$$

(27)

The ESR spectrum of this radical cation has been investigated.[300]

For two stannylated 1,2,4-triaza-3,5-diborolidines (containing SnMe$_3$ groups in position 2 or 4) (26), R = R' = Me, R" = SnMe$_3$, and (26a), R = R' = R" = Me, R''' = SnMe$_3$, a multinuclear NMR (^{1}H, ^{11}B, ^{13}C and ^{119}Sn) investigation has been carried out in C$_6$D$_6$, the chemical shifts and coupling constants being determined.[309]

The structure of pentamethyl-1,2,4-triaza-3,5-diborolidine has been investigated.[332] In both the gas phase and in solution the compound is monomeric; at −30 °C it crystallized. The ring is planar.

2.4.6   Boron–nitrogen–silicon heterocycles

1,3-Diaza-2-sila-4-boracyclobutane. Four-membered B–N–Si heterocycle derivatives (28) can be prepared from the N-lithio derivative of a diaminosilane, for example $Me_2Si(NHBu^t)_2$, by treatment with R_2BX (X = F, Cl, Br) and elimination of BR_3; the bis(diorganoboryl) derivative $Me_2Si(N-Bu^tBR_2)_2$ is an unstable intermediate.[310]

1,2,4-Triaza-3-sila-5-boracyclopentane. From the reaction of N,N'-bis-(trimethylsilyl)-N,N'-dimethylhydrazine $(Me_3Si)MeN—NMe(SiMe_3)$ and (bromomethylboryl)(bromodimethylsilyl)methylamine $MeN(SiMe_2Br)$-(BMeBr), the five-membered ring compound (29), R = R' = R'' = Me, has been obtained.

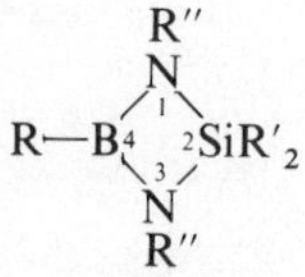

R''
|
N
|
R—B⁴ ²SiR'₂
\₃/
N
|
R''

(28)

R''   R''
| |
N—N
|₁ ₂|
R—B⁵ ₄ ³SiR'₂
\ /
N
|
R''

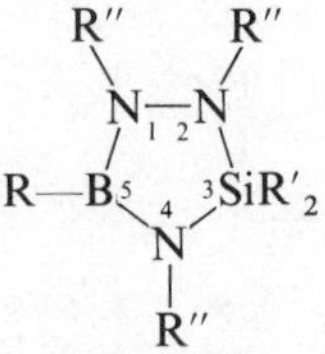

(29)

The BN_3Si ring system is also formed in the reaction of lithiated N,N'-dimethylhydrazine $LiMeN—NMeLi$, with $MeN(SiMe_2Br)_2$. It has been shown that yields depend on the nature of the solvent used. In diethyl ether or tetrahydrofuran, considerable amounts of pentamethyl-1,2,4-triaza-2,5-diborolidine (26) are formed.[311]

An excess of $MeSiCl_3$ reacts with (29), R = R' = R'' = Me, to replace the $SiMe_2$ group with a SiMeCl unit.[311]

Silaborazines. Boron in borazine can be partially replaced with silicon to give mono- and disilaborazines (30) and (31):

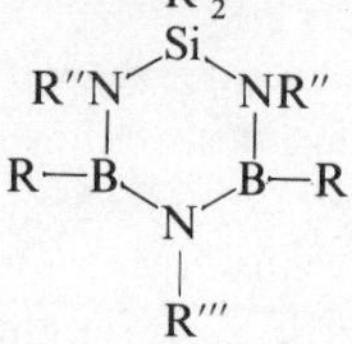

R'₂
Si
R''N NR''
| |
R—B   B—R
\ N /
|
R'''

(30)

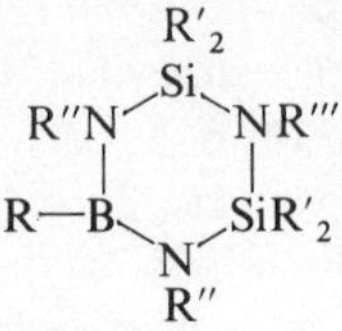

R'₂
Si
R''N NR'''
| |
R—B   SiR'₂
\ N /
|
R''

(31)

In the reaction of $MeB(NMeLi)_2$ with Ph_2SiCl_2, a four-membered ring was not obtained as expected, and instead 2-sila- and 2,4-disilaborazines

(30) and (31) were formed (R = Me, R' = Ph, R'' = R''' = Me). By-products in this reaction were $(Ph_2SiNMe)_3$ and $(MeBNMe)_3$.[73]

The reaction of 1,3-bis(methylamino)pentamethyldisilazane MeN(Si-Me$_2$NHMe)$_2$, after metallation or in the presence of triethylamine, with PhBCl$_2$ yielded the 2,4-disilaborazine (31), R = Ph, R' = R'' = R''' = Me.[73]

Another way for the synthesis of 2,4-disilaborazines (31) is based on a [3 + 3] ring condensation of RB(NLiR'')$_2$ with 1,3-dichlorotetramethyldi-silazane derivatives R'''N(SiMe$_2$Cl)$_2$. Using this procedure, compounds (31) with R = Me, Ph; R' = Me, R'' = Me, Et, Bu^s, Ph, R''' = H, Me, have been prepared.[313] A similar [3 + 3] ring condensation of Me$_2$Si(NMeLi)$_2$ with (BrMeB)$_2$NMe has also been successful in the synthesis of disila-borazines (31) (R = R' = R'' = R''' = Me). Lithiated dimethylaminoborane Me$_2$B—NMeLi reacts with 1,3-dibromopentamethylsilylamine MeN(Si-Me$_2$Br)$_2$ and Me$_2$B—NMe—SiMe$_2$Br to give in both cases a 2,4-disila-borazine (31), R = R' = R'' = R''' = Me.[68]

A 2-silaborazine derivative (30), with R = NMe$_2$, R' = Me, R'' = Me, R''' = CH$_2$Ph, results from a [5 + 1] ring condensation of Me$_2$Si[NMe—B(NMe$_2$)$_2$]$_2$ with Ph—CH$_2$NH$_2$.[313]

Attempts to synthesize silaborazines by [2 + 2 + 2] and [4 + 2] ring condensations have been started from bis(methylamino)dimethylsilane Me$_2$Si(NHMe)$_2$ and bis(dimethylamino)borane derivatives R—B(NMe$_2$)$_2$ (R = Me or Bu^n). Neither the formation of siladiazaboracyclobutane (28) nor the formation of NSiNBN or NBNSiNBN chains has been observed. The formation of (30) is favoured, and methylaminoborane derivatives are formed in side reactions. By reaction of Me—B(NMe—SiMe$_3$)$_2$ with BrMeB—NMe—SiMe$_2$Br, the permethylated silaborazine (30) was also formed, but could not be separated from the by-product, hexamethyl-borazine.[314]

The [3 + 3] cyclocondensation of Me$_2$Si(NMeLi)$_2$ with (BrBMe)$_2$N—SiMe$_3$ also produced a 2-silaborazine (30), R = R' = R'' = Me, R''' = SiMe$_3$, as expected.[314]

Phenyldichloroborane PhBCl$_2$ reacts with the cyclosilazane (Me$_2$SiNH)$_3$ to give either monosila- (30) or disila- (31) borazine derivatives (R = Ph, R' = Me, R'' = R''' = H), depending upon the molar ratio used.[71] By-products of these reactions are 2,4,6-triphenylborazine (PhBNH)$_3$, 2-chloro-1-dimethylphenylsilyl)-4,6-diphenylborazine and HN(SiMe$_2$Cl)$_2$. In a similar reaction with PhBCl$_2$, octamethylcyclotetrasilazane (Me$_2$SiNH)$_4$ contracts to a 2,4-disilaborazine (31), R = Ph, R' = Me, R'' = R''' = H.[71]

A complex reaction occurs between an unsymmetrically substituted cyclo-trisilazane Me$_6$Si$_3$N$_3$(SiMe$_2$F)LiH and BF$_3$.OEt$_2$. Among the various products formed in this reaction (linear and cyclic silazanes), there is a disilaborazine (31), R = F, R′ = Me, R″ = H, R‴ = $SiMe_2F$.[315] Various related silaborazines (31) have been prepared in good yields by treatment of unsymmetrical cyclotrisilazanes with $BF_3.OEt_2$.[315]

1,3-Diaza-4,5-disila-2-boracyclopentane. The first compound of this type, (32), R = R′ = R″ = Me, has been obtained by [3 + 2] cyclocondensation of a lithiated aminoborane $MeB(NMeLi)_2$ with 1,2-dichlorotetramethyl-disilane $ClMe_2Si$—$SiMe_2Cl$:[316]

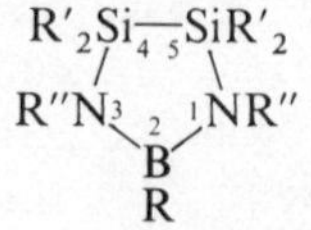

$$R'_2Si\overset{4\ 5}{\underset{}{\frown}}SiR'_2$$
$$R''N^3_{\ 2}\ \ NR''$$
$$B$$
$$R$$

(32)

By [4 + 1] cyclocondensation, starting from bis(methylamino)tetramethyl-disilane $R''HN$-$SiMe_2SiMe_2$-NHR'', dichlorophenylborane and triethyl-amine,[317] and from lithiated alkylaminodisilanes and dichlorophenyl-borane[317-319a], a series of BN_2Si_2 ring derivatives (32), R = Ph, 4-MeC_6H_4, NMe_2, $N(SiMe_3)_2$, OEt and R′ = Me, R″ = Me, Et, have been obtained.[319]

1,3,5,7-Tetraza-2-bora-4,6,8-trisilacyclo-octane. The only known eight-membered ring containing boron, silicon and nitrogen (33), R = Me, R′ = Pr^n, has been synthesized by a [7 + 1] cyclocondensation of dilithiated trisilazane-1,5-di-n-propyl derivatives $Pr^nNH(Me_2Si$—$NMe)_2SiMe_2$—$NHPr^n$ with phenyldichloroborane $PhBCl_2$.[320]

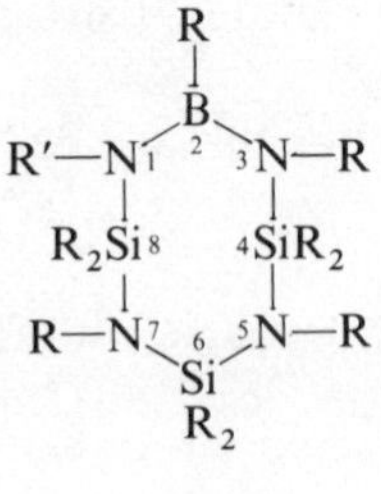

$$R$$
$$|$$
$$B$$
$$R'-N^1\ ^2\ _3N-R$$
$$|\qquad\qquad|$$
$$R_2Si^8\quad _4SiR_2$$
$$|\qquad\qquad|$$
$$R-N^7\ _6\ _5N-R$$
$$Si$$
$$R_2$$

(33)

2.4.7 Boron–nitrogen–phosphorus heterocycles

1,3,2,4-Diazaphosphaboretidine. The four-membered ring compound (34a), R = R′ = Cl, Br; R″ = $SiMe_3$, is formed in the reaction of bis(trimethyl-silyl)aminotrimethylsilyliminophosphane $(Me_3Si)_2N$—P═N-$SiMe_3$, with boron trihalides. An iminophosphane–boron-trihalide adduct in the first step of this reaction has been supposed, owing to the formation of bis(trimethyl-silyl)aminodihalogenophosphane $(Me_3Si)_2N-PX_2$ as by-product.[321,322]

(34a)   (34b)

Another method used for the synthesis of four-membered ring derivatives (34a) and (34b), R = Me, Et, F; R' = Me, R'' = Bu^t, X = electron pair, O or S, starts with N-lithiated derivatives of $MeP(NHBu^t)_2$ or $Me(X)P(NH-Bu^t)_2$ and diorganoboron halides R_2BX or BF_3.[310]

The intermediate (35) formed when X = O is rather stable owing to its intramolecular B–O coordination:

(35)

This compound decomposes on heating, with elimination of BR_3.[310]

1,2,4-Triaza-3-phospha-5-borolidines. These compounds (36a) have been prepared by replacing the silicon atom in 1,2,4-triaza-3-sila-5-borolidines (29) in reactions with phosphorus halides (R = Me, R' = Cl, Br, Me, Ph):

(29)   (36a)

$$(24)$$

The Lewis acidity of the phosphorus compound has been proved to be essential for the yield of the BN_3P ring compounds (36a). The efficiency decreases in the order $PCl_3 > PBr_3 > PhPCl_2 > MePCl_2$.[311]

The 3-chloro derivative of (36a), R' = Cl, has been the starting material for substitution reactions, occurring on treatment with LiMe, LiPh, $HNMe_2$ and NaOR, to give (36a), R = Me, R' = Me, Ph, NMe_2, OMe.[311] The phosphorus(III) derivative (36a), R = R' = Me, adds sulphur, to form a phosphorus(V) derivative (36b), R = R' = Me, X = S, while an oxygen analogue (36b), R = R' = Me, X = O, was obtained in a Michaelis–Arbusov-type rearrangement of the methoxy derivative (36a), R' = OMe, on heating with methyl iodide.[311]

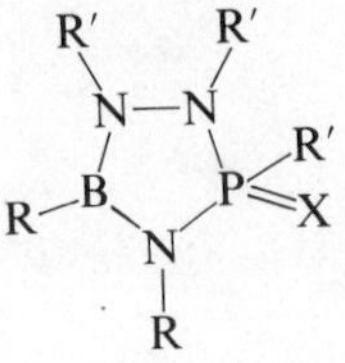

(36b)

3-Chloro-1,2,4,5-tetramethyl-1,2,4,3,5-triazaphosphaborolidine   (36a), R = Me, R' = Cl, on treatment with $AlCl_3$ in CH_2Cl_2, formed the 1,2,4,5-tetramethyl-1,2,4,3,5-triazaphosphaborolidin-3-ylium   tetrachloroaluminate (36c):

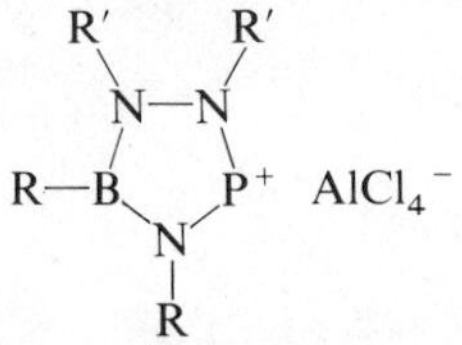

(36c)

The large change in ^{31}P NMR shifts indicates that cationic P(III) is able to compete more efficiently than neutral P(III) for the electron density of the adjacent nitrogen atoms. A degree of participation of phosphorus in the π-system leads to greater planarity of the ring.[323]

1,3,5-Triaza-2-phospha-4,6-diboracyclohexane. A six-membered B_2N_3P heterocycle (37a) results from the Si–N cleavage reaction of bisphenyl(trimethylsilylamino)borylmethylamine with halogenophosphine oxide (R = Ph, R' = R'' = R''' = Me):[324]

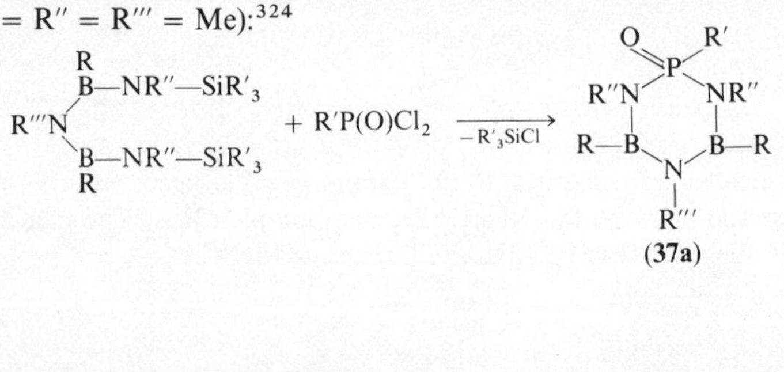

(25)

(37a)

The determination of the melting point of (37a), R = Ph, R' = R" = R''' = Me, shows a first melting point at 66–70 °C, but resolidification at 80 °C; the sample melts again reversibly at 100–103 °C. This is an indication of crystallization in two modifications. The results of the NMR spectroscopic investigation indicates two conformers for this compound, which do not interconvert until 160 °C:[324]

In silaborazines, the nitrogen atoms differ in their basicity. The nitrogen atoms attached to silicon are more basic than the two nitrogens of the BNB fragments; as a result the reaction with phosphorus halides is selective and produces 2-phospha(III)borazines (37b), R = Me, R' = Me or SiMe$_3$; X = Cl or Br:[314]

$$(26)$$

(30)   (37b)

The chlorine substituent at phosphorus can be readily replaced with OMe and NMe$_2$ groups in reactions with appropriate reagents.[314] The phosphorus(III) derivatives (37b), R = Me, R' = Me, X = Cl, NMe$_2$, add sulphur to form phosphorus(V) derivatives (37c):[314]

(37c)

The nucleophilic substitution of chlorine indicates a relatively weak P–Cl bond and the resonance stabilization of the phosphonium ion with dicoordinated phosphorus is substantiated by the reaction with anhydrous AlCl$_3$ in CH$_2$Cl$_2$ of the 2-chloro derivative (37b), R = R' = Me, X = Cl, which gives the tetrachloroaluminate of the 3-phosph(III)oniaborazine cation (37d).[323] In contrast with this, BBr_3 is added to the free electron pair at phosphorus, which is more basic than the nitrogen site, and (37e), $R = R' = Me$, is formed:[314]

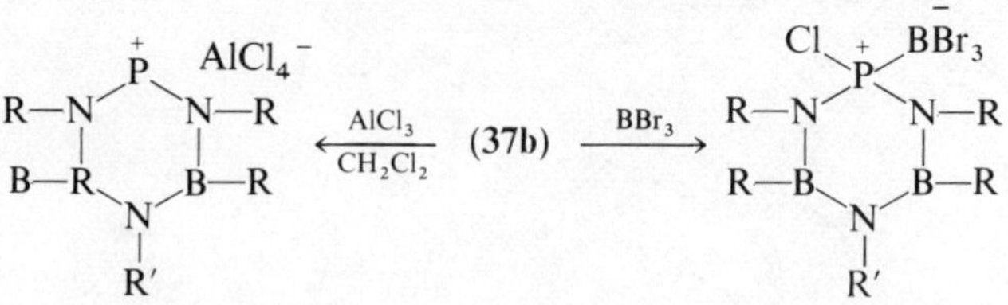

(27)

(37d)   (37e)

The six-membered ring B_2N_3P seems to be nonplanar. This is supported by NMR spectral data.[314] The planar NBNBN fragment, which permits the nitrogen and boron atoms to be sp^2 hybridized and thus to achieve $p_\pi-p_\pi$ overlap, is indicated by the fact that ^{11}B chemical shifts of (37b) are quite similar to those of planar hexamethylborazine and chloropentamethylborazine,[323] but the phosphorus atom is out-of-plane.

1,3,5-Triaza-2,4-diphospha-6-boracyclohexenes. Two, formally unsaturated, six-membered B–N–P rings (38) and (39) have been described in the literature, one of them in a patent:

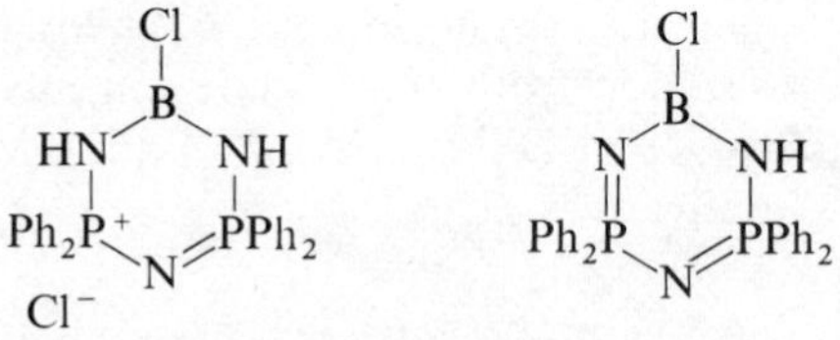

(38)   (39)

The two compounds have been investigated in relation to their potential fungicidal properties.[325,326]

2.4.8   Boron–nitrogen–arsenic heterocycles

Arsenic trichloride reacts with 1,2,4-triaza-3-sila-2-borolidine (29) to form the only known B–N–As heterocycle (40), 3-chloro-1,2,4,5-tetramethyl-1,2,4-triaza-3λ^3-arsa-5-borolidine (R = Me):[311]

$$R-B\overset{\overset{\displaystyle R\quad R}{\underset{\displaystyle N-N}{\diagdown\diagup}}}{\underset{\underset{\displaystyle R}{\displaystyle N}}{\diagdown\diagup}}SiR_2 \; + \; AsCl_3 \quad \xrightarrow[-R_2SiCl_2]{} \quad R-B\overset{\overset{\displaystyle R\quad R}{\underset{\displaystyle N-N}{\diagdown\diagup}}}{\underset{\underset{\displaystyle R}{\displaystyle N}}{\diagdown\diagup}}As-Cl \qquad (28)$$

(29)   (40)

The arsenic-containing heterocycle is extremely sensitive to hydrolysis.

2.4.9 Boron–nitrogen–oxygen heterocycles

A five-membered ring compound, 2,3,4,5-tetramethyl-1-oxa-3,4-diaza-2,5-diborolidine (41), R = Me, was formed on treatment of N,N'-bis(methyl-bromoboryl)dimethylhydrazine(BrMeB)MeN—NMe(BMeBr)with Me₃Si—OMe or NaOMe;[327,328] methyl bromide was eliminated in addition to Me₃SiBr (or NaBr):

$$R-B\overset{\overset{\displaystyle R\quad R}{\underset{\displaystyle N\underset{3\;\;4}{-}N}{\diagdown\diagup}}}{\underset{\underset{\displaystyle O}{1}}{\diagdown}}^{2\quad 5}B-R$$

(41)

The PE spectrum of (41) has been reported.[307] The five-membered ring compound (41), R = Me, has also been prepared from (MeS)MeB—NMe—NMe—BMe(SMe), N,N'-dimethylhydrazine and an equimolar amount of water.[305]

The reaction of 1,3-di-t-butyl-2,4-bis(pentafluorophenyl)-1,3,2,4-diazadiboretidine (22), R = C_6F_5, R' = Bu^t, with a nitrone leads to the formation of 1,2,4-oxadiaza-3,5-diborolidine (42):[44]

$$\overset{\displaystyle R\quad\quad R'}{\underset{\underset{\displaystyle R'\quad\; R}{\diagup\diagdown}}{\underset{\displaystyle N-B}{\overset{\displaystyle B-N}{\diagup\diagdown}}}} \;+\; \underset{\underset{\displaystyle Me}{|}}{\overset{\overset{\displaystyle C_6H_5-CH\quad O^-}{|}}{N^+}} \quad \xrightarrow[-PhHC=NBu^t]{} \quad R-B\overset{\overset{\displaystyle Me}{\underset{\displaystyle N\underset{2\;1}{-}O}{\diagdown\diagup}}}{\underset{\underset{\displaystyle R'}{\displaystyle N}}{{}_3\quad_5}}B-R \qquad (29)$$

(22)   (42)

Bis(trimethylsilyl)peroxide Me₃Si—O—O—SiMe₃ has been used as a reagent to prepare a 1,2-dioxa heterocycle (43), R = Me, by reaction with bis(bromomethylboryl)trimethylsilylamine Me₃Si-N(BRBr)₂:[328]

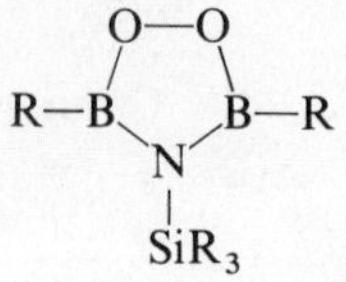

$$R-B \underset{\underset{\underset{SiR_3}{|}}{N}}{\overset{O-O}{\diagdown \diagup}} B-R$$

(43)

The He(I) photoelectron spectrum of (43), R = Me, has been discussed. Compared with pentamethyltriazadiborolidine, replacement of the Me—N—N—Me group by the peroxide group stabilizes the b_1 (π) orbital.[328]

3,5-Dimethyl-1,2,4-trithia-3,5-diborolane reacts in a complex redox reaction with sulphinylamines RNSO to give 1,2,4,3,5- and 1,3,4,2,5-dithiazadiborolidines; in addition, trimethylboroxine $(RBO)_3$, 1,3,5-oxadiaza-2,4,6-triboracyclohexane (44) and 1,3,5-dioxaza-2,4,6-triboracyclohexane (45) derivatives have been isolated (R = Me, R' = SO_2Me, cyclo-C_6H_{11}, p-FC_6H_4):[329]

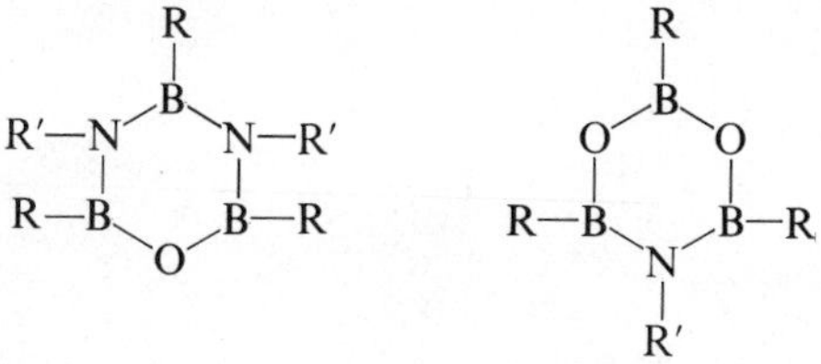

(44)   (45)

2.4.10 Boron–nitrogen–oxygen–silicon heterocycles

1,3-Dichloropentamethyldisilazane reacts with phenylboronic acid and triethylamine in petrol at −78 °C to form 1,3,5-dioxaza-2,4,6-boradisilacyclohexane (46), R = Me:[312a,330]

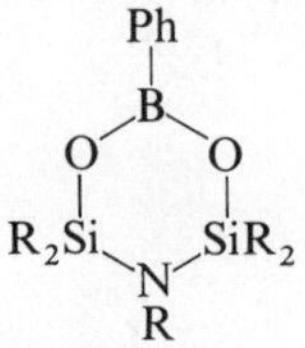

(46)

By the reaction of dilithio-1,5-bis(methylamino)hexamethyl-1,3,5-trisila-2,4-dioxapentane $Me_2Si(OSiMe_2-NMeLi)_2$ with $PhBCl_2$ an eight-membered $BN_2O_2Si_3$ ring was formed.[312a] Of the three isomeric structures (47a–c), R = Me, the first one is the most likely, if no rearrangement occurs during the reaction:

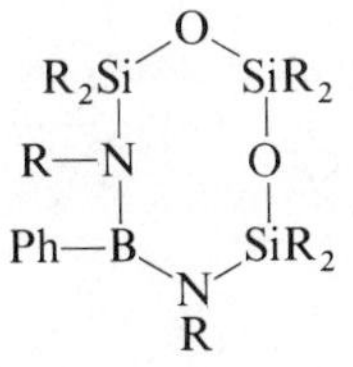

(47a) 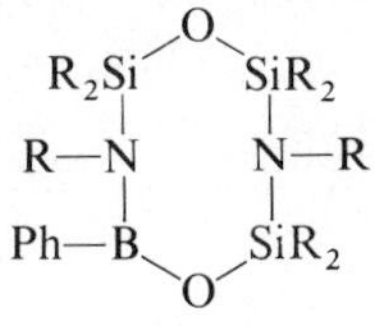   (47b) 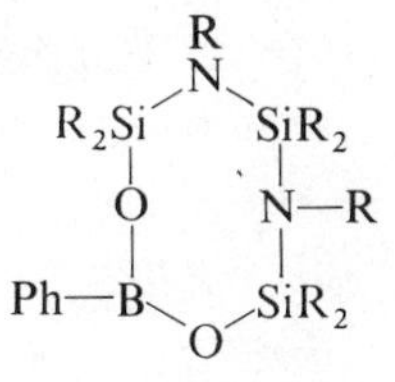   (47c)

2.4.11 Boron–nitrogen–sulphur heterocycles

An insertion reaction takes place on interaction of phenylbis(dimethyl-amino)borane and sulphonyl diisocyanate, and thiadiazaboretidine (48) is formed in quantitative yield:[331]

$$PhB(NMe_2)_2 + O_2S(NCO)_2 \longrightarrow$$

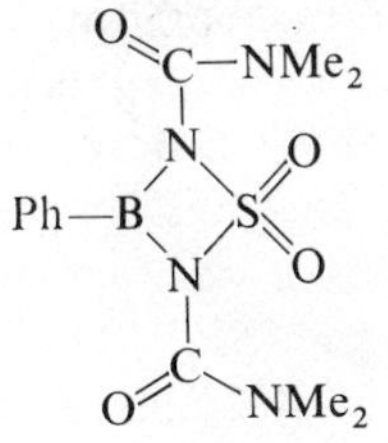

(30)

(48)

Primary amines react with 3,5-dimethyl-1,2,4,3,5-trithiadiborolidine $R_2B_2S_3$ by ring opening followed by an intramolecular cyclocondensation to yield 1,2,4,3,5-dithiazadiborolidines (49), 1,3,5-thiadiaza-2,4,6-tribora-cyclohexane (50) and borazines $(RBNR')_3$, R = Me:[76]

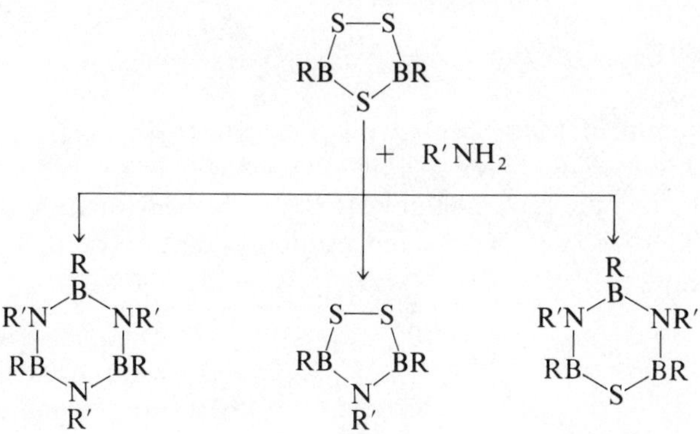

(49)   (50)

The product distribution is influenced by steric and electronic effects.

A rather complex reaction takes place between 1,2,4,3,5-trithiadiboro-lidine $R_2B_2S_3$ and N-sulphinylamines R'NSO. Among other products, derivatives of 1,2,4-dithiaza-diborolidines (51a) are formed (R = Cl, Br, Me, R' = C_6F_5, Bu^t or SiMe$_3$):[329]

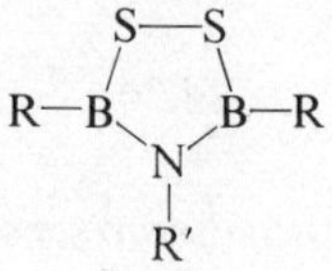

$$\begin{array}{c} S\!\!-\!\!S \\ R\!\!-\!\!B \diagdown \diagup B\!\!-\!\!R \\ N \\ | \\ R' \end{array}$$

(51a)

In the synthesis of (51a), R = Me, R' = Bu^t, 2,4,6-trimethylboroxine (RBO)$_3$, 1,3,5,2,4,6-dioxazaboracyclohexane (45) and 1,3,5,2,4,6-oxadiaza-triboracyclohexane (44) have been identified as by-products.[329] The halogen substituents in (51a), R = Cl, Br, R' = SiMe$_3$, can be replaced in reactions with LiR (R = Bu^n, Bu^t), HNR$_2''$ (R'' = Me, Et) to give (51a) with R = Bu^n, Bu^t, NMe$_2$, NEt$_2$.[329] Similarly, (51a), R = Br, R' = C_6F_5, reacts with diethylamine to give (51a), R = NEt$_2$, R' = C_6F_5.[329]

Another B–N–S five-membered ring (51b), R = Me, R' = SiMe$_3$, 2-CF$_3$C$_6$H$_4$, 2,6-Me$_2$C$_6$H$_3$, was isolated from the reaction of 3,5-dimethyl-1,2,4,3,5-trithiadiborolidine $R_2B_2S_3$ with sulphinylamines R'NSO; other cyclic by-products (44), (45) and (RBO)$_3$ were also identified:[329]

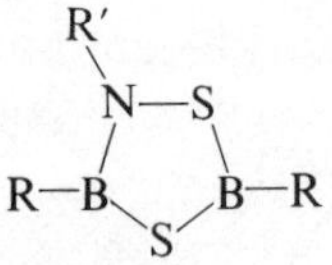

$$\begin{array}{c} R' \\ \diagdown \\ N\!\!-\!\!S \\ R\!\!-\!\!B \diagdown \diagup B\!\!-\!\!R \\ S \end{array}$$

(51b)

In the reaction of MeB(SMe)$_2$ with N,N'-dimethylhydrazine, the linear compound Me(MeS)B—NMe—NMe—BMe(SMe) is formed; this reacts with H$_2$S to eliminate methanethiol MeSH, with the formation of the five-membered heterocycle 1,3,4,2,5-thiadiazadiborolidine (51c), R = Me:[305]

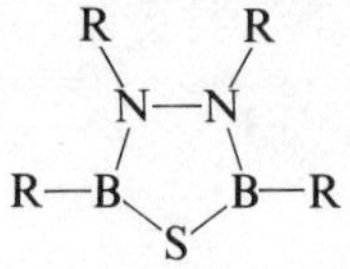

$$\begin{array}{c} R \qquad R \\ \diagdown \qquad \diagup \\ N\!\!-\!\!N \\ R\!\!-\!\!B \diagdown \diagup B\!\!-\!\!R \\ S \end{array}$$

(51c)

The B–S bonds in (51c) can be cleaved by bromide, phosphorus(III) halides (PCl_3, PBr_3) and SbF_3, to form linear N,N'-bis(methylhalogeno-boryl)-N,N-dimethylhydrazine derivatives.[328]

The crystal and molecular structure of some 6 π-electron B–N–S heterocycles has been determined.[332] The results are briefly summarized here.

3,4,5-Trimethyl-1,2,4,3,5-dithiazadiborolidine   (51a),   R = R' = Me, forms orthorhombic crystals (space group *Cmcm*, Z = 4); the bond distances are S–S 2.057 Å, B–S 1.813 Å, B–N 1.435 Å; and the bond angles are SSB 96.8°, SBC 119.3°, CBN 127.4°, SBN 113.3°, BNC 120.0°, BNB 119.9°.[332]

2,3,4,5-Tetramethyl-1,3,4,2,5-thiadiazadiborolidine (51c), R = R' = Me, forms monoclinic crystals (space group $P2_1/n$, Z = 4); the bond distances are N–N 1.414 Å, N–B 1.396 Å and 1.374 Å, B–S 1.789 Å and 1.810 Å; and the bond angles B_1N_1C 128.2°, $B_1N_1N_2$ 115.6°, N_1N_2C 116.2°, $N_1N_2B_2$ 112.9°, N_1N_2C 118.7°, N_2B_2S 111.3°, B_1SB_2 90.2°, N_1B_1S 110.0°, N_1B_1C 125.5°, SB_1C 124.4°.[332]

Both (51a) and (51c) show planar five-membered 6 π-electron systems. The sums of the angles at the threefold coordinated boron and nitrogen atoms are 360°, thus showing sp^2 hybridization. Therefore optimal B–N π-interactions are possible. The structural data are in agreement with the ring conformation of (51a) and (51c) deduced from He(I) PE spectra.[307]

Heating equimolecular amounts of $MeN(BPh—NMe—SiMe_3)_2$ with SCl_2 results in a [5 + 1] cyclocondensation, leading to 1,3,5-thiadiaza-2,4,6-diboracyclohexane, (52) R = Ph, R' = Me, with $(PhBNMe)_3$ as by-product.[52]

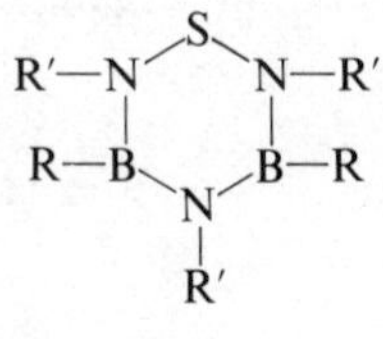

(52)

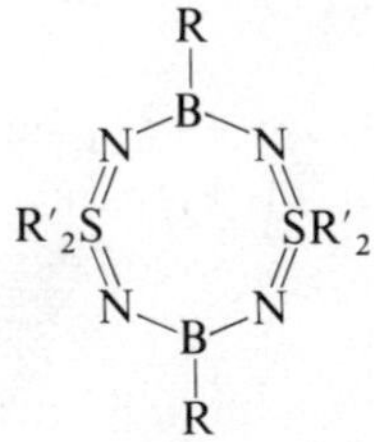

(53)

1,1,5,5-Tetramethyl-3,7-diphenyl-$1\lambda^6,5\lambda^6$,2,4,6,8,3,7,-dithiatetrazadiborocine (53), an eight-membered ring system, was synthesized by the reaction of dichlorophenylborane $PhBCl_2$ with S,S-dimethyl-N,N'-bis(trimethyl-

silyl)sulphodiimide $Me_2S(=NSiMe_3)_2$, with elimination of Me_3SiCl
(R = Ph, R' = Me).[333]

2.4.12 Boron–nitrogen–selenium heterocycles

N,N'-Dimethylhydrazine can replace the Se–Se fragment of 1,2,4-triselena-3,5-diborolidine, to form the 1,2-diaza-4-selena-3,5-diborolidine ring
(54); reactions with primary amines result in replacement of the selenium
atom from the B–Se–B fragment with formation of 1,2-diselena-4-aza-3,5-diborolidine ring (55), R = Me or Pr^n, R' = Me or Ph:[334,335]

$$\tag{31}$$

All B–Se–N ring derivatives are liquids, and consequently no X-ray diffraction structure analysis has been performed on any of them.

2.4.13 Fused-ring systems containing boron, nitrogen and a heteroatom

There are several boron–nitrogen heterocycles fused to organic ring systems
described in the literature
   Fused-ring silaborazines (56) can be obtained by [3 + 3] cyclocondensation of N,N'-dilithiotriazaboradecalin (obtained by lithiation of triazabora-
decalin) with 1,3-halogenobis(silyl)amines (R, R' = Me, Ph, Cl, R'' = H,
Me).[73, 313]
   By reaction of N,N'-dilithiotriazaboradecalin with $(Cl_3Si)_2NH$, a derivative of 2,4,6,8,9-pentaza-1,5-disila-3,7-diborabicyclo [3.3.1] nonane has been
obtained with structure (57).[313]
   The reaction of N,N'-dilithiotriazaboradecalin with 1,2-dichlorotetra-
methyldisilane $ClMe_2Si$—$SiMe_2Cl$ results in the formation of a fused
system containing a five-membered ring (58) (R = Me).[63,316]
   Starting from the same N,N'-dilithiotriazaboradecalin and bis(chloro-
phosphino)methylamines $MeN(PRCl)_2$ (R = Me, Ph, Cl) and bis(dichloro-

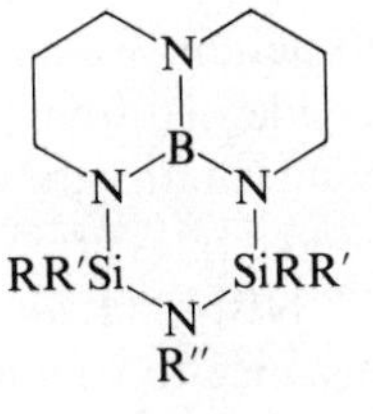

(56)

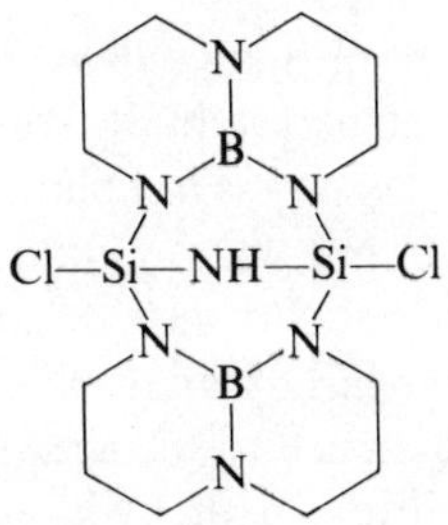

(57)

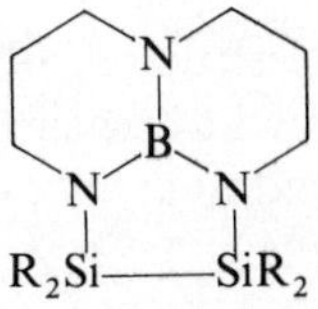

(58)

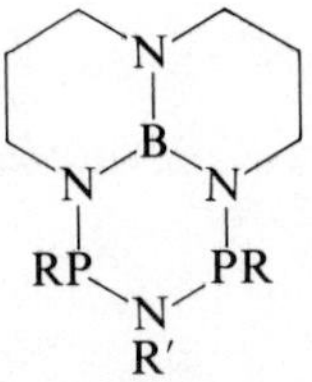

(59a)

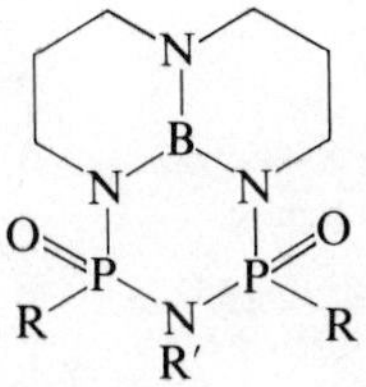

(59b)

phosphoryl)methylamine MeN(POCl$_2$)$_2$, tetrazadiphosphaboraphenalene analogues (59a) and (59b) have been obtained.[324]

Other fused-ring systems contain only inorganic rings, and two bicyclic systems (60) and (61) containing the SiN$_2$B$_2$ rings, have been reported:[313a]

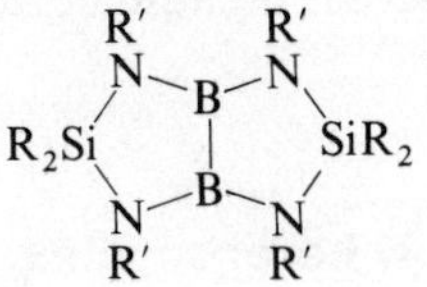

(60)

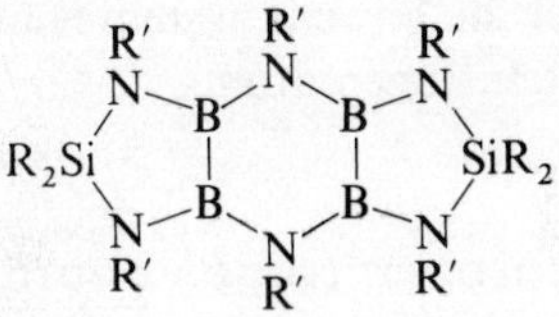

(61)

The reaction of SnCl$_2$ with Me$_3$Si—NLi—BMe$_2$ leads to a dimeric 4-methyl-1,3-bis(trimethylsilyl)-1,3,2,4-diazastannaboretidine (62), after elimination of BMe$_3$:[74]

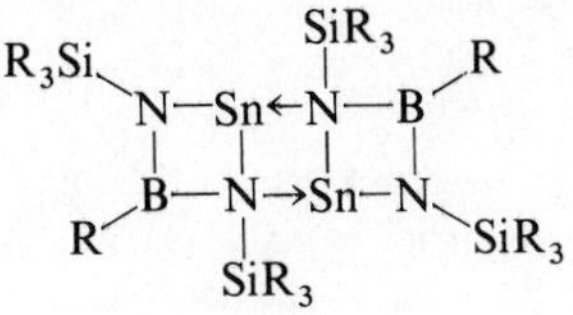

(R = Me)

(62)

The structure of (62) has been analysed by X-ray diffraction, which confirmed a tricyclic tub-form structure built from three condensed four-membered rings. The central Sn_2N_2 ring is not planar. The bond lengths are Sn–N 2.254 Å, 2.231 Å, 2.105 Å, B–N 1.422 Å, 1.484 Å; the bond angles are NSnN 82.26°, 66.90°, 94.63°, SnNB 86.76°, 115.28° and 94.36°, NBN 111.74°. The BN_2Sn rings are, however, nearly planar. The Sn–N distance to trigonal N is much shorter than the distance to tetracoordinated nitrogen.[74]

In a reaction of $MeB(SMe)_2$ with hydrazine and methylamine, a new bicyclic B–N system hexamethylbicyclo[3.3.0]-1,3,5,7-tetraza-2,4,6,8-tetraboraoctane (63), R = Me, was isolated instead of the expected 3,4,5-trimethyl-1,2,4,3,5-triazadiborolidine:[305]

(63)   (64)

Another bicyclic compound, 2,4,6,8-tetramethyl-3,7-dithia-1,5-diaza-2,4,6,8-tetraborabicyclo[3.3.0]octane (64) has been investigated by X-ray diffraction.[336] The two CBNBC planes are twisted by 18° against one another. The important bond distances and angles are B–S 1.795 Å and 1.806 Å, N–B 1.420 Å, N–N 1.450 Å; BSB 93.0°, SBN 108.7°, NNB 113.4°. Planarity of the bicyclic system is not achieved, because of the free electron pairs at the nitrogen atoms.[336]

2.5 Boron–nitrogen coordination heterocycles. Systems containing four-coordinate boron

2.5.1 Four-membered B_2N_2 rings

Dimeric aminoboranes. Dimerization of many aminoboranes through donor–acceptor bonds leads to four-membered ring derivatives (65) of general composition $(R_2B—NR_2')_2$:

(65)

A series of novel, volatile 1,3-diazonia-2,4-diboratacyclobutane derivatives (65) has been prepared from cyclic amines and diborane at elevated temperatures:[337]

$$2 \underbrace{CH_2(CH_2)_x NH} \xrightarrow{B_2H_6} 2 \underbrace{CH_2(CH_2)_x NH.BH_3} \xrightarrow[-H_2]{\Delta} \underbrace{[CH_2(CH_2)_x N.BH_2]_2}$$

$$x = 2,3,4 \tag{32}$$

The dimer of $Cl_2BNHSCF_3$ (obtained by aminolysis of BCl_3 with CF_3SNH_2) reacts with chlorine (at $-36\,°C$) to yield a sublimable solid $[Cl_2B—NCl(SCF_3)]_2$, which consists of *cis* and *trans* isomers.[338] The compound is also obtained by treating $Cl_2B—N(SCF_3)_2$ with chlorine, after cleavage of CF_3SCl.

The reaction of *N*-trimethylsilyl-2-aminopyridine with boron trichloride forms a dimeric aminoborane $[Cl_2B—NH(C_5H_4N)]_2$, probably containing a four-membered ring skeleton (65), although an alternative structure has also been considered.[339]

Aminoboration of ketene $H_2C=C=O$ with $XB(NR_2)_2$ (where R = Me, Et; X = Cl, Br, I) yielded $[(R_2N—COCH_2)XBNR_2]_2$ in 40–83% yields; the compounds contain the four-membered ring (65).[340]

If sodium tetrahydridoborate is heated with iodine in boiling tetrahydrofuran, borane is generated, and this reacts with primary and secondary amines to yield dimeric aminoboranes $(H_2BNR_2)_2$ (R = Me, Et).[341] On attempted distillation, the derivatives of primary amines $(H_2BNHR)_2$ decomposed to borazines $(HBNR)_3$ and hydrogen.[341] These reactions provide simple routes to dimeric aminoboranes and borazines, starting with readily available materials.

Dimeric $(HClBNMe_2)_2$ and $(HBrBNMe_2)_2$ have been prepared by heating $(H_2BNMe_2)_2$ with $HgCl_2$ or $HgBr_2$ at $110\,°C$. Both compounds derive from (65) and exist in two isomeric *cis* and *trans* forms.[342]

The interaction of carbon atoms from a carbon arc with dimethylaminoborane, $H_2B—NMe_2$ yields as a main product (dimethylamino)methylborane, which combines with the starting material to form the four-membered ring system (65a) (R = Me):[343]

$$R_2N—BH_2 \xrightarrow{R_2N—BR_2} 2 R_2N—BHR \rightleftharpoons (HRB—NR_2)_2$$

$$\Big\updownarrow + R_2N—BH_2$$

$$\begin{array}{c} R_2N \overset{+}{\underset{-}{\longrightarrow}} BH_2 \\ | \quad\quad | \\ H—B \overset{-}{\underset{+}{\longrightarrow}} NR_2 \\ | \\ R \end{array}$$

$$(65a)$$

(33)

The latter is a rare example of an unsymmetrically substituted B_2N_2 ring system.

2-Alkyl-1,2-azaborolidines are equilibrium mixtures of monomeric and dimeric (66) forms, (R = H, OBu^n, SBu^n; R' = Pr^n, Bu^n, CH_2Ph):

$$2 \quad \underset{\underset{R}{\overset{|}{B}}}{\overset{\overbrace{}}{}}\!\!-N-R' \quad \underset{\Longleftarrow}{\Longrightarrow} \quad \text{(dimer 66)} \tag{34}$$

(66)

These equilibrium mixtures are formed by reactions of allyl-arylamines $H_2C{=}CH{-}CH_2NR_2$ with triethylamine–borane $Et_3N.BH_3$ at 110–140 °C with elimination of H_2 and NEt_3. A second way is hydroboration of allylimines, $PhCH{=}NCH_2CH{=}CH_2$ with $Et_3N.BH_3$.[344]

The crystalline dimers of 1-alkyl-1,2-azaborolidines (66) are stable against water and alcohol at room temperature. The 1-alkyl-1,2-azaborolidines react with alcohols, mercaptans and amines to replace the hydrogen at boron sites and to form B-substituted derivatives by eliminating hydrogen.[344]

Various substituted 1,2-azaborolidines have been convenient models for the investigation of the association of aminoboranes and, in particular, the effect of different B- and N-substituents on the tendency for association to occur. While most 1,2-diorgano-1,2-azaborolidines are monomeric, the 2-chloro derivatives may exist in the dimeric form (66), R = Cl, R' = Et, Pr^n, CH_2Ph.[345]

Dimeric 2-phenyl-1,2-azaborolidine (66), R = Ph, R' = H, has been synthesized by the reaction of allylamine and diphenylborane.[345]

A comparison of alkyl-substituted 1,2-azaborolidines shows that these can be arranged in the following order with respect to their tendency to dimerize:

$$\underset{R-B}{\overset{R'-N}{}}\!\!\big] \;<\; \underset{R-B}{\overset{H-N}{}}\!\!\big] \;<\; \underset{H-B}{\overset{R-N}{}}\!\!\big]$$

Steric and electronic influences are responsible for this order.[345]

A ring contraction, with change in coordination number, leading to formation of (65) derivatives, has been observed in the reaction of 2,4,6-trichloroborazine with acetic acid:[113,114]

$$2 \text{ (ClBNH)}_3 + 12 \text{ AcOH} \longrightarrow 3 [(AcO)_2BNH_2]_2 + 6 \text{ HCl} \qquad (35)$$

In an analogous manner, several dimeric halogenoacetate derivatives $[(RCOO)_2BNH_2]_2$, were prepared with $R = CF_3$, CHF_2, CCl_3, $CHCl_2$, CH_2Cl, CBr_3, CH_2Br.[115]

Dimeric acetamidodialkylboranes $(R_2B—NHAc)_2$ have been prepared by reacting alkylmercaptoboranes $R_2B—SR'$ with acetamide at 60–90 °C, with elimination of the mercaptan $R'SH$.[346]

The reaction of LiH with $(H_2B—NMe_2)_2$ at room temperature has been observed over a period of several months and the system monitored periodically by ^{11}B NMR. From the collected data it has been concluded that the main pathway involves ring opening by reaction of LiH with aminoborane dimer:

$$(H_2B—NMe_2)_2 + LiH \longrightarrow LiMe_2N—BH_2—NMe_2.BH_3 \qquad (36)$$

In a subsequent step, $LiMe_2N.BH_3$ is formed. After two months the reaction was 90% complete.[347]

On warming the adduct $F_2P—NMe_2.BF_3$, fluorine transfer occurs from boron to phosphorus, and $(F_2B—NMe_2)_2$ and PF_3 are formed. The reaction is probably favoured by an intramolecular $N{\rightarrow}B$ interaction. The compound $FP(NMe_2)_2$ reacts with BF_3 in a similar manner, to yield $(F_2B—NMe_2)_2$ and $F_3B.F_2P—NMe_2$.[348]

A normal-coordinate analysis of $(H_2B—NMe_2)_2$, $(Cl_2B—NMe_2)_2$ and $(Br_2B—NMe_2)_2$ has been carried out. X-ray diffraction studies for $(Cl_2B—NMe_2)_2$ and $(H_2B—NMe_2)_2$ have shown D_{2h} molecular symmetry (planar rings), and infrared spectra have been recorded for these compounds and isotopically labelled analogues.[349]

First- and second-order quadrupole effects of the ^{11}B NMR resonance have been measured in polycrystalline samples of dimeric dimethylaminoboranes. The NQR spectra of $(F_2B—NMe_2)_2$ and $(Cl_2B—NMe_2)_2$ have been reported and discussed.[350]

The vibration frequencies have been calculated for the $(H_2B—NH_2)_2$ molecule using the MNDO method. Assignments have been made for the calculated and observed frequencies in the IR spectrum.[285] Gas-phase UV spectra of some monoaminoboranes have been reported. It has been shown with the aid of the PE spectrum that $F_2B—NMe_2$ vaporizes predominantly in the dimeric form. At room temperature CFH_3 is slowly formed. The dimerization equilibrium of $H_2B—NMe_2$ is reached very slowly at room temperature, and the vapour consists of monomeric and dimeric forms in a ratio of $1:1$.[351].

(Dimethylamino)diethynylborane has also been found to be dimeric, $[(HC{\equiv}C)_2B—NMe_2]_2$; its ^{1}H and ^{13}C NMR spectra have been reported.[352]

Dimeric iminoboranes. The chemistry of iminoboranes until 1972 has been reviewed.[B 28] Dimeric iminoboranes are prepared essentially in three ways:

(a) reaction of lithiated or non-lithiated imines with boranes;
(b) Si–N bond cleavage reactions by halogenoboranes;
(c) insertion reactions.

Substitution reactions on a preformed B_2N_2 skeleton of dimeric iminoboranes are also possible.

Crystalline dimeric diphenylketiminoboron dihalides (67), X = F, Cl, Br, I, Me, R = Ph, were prepared from ethereal solutions of $Ph_2C{=}NLi$ and boron trihalides:[353]

$$R_2C{=}N{\overset{+}{\underset{-}{\,}}}\overset{X\ \ X}{\underset{X\ \ X}{\underset{B}{\overset{B}{\diamondsuit}}}}{\overset{+}{\underset{-}{\,}}}N{=}CR_2$$

(67)

The methyl derivatives of (67) result from insertion of acetonitrile into the B–H bond of $HBCl_2$:[353]

$$2\ Me{-}CN + 2\ HBCl_2 \longrightarrow (Cl_2B{-}N{=}CH{-}Me)_2 \qquad (37)$$

In reactions of diphenylketiminotrimethylsilane with $PhBCl_2$ and the catechol derivative $C_6H_4O_2BCl$, the Si–N bond is split and dimeric iminoboranes, $[Ph(Cl)B{-}N{=}CPh_2]_2$ and $[1,2\text{-}C_6H_4O_2B{-}N{=}CPh_2]_2$ respectively are formed.[354] In a similar manner, diphenylmethyleneiminotrimethylsilane $Ph_2C{=}N{-}SiMe_3$ reacts with biphenyl-2,2′-yleneboron fluoride $(C_6H_4)_2BF$, and benzylideniminotrimethylsilane $PhCH{=}N{-}SiMe_3$ with $PhBCl_2$ and Ph_2BCl; the respective products are $[(C_6H_4)_2B{-}N{=}CPh_2]_2$, $[Ph(Cl)B{-}N{=}CHPh]_2$ and $(Ph_2B{-}N{=}CHPh)_2$.[355]

Dihalogeno(t-butylaldimino)boranes $(X_2B{-}N{=}CHBu^t)_2$ have been obtained from di-n-butyl derivatives $(Bu^n{}_2B{-}N{=}CHBu^t)_2$ by heating with boron trihalides at 90–130 °C (chloride or bromide); in these reactions a B–C bond is cleaved and Bu^nBCl_2 is formed as a by-product.[356]

As the ^{11}B NMR chemical shifts suggest for the n-Bu derivative, both monomeric and dimeric forms exist in the melt. The assignment of mass-spectrometric fragmentation is given.[356,356a]

By reacting variously substituted nitriles with halogenoboranes, it has been found that insertion into B–halogen bonds occurs if the CN groups are carrying electron-withdrawing substituents:[356,357]

$$2\ R{-}C{\equiv}N + 2\ BXYZ \longrightarrow [Z(Y)B{-}N{=}CRX]_2 \qquad (38)$$

where $R = Cl$, Br, I, Me, OMe, SMe; X, Y $= Cl$, Br, I, Me, Et, Pr^n, Pr^i, Ph, C_6F_5, SMe, SBu^n; Z $= CF_2CF_3$, C_6F_5, CH_2F, CH_2Cl, CH_2Br, etc., in various combinations. Some of the iminoborane derivatives form either the monomer $Z(Y)B—N{=}CRX$ or the adduct $Z(Y)XB.N{\equiv}C—R$ during attempts to sublime them in vacuum.

Thioboration of fluoroacetonitrile leads to dimeric iminoboranes $[(RS)_2B—N{=}C(SR)(CH_2F)]_2$ (R = Me, Ph).[358]

Organothiocyanates RSCN form adducts with BCl_3 which can be converted to dimeric iminoboranes $[Cl_2B—N{=}C(SR)Cl]_2$ by heating in refluxing carbon tetrachloride.[359]

Unsymmetrically C–S substituted dimeric iminoboranes (68) are also synthesized by substitution reactions of dimeric halogenoiminoboranes $[Cl_2B—N{=}CCl_2]_2$ with mercaptans $R'SH$ (R = Cl, CH_2Cl; R' = Me, Bu^n).[359]

$$\begin{array}{c}
\text{Cl}\quad\text{Cl}\\
R\diagdown\quad\diagup B\diagdown\quad R\\
\diagdown C{=}N^{+}\quad {}^{+}N{=}C\diagup\\
R'S\diagup\quad\diagdown B\diagup\quad\diagdown\text{Cl}\\
\text{Cl}\quad\text{Cl}
\end{array}$$

(68)

The B–halogeno substituents can be replaced by alkyl groups on treatment with Grignard reagents and organolithium compounds, to give $(R_2B—N{=}CCl_2)_2$.[359] A series of diazido- and tetraazido- derivatives $[R(X)B—N{=}C(Y)Z]_2$ (where R = Me, N_3; X = N_3, Me; Y = Cl, Br, C_6F_5, CH_2F, $CHCl_2$, etc, Z = Cl, Br, N_3) have been prepared by treating various bis(imino)borane derivatives with NaN_3 in acetonitrile.[360]

A dimeric (isopropylideneimino)dimethylborane $(Me_2C{=}N—BMe_2)_2$ was made by a reaction of N-chloroketimine $Me_2C{=}NCl$ with trimethylborane after elimination of methyl chloride.[361] Thioboranes also react readily with ketimines; thus elimination of n-butylmercaptan from R_2BSBu^n and $Ph(Ar)C{=}NH$ on heating gave $[Ph(Ar)C{=}N—BR_2]_2$, with R = Bu^n, Ph, Ar = Ph, $o\text{-}MeC_6H_4$.[362] The reaction of $Ph—C({=}NH)R$ (R = Et, Pr^n) with $Bu_2B\text{-}SBu$ gave a good yield of the dimeric iminoborane $[Bu_2B—N{=}CPh(CH_2R)]_2$, which is in tautomeric equilibrium with $Bu_2B—NHC(Ph)({=}CHR)$.[363] Similar tautomeric mixtures were also observed for the reaction of $R_2B—N{=}CPh(R')$ (R = Pr^n, Bu^n; R' = Et, Pr^i, CH_2Ph) with mercaptoboranes. Thus the reaction of $Pr(Ph)C{=}NH$ with $Ph_2B—SBu^n$ gave 74% $(Ph_2B—N{=}CPhPr)_2$.[364]

Allyl(dialkyl)boranes $H_2C{=}CHCH_2BR_2'$ react with nitriles RCN (R = Et, Ph) at $-50\,°C$ to yield $[R_2'B—N{=}CR(C_3H_5)]_2$ (R' = Et).[365]

The reaction of diarylketimine hydrochloride and a dialkylaminoborane also yields dimeric iminoboranes.[366] (Dipropylboryl)imino-1-methyl-1,2-dihydropyridine dimerizes on standing to form a cyclic iminoborane dimer.[367]

Trialkylboranes react smoothly with N-alkylbenzamidines Ph—C(=NH)NHR' to form amidinoboranes Ph—C(=NBR$_2$)NHR', which dimerize on standing (R = Pr^i, cyclo-C$_6$H$_{11}$, R' = Pr^n, Bu^n, Pr^i).[368]

N-Trimethylsilyltriphenylphosphinimine Ph$_3$P=N—SiMe$_3$ reacts with BX$_3$ (X = F, Cl, Br) to form dimeric phosphiniminodihalogenoboranes (X$_2$B—N=PPh$_3$)$_2$.[369]

The ^{35}Cl NQR frequencies of some chlorine-containing iminoboranes of the type [Cl$_2$B—N=C(R)Cl]$_2$ (R = CH$_2$Cl, CHCl$_2$ and CCl$_3$) have been measured, and the data are consistent with the cyclic structure of dimeric iminoboranes.[370]

2.5.2 Six-membered B$_3$N$_3$ rings (trimeric aminoboranes)

In several cases aminoboranes trimerize to form six-membered coordination rings (69) through donor–acceptor bonds:

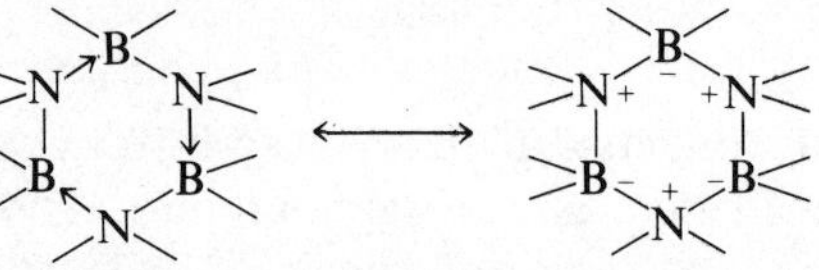

(69)

Refluxing aziridine–borane in benzene resulted in the formation of the 1,3,5-triazonia-2,4,6-triboratacyclohexane derivatives (69a):[337,371,372]

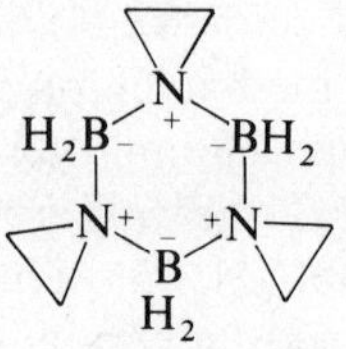

(69a)

Another trimeric aziridinyl derivative has been prepared by treating 4,4,6-trimethyl-1,3,2-dioxaborinane   with   aziridine,   or   by   treating 2-chloro-4,4,6-trimethyl-1,3,2-dioxaborinane with 1-aziridinyllithium.[373]

An unsymmetrically substituted derivative of (69) was obtained by the action of methylenedilithium on dimethylaminodichloroborane.[374] Isocyanates RNCO (R = Me, Ph) and isothiocyanates RNCS (R = Me) on reaction with B_2H_6 also give cyclic trimers $[H_2B—NR(CHX)]_3$ (X = O, S).[375, 376] Some derivatives of 1-aza-2-bora-3-oxacyclopentane also trimerize via N→B bonds, to form derivatives of (69); these are prepared by cleaving N–Si and O–Si bonds in N,O-disilylated ethanolamines, $Me_3Si—NR—CH_2CH_2OSiMe_3$ (R = H, Me) with boron trihalides BX_3 (with X = F, Cl, Br).[377] Methylaminoethanolamine reacts with diborane to form another trimeric 1-aza-2-bora-3-oxa-cyclopentane.[377]

Cis and trans isomers of trimeric alkylaminoboranes $(H_2BNHR)_3$ (R = Me, Et) have been separated by thin-layer chromatography, using dichloromethane as developing solvent.[378]

The infrared spectra (mainly the v_{N-H}) of $(H_2B—NHR)_3$ (R = Et, Pr, Bu) have been analysed.[379]

The dipole moment (3.2 ± 0.1 D) and the heat of sublimation (25 ± 3 kcal/mol) of 1,3,5-triazonia-2,4,6-triboratacyclohexane $(H_2B—NH_2)_3$ have been determined. The increased heat of sublimation compared with cyclohexane has been attributed to dipole–dipole interaction.[380–382]

The crystal and molecular structures of $(Cl_2B—N_3)_3$[383], $(Cl_2B—NH_2)_3$[384] and $(H_2B—NH_2)_3$[385] have been determined by diffraction methods. In $(Cl_2B—N_3)_3$ the ring is not planar; it has a twist conformation. The endocyclic B–N distances vary from 1.546 to 1.606 Å; the bond angles NBN of the three boron atoms with endocyclic N atoms are 99.5°, 99.8° and 102.6°. The corresponding BNB angles are 127.8°, 127.7° and 126.9°.[383]

The molecule of $(Cl_2B—NH_2)_3$ shows a nearly ideal chair conformation; interatomic distances are (mean values) B–Cl 1.845 Å and B–N 1.575 Å and bond angles NBN 109.1° and 110.3°, BNB 120.1° and 120.8°.[384]

In $(H_2B—NH_2)_3$[385] the interatomic distances are B–N 1.576 Å, B–H (equatorial) 1.11 Å and B–H (axial) 1.08 Å and bond angles NBN 107.3°, BNB 115.8° and 116.0°.

The average partial charges found on each type of atom are as follows: B +0.7e, N −0.3e, H(B) −0.3e and H(N) +0.1e. The relatively large charges at the hydrogen atoms are consistent with the high heat of sublimation. The calculated dipole moment, based upon these point charges, is 3.0 D, which compares well with the experimental value.[385] The results of a nonempirical LCAO–MO SCF study of the chair and boat forms of cyclohexane and $(H_2B—NH_2)_3$ have been reported.[386,387]

2.5.3 Dimeric and tetrameric hydrazinoboranes

Like aminoboranes, several hydrazinoboranes were found to dimerize, with formation of six-membered cyclic compounds (70):

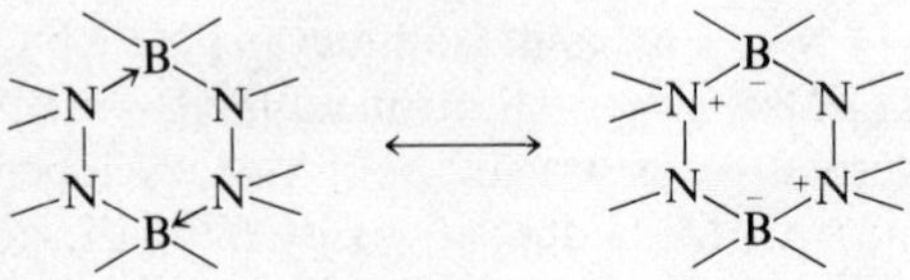

(70)

Dimerization of N,N-dimethyl-N',N'-bis(dimethylboryl)hydrazine Me_2-N—N(BMe$_2$)$_2$ has been investigated by ^{11}B NMR. Only one broad resonance has been observed instead of the expected two signals; this was attributed to a dynamic process in which B–N bonds are continuously opened and closed.[388]

Derivatives of (70) are obtained by the reaction of 1,2,4,5,3,6-tetraza-diborines with HCl to form addition compounds (hydrochlorides) (70a) which preserve the cyclic skeleton (R = Me, Ph, R' and R" = H or Me):[389]

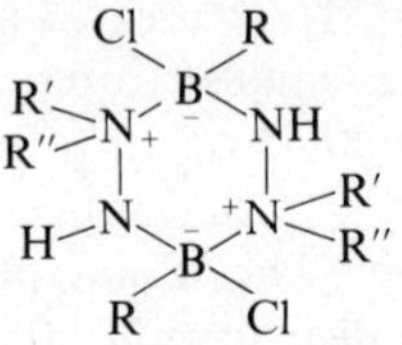

(70a)

Sometimes tetrazadiborines (25) (RBNR'NR")$_2$ dimerize through inter-cyclic N→B donor–acceptor bonds to form tetrameric cage compounds of general composition (RBNR'NR")$_4$ (71):

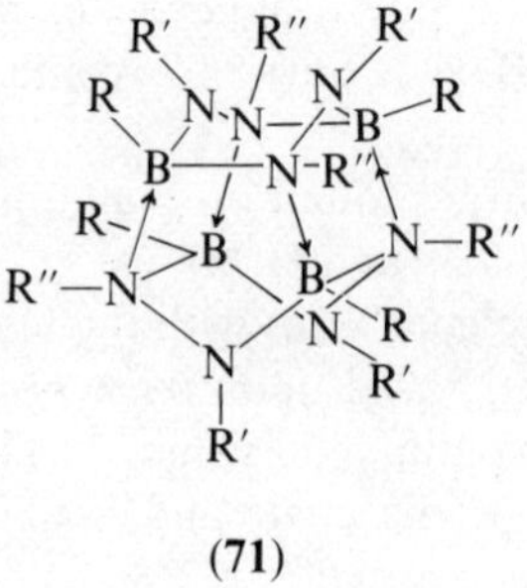

(71)

Tetrameric cage compounds (71) result from the reaction of MeNHNH$_2$ and MeNHNHMe with HB(NMe$_2$)$_2$,[296] and can also be obtained by cleavage of the exocyclic B–N bond of 1,2,4,5-tetramethyl-3,6-bis-(dimethylamino)-1,2,4,5-tetraza-3,6-diborine (25), R = NMe$_2$, R' = R" = Me, with boron trihalides BCl$_3$, BBr$_3$ and BF$_3$.OEt$_2$.[297]

The product formed is $(XB—NMe—NMe)_4$.[297] The fluoro- derivative $(FB—NMe—NMe)_4$ has also been obtained by reaction of (25), R = SMe, R' = R'' = Me, with antimony trifluoride.[299]

If hydrogen chloride is reacted in excess with $(HBNMeNMe)_4$ only the adduct $(HBNMeNMe)_4.2HCl$ is formed, for which an ionic structure has been proposed. Because of steric hindrance, adducts with BF_3 or BCl_3 are formed only by $(HBNH—NMe)_4$, but not by $(HBNMe—NMe)_4$.[389]

The dimer (71) of 3,6-di-t-butyl-1,2,4,5,3,6-tetrazadiborine (25), R = Bu^t, R' = R'' = H, reacts with two moles of MeNCO.[294] The attack of the isocyanate is expected to occur at the trigonal nitrogen sites of (71). The trisubstituted and tetrasubstituted derivatives with MeNCO have also been obtained.[294]

2.5.4 Pyrazaboles

Pyrazolylborate chemistry has been extensively investigated during the last years, and a large number of transition-metal complexes have been prepared.[B 30] These will not be treated here.

Pyrazolylboranes dimerize through intermolecular N→B donation to form so-called pyrazaboles (72), which contain six-membered B_2N_4 rings:

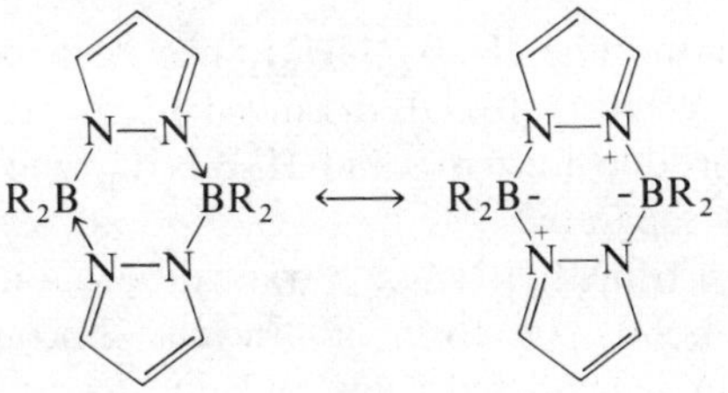

(72)

Unsymmetrically disubstituted pyrazaboles (72a) are formed in the reaction of bis(pyrazolyl)borates with boranes containing at least one leaving group (R = H, Me, Et, 1-pyrazolyl; R' = F, Me, Et, Bu):[390,391]

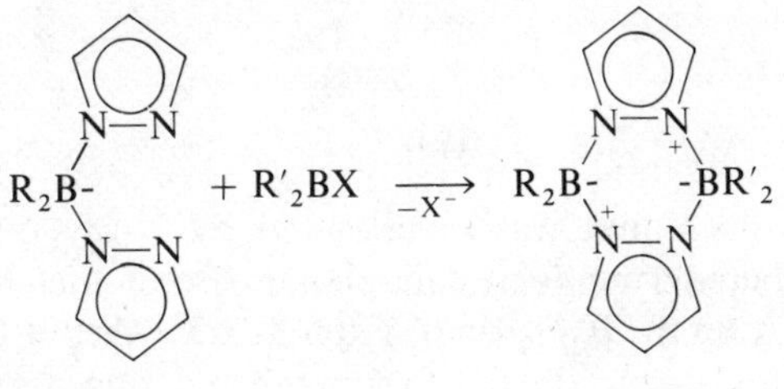

$$R_2B^- \quad + R'_2BX \xrightarrow{-X^-} R_2B^- \quad -BR'_2 \tag{39}$$

(72a)

Other pyrazaboles have been readily prepared from mercaptoboranes RB(SR′)$_2$ and pyrazole (R = Ph, R′ = Et).[392] B-Mercaptopyrazaboles (72), R = SR′, have been synthesized by substitution reactions of pyrazabole (72), R = H, with mercaptans and dithiols.[392] Similarly, (72), R = H, reacts with pyrrole to form tetrakis(pyrrol-1-yl)pyrazabole (72), R = pyrrol-1-yl.[392] Pyrazaboles (72) with R = H can be chlorinated; thus (72a), R = H, R′ = Et, was treated with chlorine to give (72a), R = Cl, R′ = Et.[390]

The reaction of 3-methylpyrazole with BMe$_3$ in a sealed tube yielded the pyrazabole [Me$_2$B(Mepz)]$_2$ (where Mepz = methylpyrazolyl).[393]

Mass spectra[394,395], ^{13}C NMR spectra[394] and UV spectra[394] of pyrazaboles have been discussed. The structures of some pyrazaboles have been analysed by X-ray diffraction.[396,397] Thus bis(3,5-dimethylpyrazolyl)borane dimer has a flattened chair conformation.[396]

Imidazole derivatives of boron are also associated. Thus tetrakis-(2-methylimidazolyl)-tetrakis(dimethylboron) [Me$_2$B(N$_2$C$_4$H$_5$)]$_4$ has been obtained from 2-methylimidazole and BMe$_3$ at 135 °C in a sealed tube and bears some resemblance to pyrazaboles.[398]

2.5.5 Other ring systems, containing only boron and nitrogen

A pentameric aminoborane (H$_2$B—NH$_2$)$_5$ has been obtained by the decomposition of [H$_2$B(NH$_3$)$_2$][BH$_4$], dissolved in glyme, in the presence of ammonia; the by-products are H$_2$ and H$_3$B.NH$_3$, and the insoluble pentamer can be easily separated.[399]

3,5-Dichloro-1,2,4-trimethyl-1,2,4,3,5-triazadiborolidine dimerizes via additional N→B bonds to form a tricyclic boron–nitrogen system, consisting of two five-membered rings and one four-membered ring (73), R = Me:[332]

(73)

The structure of this dimer was investigated by X-ray diffraction. The central four-membered ring is square-planar. Bond distances are B–N 1.617 Å and 1.612 Å (in the B$_2$N$_2$ ring), 1.465 Å, 1.372 Å and 1.444 Å (in the B$_2$N$_3$ ring); the bond angles in the five-membered ring are NBN 115.6°, BNB 109.3°, BNN 118.3° and NNB 104.1°.[332]

N,N'-Bis(chloromethylboryl)-N,N'-dimethylhydrazine   Cl$_2$B—NMe—N-Me—BCl$_2$ dimerizes to form a cubane structure with two open edges (74), R = Me:[328]

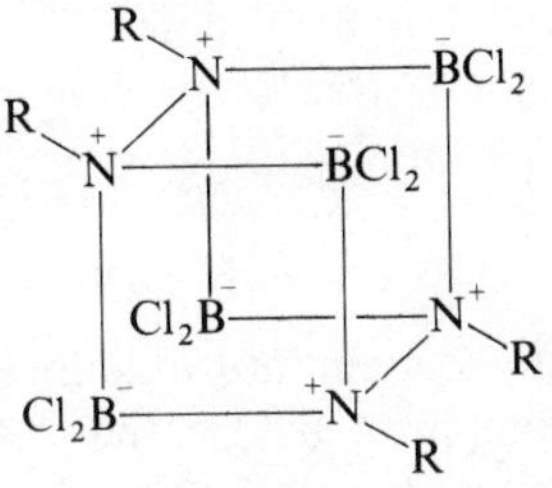

(74)

3,5-Difluoro-1,2,4-trimethyl-1,2,4,3,5-triazadiborolidine (26), R = Me, shows an interesting dimerization behaviour; although structure (75a) would be expected, the real structure of the dimer (established by X-ray diffraction) is (75b):[400]

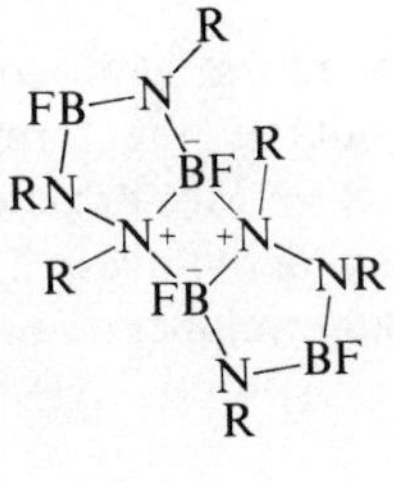

(75a) 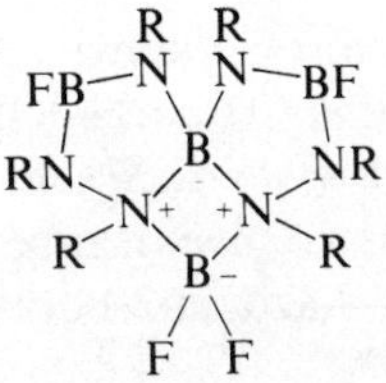   (75b)

In this dimer, two B$_2$N$_3$ rings (nearly planar) with a spirocentric tetrahedral boron atom are connected by a planar B$_2$N$_2$ ring containing four-coordinate boron and nitrogen atoms. This is the first example of an "anomalous" dimerization of an aminoborane, and a mechanism for its formation has been discussed.[400]

2.5.6   Ring systems with boron, oxygen and nitrogen

From monoximes RR′C=N—OH and bromodimethylborane some dimeric iminoxyboranes (76) have been obtained (X = Cl, Br, Me, RR′ = HMe, HPh, MeCl, –(CH$_2$)$_4^-$):[401]

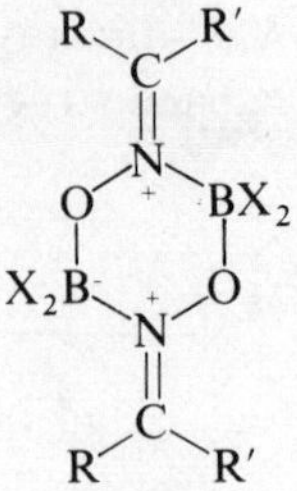

$$R\diagdown_{\overset{\displaystyle C}{\underset{}{\big\|}}}\diagup R'$$

(76)

Dimeric 1,1-dichloroethylideneiminoxydihalogenoboranes (76) (RClC= N—OBX$_2$)$_2$ have been prepared by treatment of 1,1-dichloro-1-nitroso-ethane with halogenoboranes (R = Me):[402]

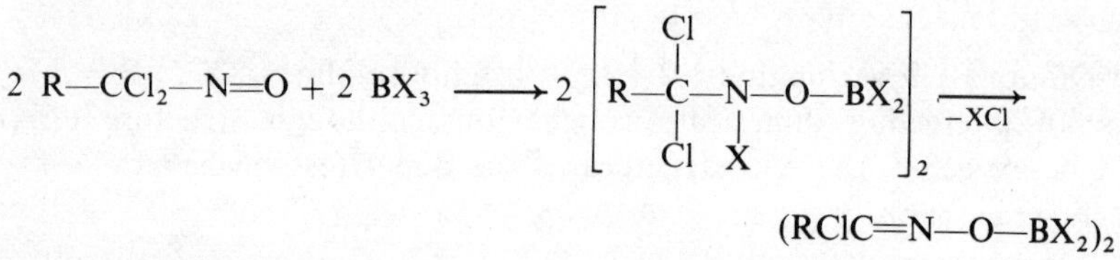

$$2\ R\text{—}CCl_2\text{—}N{=}O + 2\ BX_3 \longrightarrow 2 \left[R\text{—}\underset{\underset{\displaystyle Cl}{|}}{\overset{\overset{\displaystyle Cl}{|}}{C}}\text{—}\underset{\underset{\displaystyle X}{|}}{N}\text{—}O\text{—}BX_2 \right]_2 \xrightarrow[-XCl]{} \tag{40}$$

(RClC=N—O—BX$_2$)$_2$

The reaction of diethylhalogenoboranes with sodium salts of *aci*-nitroalkanes R'—C(R)=NOONa results in the formation of the eight-membered ring system (77), R = Et, R' = Me, Pr^n, Bu^i, Ph. On standing, (77) rearranges to five-membered ring derivatives, 2,5,5-triethyl-4-alkyli-dene-1,3-dioxa-4-azonia-2-bora-5-borata-cyclopentanes (78):[403,404]

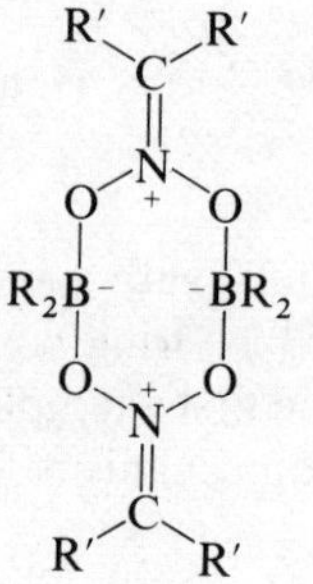

(77) 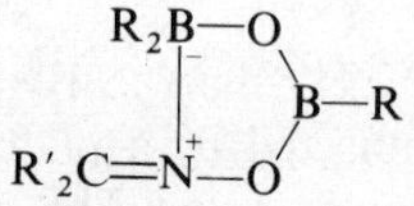   (78)

The structure of (78) has been confirmed by an independent synthesis based on the oxidation of Me$_2$C=NOBEt$_2$.[403]

2.5.7 Boron–nitrogen–phosphorus rings

Several B–N–P heterocycles (79)–(81) are known:

(79)   (80)   (81)

A dimeric phosphineaminoborane $[R_2PN(R')—BX_2]_2$ (R = Ph, R' = Me, X = Cl; R = Ph, R' = Me, X = Br) has been postulated as an intermediate in the reaction of silylaminophosphines $R_2P—NR'—SiMe_3$ with boron trihalides.[43] The compounds $Me_2P(S)—NR—BX_2$ (X = F, Cl, Br) exist in solution in an equilibrium of monomers and dimers. On the basis of NMR data, S→B coordination rather than N→B coordination has been suggested, thus indicating structure (79), R = Me.[405]

The first phosphine–diborane(4) ring complex (80) was prepared by treating bis(trifluorophosphine)diborane(4) with an equimolar amount of methylaminobis(difluorophosphine).[406]

Hexachlorodiphosphazonium tetrachloroborate $[Cl_3P\!=\!N—PCl_3][BCl_4]$ reacts with methylammonium chloride (molar ratio 1 : 2) in boiling dichloroethane, with elimination of HCl to yield the heterocycles (81), R = R' = Cl.[325] The same heterocycle resulted from the reaction of N-methyltrichlorophosphazonium-N-trichloroborate $MeN(\overset{+}{P}Cl_3)(\overset{-}{B}Cl_3)$ and methylammonium chloride.[407] By treatment with antimony trifluoride, all chlorine atoms at boron and phosphorus sites can be replaced with fluorine to give (81), R = R' = F.[325] The chlorine atoms at boron sites can be replaced in (81), R = R' = Cl, with NCS and H substituents by treatment with NH_4NCS and $NaBH_4$ respectively, to yield (81), R = NCS or H, R' = Cl. With dimethylamine, total chlorine replacement at both boron and phosphorus sites is achieved to give (81), R = R' = NMe_2.[408]

2.5.8 Boron–nitrogen–sulphur ring systems

A six-membered ring (82), containing two $H_2B—NMe_2$ and one $H_2B—SMe$ group, joined through donor–acceptor bonds, has been described. It can be formed by reacting $R_2NB_2H_5$ with RSH and R_2NH at 110 °C for 25 h in a

sealed tube, or by heating $R_2NB_2H_5$ with R_2NBH_2 and $B(SR)_3$ at
105–125 °C for 70 h in a sealed tube (R = Me):[409,410]

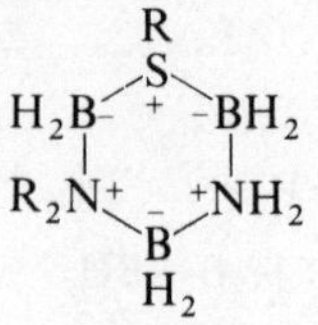

$$
\begin{array}{c}
R \\
\underset{\;\;}{\overset{S}{|}} \\
H_2B\overset{+}{\underset{|}{\diagup}}\;\;\overset{-}{\underset{|}{\diagdown}}BH_2 \\
R_2N^+\underset{\underset{H_2}{B}}{\diagdown}\;\;\overset{+}{\diagup}NH_2
\end{array}
$$

(82)

From the interaction of 3,5-dibromo-1,2,4,3,5-trithiadiborolane $Br_2B_2S_3$
and 1,2,3-trimethyl-1,3,2-diazaborolidine, an adduct of 4,6-dibromo-
1,2,3,5,4,6-tetrathiadiboracyclohexane and 1,2,3-trimethyl-1,3,2-diazaboro-
lidine with a cage structure has been isolated.[411]

2.5.9   Boron–nitrogen–aluminium ring systems

The B–N–Al ring systems reported are as follows:

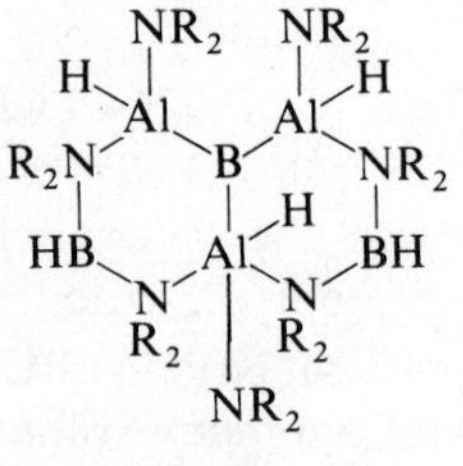

(83) 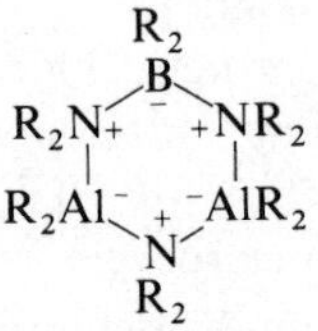   (84) 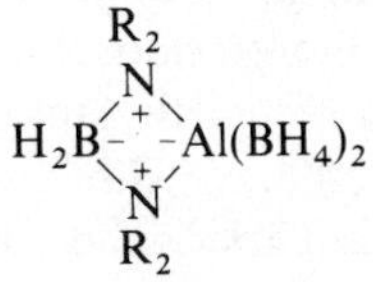   (85)

Treatment of bis(trimethylamine)alane $H_3Al.2NMe_3$ with tetrakis(dimethylamino)diborane(4) results in the formation of hydrogen, NMe_3,
$(H_2BNMe_2)_2$, $(Me_2N)_2BH$ and $H_3B.NMe_3$. In addition, a new species
$Al_3B_3(NMe_2)_7H_5$, has been isolated and the structure (83) has been suggested (R = Me).[412] It seems to be in agreement with the ^1H and ^{11}B NMR
data, although the presence of B–Al bonds has not been demonstrated.

The reaction of Al_2Me_6 with excess $HB(NMe_2)_2$ produces a sublimable
crystalline solid of composition $Al_2B(NMe_2)_6Me_4H_2$, for which the structure
(84) has been suggested on the basis of ^1H and ^{11}B NMR data, and the
probable mechanism of its formation has been discussed.[413]

In the reaction of tris(dimethylamino)alane $Al(NMe_2)_3$ with excess diborane in diethyl ether, besides $(H_2B\!-\!NMe_2)_2$, $Me_2NB_2H_5$, $[Me_2NAl(BH_4)_2]_2$ and $H_2B(NMe_2)\!-\!BH_2\!-\!NMe_2.BH_3$, the new heterocyclic compound $H_2B(NMe_2)_2Al(BH_4)_2$ was isolated, to which structure (85) (R = Me) is attributed.[414]

Bibliography

B 1 A. Finch, J. B. Leach and J. H. Morris, *Organomet. Chem. Rev.* **A4** (1969) 1.

B 2 T. Onak, *Organoborane Chemistry* (Academic Press, New York, 1975).

B 3 H. Nöth, *Prog. Boron Chem.* **3** (1970) 211.

B 4 A. Finch and P. J. Gardner, *Prog. Boron Chem.* **3** (1970) 177.

B 5 A. Meller, *Top. Curr. Chem.* **15** (1970) 146.

B 6 A. Meller, *Gmelin Handbuch der Anorganischen Chemie*, Bd. 51, Tl. 17, (Springer, Berlin, 1978).

B 6a I. Haiduc, *The Chemistry of Inorganic Ring Systems*, Part 1 (Wiley-Interscience, London, 1970), p. 124.

B 7 K. Molt and W. Sawodny, *Proc. 5th Int. Conf. on Raman Spectroscopy Freiburg i. Br.*, 1976 p. 86.

B 8 The Chemical Society, London, *Spectrosc. Prop. Inorg. Organometal. Compounds* **1** (1968)–14 (1981).

B 9 A. Serafini, J. F. Labarre and F. Gallais, *Proc. 16th Int. Conf. Coord. Chem., Dublin*, 1974, p. 2.4a; *CA* **84** (1976) 25604.

B 10 J. O. Cox and G. Pilcher, *Thermochemistry of Organic and Organometallic Compounds* (Academic Press, New York, 1970).

B 11 H. Nöth and B. Wrackmeyer, *Nuclear Magnetic Resonance Spectroscopy of Boron Compounds* (Springer, Berlin, 1978).

B 12 W. G. Henderson and E. F. Mooney, *Ann. Rep. NMR Spectrosc.* **2** (1969) 219.

B 13 D. F. Gaines and J. Borlin, *Boron Hydride Chemistry* (Academic Press, New York, 1975), p. 241.

B 14 G. R. Eaton and W. N. Lipscomb, *NMR Spectroscopy of Boron Hydrides and Related Compounds* (W. A. Benjamin, New York, 1969).

B 15 M. R. Robin, *Higher Excited States of Polyatomic Molecules*, Vol. 1 (New York, 1974).

B 16 W. Fuß and H. Bock, in *Chemical Spectroscopy and Photochemistry in the Vacuum Ultraviolet*, edited by C. Sandorf, P. J. Ausloos and M. B. Robin (Reidel, Dordrecht 1974), p. 223.

B 17 W. Fuß, Thesis, Univ. Frankfurt a.M. (1971), p. 1

B 18 H. Nöth, *Int. Union of Pure Appl. Chem., 24th Int. Congr. of Pure and Appl. Chem.*, **4** (1974) 13.

B 19 M. R. Litzow and T. R. Spalding, *Physical Inorganic Chemistry*, Monograph No. 2: *Mass Spectrometry of Inorganic and Organometallic Compounds*, (Amsterdam, Elsevier, 1973).

B 20 W. Liebscher, *Handbuch zur Anwendung der Nomenklatur organ. chem. Verbindungen* (Berlin, 1979), p. 524.

B 21 IUPAC, *Internationale Regeln f. d. Chemische Nomenklatur und Terminologie*, deutsche Ausgabe, Bd. 2: *Regeln für die Nomenklatur der Anorganischen Chemie* (Verlag Chemie, Weinheim, 1976).

B 22 IUPAC, *Internationale Regeln f. d. Chemische Nomenklatur und Terminologie*, deutsche Ausgabe, Bd. 1: *Regeln für die Nomenklatur der Organischen Chemie* (Verlag Chemie, Weinheim, 1976).

B 23 IUPAC, *International Rules for Chemical Nomenclature* (English Edition) *Pure Appl. Chem.* **28** (1971) 1.

B 24 IUPAC, *Nomenclature of Organic Chemistry*, Sections, A, B, C, D, E, F, H *(Heterocyclic Boron Compounds*, p. 440) (Pergamon Press, Oxford, 1979).

B 25 R. M. Adams, *Pure Appl. Chem.* **30** (1972) 681.

B 26 I. Haiduc, *Rev. Inorg. Chem.* **2** (1980) 219.

B 27 J. J. Lagowski, *Coord. Chem. Rev.* **22** (1977) 185.

B 28 A. Meller, *Top. Curr. Chem.* **26** (1972) 37.

B 29 K. Niedenzu, *Chem.-Ztg* **98** (1974) 487.

B 30 S. Trofimenko, *Chem. Rev.* **72** (1972) 497.

B 31 *Gmelin Handbook of Inorganic Chemistry. Boron Compounds.* Part 17. *Borazine and its Derivatives* (Springer, Berlin, 1978).

B 32 *Gmelin Handbook of Inorganic Chemistry.* 1st Suppl. Vol. 2. *Organoboron Halides. Borazines. Other Heterocyclic Compounds* (Springer, Berlin, 1980).

B 33 *Gmelin Handbook of Inorganic Chemistry.* 2nd Suppl. Vol. 1. *Organoboron–Nitrogen and –Oxygen Compounds* (Springer, Berlin, 1983).

B 34 I. B. Atkins, B. R. Currell, *Inorg. Macromol. Rev.* **1** (1971) 203 (B–N polymers).

References

1 V. V. Volkov, G. I. Bagryantsev and M. G. Myakishev, *Zh. Neorg. Khim.* **15** (1970) 2902; *Russ. J. Inorg. Chem.* **15** (1970) 1510.

2 V. V. Volkov, G. I. Bagryantsev and L. N. Ragozin, *USSR Pat.* 259 (1980); *CA* **92** (1980) 149405.

3 V. V. Volkov, G. I. Bagryantsev, M. G. Myakishev, V. V. Voskoboinikov and V. A. Grigor'ev, *Fiz. Khim. Gidridov* (1972) 173; *CA* **83** (1975) 187701.

4 V. W. Hough, R. C. Guibert and G. T. Hefferan, *US Pat.* 4 150 097 (1979); *CA* **91** (1979) 59610.

5 C. L. Bramlett and A. T. Tabereaux, *Inorg. Chem.* **9** (1970) 978.

6 A. B. Burg, *Inorg. Chem.* **12** (1973) 1448.

7 O. T. Beachley, and T. R. Durkin, *Inorg. Chem.* **12** (1973) 1128.

8 D. T. Haworth, *Inorg. Synth.* **13** (1972) 41.

9 S. Koide, K. Nakamura and K. Yoshimura, *Nihon Daigaku Bunrigaku Shizen, Kagaku Kenkyusho Kenkyo Kiyo* **12** (1977) 7; *CA* **87** (1977) 145098.

10 O. T. Beachley, *Inorg. Chem.* **8** (1969) 2665.

11 O. T. Beachley, and T. R. Durkin, *Inorg. Chem.* **13** (1974) 1768.

12 J. G. Haasnoot and W. L. Groeneveld, *Z. Naturforsch.* **29b** (1974) 52.

12a J. G. Haasnoot, Diss. Rijksuniversiteit te Leiden (1975).

13 M. B. Oertel, Diss. Cornell Univ. (1971); Diss. Abstr. Int. **B31** (1971) 7170.

14 G. Elter, H. J. Külps and O. Glemser, *Angew. Chem.* **87** (1975) 741.

14a H. J. Külps, Diss. Univ. Göttingen (1977).

15 G. Elter, Diss. Univ. Göttingen (1972).

16 G. Elter, O. Glemser and W. Herzog, *Chem. Ber.* **105** (1972) 115.

17 K. Dehnicke and V. Fernandéz, *Chem. Ber.* **109** (1976) 488.

18 A. M. Pinchuk, G. K. Bespal'ko, T. U. Khimchenko and Z. I. Kuplennik, *Zh. Obshch. Khim.* **47** (1977) 2153; *J. Gen. Chem.* **47** (1977) 1964.

19 R. H. Cragg, M. Nazery and A. F. Weston, *Inorg. Nucl. Chem. Lett.* **9** (1973) 497.

20 E. Gayoso and M. Gayoso, *Anales Quim.* **73** (1973) 1112.

21 O. T. Beachley, *J. Am. Chem. Soc.* **93** (1971) 5066.
22 O. T. Beachley, *Inorg. Chem.* **12** (1973) 2503.
23 I. B. Atkinson, D. B. Clapp, C. A. Beck and B. R. Currell, *J. Chem. Soc. Dalton Trans.* **1972**, 182.
24 O. T. Beachley, *J. Am. Chem. Soc.* **94** (1972) 4223.
25 A. F. Zhigach, V. T. Laptev, A. B. Petrunin, V. S. Nikitin and D. B. Becker, *Zh. Neorg. Khim.* **18** (1973) 2037; *Russ. J. Inorg. Chem.* **18** (1973) 1080.
26 A. F. Zhigach, V. T. Laptev, A. B. Petrunin, D. B. Becker, R. I. Fedotova and S. F. Smirnova, *USSR Pat.* 318 580 (1971); *CA* **76** (1972) 45711.
27 A. F. Zhigach, V. T. Laptev, A. B. Petrunin, V. S. Nikitin and D. B. Becker, *Zh. Neorg. Khim.* **18** (1973) 2037; *Russ. J. Inorg. Chem.* **18** (1973) 1249.
27a A. F. Zhigach, R. A. Svitsyn and E. S. Sobolev, *Zh. Obshch. Khim.* **43** (1973) 1958; *J. Gen. Chem. USSR* **43** (1973) 1941.
28 E. C. Ashby and R. A. Kovar, *Inorg. Chem.* **10** (1971) 1524.
29 T. A. Shchegoleva, E. M. Shashkova and B. M. Mikhailov, *Izv. Akad. Nauk SSSR, Ser. Khim.* **1969** 366; *Bull. Acad. Sci. USSR, Div. Chem. Sci.* **1969**, 317.
30 K. Niedenzu, I. A. Boenig and E. F. Rothgery, *Chem. Ber.* **105** (1972) 2258.
31 H. Nöth and M. J. Sprague, *J. Organomet. Chem.* **22** (1970) 11.
32 V. I. Spitsyn, I. D. Kolli and T. G. Sevastyanova, *Izv. Akad. Nauk SSSR, Ser. Khim.* **1973**, 1203; *Bull. Acad. Sci. USSR, Div. Chem. Sci.* **1973**, 1165.
33 G. Elter, O. Glemser and W. Herzog, *J. Organomet. Chem.* **36** (1972) 257.
34 J. A. Miller and G. L. Wilson, *Inorg. Chem.* **13** (1974) 498.
34a J. A. Miller and G. L. Wilson, *J. Fluorine Chem.* **4** (1974) 207.
35 J. J. Harris and B. Rudner, *Inorg. Chem.* **8** (1969) 1258.
36 D. T. Haworth, *Inorg. Synth.* **13** (1972) 43.
37 I. A. Boenig, Diss. Univ. Kentucky (1972); *Diss. Abstr. Int.* **B33** (1972) 1975.
38 I. A. Boenig and K. Niedenzu, *Synth. Inorg. Metal-Org. Chem.* **1** (1971) 159.
39 J. Cueilleron and B. Frange, *Bull. Soc. Chim. Fr.* **1972**, 107.
40 B. Frange, Diss. Lyon (1972).
41 J. R. Blackborow, J. E. Blackmore and J. C. Lockhart, *J. Chem. Soc. A* **1971**, 49.
42 J. L. Adcock, Diss. Univ. Texas (1971); *Diss. Abstr. Int.* **B32** (1972) 5669.
42a A. Meller, M. Wojnowska and H. Marecek, *Monatsh. Chem.* **100** (1969) 175.
43 H. Nöth and W. Storch, *Chem. Ber.*, **110** (1977) 2607.
43a B. Frange, *Bull. Soc. Chim. Fr.* **1973**, 1216.
43b B. Frange, *Bull. Soc. Chim. Fr.* **1973**, 2165.
44 P. Paetzold, A. Richter, Th. Thijssen and Str. Würtenberg, *Chem. Ber.* **112** (1979) 3811.
45 W. Haubold and K. Zurmühl, *Chem. Ber.* **113** (1980) 2333.
46 K. Niedenzu, E. Blick and I. A. Boenig, *Z. Anorg. Allg. Chem.* **387** (1971) 107.
47 J. R. Bowser, R. H. Neilson and R. L. Wells, *Inorg. Chem.* **17** (1978) 1882.
48 J. R. Bowser, Diss. Duke Univ. (1976); *Diss. Abstr. Int.* **B37** (1977) 3400.
49 A. Meller, W. Maringgele and K. D. Kablau, *Z. Anorg. Allg. Chem.* **445** (1978) 122.
49a K. D. Kablau, Diplomarbeit Univ. Göttingen (1978).
50 A. Meller and H. J. Füllgrabe, *Monatsh. Chem.* **106** (1975) 1407.
51 A. Meller, W. Maringgele and K. D. Kablau, *Z. Naturforsch.* **33b** (1978) 891.
52 H. Nöth and M. J. Sprague, *J. Organomet. Chem.* **23** (1970) 323.
53 A. Rizzo and B. Frange, *J. Organomet. Chem.* **76** (1974) 1.
54 A. Meller and H. Batka, *Monatsh. Chem.* **101** (1970) 648.
55 W. L. Cook and K. Niedenzu, *Synth. Inorg. Metal-Org. Chem.* **4** (1974) 53.
56 W. L. Cook, Diss. Univ. Kentucky (1973); *Diss. Abstr. Int.* **B35** (1974) 1194.
57 B. R. Cragg and K. Niedenzu, *J. Organomet. Chem.* **117** (1976) 1.
58 B. L. Therell and E. K. Mellon, *Inorg. Chem.* **11** (1972) 1137.

59   V. A. Zamyatina and V. V. Korshak, *Izv. Akad. Nauk SSSR, Ser. Khim.* **1971**, 1816; *Bull. Acad. Sci. USSR, Div. Chem. Sci.* **1971**, 1709.
60   H. Nöth, D. Reiner and W. Storch, *Chem. Ber.* **106** (1973) 1508.
61   K. Jonás, H. Nöth and W. Storch, *Chem. Ber.* **110** (1977) 2783.
62   H. Nöth and W. Storch, *Chem. Ber.* **109** (1976) 884.
63   I. Geisler and H. Nöth, *Chem. Ber.* **106** (1973) 1943.
64   H. Fußstetter, R. Kroll and H. Nöth, *Chem. Ber.* **110** (1977) 3829.
65   P. M. Johnson and E. K. Mellon, *Inorg. Chem.* **13** (1974) 2769.
66   N. A. Vasilenko, A. S. Teleshova and A. N. Pravednikov, *Zh. Obshch. Khim.* **43** (1973) 1124; *J. Gen. Chem. USSR* **43** (1973) 1114.
67   J. C. Kotz and W. J. Painter, *J. Organomet. Chem.* **32** (1971) 231.
67a W. J. Painter, Diss. Kansas State Univ. (1970); *Diss. Abstr. Int.* **B31** (1971) 6482.
68   H. Fußstetter, G. Kopitz and H. Nöth, *Chem. Ber.* **113** (1980) 728.
69   R. Köster and G. Seidel, *Liebigs Ann. Chem.* **1977**, 1837.
70   D. Enterling, U. Klingebiel and A. Meller, *Z. Naturforsch.* **33b** (1978) 527.
71   H. Nöth, W. Tinhof and T. Taeger, *Chem. Ber.* **107** (1974) 3113.
72   K. Barlos and H. Nöth, *Chem. Ber.* **110** (1977) 2790.
73   I. Geisler and H. Nöth, *Chem. Ber.* **103** (1970) 2234
74   H. Fußstetter and H. Nöth, *Chem. Ber.* **112** (1979) 3672.
75   D. Nölle and H. Nöth, *Angew. Chem.* **83** (1971) 112.
76   D. Nölle, H. Nöth and T. Taeger, *Chem. Ber.* **110** (1977) 1643.
76a H. Fußstetter and H. Nöth, *Chem. Ber.* **111** (1978) 3596.
77   M. Charkravorty and R. F. Porter, *Inorg. Chem.* **8** (1969) 1997.
78   M. Charkravorty, Diss. Cornell Univ. (1970); *Diss. Abstr. Int.* **B31** (1970) 1133.
79   H. Wille and J. Goubeau, *Chem. Ber.* **105** (1972) 2156.
80   R. H. Cragg and M. Nazery, *Inorg. Nucl. Chem. Lett.* **10** (1974) 481.
81   K. Niedenzu and C. D. Miller, *Fortschr. Chem. Forsch.* **15** (1970) 191.
82   K. Niedenzu, P. J. Busse and C. D. Miller, *Inorg. Chem.* **9** (1970) 977.
82a R. Goetze and H. Nöth, *Chem. Ber.* **109** (1976) 3247.
83   S. H. Dandegaonker and A. S. Mane, *J. Indian Chem. Soc.* **50** (1973) 622.
84   R. H. Cragg and A. F. Weston, *J. Chem. Soc. Chem. Commun.* **1972** 79.
85   R. H. Cragg and A. F. Weston, *J. Chem. Soc. Dalton Trans.* **1975**, 93
86   R. H. Cragg and A. F. Weston, *J. Chem. Soc. Dalton Trans.* **1975**, 1761.
87   R. H. Cragg and A. F. Weston, *J. Chem. Soc. Dalton Trans.* **1973**, 1054.
88   A. Meller and G. Beer, *Monatsh. Chem.* **104** (1973) 1055.
89   A. Meller and H. J. Füllgrabe, *Angew. Chem.* **87** (1975) 382.
90   A. Meller and H. J. Füllgrabe, *Z. Naturforsch.* **33b** (1978) 156.
91   A. Meller and H. J. Füllgrabe, *Chem. Ber.* **111** (1978) 819.
92   A. Meller, H. J. Füllgrabe and C. D. Habben, *Chem. Ber.* **112** (1979) 1252.
92a C. D. Habben, Diplomarbeit Univ. Göttingen, (1978).
93   M. Oertel and R. F. Porter, *Inorg. Chem.* **9** (1970) 904.
94   M. A. Neiss and R. F. Porter, *J. Phys. Chem.* **76** (1972) 2630.
95   M. P. Nadler and R. F. Porter, *Inorg. Chem.* **8** (1969) 599.
96   M. P. Nadler, Diss. Cornell Univ., (1969), *Diss. Abstr. Int.* **B30** (1970) 5453.
97   M. A. Neiss and R. F. Porter, *J. Am. Chem. Soc.* **94** (1972) 1438.
98   M. A. Neiss, Diss. Cornell Univ. (1971); *Diss. Abstr. Int.* **B32** (1972) 6957.
99   L. J. Turbini, G. M. Golenwsky and R. F. Porter, *Inorg. Chem.* **14**, (1975) 691.
100 L. J. Turbini and R. F. Porter, *Inorg. Chem.* **14** (1975) 1252.
101 L. J. Turbini, T. J. Mazanek and R. F. Porter, *J. Inorg. Nucl. Chem.* **37** (1975) 1129.
102 G. A. Kline and R. F. Porter, *Inorg. Chem.* **16** (1977) 11.
103 A. J. De Stefano and R. F. Porter, *Inorg. Chem.* **15** (1976) 2569.

104 G. A. Anderson and J. J. Lagowski, *Inorg. Chem.* **14** (1975) 1845.
105 G. A. Anderson and J. J. Lagowski, *Inorg. Chem.* **10** (1971) 1910.
106 O. T. Beachley and B. Washburn, *Inorg. Chem.* **14** (1975) 120.
107 O. T. Beachley, *Inorg. Chem.* **16** (1977) 2642.
108 J. L. Adcock and J. J. Lagowski, *Inorg. Nucl. Chem. Lett.* **7** (1971) 473.
109 L. A. Melcher, J. L. Adcock and J. J. Lagowski, *Inorg. Chem.* **11** (1972) 1247.
110 P. Powell, *Inorg. Chem.* **12** (1973) 913.
111 B. M. Mikhailov and A. F. Galkin, *Izv. Akad. Nauk SSSR, Ser. Khim.* **1969**, 604; *Bull. Acad. Sci. USSR, Div. Chem. Sci.* **1969**, 539.
112 E. Amberger and W. Stoeger, *J. Organomet. Chem.* **17** (1969) 287.
113 D. T. Haworth, L. A. Melcher and K. Niedenzu, *Inorg. Synth.* **14** (1973) 55.
114 D. T. Haworth and E. S. Matushek, *Chem. Ind. (London)* **1971**, 130.
115 G. J. Barrett and D. T. Haworth, *Inorg. Chim. Acta* **6** (1972) 504.
116 R. Clément and Y. Proux, *Bull. Soc. Chim. Fr.* **1969**, 558.
117 H. Nöth, *24th Int. Congr. Pure Appl. Chem., Hamburg 1973,* Bd. 4, p 13 (1974).
118 I. B. Atkinson, D. C. Blundell and D. B. Clapp, *J. Inorg. Nucl. Chem.* **34** (1972) 3037.
119 H. Werner, R. Prinz and E. Deckelmann, *Chem. Ber.* **102** (1969) 95.
120 J. L. Adcock and J. J. Lagowski, *Inorg. Chem.* **12** (1973) 2533.
121 R. Prinz and H. Werner, *Angew. Chem.* **79** (1967) 63.
122 M. Scotti and H. Werner, *J. Organomet. Chem.* **81** (1974) C 17.
123 K. Deckelmann and H. Werner, *Helv. Chim. Acta* **53** (1970) 139.
124 K. Deckelmann and H. Werner, *Helv. Chim. Acta* **54** (1971) 2189.
125 M. Scotti and H. Werner, *Helv. Chim. Acta* **57** (1974) 1234.
126 R. Goetze and H. Nöth, *J. Organomet. Chem.* **145** (1978) 151.
127 G. Huttner and B. Krieg, *Angew. Chem.* **83** (1971) 541.
128 G. Huttner and B. Krieg, *Chem. Ber.* **105** (1972) 3437.
129 M. Scotti, H. Werner, D. L. S. Brown, S. Cavell, J. A. Connor and H. A. Skinner, *Inorg. Chim. Acta* **25** (1977) 261.
130 K. Deckelmann and H. Werner, *Helv. Chim. Acta* **52** (1969) 892.
131 B. Herber, M. Scotti and H. Werner, *Helv. Chim. Acta* **58** (1975) 1225.
132 I. V. Kryukova and E. M. Fedneva, *Tr. Chelyabinsk Politekhn. Inst.* **91** (1971) 112; *CA* **77** (1972) 152319.
133 D. T. Haworth and E. S. Matushek, *Inorg. Nucl. Chem. Lett.* **7** (1971) 261.
134 L. A. Melcher, *Diss. Univ. Texas at Austin* (1971); *Diss. Abstr. Int.* **B32** (1972) 5674.
135 J. Witschel and B. M. Fung, *J. Chem. Phys.* **56** (1972) 5417.
136 L. J. Turbini and R. F. Porter, *Org. Magn. Reson.* **6** (1974) 456.
137 E. K. Mellon, B. M. Coker and P. B. Dillon, *Inorg. Chem.* **11** (1972) 852.
138 O. T. Beachley, *J. Am. Chem. Soc.* **92** (1970) 5372.
139 O. T. Beachley, *Inorg. Chem.* **8** (1969) 981.
140 J. L. Adcock, L. A. Melcher and J. J. Lagowski, *Inorg. Chem.* **12** (1973) 788.
141 J. Gallier, M. A. Chassoneau and J. Meinnel, *C. R. Acad. Sci. Paris* **C276** (1973) 1231.
142 M. Pasdeloup, J. P. Laurent and G. Commenges, *J. Chim. Phys.* **69** (1972) 1023.
142a G. Cros, M. Pasdeloup and J. P. Laurent, *Bull. Soc. Chim. Fr.* **1969**, 2601.
143 K. N. Scott and W. S. Brey, *Inorg. Chem.* **8** (1969) 1703.
144 W. Biffar, H. Nöth, H. Pommerening and B. Wrackmeyer, *Chem. Ber.* **113** (1980) 333.
145 J. B. Rowbotham and T. Schaefer, *Can. J. Chem.* **52** (1974) 489.
146 L. J. Turbini, *Diss. Cornell Univ.* (1974); *Diss. Abstr. Int.* **B35** (1975) 3233.
147 E. L. Burrows, *Diss. Univ. Texas* (1971), *Diss. Abstr. Int.* **B32** (1972) 6269.
148 M. Pasdeloup, G. Cros, G. Commenges and J. P. Laurent, *Bull. Soc. Chim. Fr.* **1971** 754.
149 B. R. Gragg, W. J. Layton and K. Niedenzu, *J. Organomet. Chem.* **132** (1977) 29.
150 G. Beaumelou, M. Pasdeloup and J. P. Laurent, *Org. Magn. Reson.* **5** (1973) 585.

151 J. L. Adcock and J. J. Lagowski, *J. Organomet. Chem.* **72** (1974) 323.
152 B. Wrackmeyer and H. Nöth, *Chem. Ber.* **109** (1976) 3480.
153 H. Nöth and B. Wrackmeyer, *Chem. Ber.* **114** (1981) 1150.
154 H. Bock and B. G. Ramsey, *Angew. Chem.* **85** (1973) 773.
154a W. Fuß, Diss. Univ. Frankfurt (1971).
155 H. Bock and W. Fuß, *Angew. Chem.* **83** (1971) 169.
156 D. C. Frost, F. G. Herring, C. A. McDowell and I. A. Stenhouse, *Chem. Phys. Lett.* **5** (1970) 291.
157 D. R. Lloyd and N. Lynaugh, *J. Chem. Soc. Dalton Trans.* **1971**, 125.
158 D. R. Lloyd and N. Lynaugh, *Phil. Trans. R. Soc. Lond.* **A268** (1970) 97; *CA* **74** (1971) 36637.
159 J. Kroner, D. Proch, W. Fuß and H. Bock, *Tetrahedron* **28** (1972) 1585.
160 C. R. Brundle, M. B. Robin and N. A. Kuebler, *J. Am. Chem. Soc.* **94** (1972) 1466.
161 P. H. Blustin, *Mol. Phys.* **36** (1978) 279.
162 D. R. Armstrong and D. T. Clark, *Theor. Chim. Acta* **24** (1972) 307.
163 D. R. Armstrong and D. T. Clark, *Chem. Commun.* **1970**, 99.
164 M. J. S. Dewar and M. L. McKee, *J. Am. Chem. Soc.* **99** (1977) 5231.
165 C. E. Doiron, F. Grein, T. B. McMahon and K. Vasudevan, *Can. J. Chem.* **57** (1979) 1751.
166 J. Gayoso, B. Maouche and A. Boucekkine, *C. R. Acad. Sci. Paris* **C276** (1973) 257.
167 F. Grein and K. Weiss, *Theor. Chim. Acta* **34** (1974) 315.
168 M. F. Guest, I. H. Hillier and I. C. Shenton, *Tetrahedron* **31** (1975) 1943.
169 V. M. Scherr and D. T. Haworth, *Theor. Chim. Acta* **21** (1971) 143.
170 V. M. Scherr and D. T. Haworth, *J. Inorg. Nucl. Chem.* **35** (1973) 660.
171 A. Serafini and J. F. Labarre, *J. Mol. Struct.* **26** (1975) 129.
172 K. Vasudevan and F. Grein, *Theor. Chim. Acta* **52** (1979) 219.
173 S. D. Peyerimhoff and R. J. Buenker, *Theor. Chim. Acta* **19** (1970) 1.
173a J. Kroner, H. Nöth and K. Niedenzu, *J. Organomet. Chem.* **71** (1974) 165.
174 I. Pozela and C. Radvilavicius, *Fiz. Elektron. (Kaunas)* **1** (1972) 166; *CA* **82** (1975) 35217.
175 I. Pozela and C. Radvilavicius, *Lietuvos Fiz. Rinkinys* **15** (1975) 559, *CA* **84** (1976) 50029.
176 E. Ya. Bolycheva, V. N. Mochalkin and A. B. Bolotin, *Lietuvos Fiz. Rinkinys* **17** (1977) 51; *CA* **87** (1977) 13670.
177 E. Ya. Bolycheva, V. N. Mochalkin and A. B. Bolotin, *Lietuvos Fiz. Rinkinys* **17** (1977) 41; *CA* **87** (1977) 73687.
178 L. A. Melcher, J. L. Adcock, G. A. Anderson and J. J. Lagowski, *Inorg. Chem.* **12** (1973) 601.
179 J. M. Miller and G. L. Wilson, *Inorg. Chem.* **13** (1974) 498.
180 P. Powell, P. J. Sherwood, M. Stephens and E. F. H. Brittain, *J. Chem. Soc.* A **1971**, 2951.
181 R. H. Cragg and A. F. Weston, *J. Organomet. Chem.* **96** (1975) C35.
182 R. F. Porter and J. J. Solomon, *J. Am. Chem. Soc.* **93** (1971) 56.
183 L. D. Betowski, J. J. Solomon and R. F. Porter, *Inorg. Chem.* **11** (1972) 424.
184 A. J. De Stefano and R. F. Porter, *Inorg. Chem.* **14** (1975) 2882.
185 A. J. De Stefano and R. F. Porter, *J. Phys. Chem.* **80** (1976) 2818.
186 R. G. Wilson, *J. Appl. Phys.* **44** (1973) 5056.
187 C. H. Chang, R. F. Porter and S. H. Bauer, *Inorg. Chem.* **8** (1969) 1677.
188 J. C. Huffman, H. Fußstetter and H. Nöth, *Z. Naturforsch.* **31b** (1976) 289.
189 W. Harshbarger, G. H. Lee, R. F. Porter and S. H. Bauer, *J. Am. Chem. Soc.* **91** (1969) 551.
190 H. Hess and B. Reiser, *Z. Anorg. Allg. Chem.* **381** (1971) 91.
191 D. Lux, W. Schwarz and H. Hess, *Cryst. Struct. Commun.* **8** (1979) 33.
192 W. Schwarz, D. Lux and H. Hess., *Cryst. Struct. Commun.* **6** (1977) 431.

193 W. Harshbarger, G. Lee, R. F. Porter and S. H. Bauer, *Inorg. Chem.* **8** (1969) 1683.
194 P. J. Roberts, D. J. Brauer, Y. H. Tsay and C. Krüger, *Acta Crystallogr.* **B30** (1974) 2673.
195 U. Müller, *Acta Crystallogr.* **B27** (1971) 1997.
196 J. G. Haasnoot, G. C. Verschoor, C. Romers and W. L. Groeneveld, *Acta Crystallogr.* **B28** (1972) 2070.
197 M. S. Gopinathan, M. A. Whitehead and C. A. Coulson, *Acta Crystallogr.* **B30** (1974) 731.
197a M. Kaplansky and M. A. Whitehead, *Can. J. Chem.* **48** (1970) 697.
198 K. Anton, H. Fußstetter and H. Nöth, *Chem. Ber.* **114** (1981) 2723.
198a H. Nöth, R. Ullmann and H. Vahrenkamp, *Chem. Ber.* **106** (1973) 1165.
199 J. A. S. Smith and D. A. Tong, *J. Chem. Soc.* A **1971** 178.
200 G. M. Whitesides, St. L. Regen, J. B. Lisle and R. Mays, *J. Phys. Chem.* **76** (1972) 2871.
201 F. Gallais, J. P. Laurent, G. Cros and M. Pasdeloup, *J. Chim. Phys.* **66** (1969) 122.
202 M. Pasdeloup and J. P. Laurent, *J. Chim. Phys.* **70** (1973) 199.
203 A. Serafini, J. F. Labarre and F. Gallais, *Proc. 16th Int. Conf. Coord. Chem. Dublin,* 1974, p. 2.4a; *CA* **85** (1976) 25604.
204 M. B. Robin and N. A. Kuebler, *J. Mol. Spectrosc.* **70** (1978) 472.
205 A. Kaldor, *J. Chem. Phys.* **55** (1971) 4641.
206 E. R. Bernstein and J. R. Reilly, *J. Chem. Phys.* **57** (1972) 3960.
207 I. Santucci and C. Triboulet, *J. Chem. Soc.* A **1969** 392.
208 W. F. Young, F. Grein, J. Passmore and I. Unger, *Can. J. Chem.* **49** (1971) 1.
209 Md. A. Haque, A. B. Sannigrahi and B. G. Niyogi, *J. Phys. Chem.* **83** (1979) 1348.
210 Ph. J. Vergragt and J. H. Van der Waals, *Chem. Phys. Lett.* **42** (1976) 193.
211 J. M. Van Pruyssen and St. D. Colson, *Chem. Phys.* **6** (1974) 382.
212 M. Yanase, M. Koyanagi and Y. Kanda, *Mem. Fac. Sci., Kyushu Univ.* **C8** (1972) 35; *CA,* **76** (1972) 160422.
213 C. D. Dejardins, F. Greni, M. E. Macbeath, J. Passmore and I. Unger, *J. Photochem.* **1** (1972/73) 153.
214 C. E. Doiron, M. E. Macbeath and T. B. McMahon, *Chem. Phys. Lett.* **59** (1978) 90.
215 M. Graffeuil and J. F. Labarre, *J. Chim. Phys.* **68** (1971) 1379.
216 M. Graffeuil and J. F. Labarre, *J. Chim. Phys.* **66** (1969) 177.
217 J. V. Kainnady and A. Weber, *J. Raman Spectr.* **5** (1976) 35.
218 B. Roussel, A. Chapput and G. Fleury, *J. Mol. Struct.* **31** (1976) 371.
219 R. S. Wall and J. R. Riter, *Spectrochim. Acta* **36A** (1980) 673.
220 L. Ya. Rikhter and L. M. Sverdlov, *Zh. Prikl. Spektrosk.* **17** (1972) 491; *J. Appl. Spectrosc.* **17** (1972) 1194.
221 K. E. Blick, K. Niedenzu, W. Sawodny, M. Takasuka, T. Totani and H. Watanabe, *Inorg. Chem.* **10** (1971) 1133.
221a K. E. Blick, I. A. Boenig and K. Niedenzu, *Inorg. Chem.* **10** (1971) 1917.
221b K. E. Blick and K. Niedenzu, AD-701704 (1969); *US Govt Res. Develop. Rep.* **70** (1970) 80; *CA* **73** (1970) 60825.
222 K. Hensen and K. P. Messer, *Theor. Chim. Acta* **9** (1967/68) 17.
223 E. Blick, E. B. Bradley, K. Iwatani, K. Niedenzu, T. Takusaka, T. Totani and H. Watanabe, *Z. Anorg. Allg. Chem.* **417** (1975) 19.
224 K. Molt and W. Sawodny, *Z. Anorg. Allg. Chem.* **474** (1981) 182.
225 L. Ya. Rikhter and L. M. Sverdlov, *Izv. Vyssh. Ucheb. Zaved. Fiz.* **15** (1972) 113; *Sov. Phys. J.* **15** (1972) 1178.
226 L. Ya. Rikhter and L. M. Sverdlov, *Vopr. Mol. Spektrosk* **1974,** 71; *CA* **82** (1975) 162263.
227 A. Meller and G. Beer, *Z. Anorg. Allg. Chem.* **460** (1980) 169.
228 E. Huler, E. Silbermann and E. A. Jonas, *Spectrochim. Acta* **A26** (1970) 2241.

229   E. J. Huler, Diss. Vanderbilt Univ. (1969), *Diss. Abstr. Int.* **B30** (1970) 4742.

230   L. Ya. Rikhter and L. M. Sverdlov, *Zh. Fiz. Khim.* **49** (1975) 2713; *Russ. J. Phys. Chem.*
        49 (1975) 1598.

231   L. Ya. Rikhter and L. M. Sverdlov, *Spektrosk. Ee Primen. Geofiz. Khim.* **1975** 203; *CA* **84**
        (1976) 51810.

232   L. Ya. Rikhter and L. M. Sverdlov, *Teor. Eksp. Khim.* **11** (1975) 672; *Theor. Exp. Chem.*
        11 (1975) 562.

233   L. Ya. Rikhter and L. M. Sverdlov, *Zh. Fiz. Khim.* **45** (1971) 2687; *Russ. J. Phys. Chem.*
        45 (1971) 1527.

234   L. Ya. Rikhter and L. M. Sverdlov, *Zh. Fiz. Khim.* **45** (1971) 1579; *Russ. J. Phys. Chem.*
        45 (1971) 896.

235   L. Ya. Rikhter and L. M. Sverdlov, *Izv. Vyssh. Uchebn. Zaved. Fiz* **15** (1972) 113; *Sov.
        Phys. J.* **1972**, 1178.

236   L. Ya. Rikhter and L. M. Sverdlov, *Kvant. Khim.* **1975**, R-201; *CA* **84** (1976) 97320.

237   A. Kaldor, Diss. Cornell Univ. (1970); *Diss. Abstr. Int.* **B31** (1971) 7221.

238   M. K. Subhedar and G. Thyagarajan, *Indian J. Pure Appl. Phys.* **12** (1974) 498.

239   M. K. Subhedar and G. Thyagarajan, *Indian J. Pure Appl. Phys.* **18** (1980) 396.

240   G. A. Anderson and J. J. Lagowski, *Spectrochim. Acta* **26** (1970) 2013.

241   A. Kaldor and R. F. Porter, *Inorg. Chem.* **10** (1971) 775.

242   K. E. Blick, E. B. Bradley, K. Niedenzu, M. Takasuka, T. Totani and H. Watanabe, *Z.
        Anorg. Allg. Chem.* **442** (1978) 183.

243   O. T. Beachley and R. G. Simmons, *Inorg. Chem.* **16** (1977) 2935.

244   T. Totani, H. Watanabe and M. Kubo, *Spectrochim. Acta* **A25** (1969) 585.

245   M. A. Molinari and G. J. Videla, *Arg. Com. Nacl. Energia At.* No. CNEA-302 (1971);
        CA **76** (1972) 119458.

246   B. C. Smith, L. Thakur and M. A. Wassef, *Can. J. Chem.* **48** (1970) 1936.

247   A. Ya. Chernikov, M. N. Yakovlev, V. B. Lysova, E. L. Gefter and N. N. Shmagina, *Fr.
        Demande* 2 428 655 (1980); *CA* **92** (1980) 216210.

248   A. Ya. Chernikov, M. N. Yakovlev, V. B. Lysova, E. L. Gefter and N. N. Shmagina,
        Ger. Offen. 2 825 413 (1980); *CA* **93** (1980) 8757.

249   A. Ya. Chernikov, M. N. Yakovlev, N. S. Rogov, A. P. Petrova, E. B. Martirosov and
        V. E. Gul, *Ger. Offen.* 2 907 195 (1980), *CA* **93** (1980) 240508.

250   V. V. Korshak, V. A. Zamyatina and R. M. Oganesyan, *USSR Pat.* 309 937 (1971); *CA*
        76 (1972) 86387.

251   I. B. Atkinson, D. Clapp, B. R. Currell and W. Flavell, *Brit. Polym. J.* **3** (1971) 169; *CA*
        75 (1971) 152138.

251a   I. B. Atkinson, D. B. Clapp, W. Flavell and B. R. Currell, *Polym. Prepr. Am. Chem. Soc.
        Div. Polym. Chem.* **13** (1972) 770; *CA* **81** (1974) 121072.

252   V. I. Spitzin, *Z. Chem.* **1974**, 459.

253   V. A. Zamyatina, V. V. Korshak and N. M. Gnutova, *Vasokomol. Soedin.* **B15** (1973)
        846; *CA* **80** (1974) 71167.

254   V. V. Korshak, A. F. Zhigach, I. G. Sarishvili and M. V. Sobolevski, *Prog. Polim. Khim.*
        1968, 321; *CA* **72** (1970) 13080.

255   I. Taniguchi, K. Harada and T. Maeda, *Japan. Kokai* 76-53.000 (1976); *CA* **85** (1976)
        96582.

256   A. Ya. Chernikov, M. N. Yakovlev, N. S. Rogov and A. P. Petrova, *Ger. Offen.*
        2 907 195 (1979); *CA* **94** (1981) 192722.

257   V. A. Zamyatina, V. V. Korshak and K. P. Grinevich, *Vysokomol. Soedin.* **B14** (1972)
        410; *CA* **77** (1972) 115179.

257a   V. A. Zamyatina, V. V. Korshak and R. M. Oganesyan, *Vysokomol. Soedin.* **B15** (1973)
        848; *CA* **80** (1974) 71168.

258 K. N. Karadzhyan and G. A. Kazaryan, *USSR Pat.* 519 430 (1976); *CA* **85** (1976) 109524.

259 G. P. Kazaryan, K. N. Karadzhyan, M. E. Kuimova, N. E. Nikolaeva and A. Shirinyan, *Plast. Massy* **1978**, 73; *CA* **89** (1978) 198320.

259a K. N. Karadzhyan and G. A. Kazaryan, *Arm. Khim. Zh.* **29** (1976) 1049; *CA* **86** (1977) 172213.

260 Y. Proux and R. Clement, *Bull. Soc. Chim. Fr.* **1970**, 528.

261 H. Kondo and M. Sugata, *Japan. Kokai Tokkyo Koho* 78–131.844 (1978); *CA* **91** (1979) 81565.

261a H. Kondo and M. Sugata, *Japan. Kokai Tokkyo Koho* 78–131.845 (1978); *CA* **91** (1979) 81566.

262 R. Liepins, S. Gilbert, F. Tibbets and J. Kearney, *J. Appl. Polym. Sci.* **17** (1973) 2523.

263 M. Yamamoto, M. Hirami, T. Sasaki and H. Saito, *Japan. Pat.* 73–07.847 (1973); *CA* **80** (1974) 61025.

264 G. L. Nelson and J. L. Webb, *J. Fire Flammability* **1973**, 325.

265 S. Samukawa and Sh. Ikebe, *Japan. Pat.* 71–34.900 (1971); *CA* **77** (1972) 76068.

266 M. A. Molinari, A. E. Rojo and G. J. Videla, CNEA-206 (1967); *Nucl. Sci. Abstr.* **22** (1968) 37974; *CA* **70** (1969) 24994.

267 M. A. Molinari and G. J. Videla, *Acta Cient.* **1** (1968) 13.

268 T. Hirata, *US Nat. Tech. Inform. Serv.* AD-729339 (1971); from *Govt Rep. Announce (US),* **71** (1971) 66; *CA* **76** (1972) 47910.

268a T. Hirata, *US Nat. Tech. Inform. Serv.* AD-728806 Avail NTIS, (1971); from *Govt Rep. Announce (US)* **71** (1971) 19; *CA* **76** (1972) 20828.

269 L. N. Akimova, A. Ya. Apin and L. N. Stesik, *Fiz. Goreniya Vzryva* **8** (1972) 475; *CA* **78** (1973) 161670.

270 A. A. Savel'ev, A. A. Pukhov, B. A. Vishnyakov, A. D. Sulimin and A. P. Ishchenko, *Fiz. Chim. Obrab. Mater.* **1981**, 85; *CA* **94** (1981) 201401.

271 V. Liepins, I. G. Kuznetsova, A. A. Maier and O. E. Shumilina, *Mosk. Khim.-Tekhnol. Inst. im. V. I. Mendeleeva* **108** (1979) 89; *CA* **95** (1981) 11388.

272 R. L. Hough, *US Pat.* 3 451 840 (1969); *CA* **71** (1969) 41724.

273 J. F. Clarke, *US Pat.* 3 476 080 (1969); *CA* **72** (1970) 15215.

274 R. Feurer, G. Cros and G. Constant, *3rd Eur. Conf. Chem. Vap. Deposition,* 1980, p. 49; *CA* **95** (1981) 174092.

275 Sh. Koide, K. Nakamura and K. Yoshimura, *Kenkyu Kiyo Nihon Daigaku Buringakubu Shizen Kagaku Kenkyusho* **14** (1979) 9; *CA* **94** (1981) 19465.

276 L. E. Branovich, W. B. Fitzpatrick and M. L. Long, *US Pat.* 3 692 566 (1972); *CA* **77** (1972) 168055.

277 R. Haberecht and R. J. Patterson, *US Pat.* 3 564 565 (1971); *CA* **74** (1971) 90485.

278 E. A. Balabanova, A. I. Vargunin, E. M. Orlowa, I. Kolli, E. B. Sokolov and L. S. Sukhanova, *Zh. Neorg. Khim.* **25** (1980) 1793; *Russ. J. Inorg. Chem.* **25** (1980) 995.

279 A. C. Adams, *J. Electrochem. Soc.* **128** (1981) 1378.

280 A. Borkovec, J. A. Settepani, G. C. La Breque and R. L. Fye, *J. Econ. Entomol.* **62** (1969) 1472; *CA* **70** (1969) 105408.

281 H. Neilson and R. L. Wells, *Synth. Inorg. Metal-Org. Chem.* **3** (1973) 283.

282 H. Hess, *Acta Crystallogr.* B25 (1969) 2342.

282a H. A. Steuer, Staatsexamensarbeit Univ. Göttingen (1980).

282b A. F. Weston, Diss. Univ. Kent, Canterbury (1973).

283 P. I. Paetzold and G. Maier, *Chem. Ber.* **103** (1970) 281.

284 N. C. Baird, *Inorg. Chem.* **12** (1973) 473.

285 M. J. S. Dewar and M. L. McKee, *J. Mol. Struct.* **68** (1980) 105.

286 B. Hesset, J. B. Leach, J. H. Morris and P. G. Perkins, *J. Chem. Soc. Dalton Trans.* **1972**, 131.

287 J. B. Leach, J. H. Morris and P. G. Perkins, *J. Chem. Soc.* A **1970**, 1077.

288 D. R. Armstrong, P. G. Perkins, J. M. Scott and J. J. P. Stewart, *Theor. Chim. Acta (Berlin)* **26** (1972) 237.

289 D. C. Rajwar and S. C. Tiwari, *Thermochim. Acta* **34** (1979) 383.

290 B. Hesset, J. H. Morris and P. G. Perkins, *J. Chem. Soc.* A **1971**, 2466.

291 B. Hesset, J. H. Morris and P. G. Perkins, *J. Chem. Soc.* A **1971** 2056.

292 J. H. Morris, *J. Inorg. Nucl. Chem.* **36** (1974) 2439.

293 N. Falla, C. H. J. Wells and J. H. Morris, *Chem. Commun.* **1969**, 1224.

294 J. J. Miller and F. A. Johnson, *Inorg. Chem.* **9** (1970) 69.

295 C. Ungurenasu, St. Cihodaru and I. Popescu, *Tetrahedron Lett.* **1969**, 1435.

296 H. Nöth and W. Regnet, *Chem. Ber.* **102** (1969) 167.

297 D. Nölle and H. Nöth, *Chem. Ber.* **111** (1978) 469.

298 D. Nölle, H. Nöth and W. Winterstein, *Chem. Ber.* **111** (1978) 2465.

299 H. Nöth and W. Winterstein, *Chem. Ber.* **111** (1978) 2469.

299a H. Nöth and W. Regnet, *Chem. Ber.* **102** (1969) 2241.

299b N. I. Bekasova, V. V. Korshak and M. P. Prigozhina, *Vysokomol. Soedin.* **B11** (1969) 366; *CA* **71** (1969) 39505.

299c J. J. Miller, *J. Organomet. Chem.* **24** (1970) 595.

300 H. Nöth, W. Winterstein, W. Kaim and H. Bock, *Chem. Ber.* **112** (1979) 2494.

301 H. Bock, W. Kaim, H. Nöth and A. Semkow, *J. Am. Chem. Soc.* **102** (1980) 4421.

302 H. Bock, W. Kaim, A. Semkow and H. Nöth, *Angew. Chem.* **90** (1978) 308, *Angew. Chem. Int. Ed. Eng.* **17** (1978) 286.

303 W. Winterstein, Diplomarbeit Univ. München (1973).

304 J. Kroner, D. Nölle, H. Nöth and W. Winterstein, *Z. Naturforsch.* **29b** (1974) 476.

304a J. D. Odom, T. F. Moore, R. Goetze, H. Nöth and B. Wrackmeyer, *J. Organomet. Chem.* **173** (1979) 15.

305 D. Nölle and H. Nöth, *Z. Naturforsch.* **27b** (1972) 1425.

306 D. Nölle, H. Nöth and W. Winterstein, *Z. Anorg. Allg. Chem.* **406** (1974) 235.

307 J. Kroner, D. Nölle, H. Nöth and W. Winterstein, *Chem. Ber.* **108** (1975) 3807.

308 H. Nöth, W. Reichenbach and W. Winterstein, *Chem. Ber.* **110** (1977) 2158.

309 W. Biffar, H. Nöth, H. Pommerening, R. Schwerthöffer, W. Storch and B. Wrackmeyer, *Chem. Ber.* **114** (1981) 49.

310 W. Storch, W. Jackstiess, H. Nöth and G. Winter, *Angew. Chem.* **89** (1977) 494.

310a E. Bachholz, Staatsexamensarbeit Univ. Göttingen (1976).

311 K. Barlos and H. Nöth, *Z. Naturforsch.* **35b** (1980) 407.

312 U. Wannagat, E. Bogusch and R. Braun, *J. Organomet. Chem.* **19** (1969) 367.

312a F. Rabet and U. Wannagat, *Z. Anorg. Allg. Chem.* **384** (1971) 115.

313 H. Nöth and W. Tinhof, *Chem. Ber.* **108** (1975) 3109.

313a R. W. Nelson, Diss. Duke Univ. (1968); *Diss. Abstr. Int.* **B29** (1969) 4578.

314 K. Barlos and H. Nöth, *Z. Naturforsch,* **35b** (1980) 415.

315 M. Hesse, U. Klingebiel and L. Skoda, *Chem. Ber.* **114** (1981) 2287.

316 I. Geisler and H. Nöth, *Chem. Commun.* **1969**, 775.

317 U. Wannagat and M. Schlingmann, *Abhandl. Braunschweig Wiss. Ges.* **24** (1974) 79.

318 U. Wannagat, G. Eisele and M. Schlingmann, *Z. Anorg. Allg. Chem.* **429** (1977) 83.

319 U. Wannagat, M. Schlingmann and H. Autzen, *Chem.-Ztg.* **98** (1974) 372.

319a H. Nöth, W. Tinhof and B. Wrackmeyer, *Chem. Ber.* **107** (1974) 518.

320 U. Wannagat and L. Gerschler, *Z. Anorg. Allg. Chem.* **383** (1971) 249.

321 E. Niecke and W. Bitter, *Angew. Chem.* **87** (1975) 34.

322 E. Niecke and W. Bitter, *Chem. Ber.* **109** (1976) 415.
323 K. Barlos, H. Nöth, B. Wrackmeyer and W. McFarlane, *J. Chem. Soc. Dalton Trans.* **1979**, 801.
324 H. Nöth and W. Tinhof, *Chem. Ber.* **107** (1974) 3806.
325 H. Binder, *Phosphorus* **1** (1972) 287.
326 C. D. Schmulbach, *US Pat.* 3 538 155 (1970); *CA* **74** (1971) 23006.
327 D. Nölle, Diss. Univ. München (1975).
328 K. Barlos and H. Nöth, *Z. Naturforsch.* **35b** (1980) 125.
329 A. Meller and C. Habben, *Monatsh. Chem.* **113** (1982) 139.
330 U. Wannagat and G. Eisele, *Z. Naturforsch.* **33b** (1978) 475.
331 H. W. Roesky and S. K. Mehrotra, *Angew. Chem.* **90** (1978) 626; *Angew. Chem. Int. Ed. Engl.* **17** (1978) 599.
332 H. Fußstetter, H. Nöth, K. Peters, H. G. v. Schnering and J. C. Huffman, *Chem. Ber.* **113** (1980) 3881.
333 H. W. Roesky, S. K. Mehrotra and S. Pohl, *Chem. Ber.* **113** (1980) 2063.
334 F. Riegel, Diss. Univ. Würzburg (1973).
335 W. Siebert and F. Riegel, *Chem. Ber.* **106** (1973) 1012.
336 H. Nöth and R. Ullmann, *Chem. Ber.* **108** (1975) 3125.
337 A. Storr, B. S. Thomas and A. D. Penland, *J. Chem. Soc. Dalton Trans.* **1972**, 326.
338 A. Haas and M. Häberlein, *J. Fluorine Chem.* **7** (1976) 123.
339 W. Haubold and K. Stanzl, *Liebigs Ann. Chem.* **1980**, 1659.
340 P. Paetzold and S. Kosma, *Chem. Ber.* **112** (1979) 654.
341 K. Niedenzu, I. A. Boenig and E. F. Rothgery, *Chem. Ber.* **105** (1972) 2258.
342 O. T. Beachley and B. Washburn, *Inorg. Chem.* **15** (1976) 725.
343 W. Haubold and R. Schaeffer, *Chem. Ber.* **104** (1971) 513.
344 V. A. Dorokhov, O. G. Boldyreva and B. M. Mikhailov, *Zh. Obshch. Khim.* **40** (1970) 1528; *J. Gen. Chem. USSR* **40** (1970) 1515.
345 B. M. Mikhailov, V. A. Dorokhov, N. V. Mostovoi, O. G. Boldyreva, M. N. Boldyreva, *Zh. Obshch. Khim.* **40** (1970) 1817; *J. Gen. Chem. USSR* **40** (1970) 1801.
346 B. M. Mikhailov, V. A. Dorokhov, V. S. Bogdanov, I. P. Yakovlev and A. D. Naumov, *Dokl. Akad. Nauk SSSR* **194** (1970) 595; *Dokl. Chem.* **194** (1970) 690.
347 Ph. C. Keller, *Inorg. Chem.* **14** (1975) 440.
348 S. Fleming and R. W. Parry, *Inorg. Chem.* **11** (1972) 1.
349 H. E. Clark, Diss. Univ. Michigan (1968); *Diss. Abstr. Int.* **B30** (1969) 150.
350 K. Wiedemann and J. Voigtländer, *Z. Naturforsch.* **24a** (1969) 566.
351 W. Fuß, *Z. Naturforsch.* **29b** (1974) 514.
352 B. Wrackmeyer and H. Nöth, *Chem. Ber.* **110** (1977) 1086.
353 J. R. Jennings, I. Pattison and K. Wade, *J. Chem. Soc. A* **1969**, 565.
354 C. Summerford and K. Wade, *J. Chem. Soc. A* **1969**, 1487.
355 C. Summerford and K. Wade, *J. Chem. Soc. A* **1970**, 2010.
356 A. Meller and W. Maringgele, *Monatsh. Chem.* **101** (1970) 753.
356a V. A. Dorokhov and M. F. Lappert, *J. Chem. Soc. A* **1969**, 433.
357 A. Meller and A. Ossko, *Monatsh. Chem.* **100** (1969) 1187.
358 A. Meller and A. Ossko, *Monatsh. Chem.* **101** (1970) 1104.
359 A. Meller and A. Ossko, *Monatsh. Chem.* **102** (1971) 131.
360 A. Meller and W. Maringgele, *Monatsh. Chem.* **101** (1970) 387.
361 F. Weller and K. Dehnicke, *Chem. Ber.* **110** (1977) 3935.
362 B. M. Mikhailov, G. S. Ter-Sarkisyan, N. N. Govorov and N. A. Nikolaeva, *Izv. Akad. Nauk SSSR, Ser. Khim.* **1976**, 1820; *Bull. Acad. Sci. USSR, Div. Chem. Sci.* **25** (1976) 1715.

363 B. M. Mikhailov, G. S. Ter-Sarkisyan, N. N. Govorov and N. A. Nikolaeva, *Khim. Elementoorg. Soedin.* **1976**, 3; *CA* **86** (1977) 16721.

364 B. M. Mikhailov, G. S. Ter-Sarkisyan and N. N. Govorov, *Izv. Akad. Nauk SSSR, Ser. Khim.* **1976**, 1823; *Bull. Acad. Sci. USSR, Div. Chem. Sci.* **25** (1976) 1717.

365 Yu. N. Bubnov, A. V. Tsyban and B. M. Mikhailov, *Izv. Akad. Nauk SSSR, Ser. Khim.* **1976**, 2842; *Bull. Acad. Sci. USSR, Div. Chem. Sci.* **25** (1976) 2653

366 B. M. Mikhailov, G. S. Ter-Sarkisyan, N. N. Govorov, N. A. Nikolaeva and V. G. Kiselev, *Izv. Akad. Nauk SSSR, Ser. Khim.* **1976**, 870; *Bull. Acad. Sci. USSR, Div. Chem. Sci.* **25** (1976) 848.

367 V. A. Dorokhov, V. I. Seredenko and B. M. Mikhailov, *Izv. Akad. Nauk SSSR, Ser. Khim.* **1977**, 1593; *Bull. Acad. Sci. Ussr, Div. Chem. Sci.* **26** (1977) 1464.

368 V. A. Dorokhov, I. P. Yakovlev and B. M. Mikhailov, *Izv. Akad. Nauk SSSR, Ser. Khim.* **1980**, 663; *Bull. Acad. Sci. USSR, Div. Chem. Sci.* **29** (1980) 485.

369 W. Maringgele, A. Meller, H. Nöth and R. Schroen, *Z. Naturforsch.* **33b** (1978) 673.

370 S. Ardjomand and E. A. C. Lucken, *Helv. Chim. Acta* **54** (1971) 176.

371 S. Akerfeld, K. Wahlberg and M. Hellström, *Acta Chem. Scand.* **23** (1969) 115.

372 R. L. Williams, *Acta Chem. Scand.* **23** (1969) 149.

373 H. D. Smith and R. J. Brotherton, *Inorg. Chem.* **9** (1970) 2443.

374 P. Krohmer and J. Goubeau, *Chem. Ber.* **104** (1971) 1347.

375 R. Molinelli, S. R. Smith and J. Tanaka, *J. Chem. Soc. Dalton Trans.* **1972**, 1363.

376 R. P. Molinelli, *Diss. Univ.* Connecticut (1971); *Diss. Abstr. Int.* **B33** (1972) 2517.

377 H. G. Köhn and A. Meller, *Z. Naturforsch.* **35b** (1980) 447.

378 G. B. Kauffman, B. H. Gump, B. J. Stedjee and R. A. Houghter, *J. Chromatogr.* **123** (1976) 448.

379 M. P. Brown, R. W. Heseltine, P. A. Smith and P. J. Walker, *J. Chem. Soc.* A **1970**, 410.

380 D. R. Leavers, J. R. Long, S. G. Shore and W. J. Taylor, *J. Chem. Soc.* A **1969**, 1580.

381 D. R. Leavers and W. J. Taylor, *J. Phys. Chem.* **81** (1977) 2257.

382 D. R. Leavers, *Diss.* Ohio State Univ. (1971); *Diss. Abstr. Int.* **B32** (1972) 6329.

383 U. Müller, *Z. Anorg. Allg. Chem.* **382** (1971) 110.

384 H. Hess, D. Lux and W. Schwarz, *Z. Naturforsch.* **32b** (1977) 982.

385 P. W. R. Corfield and S. G. Shore, *J. Am. Chem. Soc.* **95** (1973) 1480.

386 R. H. Findley, *J. Chem. Soc. Chem. Commun.* **1975**, 98.

387 R. H. Findley, *J. Chem. Soc. Dalton Trans.* **1976**, 851.

388 H. Fußstetter and H. Nöth, *Liebigs Ann. Chem.* **1981**, 633.

389 H. Nöth and W. Regnet, *Chem. Ber.* **102** (1969) 2241.

390 S. Trofimenko, *Inorg. Chem.* **8** (1969) 1714.

391 S. Trofimenko, *Acc. Chem. Res.* **4** (1971) 17.

392 T. G. Hodgkins, K. Niedenzu, K. S. Niedenzu and S. S. Seelig, *Inorg. Chem.* **20** (1981) 2097.

393 L. K. Peterson and K. I. Thé, *Can. J. Chem.* **57** (1979) 2520.

394 C. E. May, K. Niedenzu and S. Trofimenko, *Z. Naturforsch.* **33b** (1978) 220.

395 C. E. May, K. Niedenzu and S. Trofimenko, *Z. Naturforsch.* **31b** (1976) 1662.

396 N. W. Alcock and J. F. Sawyer, *Acta Crystallogr.* **B30** (1974) 2899.

397 E. M. Holt, S. L. Holt, K. J. Watson and B. Olsen, *Cryst. Struct. Commun.* **7** (1978) 613.

398 K. E. Breakall, D. F. Rendle, A. Storr and J. Trotter, *J. Chem. Soc. Dalton Trans.* **1975**, 1584.

399 E. Mayer, *Inorg. Nucl. Chem. Lett.* **9** (1973) 343.

400 H. Fußstetter, H. Nöth and W. Winterstein, *Chem. Ber.* **110** (1977) 1931.

401 W. Maringgele and A. Meller, *Monatsh. Chem.* **106** (1975) 1369.

402 A. Meller, W. Maringgele and H. G. Köhn, *Monatsh. Chem.* **107** (1976) 89.

403 O. P. Shitov, S. L. Ioffe, L. M. Leont'eva and V. A. Tartarovskii, *Zh. Obshch. Khim.* **43** (1973) 1266; *J. Gen. Chem. USSR* **43** (1973) 1257.

404 O. P. Shitov, L. M. Leont'eva, S. L. Ioffe, B. N. Khasanov, V. M. Novikow, A. U. Stepanyants and V. A. Tartarovskii, *Izv. Akad. Nauk, Ser. Khim.* **1974**, 2782; *Bull. Acad. Sci. USSR, Div. Chem. Sci.* **23** (1974) 2684.

405 G. Muckle, H. Nöth and W. Storch, *Chem. Ber.* **109** (1976) 2572.

406 R. T. Paine, *J. Am. Chem. Soc.* **99** (1977) 3884.

407 H. Binder, *Z. Naturforsch.* **26b** (1971) 616.

408 H. Binder and J. Palmtag, *Z. Naturforsch.* **34b** (1979) 179.

409 A. B. Burg, *Inorg. Chem.* **11** (1972) 2283.

410 A. B. Burg, *Govt Rep. Announce (US)* **72** (1972) A.D. -739712; *CA* **77** (1972) 69556.

411 H. Nöth and R. Staudigl, *Chem. Ber.* **115** (1982) 813.

412 R. E. Hall and P. E. Schram, *Inorg. Chem.* **8** (1969) 270.

413 R. E. Hall and P. E. Schram, *Inorg. Chem.* **10** (1971) 192.

414 P. C. Keller, *J. Am. Chem. Soc.* **94** (1972) 4020.

3

Boron–Phosphorus and Boron–Arsenic Heterocycles

D. B. SOWERBY

University of Nottingham, England

3.1 Introduction

The most common cyclic boron–phosphorus compounds are those in which the atoms alternate, and can be considered as arising from polymerization of R_2BPR_2' units through donation of the lone pair on phosphorus into an empty boron p-orbital. In these cases, each of the ring atoms becomes fourfold coordinate and dimeric (1) trimeric (2) and tetrameric (3) derivatives are known:

$$(1) \qquad (2) \qquad (3)$$

In addition, condensed species such as (4) are possible. Less well-known are compounds similar to these except that both ring atoms are threefold coordinate. One such compound (5, R = R^1 = Ph, was described in 1961,[1] but in principle three-coordinate analogues of (2)–(4) could also exist. It is also possible to envisage complete series of compounds in which the phosphorus atoms in (1)–(5) are replaced by arsenic atoms.

THE CHEMISTRY OF INORGANIC HOMO-
AND HETEROCYCLES VOL 1 ISBN 0-12-655775-6

$$\begin{array}{ccc} & R_2{}'\ \ R\ \ R_2{}' & \\ & \underset{|}{P}\ \ \underset{|}{P} & \\ R_2B\underset{+}{\overset{-}{\diagup}}\ \underset{|}{B}\ \overset{+}{\underset{|}{\diagdown}}\ \overset{-}{-}BR_2 & & RB{-\!\!-}PR' \\ R_2{}'\overset{+}{\underset{|}{P}}\underset{|}{\diagdown}\ \overset{+}{\underset{|}{P}}\underset{-}{\diagup}{}^{+}PR_2{}' & & R'P{-\!\!-}BR \\ & \underset{|}{B}\ \ \underset{|}{B} & \\ & R_2\ \ R'\ R_2 & \end{array}$$

$$(4) \qquad\qquad\qquad (5)$$

3.2 Cyclodiphosphinoborines $(R_2BPR_2')_2$ and $(RBPR')_2$

The synthetic route to four-coordinate compounds recently exploited is the reaction of a lithium phosphide with a boron halide.[2] At room temperature, the reaction between $LiPEt_2$ and BCl_3 varies with the molar ratio of reactants, yielding as major products compounds (6), (7), X = Cl, and $LiB(PEt_2)_4$ respectively as the ratio increases from 1 : 1 to 3 : 1:

$$\begin{array}{cc} Cl_2B{-}_{-}^{+}PEt_2 & \overset{X}{\diagdown}B{-}_{-}^{+}PEt_2 \\ {}_{|}^{+}\ {}_{-}^{\ }| & Et_2P{\diagdown}{}_{|}^{\ }\ {}_{-}^{|}{\diagup}X \\ Et_2\overset{+}{P}{-\!\!-}BCl_2 & Et_2\overset{+}{P}{-}B \\ & {\diagdown}PEt_2 \end{array}$$

$$(6) \qquad\qquad\qquad (7)$$

At $-78\ ^\circ C$ in ether solution a 1 : 1 ratio of reactants gives compound (6) as a minor product, the major components being $BCl_3.PEt_3$ and $BCl_2(OEt).$ PEt_3, which arise from reaction with the solvent. When the 3 : 1 reaction was carried out at $-50\ ^\circ C$, the product was the dimer $[B(PEt_2)_3]_2$ (7), X = PEt_2, and this compound also results on treating $LiB(PEt_2)_4$ with BCl_3. Reactions of the ethoxy derivative $BCl_2(OEt).PEt_3$, mentioned above, also yield the phosphinoborines (6) and (7), X = Cl or PEt_2, with $LiPEt_2$ in the appropriate molar ratio.

NMR and mass-spectrometric data are available for all these compounds[2] and the structure of an analogue $(Ph_2PBI_2)_2$, previously obtained via the alternative preparative route from Ph_2PH and BI_3,[3] has been determined.[4] The B_2P_2 ring is nonplanar, with c. mm2 symmetry; endocyclic angles at boron and phosphorus are 88.2° and 87.9° respectively and one boron–phosphorus distance (194 pm) is significantly shorter than the other three (mean 201 pm).

Few reactions of these compounds have been reported but with $LiAlH_4$, compounds (6) and (7), X = Cl, are converted to the trimeric hydrides $(H_2BPEt_2)_3$ and $[HB(PEt_2)_2]_3$ respectively.[2] The former is obtained only at 150 °C.

Reactions between a lithium phosphide and an aminochloroborane give monomeric products, as shown in (1)–(3),[5] where it is presumed that polymerization does not occur because of intramolecular N → B interaction:

$$(Me_{3-n}H_nSi)_2PLi.2THF + (Me_2N)_2BCl \longrightarrow$$
$$n = 1-3 \qquad LiCl + (Me_{3-n}H_nSi)_2PB(NMe_2)_2 \qquad (1)$$

$$(Me_3Si)_2PLi.2THF + Me_2NBCl_2 \longrightarrow LiCl + (Me_3Si)_2PB(NMe_2)Cl \qquad (2)$$

$$2(Me_3Si)_2PLi.2THF + Me_2NBCl_2 \longrightarrow LiCl + [(Me_3Si)_2P]_2BNMe_2 \qquad (3)$$

The product from reaction (3) loses $(Me_3Si)_3P$ on heating to 150 °C, giving the second example of a cyclic diphosphadiboretane (8) in near quantitative yield:

$$[(Me_3Si)_2P]BNMe_2 \longrightarrow 2(Me_3Si)_3P + \begin{array}{c} Me_2NB\!-\!\!-\!PSiMe_3 \\ | \qquad\quad | \\ Me_3SiP\!-\!\!-\!BNMe_2 \end{array} \qquad (4)$$

$$(8)$$

Similar loss of $(Me_3Si)_3P$ occurs when $[(Me_3Si)_2P]_2BPh$, obtained from $PhBCl_2$ and two moles of $(Me_3Si)_2PLi.2THF$, is distilled to give the B-phenyl derivative of (8), $(Me_3SiBPPh)_2$, as a yellow solid.[5]

3.3 Cyclotriphosphinoborines $(R_2BPR_2')_3$

Since 1969, most new investigations in this area have been concerned with the spectroscopy and reactivity of previously prepared compounds, but one new compound $(Cl_2BPHPh)_2$ (m.p. 257 °C dec) has been isolated by heating a toluene solution of boron trichloride and $PhPH_2$.[6]

The unusual stability of these compounds has been probed by CNDO calculations using the hydrogen derivative $(H_2BPH_2)_3$ as an example.[7] Unexpectedly large cross-ring P...P interactions are present, and these, associated particularly with phosphorus d-orbitals, probably have a significant stabilizing effect.

IR and Raman data for $(X_2BPMe_2)_3$, where X = H,[8] F,[9] Cl, Br, I and Me, can be interpreted on the basis of a C_{3v} (chair) model with ring vibrations at 685, 669, 578 and 547 cm^{-1} for the hydride.[8] For the halides, the highest energy mode decreases from 683 cm^{-1} for $(Me_2PBF_2)_3$ to 593 cm^{-1} for the corresponding iodide.[9] Spectra of the B-monosubstituted derivatives, $(Me_2P)_3B_3XH_5$ for X = Br, I or CN, are similar to those of the parent compound, with the addition of three extra bands due to B–X motions.[9]

A total of 14 compounds from the $(Me_2PBX_2)_3$ and $(Me_2P)_3B_3XH_5$ series have been examined by multinuclear NMR spectroscopy.[10] Boron chemical shifts vary for homogeneously substituted compounds from -3.79 ppm for $X = F$ to 44.17 ppm (from $BF_3.OEt_2$) for $X = I$, and, with the exception of the monocyano derivative, the monosubstituted compounds show the expected two boron signals. Variable-temperature measurements point to ring flipping, which equilibrates the equatorial and axial positions, but such motion is very rapid in symmetrically substituted compounds and cannot be observed on the NMR time scale.

The structure of the triclinic α-form of $(H_2BPPh_2)_3$ (m.p. 161 °C) has been determined, confirming the chair conformation (c. C_s symmetry) of the ring;[11] mean values of the important ring parameters are B–P 194.8 pm, BPB 114.3° and PBP 112.6°. There is also a β-form (m.p. 179 °C), which crystallizes in the monoclinic system.

Phosphinoborines containing B–H bonds are unexpectedly stable towards hydrolysis, oxidation and thermal decomposition, but they can be halogenated with N-halogenosuccinimides,[12,13] ICl,[12] IBr[12] or bromine[13] itself. With N-bromosuccinimide, for example, $(H_2BPPh_2)_3$ was converted to all products in the series $H_{6-n}Br_nB_3P_3Ph_6$, while with ICl the three products isolated were $H_{6-n}I_nB_3P_3Ph_3$, with $n = 1$–3.[12] When $(H_2BPMe_2)_3$ was the substrate, product mixtures were generally obtained, but there was an increase in selectivity as the atomic weight of the halogen was increased from N-chloro- to N-iodosuccinimide.[13] The products can be separated using the normal fractionation and chromatographic techniques. Halogenation follows a nongeminal substitution path according to IR (absence of BH_2 modes)[12] and NMR[13] evidence, and this is supported by the isolation of (cis–trans) pairs of isomers for $H_3Br_3B_3P_3Ph_6$,[12] $X_2H_4B_3P_3Me_6$, where $X = Br$ or I,[13] $Cl_3H_3B_3P_3Me_6$ and $Cl_4H_2B_3P_3Me_6$.

The halogen atoms in the monosubstituted compounds $XH_5B_3P_3Me_6$ undergo nucleophilic displacements with amides in the presence of water at 100 °C:[14]

$$XH_5B_3P_3Me_6 + RC(O)NR_2' \longrightarrow (RCOO)H_5B_3P_3Me_6 + R_2'NH_2X \quad (5)$$

$X = Br$ or I; R, $R' = H$ or Me.

With silver formate and acetate, the monoiodide is converted to the same formyl and acetoxy products.

The B-halogeno derivatives can be alkylated with boron, aluminium or zinc derivatives.[15] With the fluoride $(F_2BPMe_2)_3$, the best yield of the methyl derivative was obtained with BMe_3, and similarly BEt_3 gave a good yield of $(Et_2BPMe_2)_3$. The heavier halide derivatives, however, reacted most readily with $AlMe_3$, but even with $AlEt_3$, the chloro-, bromo-, and iodo-compounds generally gave mixtures of partially ethylated products.

The B–P ring in $(H_2BPEt_2)_3$ does not react with $LiPEt_2$, in contrast with the ready decomposition of the aluminium analogue $(H_2AlPEt_2)_3$ to $LiAlH_2(PEt_2)_2$.[16]

An excess of Me_3N or Me_3P on the other hand, cleaves the ring in $[H_2BP(CF_3)_2]_3$:[17]

$$[H_2BP(CF_3)_2]_3 + Me_3N \longrightarrow Me_3NBH_2P(CF_3)_2$$
$$+ [(Me_3N)BH_2]^+ [(CF_3)_2PBH_2P(CF_3)_2]^- \tag{6}$$

Products such as $(CF_3)_2PBH_2P(CF_3)_2BH_2P(CF_3)_2^-$ and (base) $BH_2P(CF_3)_2BH_2P(CF_3)_2$ can be identified, if there is a deficiency of base, suggesting that in the initial step one ring bond is cleaved to give an open-chain trimer complex. Ring cleavage also occurs in the borohydride reduction of $(Cl_2BPHPh)_3$ giving a salt considered to be $Na_3[Ph(H_3B)PBCl_2]_3$.[6]

Few uses have been reported for these compounds, but they may be of value as stabilizers for nylon.[18]

3.4 Cyclotetra- and condensed phosphinoborines

Analysis of IR and Raman spectra[19] for $(H_2BPMe_2)_4$ shows consistency with the D_{2d} symmetry found earlier by X-ray crystallography, and a full multinuclear NMR study has been reported for the compound.[10]

The condensed ring compound $H_8BrB_5P_5Me_5$ (see (4)), first reported in 1962, has a decalin-like structure of alternating boron and phosphorus atoms, with chair conformations for the two *cis*-fused cyclohexane rings.[20] The single bromine atom occupies the equatorial site at the 4-position, slightly distorting the ring bonding. Distances associated with this atom are 191.1 and 197.4 pm compared with a mean B–P distance of 194.3 pm. The PBP and BPB angles are 112.2° and 113.6° respectively, indicating a basically σ-bonded system with little electron delocalization.

3.5 Cycloarsinoborines

This area has been little investigated recently. The trimeric compound (9), R = Ph, R' = Me, results when Me_3SiCl is evolved from a mixture of Ph_2BCl and $Me_2AsSiMe_3$ heated to 50 °C,[21] and the analogue with R = Me (m.p. 65 °C) from a reaction between Me_2BBr and $Me_2AsSnMe_3$ in hexane:[22]

$$
\begin{array}{ccc}
\underset{R_2'As}{\overset{R_2}{\underset{\underset{As}{\overset{+}{|}}}{\overset{\overset{B}{|}}{\diagdown}}}}\overset{AsR_2'}{\underset{R_2B}{\overset{+}{\diagup}}BR_2} & \underset{Me_2As}{\overset{Me}{\underset{\underset{B}{\overset{+}{|}}}{\overset{\overset{B}{|}}{\diagdown}}}}\overset{Br}{\underset{AsMe_2}{\overset{+}{\diagup}Me}} & \underset{Me_2As+}{\overset{Me_2}{\overset{H_2B-As-BH_2}{|}}}\overset{+AsMe_2}{\underset{H_2B-As-BH_2}{\overset{|}{Me_2}}}
\end{array}
$$

$$
\begin{array}{ccc}
\quad R'_2 & Br \quad As \quad Br & \\
& Me_2 &
\end{array}
$$

$$
\begin{array}{ccc}
\textbf{(9)} & \textbf{(10)} & \textbf{(11)}
\end{array}
$$

Compound **(10)**, melting at c. 230 °C with decomposition, is the product with methyl dibromoborane as a reactant. Both trimeric **(9)**, R = H, R' = Me (m.p. 49 °C) and tetrameric **(11)** (m.p. 145 °C) compounds, which can be separated by fractional sublimation, are produced during thermal decomposition of the borine adduct $Me_4As_2BH_3$.[22]

References

1 G. E. Coates and J. G. Livingston, *J. Chem. Soc.* **1961**, 5053.
2 G. Fritz and E. Sattler, *Z. Anorg. Allg. Chem.* **413** (1975) 193.
3 W. Gee, R. A. Shaw and B. C. Smith, *Inorg. Syn.* **9** (1967) 19.
4 G. J. Bullen and P. R. Mallinson, *J. Chem. Soc. Chem. Commun.* **1969**, 132; *J. Chem. Soc. Dalton Trans.* **1972** 1143.
5 G. Fritz and W. Hölderich, *Z. Anorg. Allg. Chem.* **431** (1977) 61.
6 E. Mayer and A. W. Laubengayer, *Monatsh. Chem.* **101** (1970) 1138.
7 P. G. Perkins, *Phosphorus* **5** (1974) 31
8 J. D. Odom, M. A. Sens, V. F. Kalasinsky and J. R. Durig, *Spectrochim. Acta* **33A** (1977) 347.
9 J. R. Durig, V. F. Kalasinsky, M. A. Sens and J. D. Odom, *J. Raman Spectrosc.* **5** (1976) 391.
10 M. A. Sens, J. D. Odom and M. H. Goodrow, *Inorg. Chem.* **15** (1976) 2825.
11 G. J. Bullen and P. R. Mallinson, *J. Chem. Soc. Dalton Trans.* **1973**, 1295.
12 W. Gee, J. B. Holden, R. A. Shaw and B. C. Smith, *J. Chem. Soc. A* **1967**, 1545.
13 M. H. Goodrow and R. I. Wagner, *Inorg. Chem.* **15** (1976) 2830.
14 M. H. Goodrow and R. I. Wagner, *Inorg. Chem.* **15** (1976) 2836.
15 M. H. Goodrow and R. I. Wagner, *Inorg. Chem.* **17** (1978) 350.
16 G. Fritz and F. Pfannerer, *Z. Anorg. Allg. Chem.* **373** (1970) 30.
17 A. B. Burg. *Inorg. Chem.* **17** (1978) 593.
18 G. W. Follows, *Ger. Offen.* 2 017 897 (1970) (*CA* **74** (1971) 14126); *US Patent* 3 839 392 (1974) (*CA* **83** (1975) 61452); *US Patent* 3 907 746 (1975) (*CA* **83** (1975) 207481).
19 J. R. Durig, M. A. Sens, V. F. Kalasinsky and S. C. Columbia, *Spectrochim. Acta* **33A** (1977) 893.
20 G. R. Clark and G. J. Palenik, *Aust. J. Chem.* **28** (1975) 1187.
21 E. W. Abel and S. M. Illingworth, *J. Chem. Soc. A* **1969**, 1094.
22 R. Goetze and H. Nöth, *Z. Naturforsch.* **30b** (1975) 875.

4

Boron–Oxygen Heterocycles

IONEL HAIDUC

Babes–Bolyai University, Cluj-Napoca, Romania

4.1 Introduction

Boron–oxygen heterocycles of various ring sizes are known; the regular six-membered B_3O_3 ring shows the greatest tendency of formation and it has the most derivatives known. The boron–oxygen rings reported in the literature are illustrated below (only the ring types are shown):

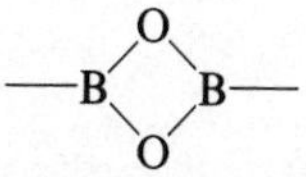

(1) 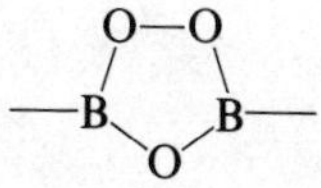 (2) 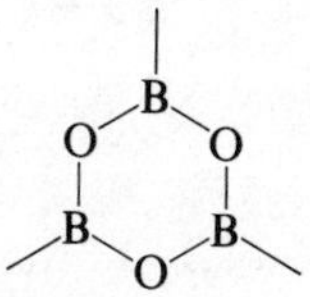 (3) 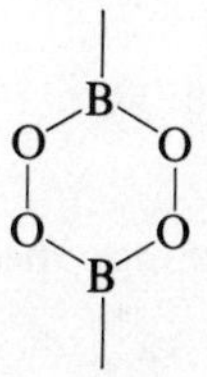 (4)

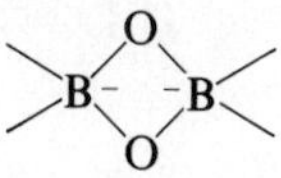

$(1a)$ 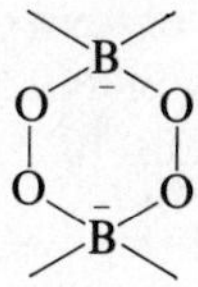 $(4a)$

In real compounds boron may become four-coordinate by addition of supplemental groups, for example OH. In particular, the four-membered

THE CHEMISTRY OF INORGANIC HOMO-
AND HETEROCYCLES VOL 1 ISBN 0-12-655775-6

ring (1) and the six-membered ring (4) are in fact known only with four-coordinate boron.

The boron–oxygen heterocycles have been reviewed.[B 1 – B 22]

The recent Gmelin Handbook volumes[B 2 – B 4] should be consulted for comprehensive data about individual compounds. This chapter supplements the review of boron–oxygen heterocycles published in ref. B 1 of the Bibliography, with the literature published after 1968, including some earlier references that were omitted in that book.

4.2 Cyclodiboroxanes

The B_2O_2 ring has been mentioned in the literature, but its chemistry is very limited. It seems to be present only in some sterically crowded, dimeric molecules, derived from heteroborahomoadamantane (5), containing four-coordinate boron:[1]

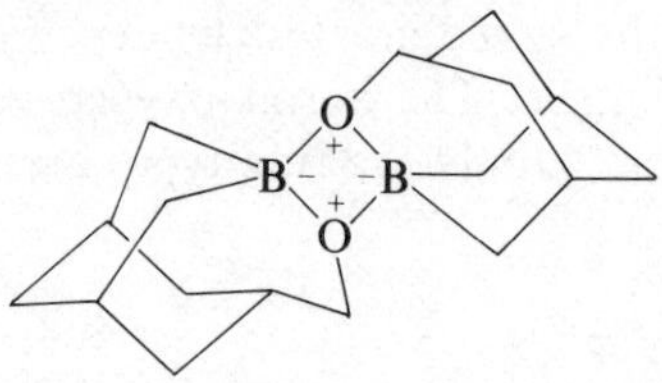

(5)

Apparently, there is no B_2O_2 ring derivative containing three-coordinate boron.

4.3 Cyclodiboratrioxanes

The parent compound of this class (6), R = H, has been observed as an intermediate in the gas-phase oxidation of various boranes, for example diborane(6),[2] tetraborane(10),[3,4] pentaborane(9),[5] borane adducts like $H_3B.PF_3$[4] or $H_3B.CO$:[3,4,6]

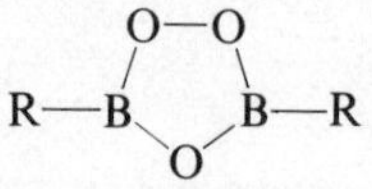

(6)

Oxidation of cyclotriboroxane $(HBO)_3$ also leads to $H_2B_2O_3$.[3,7] The kinetics of the reaction

$$(HBO)_{3(g)} + \tfrac{1}{2} O_2 \longrightarrow H_2B_2O_{3(g)} + \tfrac{1}{3} B_2O_{3(s)} + \tfrac{1}{6} B_2H_{6(g)} \qquad (1)$$

has been investigated.[7]

The compound $H_2B_2O_3$ is unstable in condensed phases and decomposes (sometimes explosively) with formation of B_2O_3 and hydrogen.[5,8] The ultraviolet[9] and infrared spectra (vibrational analysis)[8,10] have been investigated. The molecular structure was determined by microwave spectroscopy:[11] B–O 1.380 Å (in the B–O–B fragment) and 1.365 Å (in the B–O–O fragment); the bond angles are BOB 104°, OBO 113.5°, BOO 105°.[11] The bonding in $H_2B_2O_3$ has been discussed in terms of π-electron delocalization, on the basis of molecular-orbital calculations.[12] Aromaticity (π-electron delocalization) in the B_2O_3 ring was suggested on the basis of NMR and mass-spectral data.[13] Other theoretical calculations on $H_2B_2O_3$[14–16] and its dichloro (6), R = Cl, and dimethyl (6), R = Me, derivatives[15] have also been performed; thus the energies and oscillator strengths[14], ionization potentials,[14] dipole moments,[14,16] orbital energies,[15] charge distributions[15] and bond orders[15,16] were calculated using semiempirical CNDO/2[15,16] and INDO[16] methods. The results suggest π-electron delocalization, interrupted over the O–O bond.[16]

The photolysis of $H_2B_2O_3$ in a low-temperature argon matrix with 1480 Å radiation yields monomeric HBO, B_2O_3 and another unidentified compound.[17]

The dimethyl derivative (6), R = Me, is formed in the gas-phase oxidation of trimethylborane[18] and 1,1-dimethyldiborane(6) at 80 °C.[19] The compound was isolated and characterized by infrared, mass-spectrometry, NMR, vapour density, vapour pressure, heat of vaporization and melting point.[19] The mass and NMR spectra suggest aromatic character for the B_2O_3 ring.[19]

An unsymmetrical derivative $HMeB_2O_3$ was identified by mass and infrared spectroscopy in the oxidation product of methyldiborane(6) MeB_2H_5.[20]

4.4 Cyclotriboroxanes

The six-membered B_3O_3 ring (3) is the parent skeleton for an extensive class of boron–oxygen heterocyclic derivatives (7):

(7)

The basics of their chemistry were established before 1970 (see ref. B 1 of the Bibliography); most of the progress achieved in the following decade is related to the structural chemistry of cyclic borates, derived from the B_3O_3 ring, and to a lesser extent to the chemistry of organic derivatives of this ring. Although IUPAC accepted the trivial name *boroxine* for the B_3O_3 heterocycle,[21] we shall continue to use here the term *cyclotriboroxane*, suggested earlier (see ref. B 1).

4.4.1 Inorganic derivatives

Cyclotriboroxane $(HBO)_3$, the parent hydride, is stable only in the gaseous phase, in the presence of a diluting inert gas. In the condensed state it is thermodynamically unstable at room temperature and disproportionates into B_2O_3 and B_2H_6. The compound is formed in the oxidation of pentaborane(9) with oxygen deficiency,[22,23] in the photolytic reaction between diborane(6) and $H_2B_2O_3$,[2,9,23] in the pyrolysis of the $H_3B.CO$ adduct,[6] in the high-temperature reaction between water vapour and solid boron at 850 °C[22] or between water vapour and a mixture of boron and boron trioxide melt at 1400 K.[2] In the formation of $(HBO)_3$ from B_5H_9, the intermediate compounds HBO and H_2BOH were detected by mass spectrometry.[3] In the hydrogen–boron–oxygen system, at temperatures above 800 °C, the compounds $H_4B_4O_4$ and $H_4B_6O_7$ were also identified by mass spectrometry.[24] It has been suggested that these are derivatives of the B_3O_3 ring:

$H_4B_4O_4$ $H_4B_6O_7$

(8)   (9)

The molecular structure of $(HBO)_3$, containing a planar B_3O_3 ring (D_{3h} symmetry; B–O 1.376 Å) has been confirmed by electron diffraction in the gaseous phase.[25] The infrared spectrum of $(HBO)_3$ has been investigated in detail, using deuterium, ^{10}B and ^{18}O isotopically enriched species, both in the gaseous phase[22] and in an argon matrix[9,26]. Force constants and the potential-energy distribution have been calculated ($k_{BO} = 5.63$ mdyn/Å; $k_{BH} = 3.68$ mdyn/A).[27] The free-energy function $(F\text{-}H)/T$ and enthalpies of $(HBO)_3$ have been calculated for a large range of temperatures.[29] For the Raman spectrum of $(HBO)_3$ see ref. 28.

The heat of formation of $(HBO)_3$ has been determined as $\Delta H_f^{\circ} = -307 \pm 8$ kcal/mol by mass spectrometry.[30,31]

Cyclotriboroxane reacts with boron halides to give HBX_2 and boron trioxide[32,33] and with phosphorus trifluoride in the gas phase to form the adduct $H_3B.PF_3$:[34]

$$(HBO)_3 + BX_3 \longrightarrow HBX_2 + B_2O_3 + HX + \tfrac{1}{2} H_2 \tag{2}$$

$$(HBO)_3 + PF_3 \longrightarrow H_3B.PF_3 + B_2O_3 \tag{3}$$

With dialkyldiazo derivatives, cyclotriboroxane forms $R_2N_2.BH_3$ and diboron trioxide.[35]

Halogen derivatives. The halogenocyclotriboroxanes $(XBO)_3$ ($X = F$, Cl, Br) are unstable at room temperature and are in equilibrium with their starting materials:[36]

$$B_2O_3 + BX_3 \rightleftharpoons (XBO)_3 \tag{4}$$

Kinetic stabilization can be achieved at low temperature, for example $(FBO)_3$ can be studied in the solid state below $-135\ ^\circ C$. On the other hand, mass spectrometry can be used to confirm the molecular size at higher temperatures.[36,37] The heat of formation of $(FBO)_3$ has been calculated as -525.8 kcal/mol with the aid of an MNDO method.[38] A direct experimental determination from equilibrium data gave a value of $\Delta H_f^{\circ} = -567.8$ kcal/mol.[39] The compound $(FBO)_3$ is also formed in the reaction of SiO_2 with boron trifluoride at $450\ ^\circ C$.[40] A vibrational analysis was performed on the spectroscopic data obtained for $(FBO)_3$ in the gas phase[41] and in the low-temperature solid state[42,43], and Urey–Bradley force constants were calculated. Vibrational assignments and force constants for $(ClBO)_3$, $F_2ClB_3O_3$ and $FCl_2B_3O_3$ have also been reported.[42] The coefficients for calculating the free-energy functions for $(FBO)_3$ have been tabulated.[29] For mass spectra of $(XBO)_3$ see ref. 23.

Trihydroxocyclotriboroxane $(HO)_3B_3O_3$ is the trimer of the so-called (hypothetical as monomer) "metaboric acid" HBO_2. It exists in three modifications: orthorhombic α-HBO_2 (or HBO_2-III) is trihydroxocyclotriboroxane $(HO)_3B_3O_3$; monoclinic β-HBO_2 (or HBO_2-II) is a chain polymer of composition $\{H[B_3O_4(OH)_2]\}_n$, and cubic γ-HBO_2 (or HBO_2-I) is a tridimensional polymer $[B_3O_3(OH)_3]_x$, consisting of linked BO_4 tetrahedra (see ref. B 4 of the Bibliography for details). The stable polymorph is γ-HBO_2.[44] The anion $B_3O_6{}^{3-}$, derived from trihydroxocyclotriboroxane is present in some cyclic metaborates (see below).

Trihydroxocyclotriboroxane is formed on heating orthoboric acid $B(OH)_3$ at $100\ ^\circ C$[45] or in refluxing toluene with azeotropic removal of water[46] and in the pyrolysis of $(RO)_3B_3O_3$ derivatives[45]. The thermal decomposition of $B(OH)_3$ has been investigated by thermogravimetric and differential thermal analysis; the "metaboric acid" HBO_2 was an intermediate, the final product being $H_2B_4O_7$.[47] The compound $(HO)_3B_3O_3$ was detected by ^{1}H NMR spectroscopy in acetone–water–$B(OH)_3$ solutions[48] and in the reaction products of boric acid with urea[49].

The formation enthalpy has been determined as $\Delta H_f^o = -189$ kcal/mol.[50] The deuterated compound $(DO)_3B_3O_3$ is formed by recrystallization of $B(OH)_3$ from D_2O, followed by dehydration between 95 and 140 °C.[51]

The vibration spectra of $(HO)_3B_3O_3$ and its isotopomers $^{10}B_3O_3(OH)_3$ and $(DO)_3B_3O_3$ in the solid state have been investigated,[52–54] and a potential-force-field calculation based on vibration spectroscopic data has been performed[55].

The mass spectrum of $(HO)_3B_3O_3$ has also been reported.[24]

Cyclic $(HO)_3B_3O_3$ reacts with nickel(II) dimethylglyoxime to form a complex polycyclic structure in which the nickel chelate is coupled with cyclotriboroxane rings.[56]

Fluoroborates. Cyclic anions containing fluorine have been little investigated. The reaction of boric acid with potassium and ammonium fluorides was reported to give a fluoroborate containing the anion $[B_3O_3F_4(OH)]^{2-}$ (10) among other products.[57] The anion $[B_3O_3F_6]^{3-}$ (11) has been identified in solution, and some of its salts have been prepared.[58] The hydrolytic stability of this anion has also been investigated.[58] A bicyclic fluoroborate anion (12) was mentioned as a condensation product of fluorohydroxoborates.[59]

(10)   (11)   (12)

4.4.2 Cyclic borates

Crystal structures. A large number of borate structures have been determined by single-crystal X-ray diffraction, and most of them were found to contain the B_3O_3 ring, either in monocyclic anions, in bi- and tricyclic anions, or in polymeric structures containing an indefinite number of rings linked in chains, sheets or networks. The principles governing the structures of borates, first formulated by C. L. Christ in 1960[60] were revised and extended several times in recent years (see refs. B 8–B 10, B 17, B 18 in the Bibliography). These principles afford the understanding of borate structures, but not the prediction of the structure when the composition is known (except for the simplest cases, mostly hydrated borates). As a consequence, the borates are usually formulated as empirical ratios of oxides, and only when the crystal structure has been determined can a structural formula be used. Thus meyerhofferite can be written as $2CaO.3B_2O_3.7H_2O$ (empirical oxide formula) or as $Ca[B_3O_3(OH)_5].H_2O$ (structural formula); similarly, borax $Na_2B_4O_7.10H_2O$ (meaningless trivial formula) can be written either as $Na_2O.2B_2O_3.10H_2O$ (empirical oxide formula) or as $Na_2[B_4O_5(OH)_4].8H_2O$ (structural formula). The composition of the two compounds cited can be further abbreviated by classifying them as $2:3:7$ and $1:2:10$ borates respectively,[B 5] thus indicating the ratio between metal oxide, boron trioxide and water. Obviously, the structural formulae are preferable, but their use is possible only for insular species (containing isolated anions), usually hydrated borates, and it is rather difficult for polymeric (anhydrous) borates.

The principles governing the structures of borates are outlined below, and illustrated with examples established after 1970. Literature on noncyclic species is not cited; for references see B 5.

(a) The basic structure unit in borates is the BO_3 (trigonal planar unit), present in boric acid $B(OH)_3$.

(b) The boron atom can achieve four-coordination by addition of an extra OH group, thus gaining a negative charge for the tetrahedral species formed $[B(OH)_4]^-$. This anion is known in natural borates, like hexahydroborite $Ca[B(OH)_4]_2.2H_2O$, and frolovite $Ca[B(OH)_4]_2$.

(c) Negative charges in borates can appear either by proton dissociation (from B–OH groups) as in $Ca_3[BO_3]_2$ (which contains the BO_3^{3-} anion), by formation of tetrahedral units as shown above under (b), or by both ways as in sinhalite $MgAl[BO_4]$ (which contains the BO_4^{5-} anion).

(d) Both trigonal and tetrahedral units can undergo condensation (with H_2O elimination) to form polynuclear species with triangles or tetrahedra sharing edges. The simplest case is the formation of diborate species:

$$B(OH)_3 + B(OH)_3 \longrightarrow [B_2O(OH)_4] + H_2O \qquad (5)$$

$$B(OH)_3 + [B(OH)_4]^- \longrightarrow [B_2O(OH)_5]^- + H_2O \qquad (6)$$

$$[B(OH)_4]^- + [B(OH)_4]^- \longrightarrow [B_2O(OH)_6]^{2-} + H_2O \qquad (7)$$

The anion $[B_2O(OH)_6]^{2-}$ was found in pentahydroborite $Ca[B_2O(OH)_6].2H_2O$.

Proton dissociation from these species leads to diboron anions like $[B_2O_4(OH)]^{3-}$, present in acharite $Mg_2(OH)[B_2O_4(OH)]$, and $[B_2O_5]^{4-}$, present in saunite $Mg_2[B_2O_5]$ and kurchatovite $CaMg[B_2O_5]$.

$$B(OH)_3 \qquad [B(OH)_4]^- \qquad [BO_3]^{3-} \qquad [BO_4]^{5-}$$

$$[B_2O(OH)_4] \qquad [B_2O(OH)_5]^- \qquad [B_2O(OH)_6]^{2-}$$

$$[B_2O_4(OH)]^{3-} \qquad [B_2O_5]^{4-}$$

(e) Condensation of three trigonal or/and tetrahedral units results in the formation of cyclic borates, (13)–(20), containing the B_3O_3 ring:

$$B_3O_3(OH)_3 \qquad [B_3O_3(OH)_4]^- \qquad [B_3O_3(OH)_5]^{2-}$$
$$(13) \qquad\qquad (14) \qquad\qquad (15)$$

$$[B_3O_3(OH)_6]^{3-} \qquad [B_3O_6]^{3-} \qquad [B_3O_5(OH)_2]^{3-}$$
$$(16) \qquad\qquad (17) \qquad\qquad (18)$$

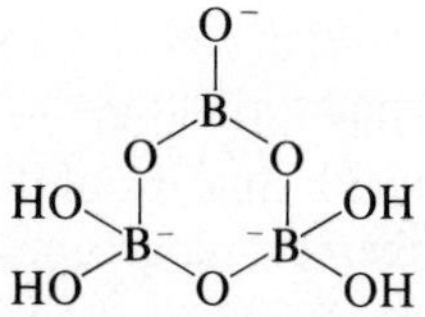

$[B_3O_4(OH)_4]^{3-}$

(19)

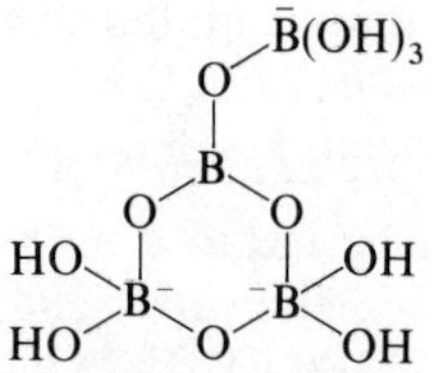

$[B_3O_3(OH)_4OB(OH)_3]^{3-}$

(20)

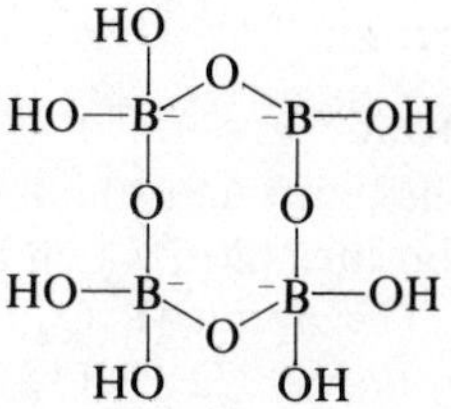

$[B_4O_4(OH)_7]^{3-}$

(21)

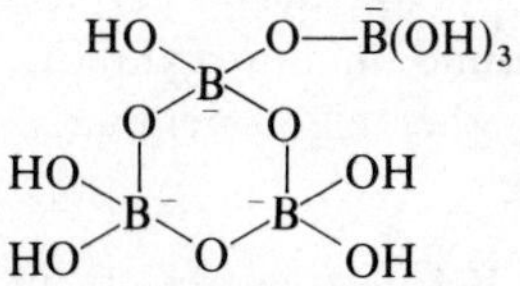

$[B_3O_3(OH)_5OB(OH)_3]^{4-}$

(22)

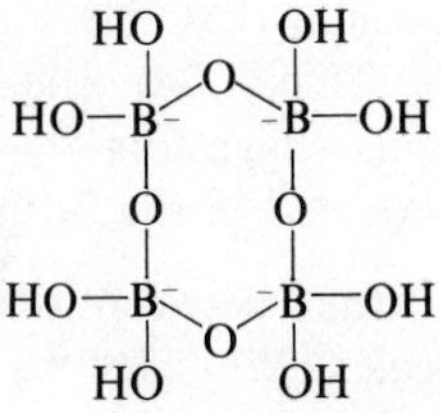

$[B_4O_4(OH)_8]^{4-}$

(23)

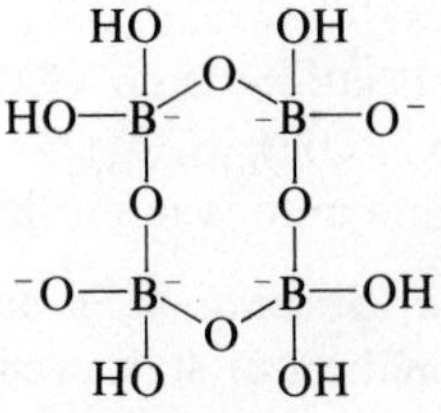

$[B_4O_6(OH)_6]^{6-}$

(24)

The $B_3O_3(OH)_3$ molecule is that of α-metaboric acid, discussed above. The anion $[B_3O_6]^{3-}$ **(17)**, derived from it by proton dissociation as shown under (c), was found in potassium metaborate $K_3[B_3O_6]$,[61] caesium metaborate $Cs_3[B_3O_6]$,[62] and barium metaborate $Ba_3[B_3O_6]_2$.[63]

The anion $[B_3O_4(OH)_4]^{3-}$ **(14)** is present in ameghinite $Na[B_3O_3\text{-}(OH)_4]$[64,65] and the anion $[B_3O_5(OH)_2]^{3-}$ **(18)**, derived from **(14)** by proton dissociation, was established in a synthetic borate $Na_3[B_3O_5(OH)_2]$.[66]

The anion $[B_3O_3(OH)_5]^{2-}$ **(15)**, previously discovered in inyioite and meyerhofferite,[B1] was also found in inderite $Mg[B_3O_3(OH)_5].5H_2O$ (monoclinic),[67,68] kurnakovite $Mg[B_3O_3(OH)_5].5H_2O$ (triclinic),[69,70] inderborite $CaMg[B_3O_3(OH)_5]_26H_2O$[71] and synthetic $Zn[B_3O_3(OH)_5].H_2O$.[72,73]

The anion $[B_3O_4(OH)]_4]^{3-}$ **(19)**, derived from **(15)** by proton dissociation, has been identified in synthetic $K_3[B_3O_4(OH)_4].2H_2O$,[74–76] $Rb_3[B_3O_4(OH)_4].2H_2O$[77,78] and in the mineral solongoite $Ca_2[B_3O_4(OH)_4]Cl$.[79–81]

The anion $[B_3O_3(OH)_6]^{3-}$ **(16)** was discovered in nifontovite

$Ca_3[B_3O_3(OH)_6]_2 2H_2O$, a compound thoroughly investigated by X-ray diffraction.[80,82-86]

(*f*) Side groups of the type $-O-\bar{B}(OH)_3$ (tetrahedral units) can be attached through an oxygen bridge to a six-membered ring, resulting in tetraborates (**20**), which are isomeric with eight-membered ring compounds (**21**). The anion $[B_3O_3(OH)_4 \cdot OB(OH)_3]^{3-}$ (**20**) was discovered in hydrochloroborite $Ca_2[B_3O_3(OH)_4 \cdot OB(OH)_3]Cl \cdot 7H_2O$.[87-89] The anion $[B_3O_3(OH)_5 \cdot OB(OH)_3]^{4-}$ (**22**), found in uralborite $Ca_2[B_3O_3(OH)_5 \cdot OB(OH)_3]$, is formed in a similar manner;[90,91] this anion is isomeric with an eight-membered ring anion (**23**), which has yet to be discovered in any borate.

(*g*) Eight-membered rings B_4O_4 are not usually present in borate structures. An exception is the anion $[B_4O_6(OH)_6]^{6-}$ (**24**), derived by proton dissociation from (**23**). The anion (**24**) was found in borcarite Ca_4Mg-$[B_4O_6(OH)_6](CO_3)_2$,[92,93] thus suggesting that other eight-membered-ring borate anions might be discovered in the future.

(*h*) Sharing of a boron atom between two B_3O_3 rings results in the formation of spirocyclic species (**25**)–(**28**):

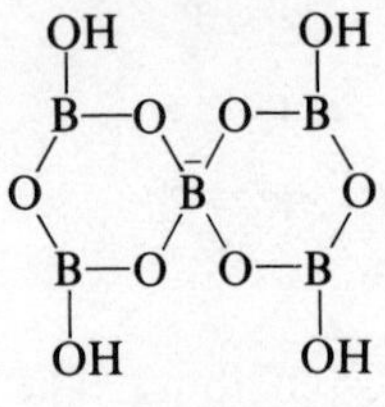

$[B_5O_6(OH)_4]^-$

(**25**)

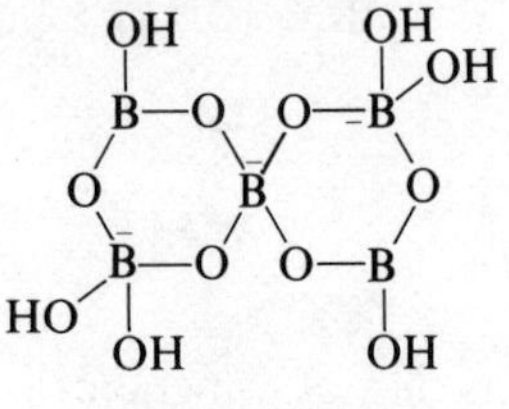

$[B_5O_6(OH)_6]^{3-}$

(**26**)

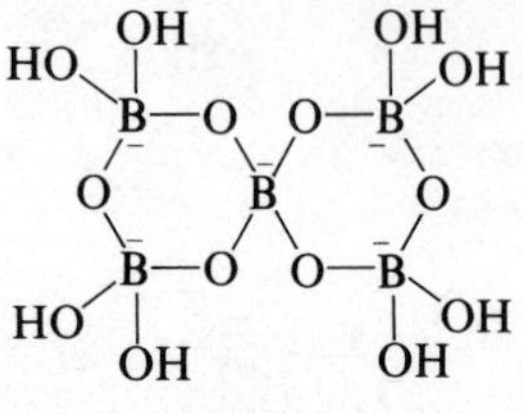

$[B_5O_6(OH)_8]^{5-}$

(**27**)

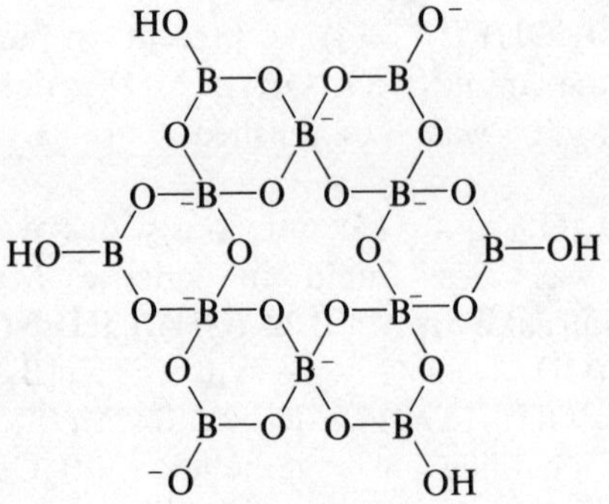

$[B_{12}O_{20}(OH)_4]^{8-}$

(**28**)

The basic structure of this type is that of the anion $[B_5O_6(OH)_4]^-$ (25), but addition of OH groups to boron atoms (converting them from trigonal to tetrahedral units) results in the formation of anions $[B_5O_6(OH)_{4+n}]^{(n+1)-}$ (where $n = 1, 2, 3$ or 4), for example the anions (26) and (27).

The anion $[B_5O_6(OH)_4]^-$ (25) has been found in many pentaborates, including synthetic $Na[B_5O_6(OH)_4]$,[94] sborgite $Na[B_5O_6(OH)_4].3H_2O$,[95] santite $K[B_5O_6(OH)_4].2H_2O$,[96,97] synthetic $Cs[B_5O_6(OH)_4].2DMSO$,[98] synthetic β-$NH_4[B_5O_6(OH)_4].2H_2O$[99] and synthetic $Tl[B_5O_6(OH)_4].$ $2H_2O$.[100] The anion $[B_5O_6(OH)_6]^{3-}$ (26) was found in ulexite $NaCa[B_5O_6(OH)_6].5H_2O$.[101]

An amazing example of spiro-condensation was found in a synthetic dodecaborate $Na_8[B_{12}O_{20}(OH)_4]$,[102] which contains six B_3O_3 rings joined by common tetrahedral boron atoms (28).

(i) Fused B_3O_3 rings with two joint oxygen-bridged boron atoms, can form bicyclic anions. The prototype is the $[B_4O_5(OH)_4]^{2-}$ anion (29), present in borax $Na_2[B_4O_5(OH)_4].8H_2O$[103-105] and also identified in the following borates: tincalconite $Na_2[B_4O_5(OH)_4].3H_2O$,[106] synthetic $K_2[B_4O_5(OH)_4].2H_2O$,[107] synthetic $(NH_4)_2[B_4O_5(OH)_4].2H_2O$,[108] hungchaoite $Mg[B_4O_5(OH)_4].7H_2O$[88,109] and synthetic $K_2Ca[B_4O_5(OH)_4]_2.$ $8H_2O$,[110,111] $(NH_4)_2Ca[B_4O_5(OH)_4]_2.8H_2O$,[112] $Rb_2Sr[B_4O_5(OH)_4]_2.$ $8H_2O$[113,114] and $Mn[B_4O_5(OH)_4].7H_2O$.[115-117]

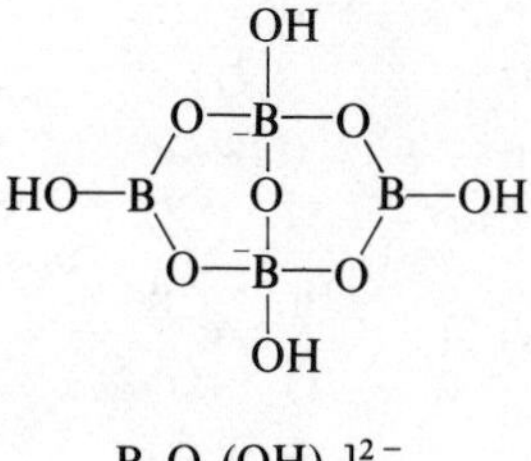

$B_4O_5(OH)_4]^{2-}$

(29)

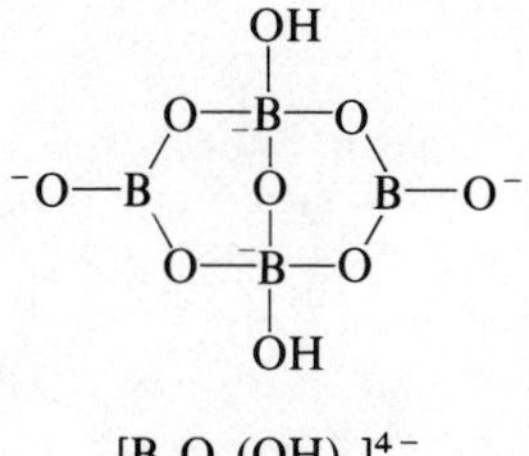

$[B_4O_7(OH)_2]^{4-}$

(30)

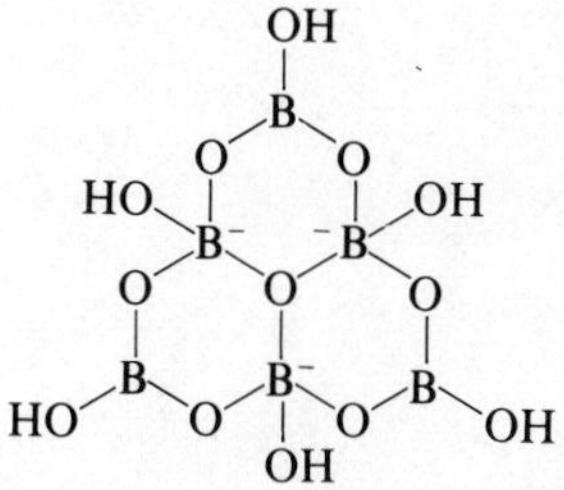

$[B_6O_7(OH)_6]^{2-}$

(31)

Other anions may be derived from (29) by transformations described under (c) (proton dissociation) or (d) (OH addition). An example is the anion $[B_4O_7(OH)_2]^{4-}$ (30), which is present in roweite $Ca_2Mn_2(OH)_4$-$[B_4O_7(OH)_2]$.[118,119]

(j) Fused tricyclic systems, containing three B_3O_3 rings sharing three boron atoms (four-coordinate) and an oxygen atom (three-coordinate) can be formed. The tricyclic anion $[B_6O_7(OH)_6]^{2-}$ (31) thus formed was found in the following crystalline hexaborates:

rivadavite	$Na_6Mg[B_6O_7(OH)_6]_4.10H_2O^{120}$
aksaite	$Mg[B_6O_7(OH)_6].2H_2O^{121,122}$
synthetic	$Mg[B_6O_7(OH)_6].3H_2O^{123,124}$
synthetic	$Mg[B_6O_7(OH)_6].4H_2O^{125}$
macallisterite	$Mg[B_6O_7(OH)_6].4.5H_2O^{126}$
synthetic	$Co[B_6O_7(OH)_6].7H_2O^{127}$
synthetic	$CoK_2[B_6O_7(OH)_6]_2.4H_2O^{128}$
synthetic	$Ni[B_6O_7(OH)_6].5H_2O^{129,130}$
synthetic	$Ni[B_6O_7(OH)_6].4.5H_2O^{131}$
synthetic	$Ni[B_6O_7(OH)_6].7H_2O^{127,128,132}$
terrugite	$Ca_4Mg(AsO_4)_2[B_6O_7(OH)_6]_2.14H_2O^{133}$

(*k*) Any of the insular (isolated) mono-, bi- and tricyclic anions listed above can undergo further condensation to form polymeric structures in which the cyclic units are linked through oxygen bridges:

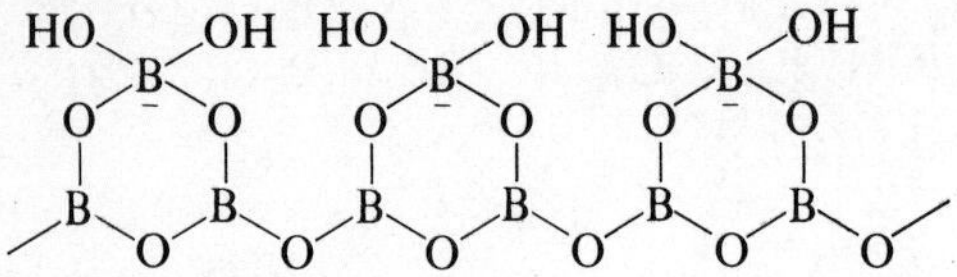

(32)

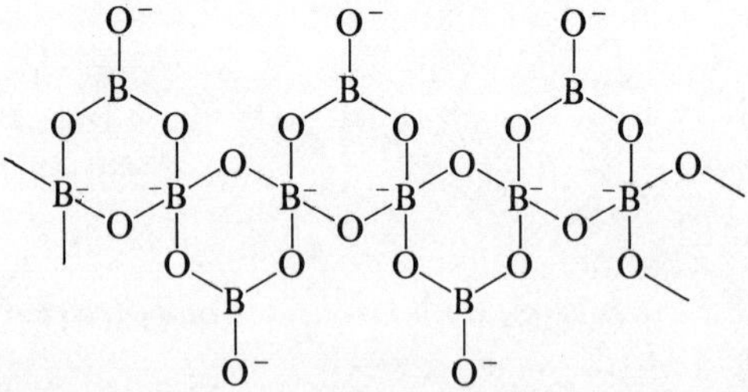

(33)

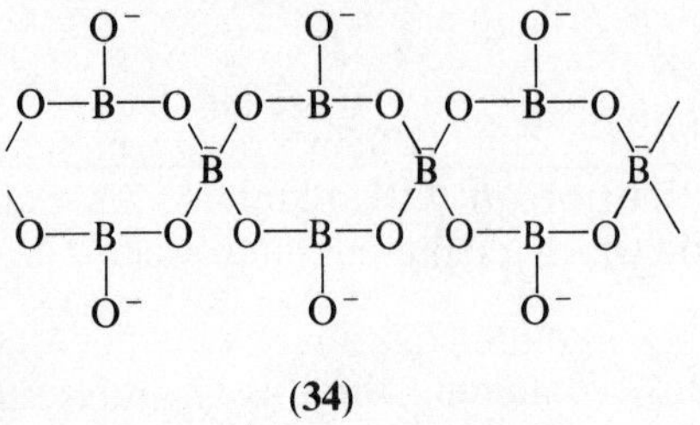

(34)

The simplest example would be a chain consisting of B_3O_3 rings (32), but connection through spiranic boron atoms (tetrahedral) is also possible, as in calciborite $Ca_2[B_4O_8]_2$ (33).[134] Another example of a spirocyclic polymeric structure, containing a rare eight-membered ring B_4O_4, is the compound $La[B_3O_6]$ with the atomic arrangement shown in (34).[135]

Much more complicated structures have been discovered by X-ray diffraction studies of many (mostly anhydrous) borates. These may contain identical or different cyclic units, linked in monodimensional (chains), bidimensional (layers) or tridimensional (networks) polymeric structures; individual examples can be found in the review literature.[B 2 – B 7] There is little room available here to deal in more detail with these polycyclic structures.

In some cases the number of connected ring units is limited to a small value. Thus in ammonioborite $(NH_4)_3[(B_5O_6)_3O_2(OH)_8].4H_2O$, which contains an isolated anion $[B_{15}O_{20}(OH)_8]^{3-}$, there are three spirocyclic groups as building units (35).[136]

(35)

The combination of rules (a)–(k) may lead to a considerable number of new borate structures, and none should be taken as unexpected, although some of the recent discoveries have indeed been surprising.

Cyclic borates in aqueous solutions. In aqueous solution boric acid undergoes complex equilibria, involving addition of hydroxo groups and condensation, with the formation of cyclic borates, for example

$$B(OH)_3 + OH^- \rightleftharpoons [B(OH)_4]^- \tag{8}$$

$$[B(OH)_4]^- + 2\ B(OH)_3 \rightleftharpoons [B_3O_3(OH)_4]^- + 3\ H_2O \tag{9}$$

$$[B(OH)_4]^- + 4\ B(OH)_3 \rightleftharpoons [B_5O_6(OH)_4]^- + 6\ H_2O \tag{10}$$

The exact composition of such a solution is a function of concentration, pH, state of ageing, presence of other cations and anions, and other factors. A great deal of research effort has been spent in investigating these equilibria, using a wide range of methods to identify the species present in solution. The data have been correlated with solid-state studies. These include acidity measurements and solubility data,[136-139] salt cryoscopy,[140] nuclear magnetic resonance[141-149] and vibrational spectroscopy (infrared and Raman).[150-165] The results of these investigations showed that the aqueous solutions contain the same anions as the hydrated (insular) crystalline borates. For example, Raman spectroscopy clearly demonstrated the presence of $[B_3O_3(OH)_4]^-$, $[B_5O_6(OH)_4]^-$ at pH = 6 and of $[B_3O_3(OH)_4]^-$, $[B_4O_5(OH)_4]^{2-}$ and $[B(OH)_4]^-$ at pH = 9.[58,165]

Infrared and other spectroscopic methods, sometimes associated with X-ray diffraction, were used to identify the cyclic borates obtained by hydrothermal preparations[166-169] or by hydrolysis of boric-acid esters $B(OR)_3$ in the presence of quaternary ammonium bases.[170-173]

4.4.3 Organic derivatives (RBO)₃

Although no essentially new reactions for the synthesis of cyclotriboroxanes have been described, much work related to their preparation has been published.

The Grignard reagents continue to be much used in the preparation of cyclotriboroxanes. Thus Grignard reagents react with trialkoxyboranes[174-182] to give cyclotriboroxanes. Aryl derivatives were found in the reaction of refluxing alkyl or aryl halides with magnesium in THF in the presence of borane.[183] The reaction of Grignard reagents with the BH_3.THF adduct also produced organocyclotriboroxanes $(RBO)_3$.[184,185]

Organolithium reagents can be used instead of Grignard reagents in the synthesis of cyclotriboroxanes.[186]

The dehydration of boric acids, $RB(OH)_2$, prepared by various procedures, is a convenient method for the preparation of organocyclotriboroxanes.[187-190] The products of hydrolysis reactions of many functional organoboron derivatives are boronic acids, which are converted (often without isolation) into their anhydrides, i.e. organocyclotriboroxanes. This reaction was used with organoboron halides $RBCl_2$ (e.g. $R = n\text{-}C_5H_{11}$,[191] ferrocenyl,[192] 2-HS—C_6H_4—C_6H_4 [193]); borazanes $(PhBNPh)_n$[194], and cyclodiboratetrazanes.[195]

Phenylboron diiodide reacts with dimethylsulphoxide in carbon disulphide to give a 98% yield of $(PhBO)_3$.[196]

Mercapto derivatives, for example $PhB(SR)_2$, extract oxygen from aldehydes to give $(PhBO)_3$.[197]

Some organoboron compounds containing B–H bonds can be converted to cyclotriboroxanes. Thus substituted diboranes $(RBH_2)_2$ hydrolyse to

$(RBO)_3$ (R = –CHCl—CHMe$_2$).[198] Refluxing of 1,1-dimethyl-1,2-azabor-olidine with t-butanol or in wet THF gave a cyclotriboroxane:[199]

$$3 \quad \underset{Me_2N{\rightarrow}BH_2}{\ce{\bigcirc}} \quad \xrightarrow[\text{or } H_2O/THF]{Bu^tOH} \quad (RBO)_3 + R = CH_2CH_2CH_2NMe_2 \quad (11)$$

Alkenylcyclotriboroxanes have been obtained via hydroboration of acetylenes, followed by hydrolysis and dehydration:[200]

$$\text{(12)}$$

Decaborane forms adducts with several carbonyl compounds (aldehydes and ketones) which are cleaved in the presence of water into alcohols, boric acid and hydrogen. Olefins and cyclotriboroxanes are formed as by-products.[201]

Thermal reactions (pyrolysis) of various organoboron–oxygen compounds, for example 1,3,2-dioxaborolane-4-ones,[202] 2-methyl-1,3,2-dioxaborolandione[203] and 1,3-dioxaborinanes,[204] yield organocyclotriboroxanes. Gas-phase thermolysis of N-oxide derivatives PhRCH—N(OBR$_2$)Me (R = Et) gives (EtBO)$_3$, PhCH=NMe and butane.[205] The diboroxanes R$_2$B—O—BR$_2$ readily undergo symmetrization, with formation of cyclotriboroxanes (RBO)$_3$ and triorganoboranes BR$_3$ (e.g. R = Bu).[206]

Triorganoboranes react with carbon monoxide in diglyme on heating to form cyclotriboroxanes.[207–209] In the reaction of triorganoboranes with isonitriles in the presence of aldehydes, cyclotriboroxanes are also formed.[210,211]

A complex boracyclohexene derivative treated with a MeOH–AcOH mixture at 80 °C gave a trimeric derivative, containing the B$_3$O$_3$ ring, in addition to other products (R = Me, Bu).[212–215]

Some organoboron derivatives of hydrazine react with aldehydes and ketones to form cyclotriboroxanes:[216]

$$3 \ R_2B—NHNH—BR_2 + 6 \ R'R''CO \longrightarrow$$
$$2 \ (RBO)_3 + 6 \ RH + 3 \ (R'R''C{=}N—)_2 \quad (13)$$

4.4.4 Functional derivatives

Tris(alkoxy)cyclotriboroxanes (ROBO)$_3$ are conveniently prepared by heating diboron trioxide with trialkoxy(aroxy)boranes B(OR)$_3$ in an organic solvent (R = Me, Et, Pr^n, Pr^i, Bu^n, alkenyl)[217-221] or by directly reacting diboron trioxide with alcohols, under controlled conditions (to avoid formation of trialkoxyborates) (R = Et, Pr^i, Pr$_2^i$CH, PhCH=CHCH$_2$, (C$_6$H$_{11}$)MeCH).[222-225]

Refluxing boric acid with high-boiling alcohols such as cyclobutylcarbinol, cyclohexylcarbinol,[226] 3,3-dimethyl-2-butanol,[226] 2-chloroethanol[227] and tridecanol-2,[228] also yields trialkoxycyclotriboroxanes.

Trialkoxycyclotriboroxanes are also formed during the oxidation of paraffins in the presence of boric acid,[229,230] in the oxidation of trialkylboranes[231] and by treatment of the BH$_3$.THF adduct with carboxylic acids, diacyloxyboranes, or triacyloxyboranes[232] and with 4-cyanobenzoic acid.[233]

Tris(germyloxy)cyclotriboroxanes (R$_3$GeO)$_3$B$_3$O$_3$ and *tris(stannyloxy)-cyclotriboroxanes* (R$_3$SnO)$_3$B$_3$O$_3$ have been prepared by heating boron trioxide with B(OER$_3$)$_3$ derivatives (E = Ge, R = Bu;[235] E = Sn, R = Me, Et, Pr, Bu, Bu^i, Ph[236]).

(36)

Tris(germyloxy)cyclotriboroxanes (R$_3$GeO)$_3$B$_3$O$_3$ and *tris(stannyloxy)-cyclotriboroxanes* (R$_3$SnO)$_3$B$_3$O$_3$ have been prepared by heating boron trioxide with B(OER$_3$)$_3$ derivatives (E = Ge, R = Bu;[235] E = Sn, R = Me, Et, Pr, Bu, Bu^i, Ph[236]).

Tris(amino)cyclotriboroxanes (R$_2$N)$_3$B$_3$O$_3$ are formed in the reaction of B$_4$(NMe$_2$)$_6$ with oxygen and in the reaction of (R$_2$N)$_2$BCl with mercury(II) oxide (R = Et, Pr):[237]

$$2\ (R_2N)_2BCl + HgO \longrightarrow \tfrac{1}{3}(R_2N)_3B_3O_3 + B(NR_2)_3 + HgCl_2 \quad (14)$$

4.4.5 Physical properties and structure

Much structural data on inorganic derivatives of the B$_3$O$_3$ ring (mainly cyclic borates) has been cited in Section 4.4.2. Here emphasis will be on organoderivatives. Only the bonding is discussed for both inorganic and organic derivatives together.

Thermodynamic data. The heat of formation for $(PhBO)_3$ has been determined as $\Delta H_f^o = -301.2 \pm 2.3$ kcal/mol,[238] and the heats of atomization for $(PhBO)_3$ and $(p\text{-}H_2C{=}CHC_6H_4BO)_3$ have been calculated using the IOC–ω technique.[239]

Viscosity. The dynamic viscosity of $(BuOBO)_3$ mixtures with all isomeric butylamines, in $1:1$, $1:2$ and $1:3$ molar ratios, has been measured.[240]

Vibration spectra (infrared and Raman). Infrared spectroscopy is used routinely for characterization of cyclotriboroxane derivatives, but some detailed spectral data and analyses have sometimes been published. Thus the infrared spectra of several organo-cyclotriboroxanes $(RBO)_3$ have been tabulated,[241] and a thorough analysis of infrared and Raman spectra of $(MeBO)_3$ and its deuterated analogues $(CD_3BO)_3$ and $(CD_2HBO)_3$ has been published.[242] The spectra of tris(alkoxy)- and (aroxy)cyclotriboroxanes[243,244] and tris(amino)cyclotriboroxanes[245] have also been analysed.

Ultraviolet spectra. The electronic spectra of triphenyl- and tris(p-bromophenyl)cyclotriboroxane have been compared with those of related compounds, and the results suggest strong interactions between the phenyl groups and the B_3O_3 ring.[246]

Magnetic, optical and magnetooptical properties. The molar refraction (Lorenz–Lorentz) for a number of organocyclotriboroxanes has been investigated.[247] The molar Kerr constant of 116×10^{-12} for $(PhBO)_3$ is in agreement with a planar ring structure; the molar polarizability and susceptibility indicate a 45% aromatic character for $(MeBO)_3$.[248] Magnetooptical data (exaltation of magnetorotation or Faraday effect) measured for several cyclotriboroxanes $(RBO)_3$ and $(ROBO)_3$ suggest π-electron delocalization in the B_3O_3 ring,[249,250] but the aromatic character is less than in borazine.

Nuclear magnetic resonance. NMR spectra of many cyclotriboroxanes have been reported, including ^1H NMR of organo and alkoxy derivatives,[13,181,199,212,232,251–254] ^{11}B NMR of organo derivatives,[255–257] alkoxycyclotriboroxanes[258] and dimethylaminocyclotriboroxane,[259] ^{13}C NMR of organo derivatives[260] and ^{17}O NMR of various derivatives.[261] NMR data correlated with mass-spectral data have been presented as significant evidence for the aromatic character of cyclotriboroxanes.[264]

Mass spectra. Fragmentation of cyclotriboroxanes under electron impact has been studied for several derivatives.[183,192,262-264] In most cases the spectra exhibit a molecular peak and the B_3O_3 ring survives fragmentation of the organic side chain and B–C bond cleavage. Mass spectra established that 2,4,6-$Me_3C_6H_2BO$ (mesityl derivative) is a trimer,[183] rather than a dimer as previously reported by Snyder in 1960.

Bonding. The B_3O_3 ring is an interesting topic for theoretical investigations, because of the presence of an electron sextet and possible aromatic delocalization. It has been subjected to numerous MO calculations, of various degrees of sophistication.[38,246,265-277] Hückel MO calculations show that in the $B_3O_6^{3-}$ anion the exocyclic B–O bonds have a more pronounced π-bond character than the endocyclic B–O bonds.[265] The charge distribution of π-electrons in cyclic borates has been calculated and the electron conjugation discussed on this basis.[266] Iterative HMO–LCAO calculations suggest that the aromatic character in the boroxine ring is weaker than in borazine, the lone pair being more localized at oxygen.[268,269] Energy levels, ionization potentials and B–N bond properties have been determined by *ab initio* calculations[271,272], which confirm that the π-distribution along the boroxine ring is much more polarized than in borazine. All valence electron calculations of the Mulliken–Wolfsberg–Helmholz type have been reported for boroxine $(HBO)_3$ and its derivatives $(XBO)_3$ (where X = OH, F, Cl), as well as for $F_2HB_3O_3$ and $FH_2B_3O_3$; the σ-electron drift in the B_3O_3 ring is towards oxygen, whereas the π-electron drift is toward boron. The π-electron delocalization from oxygen is not as large as that from nitrogen of borazine. Substitution of O, Cl, or F for hydrogen increases the negative σ-charge at oxygen.[277] Heat of formation ($\Delta H_f^\circ = -525.8$ kcal/mol), molecular geometry, first ionization potential and dipole moment have been calculated for $(FBO)_3$ by the MNDO method.[38] Binding energies have been correlated with the results of ESCA determinations.[276]

4.4.6 Chemical properties

There are two types of transformations of cyclotriboroxanes: reactions in which the B_3O_3 ring is preserved, and reactions in which the ring is cleaved. Each includes several particular cases.

Addition reactions occur between cyclotriboroxanes and various nucleophiles,[278,279] by donation from nitrogen, oxygen, phosphorus, sulphur to boron. An increase in the size of the organic groups reduces the acceptor properties of the boron atom. Usually 1 : 1 adducts are formed.

Insertion reactions take place during oxidation with various reagents, when alkyl- and arylcyclotriborazanes are converted to alkoxy and aroxy derivatives:

$$(RBO)_3 \xrightarrow{[O]} (RO\cdot BO)_3 \qquad (15)$$

This reaction has been investigated in considerable detail.[181,231,280–284] Kinetic data indicate a radical mechanism, via peroxo groups ROO—B.

The inorganic ring B_3O_3 can stand, without cleavage, several *reactions in the organic side groups*. Thus the (*p*-bromomethyl)phenyl derivative $(4\text{-BrCH}_2C_6H_4BO)_3$ can be converted into the corresponding amine by treatment with urotropine in chloroform, followed by acidification.[285] Similarly, the (*o*-bromomethyl)phenyl derivative $(2\text{-BrCH}_2C_6H_4BO)_3$ on treatment with potassium cyanide gives $(2\text{-NC—CH}_2C_6H_4BO)_3$.[286] Schiff bases are obtained from *o*-formyl derivatives, for example $(2\text{-OCH—C}_6H_4BO)_3$ with various amines (aniline, *p*-toluidine, benzylamine and propylamine).[287]

Ring-cleavage reactions are numerous and are caused by many reagents. The inorganic cyclotriboroxane $(HBO)_3$ is cleaved by boron halides[32] and phosphorus trifluoride[34] (equations 2 and 3).

Trialkoxycyclotriboroxanes are cleaved by alcohols in the presence of molecular sieves, to give trialkoxyboranes.[288,289] The silanols, which are mild reagents, also cleave the B_3O_3 ring:[290]

$$(RBO)_3 + 6\ Me_3SiOH \longrightarrow 3\ R—B(OSiMe_3)_2 + 3\ H_2O \qquad (16)$$
$$R = m\text{-CF}_3C_6H_4$$

The cleavage of the B_3O_3 ring with organic functional derivatives containing mobile hydrogen can be used for preparative purposes in the synthesis of organoboron compounds. Thus, the reaction with amidoximes[291] gives five-membered rings:

$$\tfrac{1}{3}(PhBO)_3 + \underset{\underset{NHR''}{|}}{R'C}{=}NOH \longrightarrow \quad \text{(17)}$$

The reaction of $(PhBO)_3$ with vicinal *cis*-aminoalcohols give five-membered oxazaborolidine rings[292] or seven-membered rings[293]:

With di(ethanol)amines triphenylcyclotriboroxane produces *diptych*-borox-azolidine:[294,295]

$$\tfrac{1}{3}(PhBO)_3 + RN(CH_2CH_2OH)_2 \xrightarrow[-H_2O]{} \qquad\qquad\qquad (18)$$

Organocyclotriboroxanes, for example $(PhBO)_3$, react with *cis*-diols and serve as a source of RBO_2 groups. Their reaction with sugars (glucopyranosides, glucofuranosides, etc.) have been extensively investigated[293,296-307] as $R—BO_2$ units can serve as protecting groups that assist in separation and purification of oligo- and polyhydroxy compounds.[307]

Alkoxycyclotriboroxanes, for example $[(Pr^iO)BO]_3$, in diglyme at 100–130 °C may serve as reducing agents for substituted benzaldehydes.[308] A complex reaction occurs between $(PhBO)_3$ and diethylethylidene malonate in the presence of bis(ethoxycarbonylacetonato)copper(II), in pyridine, to give Cu powder, Cu(I) pentaborate–pyridine adduct, benzene (70%), biphenyl (14%) and 1-phenylethylmalonic acid (7.5%).[309]

Analysis. The gas-chromatographic analysis of some organocyclotriboroxanes has been investigated and detection limits determined.[310,311]

4.4.7 Uses

There is a voluminous literature, especially patent sources, mentioning possible uses of various cyclotriboroxane derivatives. Only organic and organofunctional compounds are considered here, since the innumerable uses of borates, for example borax and some other cycloborates, are far beyond the scope of this book.

Organic cyclotriboroxanes or alkoxy derivatives have been recommended for use as curing agents for epoxy resins[312-319] and polysiloxanes,[320,321] heat stabilizers for polymers,[322-327] colour stabilizers for fatty acids,[328] fuel- and lubricant-oil additives,[191,220,329-333] anticorrosive agents for lubricants and other organic fluids,[334-339] hydraulic fluids,[339-342] welding fluxes and antioxidants for metal-surface protection,[343-347] catalysts for oxidation of hydrocarbons to alcohols,[230,348-366] epoxidation of olefins,[366] transposition of ketoximes,[367] isomerization of hydrocarbons,[368,369] Friedel–Crafts alkylations,[369] decomposition of alkyl peroxides,[360,370] hydrogenation of olefins,[355] N-acylation of aminophenols with carboxylic acids,[371] polymerization of epoxides,[312,372-374] polymerization of cyclic ethers[374,375] and other monomers,[376] polycondensation reactions,[334,377] and oxidation of cyclic ketones to lactones.[354,378] Cyclotriboroxanes have also been suggested for use in polymer synthesis,[379-381] polymer modification,[382-390] and as plasticizers,[391] flame retardants,[392-394] adhesives,[395] binders,[338,396] metal-surface protecting agents[397-399] and components in cosmetics[400].

Biological activity. The biological activity of several cyclotriboroxane derivatives was investigated many years ago.[401,402] For trimethoxycyclotriboroxane (MeOBO)$_3$ a low inhalation toxicity was found, but the compound produces slight lethargy.[403] Several derivatives were found to possess chemosterilant properties against the house fly (*Musca domestica*) and screwworm (*Cochliomyia hominivorax*),[404,405] while others exhibit bacteriostatic activity against *Staphylococcus aureus* and *Aspergillus niger*.[406] The reaction product between triphenylcyclotriboroxane and *d-threo*-chloramphenicol is active against *Staphylococcus aureus*, *Escherichia coli* and *Bacillus subtilis*.[407] So far no actual use based on these properties, in medicine or otherwise, is known.

4.5 Cyclodiboratetroxanes

The B_2O_4 ring is known only in anionic form (37) in peroxoborates, made up of two tetrahedra:

(37)

The structure of this anion was confirmed by single crystal X-ray diffraction[408] of sodium peroxoborate, formulated alternatively as $NaBO_3.4H_2O$ or $Na_2O.B_2O_3.2H_2O_2.6H_2O$. The same anion is probably present in other so-called peroxoborates, like $NaBO_2.H_2O_2.H_2O$,[409] $NaBO_2.H_2O_2$,[410] $KBO_2.H_2O_2$ or $KBO_3.H_2O$,[410,411] $LiBO_2.H_2O_2$,[412] $LiBO_2.H_2O_2.H_2O$[413] and $CsBO_2.H_2O_2$.[412] The peroxoborates are prepared by electrochemical methods[414,415] or by reactions of borates with hydrogen peroxide.[416,417] Since peroxoborates are used as whitening agents for textiles, their thermal decomposition (with evolution of oxygen) has been studied in some detail.[418,419]

Anions containing hydroperoxo units as exocyclic groups, like $[B_2O_4-(OH)_2(OOH)_2]^{2-}$ [412,420] and $[B_2O_4(OOH)_4]^{2-}$ [412] have also been reported in the literature.

Bibliography

B 1   I. Haiduc, *The Chemistry of Inorganic Ring Systems* (Wiley-Interscience, London, 1970), Part. 1, pp. 273–334 (boron–oxygen heterocycles).

B 2   *Gmelin Handbook of Inorganic Chemistry. Boron Compounds.* 1st Suppl. (Springer, Berlin), Vol. 1 (1980), pp. 113–258 (boron–oxygen compounds, borates, boroxines); Vol. 2 (1980), p. 131 (B–O–N rings), p. 286 (trifluoroboroxine).

B 3   *Gmelin Handbook of Inorganic Chemistry. Boron Compounds.* 2nd Suppl. (Springer, Berlin), Vol. 1 (1983) pp. 205–303 (boron–oxygen compounds, borates, boroxines).

B 4   *Gmelin Handbook der Anorganischen Chemie. Borverbindungen* (Springer, Berlin), Teil 7 (1975) (boron oxides, borates); Teil 13 (1977), pp. 115–239 (B–O rings, boroxines).

B 5   J. B. Farmer, *Adv. Inorg. Chem. Radiochem.* **25** (1982) 187 (metal borates, including cyclic borates).

B 6   G. Heller, *Top. Curr. Chem.* **15** (1970) 206 (synthesis and systematization of borates and polyborates).

B 7   R. Thompson (editor), *Mellor's Comprehensive Treatise of Inorganic and Theoretical Chemistry* (Longmans-Green, New York, 1980), Vol. 5, Part A, pp. 224–320, 343–501, 520–650 (borate chemistry).

B 8   J. R. Clark, *Am. Mineralogist* **56** (1971) 1952 (principles of borate structures extended to include new types).

B 9   J. R. Clark, *Am. Chem. Soc. Div. Water Air Waste Chem., Gen. Papers* **11**, No. 2 (1971) 187; *CA* **78** (1973) 141322 (extension of the principles of borate structures).

B 10   C. L. Christ and J. R. Clark, *Phys. Chem. Miner.* **2** (1977) 59 (borate structures, review; amended rules on polyborate ion formation).

B 11   J. L. T. Waugh, *Struct. Chem. Mol. Biol.* **1968** 731 (borate structures).

B 12   M. L. Huggins, *Inorg. Chem.* **10** (1971) 791; **7** (1968) 2108 (structon theory of borate structures).

B 13   *Boraty i boratnye sistemy* [*Borates and Borate Systems*] (Zinatne Publ. House, Riga, 1978) (collection of reviews and articles on borate chemistry).

B 14   M. A. Simonov and N. V. Belov, in *Kristalicheskie Struktury Neorganicheskikh Soedinenii* [*Crystal Structures of Inorganic Compounds*], edited by T. I. Malinovskii (Stiinta Publ. House, Kishinev, 1974), p. 166 (crystal structures of calcium metaborates).

B 15   M. A. Simonov and N. V. Belov, *Mineral. Zh.* **1** (1979) 19 (structural principles of borate structures).

B 16   S. D. Ross, *Infrared Spectra Miner.* **4** (1974) 205; *CA* **83** (1975) 45800 (review of borate structures).

B 17   M. A. Simonov, *Vestn. Mosk. Univ.* Ser. 4, *Geology* **30** (1975) Nr. 3, 15; *CA* **83** (1975) 186426.

B 18   C. L. Christ, *J. Geol. Educ.* **20** (1972) 235 (borate structures)

B 19   H. Gode (editor), *Issledovanie Sinteticheskikh Boratov* [*Investigation of Synthetic Borates*] (Latv. Gos. Univ., Riga, 1981) (collection of articles).

B 20   N. V. Belov, *Mineral. Sb. Lvovskogo Gos. Univ.* **29** (1975) No. 4, 3; *CA* **85** (1976) 163, 463 (borate structures).

B 21   J. L. T. Waugh, in *Isopolyborates in Structural Chemistry and Molecular Biology*, edited by A. Rich and N. Davidson (Freeman, San Francisco, 1968), p. 731 (classification of borates).

B 22   B. Bokii and V. B. Kravchenko, *Zh. Strukt. Khim.* 7 (1966) 920 (systematization of borate structures).

References

1   B. M. Mikhailov, L. S. Vasil'ev and V. V. Veselovskii, *Izv. Akad. Nauk SSSR, Ser. Khim.* **1980**, 1106; *CA* **93** (1980) 114589.
2   L. Barton, S. K. Wason and R. F. Porter, *J. Phys. Chem.* **69** (1965) 3160.
3   L. Barton, C. Perrin, and R. F. Porter, *Inorg. Chem.* **5** (1966) 1446.
4   L. Barton, *J. Inorg. Nucl. Chem.* **30** (1968) 1683.
5   J. F. Ditter and I. Shapiro, *J. Am. Chem. Soc.* **81** (1959) 1022.
6   A. Kaldor and R. F. Porter, *J. Am. Chem. Soc.* **93** (1971) 2140.
7   G. H. Lee and R. F. Porter, *Inorg. Chem.* **5** (1966) 1329.
8   F. A. Grimm and R. F. Porter, *Inorg. Chem.* **8** (1969) 731.
9   F. A. Grimm and R. F. Porter, *Inorg. Chem.* **7** (1968) 706.
10   F. A. Miller, *Appl. Spectrosc.* **29** (1975) 461.
11   W. V. Brooks, C. C. Costain and R. F. Porter, *J. Chem. Phys.* **47** (1967) 4186.
12   C. A. Coulson, *Acta Crystallogr.* **25B** (1969) 807.
13   L. Barton, D. Brinza, R. A. Frease and F. L. Longcor, *J. Inorg. Nucl. Chem.* **39** (1977) 1845.
14   R. Zahradnik and I. Matousek, *Coll. Czech. Chem. Commun.* **39** (1974) 57.
15   O. Gropen and P. Vassbotn, *Acta Chem. Scand.* **27** (1973) 3079.
16   C. Leibovici, *J. Mol. Struct.* **11** (1972) 141.
17   E. R. Lory and R. F. Porter, *J. Am. Chem. Soc.* **93** (1971) 6301.
18   L. Barton and G. T. Bohn, *J. Chem. Soc. Chem. Commun.* **1971**, 77.
19   L. Barton and M. Crump, *Inorg. Chem.* **12** (1973) 2252, 2506.
20   L. Barton, M. Crump and J. B. Wheatley, *J. Organomet. Chem.* **72** (1974) C1.
21   IUPAC, *Nomenclature of Inorganic Boron Compounds: IUPAC Information Bulletin*, No. 8 (1970) 19.
22   S. K. Wason and R. F. Porter, *J. Chem. Phys.* **68** (1964) 1443.
23   R. F. Porter and F. A. Grimm, *Adv. Chem. Ser.* **72** (1968) 94.
24   S. K. Gupta and R. F. Porter, *J. Chem. Phys.* **70** (1966) 871.
25   C. H. Chang, R. F. Porter and S. H. Bauer, *Inorg. Chem.* **8** (1969) 1689.
26   A. Kaldor and R. F. Porter, *Inorg. Chem.* **10** (1971) 775.
27   F. A. Grimm, L. Barton and R. F. Porter, *Inorg. Chem.* **7** (1968) 1309.
28   K. Molt and W. Sawodny, *Proc. 5th Int. Conf. Raman Spectr.*, 1976, p. 86; *CA* **87** (1977) 191304f.
29   I. B. Rozhdestvenskii, V. N. Gutov and N. A. Zhigulskaya, *Sb. Trudov-Glavniiproekt Energ. Inst.* **7** (1973) 88; *CA* **85** (1976) 83876.
30   National Bureau of Standards, *Ref. Data Ser. NBS* **1971**, No. 37.
31   A. Finch and P. J. Gardner, *Prog. Boron Chem.* **3** (1970) 177.
32   M. P. Nadler and R. F. Porter, *Inorg Chem.* **6** (1967) 1192.
33   M. P. Nadler, Diss. Cornell Univ. (1969); *Diss. Abstr. Int.* **30B** (1970) 5453.
34   L. Barton, *J. Inorg. Nucl. Chem.* **30** (1968) 1683.
35   A. Kaldor, I. Pines and R. F. Porter, *Inorg. Chem* **8** (1969) 1419.

36 A. W. Laubengayer, K. Watterson, D. R. Bidinosti and R. F. Porter, *Inorg. Chem.* **2** (1963) 519.
37 D. R. Bidinosti and L. I. Coatsworth, *Can. J. Chem.* **48** (1970) 2484.
38 M. J. S. Dewar and H. S. Rzepa, *J. Am. Chem. Soc.* **100** (1978) 58.
39 M. Farber, H. L. Peterson, J. A. Blauer, D. Brown and J. Davies, AD 284 428; *US Govt Res. Dev. Rep.* **40** (1965) No. 3, S 5.
40 F. H. J. Thevenot, P. M. V. Goeuriot, J. H. Driver and J. P. R. Lebrun, *Eur. Pat. Appl.* 15 813 (1980); *CA* **94** (1981) 34761.
41 C. J. Ultee, *J. Chem. Phys.* **40** (1964) 3746.
42 B. Latimer and J. P. Devlin, *Spectrochim. Acta* **23A** (1967) 81.
43 B. Latimer, Diss. Oklahoma State Univ. (1967): *Diss. Abstr. Int.* **B27** (1967) 4346.
44 E. Schwarzmann and G. Christoph, *Z. Naturforsch.* **24b** (1969) 965.
45 D. M. Young and C. D. Anderson, *Inorg. Chem.* **1** (1962) 429.
46 C. G. Spike, *US Pat.* 2 961 459 (1961); *CA* **55** (1961) 6379.
47 J. G. Alvarez-Valdez, C. Pico and R. E. Gutierez, *An. Quim.* **74** (1978) 424; *CA* **89** (1978) 156654.
48 A. Kamars, *Tezisy Dokladov 6th Konf. Molodykh Rabotnikov, Inst. Neorg. Khim. Akad. Nauk Latv. SSR, Riga,* 1977, p. 19; *CA* **90** (1979) 64030.
49 R. S. Tsekhanskii, V. G. Skvortsov, L. I. Vinogradov, *Sb. Nauch. Tr. Yaroslav. Gos. Ped. Inst.* **1975**, 151; *CA* **85** (1976) 201363; *Ref. Zh. Khim.* **1976**, 16 V 28.
50 N. D. Sokolova, S. M. Skuratov, A. M. Shemonaeva and V. M. Yuldasheva, *Russ. J. Inorg. Chem.* **6** (1961) 395.
51 E. Schwartzmann and G. Christoph, *Z. Naturforsch.* **23b** (1968) 1542.
52 P. Broadhead and G. A. Newman, *J. Mol. Struct.* **10** (1971) 157.
53 A. Bertoluzza, C. Castellari, M. A. Bertoluzza-Morelli, *Atti Accad. Naz. dei Lincei, Cl. Sci. Fis. Mat. Nat., Rend.* [8] **44** (1968) 554; *C. Zbl.* **1969**, 220366.
54 A. Bertoluzza, P. Monti, M. A. Battaglia and S. Bonova, *J. Mol. Struct.* **64** (1980) 123.
55 G. Nagarajan, *Bull. Soc. Chim. Belg.* **71** (1962) 431.
56 W. Fedder, F. Umland and E. Hohaus, *Z. Anorg. Allg. Chem.* **471** (1980) 77.
57 L. R. Batsanova and V. A. Egorov, *Zh. Neorg. Khim.* **12** (1967) 319.
58 L. Maya, *Inorg. Chem.* **15** (1976) 2179.
59 L. Kolditz and S. L. Cheng, *Z. Chem.* **7** (1967) 469.
60 C. L. Christ, *Am. Mineralogist* **45** (1960) 334.
61 W. Schneider and G. B. Carpenter, *Acta Crystallogr.* **26B** (1970) 1189.
62 J. Krogh-Moe, *Acta Crystallogr.* **23** (1967) 427.
63 A. D. Mighell, A. Perloff and S. Block, *Acta Crystallogr.* **20** (1966) 819.
64 A. Dal Negro, J. M. Martin Pozas and L. Ungaretti, *Am. Mineral.* **60** (1975) 879.
65 L. F. Aristarain and C. S. Hurlbut, *Am. Mineralogist* **52** (1967) 935.
66 E. Corazza, S. Menchetti and C. Sabelli, *Acta Crystallogr.* **31B** (1975) 1993.
67 E. Corazza, *Acta Crystallogr.* **32B** (1976) 1329.
68 I. M. Rumanova and A. Ashirov, *Kristallografiya (USSR)* **8** (1963) 517.
69 E. Corazza, *Acta Crystallogr.* **30B** (1974) 2194.
70 Z. P. Razmanova, I. M. Rumanova and N. V. Belov, *Dokl. Akad. Nauk SSSR* **189** (1969) 1003.
71 I. M. Rumanova and E. N. Kurkutova, *Probl. Kristallologii* **1971**, 102; *CA* **78** (1973) 21344.
72 J. Ozols, I. Tetere and A. Ievins, *Latv. PSR Zinat. Akad. Vestis, Kim. Ser.* **1973**, 3.
73 J. Ozols, I. Tetere and I. Berzina, *Boraty Boratnye Sist.* **1978**, 78; *CA* **90** (1979) 130736.
74 J. Ozols, *Latv. PSR Zinat. Akad. Vestis, Kim. Ser.* **1977**, 356.
75 I. Zviedre, J. Ozols and A. Ievins, *Latv. PSR Zinat. Akad. Vestis, Kim. Ser.* **1972**, 497.

76 I. Zviedre, J. Ozols and A. Ievins, *Latv. PSR Zinat. Akad. Vestis, Kim. Ser.* **1974**, 387.

77 I. Zviedre and A. Ievins, *Latv. PSR Zinat. Akad. Vestis, Kim. Ser.* **1974**, 395.

78 I. Zviedre and A. Ievins, *Latv. PSR Zinat. Akad. Vestis, Kim. Ser.* **1973**, 511.

79 N. A. Yamnova, Yu. K. Egorov-Tismenko, M. A. Simonov and N. V. Belov, *Dokl. Akad. Nauk SSSR* **216** (1974) 1281.

80 N. A. Yamnova, Yu. K. Egorov-Tismenko, M. A. Simonov and N. V. Belov, *Kristallokhim. Strukt. Osobov Mineralov* **1976**, 104; *CA* **87** (1977) 70993.

81 N. A. Yamnova, M. A. Simonov and N. V. Belov, *Kristallografiya* **22** (1977) 624.

82 Yu. K. Egorov-Tismenko, M. A. Simonov and N. V. Belov, *Dokl. Akad. Nauk SSSR* **210** (1973) 678.

83 M. A. Simonov, *Vest. Mosk. Univ., Geol.* **30** (1975) No. 3, 15; *CA* **83** (1975) 186426.

84 M. A. Simonov, Yu. K. Egorov-Tismenko, E. V. Kazanskaya, E. L. Belokoneva and N. V. Belov, *Dokl. Akad. Nauk SSSR* **239** (1978) 326; *Sov. Phys. Dokl.* **23** (1978/9) 159.

85 M. A. Simonov, Yu. K. Egorov-Tismenko and N. V. Belov, *Kristallografiya* **24** (1979) 439.

86 N. V. Belov, and E. N. Belova, *Mineral. Sb. (Lvov)* **34** (1980) 3; *CA* **95** (1981) 27947; *Mineral. Sb. (Lvov)* **34** (1980) 10; *CA* **95** (1981) 159950.

87 C. L. Christ and J. R. Clark, *Phys. Chem. Miner.* **2** (1977) 59.

88 Wuhan Geologic College, *K'o Hsueh T'ung Pao* **22** (1977) 451; *CA* **88** (1978) 57176.

89 G. E. Brown and J. R. Clark, *Am. Mineralogist,* **63** (1978) 814.

90 D. P. Shashkin, M. A. Simonov and N. V. Belov, *Dokl. Akad. Nauk SSSR,* **189** (1969) 532; *Sov. Phys. Dokl.* **13** (1969) 1044.

91 M. A. Simonov. Yu. K. Egorov-Tismenko and N. V. Belov, *Dokl. Akad. Nauk SSSR* **234** (1977) 822.

92 N. A. Yamnova, M. A. Simonov, E. V. Kazanskaya and N. V. Belov, *Dokl. Akad. Nauk SSSR* **225** (1975) 823.

93 L. P. Solov'eva and V. V. Bakakin, *Dokl. Akad. Nauk SSSR* **180** (1968) 1453.

94 E. S. Menchetti and C. Sabelli, *Acta Crystallogr.* **34B** (1978) 45.

95 S. Merlino and F. Sartori, *Acta Crystallogr.* **28B** (1972) 3559.

96 J. P. Ashmore and H. E. Petch, *Can. J. Phys.* **48** (1970) 1091.

97 S. Merlino and F. Sartori, *Contrib. Mineral. Petrogr. (Berlin)* **27** (1970) 159.

98 J. Frohnecke, H. Hartl and G. Heller, *Z. Naturforsch.* **32b** (1977) 268.

99 S. Merlino, *Atti Accad. Naz. Lincei, Cl. Sci. Fis. Mat. Nat., Rend.* [8] **47** (1969) 85.

100 D. Amoussou, R. Wandji and M. Touboul, *C. R. Acad. Sci. Paris,* **290C** (1980) 391.

101 S. Ghose, C. Wan and J. R. Clark, *Am. Mineralogist* **63** (1978) 160.

102 S. Menchetti and C. Sabelli, *Acta Crystallogr.* **35B** (1979) 2488.

103 H. A. Levy and G. C. Lysensky, *Acta Crystallogr.* **34B** (1978) 3502.

104 G. Thomas and M. Soustelle, *Bull. Soc. Chim. Fr.* **1970** 4202.

105 R. F. Giese, *Can. Mineralogist* **9** (1967/8) 573.

106 C. Giacovazzo, S. Menchetti and F. Scordari, *Am. Mineralogist* **58** (1973) 523.

107 M. Marezio, H. A. Plettinger and W. H. Zachariasen, *Acta Crystallogr.* **16** (1963) 975.

108 R. Janda, G. Heller and J. Pickardt, *Z. Kristallogr.* **154** (1981) 1.

109 C. Wan and S. Ghose, *Am. Mineralogist* **62** (1977) 1135.

110 G. K. Gode, I. W. Majore, M. V. Borisov and D. S. Poryvkin, *Latv. PSR Zinat. Akad. Vestis, Kim. Ser.* **1976**, 46.

111 E. V. Sokolov, N. A. Yamnova, M. A. Simonov and N. V. Belov, *Kristallografiya* **24** (1979) 1169.

112 I. Berzina, *Tezisy Dokl. Konf. Molodykh Uchenykk Rab. Inst. Neorg. Khim. Akad. Nauk Latv. SSR, 6th,* 1977, p. 18; *CA* **90** (1979) 46855.

113 N. P. Ivchenko, and E. N. Kurkutova, *Kristallografiya* **20** (1975) 533.

114 E. N. Kurkutova, N. P. Ivchenko and V. V. Ilyukhin, *Zh. Strukt. Khim.* **17** (1976) 950.
115 I. Berzina, J. Ozols and A. Ievins, *Kristallografiya* **20** (1975) 419.
116 I. Berzina, J. Ozols and A. Ievins, *Latv. PSR Zinat. Akad. Vestis, Kim. Ser.* **1974**, 648; *CA* **82** (1975) 79111.
117 I. Berzina, *Tezisy Dokl. Konf. Molodykh Uchenykh* **3** (1973) 3; *CA* **86** (1976) 163827.
118 P. B. Moore and T. Araki, *Am. Mineralogist* **59** (1974) 60.
119 N. A. Yamnova, M. A. Simonov and N. V. Belov, *Dokl. Akad. Nauk SSSR* **221** (1975) 1326.
120 A. Dal Negro, L. Ungaretti and C. Sabelli, *Naturwiss.* **60** (1973) 350.
121 A. Dal Negro, L. Ungaretti and C. Sabelli, *Am. Mineralogist* **56** (1971) 1553.
122 F. Hanic, O. Lindquist, J. Nyborg and A. Zedler, *Coll. Czech. Chem. Commun.* **36** (1971) 3678.
123 A. Dal Negro, L. Ungaretti and A. Giusta, *Cryst. Struct. Commun.* **5** (1976) 427.
124 E. A. Genkina, I. M. Rumanova and N. V. Belov, *Kristallografiya* **21** (1976) 209.
125 A. Dal Negro, L. Ungaretti and R. Basso, *Cryst. Struct. Commun.* **5** (1976) 433.
126 A. Dal Negro, C. Sabelli and L. Ungaretti, *Atti Accad. Naz. Lincei, Cl. Sci. Fis. Mat. Nat., Rend.* [8] **47** (1969) 353.
127 E. J. Silins, J. Ozols and A. Ievins, *Latv. PSR Zinat. Akad. Vestis, Kim. Ser.* **1974**, 115.
128 E. Silins, J. Ozols, A. Ievins, *Kristallografiya* **18** (1973) 503; *Latv. PSR Zinat. Akad. Vestis, Kim. Ser.* **1972**, 242.
129 E. J. Silins and A. F. Ievins, *Latv. PSR Zinat. Akad. Vestis, Kim. Ser.* **1975**, 747.
130 E. J. Silins and A. F. Ievins, *Kristallografiya* **22** (1977) 505.
131 E. J. Silins and A. F. Ievins, *Latv. PSR Zinat. Akad. Vestis, Kim. Ser.* **1976**, 357.
132 A. E. Dzene, E. M. Svarcz and A. F. Ievins, *Latv. PSR Zinat. Akad. Vestis, Kim. Ser.* **1972**, 515.
133 A. Dal Negro and I. Kumbasar, *Am. Mineralogist* **58** (1973) 1034.
134 D. P. Shashkin, M. A. Simonov and N. V. Belov, *Dokl. Akad. Nauk SSSR* **195** (1970) 345.
135 J. S. Ysker and W. Hoffmann, *Naturwiss.* **57** (1970) 129.
136 S. Merlino and F. Sartori, *Science* **171** (1971) 377.
137 R. E. Mesmer, C. F. Baes and F. H. Sweeton, *Inorg. Chem.* **11** (1972) 537.
138 A. S. Shubin, B. A. Nikolskii, and Yu.S. Plyashevskii, *Tr. Uralsk. Nauchn. Issled. Khim. Inst.* **40** (1976) 72; *CA* **91** (1979) 79645.
139 F. Peterka, *Ustav Jad. Vyzk. (Rep.), 1979,* UJV 5089; *CA* **93** (1980) 157412.
140 D. V. S. Jain and C. M. Jain, *Indian J. Chem.* **11** (1973) 1281; *CA* **81** (1974) 17440.
141 R. K. Momii and N. H. Nachtrieb, *Inorg. Chem.* **6** (1967) 1189.
142 K. Wegener, *J. Inorg. Nucl. Chem.* **29** (1967) 1847.
143 K. F. Jahr, K. Wegener, G. Heller and K. Worm, *J. Inorg. Nucl. Chem.* **30** (1968) 1677.
144 H. D. Smith and R. J. Wiersema, *Inorg. Chem.* **11** (1972) 1152.
145 A. K. Covington and K. E. Newman, *J. Inorg. Nucl. Chem.* **35** (1973) 1152.
146 B. W. Epperlein, O. Lutz and A. Schwenk, *Z. Naturforsch.* **30a** (1975) 955.
147 R. Janda and G. Heller, *Z. Naturforsch.* **34b** (1979) 1078.
148 V. P. Kochergin, O. G. Garaeva and R. N. Pletnev, *Russ. J. Inorg. Chem.* **24** (1979) 1641; *Zh. Neorg. Khim.* **24** (1979) 2952.
149 P. J. Bray, J. O. Edwards, J. G. O'Keefe, V. F. Ross and T. Tatsuzaki, *J. Chem. Phys.* **33** (1961) 435.
150 I. Sato, F. Watari and K. Aida, *Asahi Garasu Kogyo Gijutsu Shorei-kai Kenkyu Hokoku* **13** (1967) 155; *CA* **69** (1968) 81917.
151 H. Gode and I. Zuika, *Latv. PSR Zinat. Akad. Vestis, Kim. Ser.* **1970**, 724.
152 K. Frey and E. Funck, *Z. Naturforsch.* **27b** (1972) 101.

153 G. Heller and A. Giebelhausen, *J. Inorg. Nucl. Chem.* **35** (1973) 3511.
154 V. S. Suknev and T. A. Grigor'eva, *Zh. Prikl. Spektrosk.* **18** (1973) 495; *CA* **78** (1973) 166544.
155 Z. O. Fillippova, R. E. Zhaimina, T. L. Markina and A. I. Mun, *Izv. Akad. Nauk Kaz. SSR, Ser. Khim.* **25** (1975) No. 3, 5; *CA* **83** (1975) 104231.
156 I. I. Kondilenko, V. A. Pasechnyi and Yu. P. Tsyashchenko, *Zh. Prikl. Spektrosk.* **23** (1975) 1049; *CA* **84** (1976) 97289.
157 V. I. Borisenko, *Boraty Boratnye Sist.* **1978**, 142; *CA* **90** (1979) 177344.
158 A. Bertoluzza, C. Castellari, M. A. Bertoluzza-Morelli, *Atti Accad. Naz. Lincei, Cl. Sci. Fis. Mat. Nat., Rend.* [8] **44** (1968) 554; *Chem. Zbl.* (1969) 22-0366.
159 V. Devarajan, E. Graefe and E. Funck, *Spectrochim. Acta* **32A** (1976) 1225.
160 T. W. Brill, *Philips Res. Rep. Suppl.* (1976) No. 2; *CA* **85** (1976) 101580.
161 T. W. Brill, D. L. Vogel and E. Strijks, *Proc. 5th Int. Conf. Raman Spectrosc., Freiburg, 1976*, p. 656; *CA* **87** (1977) 192267.
162 J. P. Bronswijk, *Phys. Non-cryst. Solids, 4th Int. Conf.*, 1976, p. 101; *CA* **87** (1977) 108926.
163 J. P. Bronswijk and E. Strijks, *J. Noncryst. Solids*, **24** (1977) 145; *CA* **87** (1977) 75950.
164 M. Maeda, T. Hirato, M. Kotaka and H. Kakihana, *J. Inorg. Nucl. Chem.* **41** (1979) 1217.
165 R. Janda and G. Heller, *Spectrochim. Acta* **36A** (1980) 585, 997.
166 P. Kuka and H. Gode, *Uch. Zap. Latv. Univ.* (1970) No. 117, 52; *CA* **77** (1972) 118867; *Ref. Zh. Khim.* (1971) 12 B 853.
167 P. Kuka and G. K. Gode, *Zh. Vses. Khim. Ob-va* **19** (1974) 477; *CA* **81** (1974) 144822.
168 H. Gode and A. Veveris, *Latv. PSR Zinat. Akad. Vestis, Kim. Ser.* **1975**, 628; *CA* **84** (1976) 11606.
169 H. Gode and I. Majore, *Latv. PSR Zinat. Akad. Vestsis, Kim. Ser.*, **1976**, 344; *CA* **85** (1976) 86527.
170 G. Heller, *J. Inorg. Nucl. Chem.* **29** (1967) 2181; **30** (1968) 2743.
171 J. Frohnecke and G. Heller, *J. Inorg. Nucl. Chem.* **34** (1972) 69.
172 G. Heller and B. Bichowski, *Z. Naturforsch.* **33b** (1978) 20.
173 G. Heller and A. Giebelhausen, *Z. Anorg. Allg. Chem.* **460** (1980) 228.
174 J. Braun and H. Normant, *Bull. Soc. Chim. Fr.* **1966** 2557.
175 A. N. Nesmeyanov, A. E. Borisov and M. A. Osipova, *Dokl. Akad. Nauk SSSR* **169** (1966) 602; *Dokl. Chem. Proc. Acad. Sci. USSR (Engl. Transl.)* **166/171** (1966) 730.
176 E. C. J. Coffee and A. G. Davies, *J. Chem. Soc.* C **1966**, 1493.
177 Cheng-Hsun Lu, Hsiao-Shu Li, Wei-Fu Li and Hsin-Te Fung, *K'o Hsueh T'ung Pao* **17** (1966) 410; *CA* **66** (1967) 10511.
178 P. G. Allies and P. B. Brindley, *J. Chem. Soc.* B **1969** 1126.
179 G. D. Schaumberg and S. Donovan, *J. Organomet. Chem.* **20** (1969) 261.
180 H. A. Staab and B. Meissner, *Liebigs Ann. Chem.* **753** (1971) 80.
181 S. Korcek, G. B. Watts and K. U. Ingold, *J. Chem. Soc. Perkin Trans. II* **1972**, 242.
182 R. Van Veen and F. Bickelhaupt, *J. Organomet. Chem.* **43** (1972) 214.
183 S. W. Breuer and F. A. Broster, *Tetrahedron Lett.* **1972** 2193.
184 J. Solie and P. Cadiot, *Bull. Soc. Chim. Fr.* **1966**, 1981.
185 S. W. Breuer and F. A. Foster, *Tetrahedron* **1972**, 2194.
186 R. P. Dickinson and B. Iddon, *J. Chem. Soc.* C **1970**, 1926.
187 D. W. Wester and L. Barton, *Org. Prep. Proc. Int.* **1971**, 191.
188 C. J. W. Brooks and D. J. Harvey, *J. Chromatogr.* **54** (1971) 193.
189 R. T. Hawkins and D. B. Stroup, *J. Org. Chem.* **34** (1969) 1173.
190 R. A. Bowie and O. C. Musgrave, *J. Chem. Soc.* C **1966**, 566.

191   P. S. Timms, *Brit. Pat.* 1 031 533 (1966); *CA* **65** (1966) 10404.
192   E. W. Post, R. G. Cooks and J. C. Kotz, *Inorg. Chem.* **9** (1970) 1670.
193   F. A. Davies and M. J. S. Dewar, *J. Am. Chem. Soc.* **90** (1968) 3511.
194   P. I. Paetzold, P. P. Habereder and R. Müllbauer, *J. Organomet. Chem.* **7** (1967) 51.
195   J. J. Miller, *J. Organomet. Chem.* **24** (1970) 595.
196   M. Schmidt, F. R. Rittig and W. Siebert, *Z. Naturforsch.* **25b** (1970) 1344.
197   R. H. Cragg and J. P. N. Husband, *Inorg. Nucl. Chem. Lett.* **6** (1970) 773.
198   D. J. Pasto, J. Hickman and T. C. Cheng, *J. Am. Chem. Soc.* **90** (1968) 6259.
199   Z. Polivka and M. Ferles, *Coll. Czech. Chem. Commun.* **1969**, 34.
200   H. C. Brown and S. K. Gupta, *J. Am. Chem. Soc.* **94** (1972) 4370.
201   T. Tanaka, T. Matsuda, K. Kimijima and Y. Iwasaki, *Bull. Soc. Chem. Jpn* **51** (1978) 1259.
202   P. Paetzold, P. Bohm, A. Richter and E. Scholl, *Z. Naturforsch.* **31b** (1976) 754.
203   P. I. Paetzold, W. Scheibitz and E. Scholl, *Z. Naturforsch.* **26b** (1971) 646.
204   P. B. Brindley, R. Davis, B. L. Horner and D. I. Ritchie, *J. Organomet. Chem.* **88** (1975) 321.
205   P. Paetzold and G. Schimmel, *Z. Naturforsch.* **35b** (1980) 568.
206   J. P. Tuchagues, *Bull. Soc. Chim. Fr.* **1968**, 2009.
207   H. C. Brown and M. W. Rathke, *J. Am. Chem. Soc.* **89** (1967) 2737.
208   H. C. Brown and M. W. Rathke, *J. Am. Chem. Soc.* **89** (1967) 4528.
209   K. V. Puzitskii, S. D. Piroshkov, K. G. Ryabova, I. V. Pastukhova and Ya. T. Eidus, *Bull. Acad. Sci. USSR, Div. Chem. Sci.* **1973**, 1760.
210   H. Witte, W. Gulden and G. Hesse, *Liebigs Ann. Chem.* **716** (1968) 1.
211   H. Witte, P. Mischke and G. Hesse, *Liebigs Ann. Chem.* **722** (1969) 21.
212   S. I. Frolov, Yu. N. Bubnov and B. M. Mikhailov, *Izv. Akad. Nauk SSSR, Ser. Khim.* **1969**, 1996; *Bull. Acad. Sci. USSR, Div. Chem. Sci.* **1969**, 1846.
213   Yu. N. Bubnov, S. I. Frolov, V. G. Kiselev, V. S. Bogdanov and B. M. Mikhailov, *Zh. Obshch. Khim.* **40** (1970) 1311; *CA* **74** (1971) 53885.
214   Yu. N. Bubnov, S. I. Frolov, V. G. Kiselev, V. S. Bogdanov and B. M. Mikhailov, *J. Gen. Chem. USSR* **40** (1970) 1302.
215   Yu. N. Bubnov, S. I. Frolov, V. G. Kiselev, V. S. Bogdanov and B. M. Mikhailov, *Organometal. Chem. Synth.* **1** (1970/71) 37.
216   T. T. Wang and K. Niedenzu, *J. Organomet. Chem.* **35** (1972) 231.
217   R. S. Towers, *Belg. Pat.* 662.085 (1965); *CA* **65** (1966) 7058.
218   A. Salvemini, F. Smai and G. Leofanti, *Fr. Pat.* 1 501 418 (1967); *CA* **70** (1969) 3269.
219   S. C. Goel and R. C. Mehrotra, *Synth. React. Inorg. Met.-Org. Chem.* **10** (1980) 591.
220   D. L. Hunter and H. Steinberg, *US Pat.* 3 359 298 (1967); *CA* **70** (1969) 3535.
221   Y. Yuzawa and T. Tsuchiya, *Japan Pat.* 72-10710 (1972); *CA* **77** (1972) 33927.
222   T. Yamaguchi, K. Sonobe and T. Shimizu, *Japan Pat.* 66-6751 (1966); *CA* **65** (1966) 7058.
223   Chien-Wei Liao, *US Pat.* 3 303 208 (1967); *CA* **66** (1967) 94696.
224   S. C. Goel, S. K. Mehrotra and R. C. Mehrotra, *Synth. React. Inorg. Met.-Org. Chem.* **7** (1977) 519.
225   L. I. Gamaga and V. S. Markevich, *J. Org. Chem. USSR* **9** (1973) 890.
226   O. L. Chapman and G. W. Bordon, *J. Org. Chem.* **26** (1961) 4193.
227   A. Finch and I. J. Hyams, *Rec. Trav. Chim.* **84** (1965) 683.
228   N. Takamitsu, T. Hamamoto and T. Nishi, *Nippon Kagaku Kaishi* **1972**, 254; *CA* **76** (1972) 98849.
229   A. W. Dawkins, *Brit. Pat.* 1 302 703 (1973); *CA* **78** (1973) 97103.
230   B. N. Tyutyunnikov and V. M. Yakubov, *Neftepererab. Neftekhim. (Moscow)* **1971**, No. 8 32; *CA* **75** (1971) 119773.

231 A. G. Davies, K. U. Ingold, B. P. Roberts and T. Tudor, *J. Chem. Soc.* B **1971**, 698.
232 H. C. Brown and T. P. Stocky, *J. Am. Chem. Soc.* **99** (1977) 8218.
233 N. M. Yoon, C. S. Pak, H. C. Brown, S. Krishnamurty and T. P. Stocky, *J. Org. Chem.* **38** (1973) 2786.
234 A. S. Shapatin, K. K. Popkov, Z. N. Nudel'man, S. A. Golubtsov and R. F. Markina, *Zh. Obshch. Khim.* **39** (1969) 554.
235 S. K. Mehrotra, G. Srivastava and R. C. Mehrotra, *J. Organomet. Chem.* **73** (1974) 277.
236 S. K. Mehrotra, G. Srivastava and R. C. Mehrotra, *J. Organomet. Chem.* **65** (1974) 361.
237 R. H. Cragg, *J. Inorg. Nucl. Chem.* **30** (1968) 711.
238 A. Finch and P. J. Gardner, *Trans. Faraday Soc.* **62** (1966) 3314.
239 D. C. Rajwar and S. C. Tiwari, *Nat. Acad. Sci. Lett.* **3** (1980) 33.
240 D. T. Haworth and R. V. Jensen, *Ind. Engng Chem., Prod. Res. Develop.* **9** (1970) 46.
241 B. Serafinowa and M. Makosza, *Roczniki Chem.* **35** (1961) 937.
242 D. C. McKean, H. J. Becher and F. Bramsiepe, *Spectrochim. Acta* **33A** (1977) 951.
243 A. Meller and E. Schaschel, *Monatsh. Chem.* **98** (1967) 390.
244 R. Janda and G. Heller, *Z. Naturforsch.* **34B** (1979) 585.
245 A. Meller, *Monatsh. Chem.* **98** (1967) 2014.
246 L. Santucci and C. Triboulet, *J. Chem. Soc.* A **1969**, 392.
247 R. Sayre, *J. Chem. Engng Data.* **8** (1963) 244.
248 M. J. Aroney, R. J. W. Le Fèvre, D. N. S. Murthy and J. D. Saxby, *J. Chem. Soc.* B **1966**, 1066.
249 J. P. Laurent and M. Pasdeloup, *Bull. Soc. Chim. Fr.* **1966**, 908.
250 J. P. Laurent, G. Cros and M. Pasdeloup, *Bull. Soc. Chim. Fr.* **1970**, 836.
251 H. C. Beachell and D. W. Beistel, *Inorg. Chem.* **3** (1964) 1028.
252 V. S. Bogdanov, G. V. Lagodzinskaya, V. F. Pozdnev and B. M. Mikhailov, *Bull. Acad. Sci. USSR, Div. Chem. Sci.* **1966**, 908.
253 V. S. Bogdanov, V. F. Pozdnev, G. V. Lagodzinskaya and B. M. Mikhailov, *Theor. Exptl Chem.* **3** (1967) 282.
254 H. Nöth and H. Vahrenkamp, *J. Organomet. Chem.* **12** (1968) 23.
255 J. E. De Moor, and G. P. Van der Kelen, *J. Organomet. Chem.* **6** (1966) 235.
256 F. A. Davies and M. J. S. Dewar, *J. Am. Chem. Soc.* **90** (1968) 3511.
257 K. Niedenzu, K. D. Moeller, W. J. Layton and L. Komorowski, *Z. Anorg. Allg. Chem.* **439** (1978) 112.
258 T. P. Onak, H. Landesman, R. E. Williams and I. Shapiro, *J. Phys. Chem.* **63** (1959) 1533.
259 H. Nöth and B. Wrackmeyer, *Chem. Ber.* **106** (1973) 1145.
260 H. Nöth and B. Wrackmeyer, *Nuclear Magnetic Resonance Spectroscopy of Boron Compounds* (Springer, Berlin 1978).
261 W. Biffar, H. Nöth, H. Pommerening and B. Wrackmeyer, *Chem. Ber.* **113** (1980) 333.
262 W. J. Lehmann, C. O. Wilson and I. Shapiro, *J. Inorg. Nucl. Chem.* **21** (1961) 25.
263 C. J. W. Brooks, D. J. Harvey and B. S. Middleditch, *Organic Mass Spectrom.* **3** (1970) 231.
264 R. H. Cragg, J. F. J. Todd and A. F. Weston, *Org. Mass. Spectrom.* **6** (1972) 1077.
265 C. A. Coulson and T. W. Dingle, *Acta Crystallogr.* **24B** (1968) 153.
266 S. E. Svanson, *Acta Chem. Scand.* **23** (1969) 2005.
267 J. F. Labarre, M. Grafeuil, J. P. Fauscher, M. Pasdeloup and J. P. Laurent, *Theor. Chim. Acta* **11** (1968) 324.
268 J. F. Labarre, M. Grafeuil, J. P. Faucher, M. Pasdeloup and J. P. Laurent, *Theor. Chim. Acta* **11** (1968) 423.
269 J. F. Labarre, M. Grafeuil and F. Galais, *J. Chim. Phys.* **65** (1968) 638.
270 M. Grafeuil and J. F. Labarre, *J. Chim. Phys.* **68** (1971) 1379.

271 A. Serafini, J. F. Labarre and F. Galais, *Proc. 16th Int. Conf. Coord. Chem.*, 1974, p. 2.4a; *CA* **85** (1976) 25604.

272 A. Serafini and J. P. Labarre, *J. Mol. Struct.* **26** (1975) 129.

273 D. R. Armstrong and P. G. Perkins, *J. Chem. Soc.* A **1967**, 790.

274 D. R. Armstrong, P. G. Perkins and J. J. P. Stewart, *J. Chem. Soc., Dalton Trans.* **1973**, 2273.

275 M. Bossa and F. Maraschini, *J. Chem. Soc.* A **1970**, 1416.

276 D. N. Hendrickson, J. M. Hollander and W. J. Jolly, *Inorg. Chem.* **9** (1970) 612.

277 D. T. Haworth and V. M. Scherr, *J. Inorg. Nucl. Chem.* **37** (1975) 2010.

278 J. M. Ritchey, *PhD thesis, Univ. Colorado* (1968); *Diss. Abstr. Int.* **29B** (1969) 3251.

279 V. G. Tsvetkov, V. N. Alyasov, N. V. Balakshina, V. P. Maslennikov, and Yu. A. Aleksandrov, *Zh. Obshch. Khim.* **51** (1981) 269.

280 A. G. Davies and B. P. Roberts, *Acc. Chem. Res.* **5** (1972) 387.

281 K. U. Ingold, *Chem. Commun.* **1969** 911.

282 A. G. Davies, D. Griller and B. P. Roberts, *J. Chem. Soc.* B **1971**, 1823.

283 G. I. Makin, V. P. Maslennikov and V. A. Shushunov, *Zh. Obshch. Khim.* **42** (1972) 834.

284 G. I. Makin, V. P. Maslennikov and V. A. Shushunov, *Tr. Khim. Khim. Tekhnol.* **1973**, 87.

285 E. I. Pichuzhkina, I. I. Kolodkina and A. M. Yurkevich, *Zh. Obshch. Khim.* **43** (1973) 2275.

286 J. C. Catlin and H. R. Snyder, *J. Org. Chem.* **34** (1969) 1660.

287 H. E. Dunn, J. C. Catlin and H. R. Snyder, *J. Org. Chem.* **33** (1968) 4483.

288 W. V. Hough, C. R. Guibert and G. T. Heffernan, *US Pat.*, 3 853 941 (1974); *CA* **82** (1975) 155299.

289 W. V. Hough, C. R. Guibert and G. Hefferan, *Ger. Offen.*, 2 209 047 (1972); *CA* **77** (1972) 164133.

290 V. F. Gridina, A. L. Klebanskii, V. A. Bartashev, L. P. Dorofeenko, N. V. Kozlova, and L. E. Krupnova, *Zh. Obshch. Khim.* **36** (1966) 1283.

291 A. Dornow and K. Fischer, *Chem. Ber.* **99** (1966) 68.

292 I. R. McKinley, H. Weigel, C. B. Barlow and R. D. Guthrie, *Carbohyd. Res.* **32** (1974) 187.

293 E. J. Bourne, I. R. McKinley and H. Weigel, *Carbohyd. Res.* **25** (1972) 516.

294 H. C. Fu, T. G. Psarras, H. Weidmann and H. K. Zimmermann, *Liebigs Ann. Chem.* **641** (1961) 116.

295 T. G. Psarras, H. K. Zimmermann, Y. Rasiel and H. Weidmann, *Liebigs Ann. Chem.* **655** (1962) 48.

296 R. J. Ferrier, *J. Chem. Soc.* **1961**, 2325.

297 A. M. Yurkevich, S. G. Verenikina, E. G. Chauser, and N. A. Preobrazhenskii, *J. Gen. Chem USSR* **36** (1966) 1741.

298 J. C. Lockhart, *J. Chem. Soc.* A **1968**, 869.

299 G. Wulff, W. Vesper, R. Grobe-Einsle and A. Sarhan, *Makromol. Chem.* **178** (1977) 2799.

300 R. Köster and W. V. Dahlhoff, *Ger. Offen.* 2 712 112 (1977); *CA* **88** (1978) 7301.

301 W. V. Dahlhoff and R. Köster, *J. Org. Chem.* **42** (1977) 3151.

302 W. V. Dahlhoff, P. Idelmann and R. Köster, *Angew. Chem.* **92** (1980) 552.

303 R. Köster, P. Idelmann and W. V. Dahlhoff, *Angew. Chem.* **92** (1980) 553.

304 W. Fenzl, W. V. Dahlhoff and R. Köster, *Liebigs Ann. Chem.* **1980**, 1179.

305 C. J. Griffiths and H. Weigel, *Carbohyd. Res.* **81** (1980) 7.

306 R. A. Sharma, M. Bobek and A. Bloch, *J. Med. Chem.* **18** (1975) 473.

307 Studiengesellschaft Kohle m.b.H., *Belg. Pat.* 852 901 (1977); *CA* **88** (1978) 152921.

308 H. Iida, H. Kitamura, K. Seita and K. Yamada, *Bull. Chem. Soc. Jpn* **47** (1974) 1028.

309 T. Okushi and O. Manabe, *Nippon Kagaku Kaishi* **1975**, 1435.

310 M. Giorgini, A. Lucchesi, G. Masetti and P. Santini, *Ann. Chim. (Rome)* **58** (1968) 1446.

311 R. Greenhalgh and P. J. Wood, *J. Chromatogr.* **82** (1973) 410.

312 J. W. Shepherd, *US Pat.* 3 259 591 (1966); *CA* **65** (1966) 9131.

313 J. W. Shepherd, *U.S. Pat.* 3 310 507 (1967); *CA* **67** (1967) 22590.

314 A. J. Krol, *US Pat.* 3 356 306 (1967); *CA* **68** (1968) 13740.

315 D. S. Bowering, N. B. Graham and J. D. Murdoch, *DDR Pat.* 44 945 (1966); *CA* **66** (1967) 3409.

316 C. A. Braithwaite, *US Pat.* 3 324 053 (1967); *CA* **67** (1967) 44407.

317 H. L. Lęe, *US Pat.* 3 378 504 (1968); *CA* **69** (1968) 3482.

318 L. E. Brown and J. D. Nutter, *US AEC Rep.* (1969) MLM-1641; *Nucl. Sci. Abstr.* **23** (1969) 33127; *CA* **71** (1969) 125381.

319 S. C. Lin and E. M. Pearce, *J. Polym. Sci., Polym. Chem. Ed.* **17** (1979) 3095.

320 R. J. Boot, *US Pat.* 3 213 048 (1965); *CA* **64** (1966) 2264.

321 J. H. Wright, *Brit. Pat.* 1 130 683 (1968); **70** (1969) 12268.

322 N. V. Polychemie-AKU G. E., *Neth. Appl.* 6 606 803 (1967); *CA* **68** (1968) 50637.

323 H. A. Cyba, *US Pat.* 3 428 469 (1969); *CA* **71** (1969) 82160.

324 K. Schmoll and G. Klein, *Ger. Offen.* 2 064 268 (1972); *CA* **77** (1972) 127484.

325 T. Morikawa and T. Amano, *Kobunshi Ronbunshu* **35** (1978) 471; *CA* **89** (1978) 180752.

326 S. C. Lin, *PhD thesis, Polytechnic Inst. New York* (1978); *CA* **89** (1978) 180761; *Diss. Abstr. Int.* **B39** (1978) 1320.

327 Y. Kobayashi, J. Nakamura, K. Murayama and T. Veshima, *Jpn Kokai Tokkyo Koho* 79-117 557 (1979); *CA* **92** (1980) 59721.

328 L. Mannes and H. Hennig, *Ger. Pat.* 1 214 212 (1966); *CA* **64** (1966) 20016.

329 Monsanto Co., *Neth. Appl.* 6 504 746 (1965); *CA* **64** (1966) 9735.

330 G. E. Irish, J. B. Hinkamp and J. D. Bartleson, *US Pat.* 3 254 975 (1966); *CA* **65** (1966) 6979.

331 J. B. Hinkamp, J. D. Bartleson and G. E. Irish, *US Pat.* 3 356 707 (1965); *CA* **68** (1968) 61493.

332 J. B. Hinkamp, J. D. Bartleson and G. E. Irish, *US Pat.* 3 509 054 (1961); *CA* **73** (1970) 27332.

333 V. Brantl, *Ger. Pat.* 2 500 683 (1977); *CA* **88** (1978) 39662.

334 T. Shima, T. Urasaki and I. Oka, *Adv. Chem. Ser.* **128** (1973) 183.

335 M. Yoshida and S. Ikeda, *Japan Pat.* 77-20 413 (1977); *CA* **88** (1978) 77385.

336 M. Yoshida, *Japan Pat.* 77–34 251 (1977); *CA* **88** (1978) 156657.

337 B. Swinney, D. A. Kinghan and R. W. Humphrey, *Brit. Pat.* 1 474 048 (1977); *CA* **87** (1977) 204147.

338 R. W. Lerner, H. Hayati and J. R. Flasch, *Ger. Offen.* 2 258 089 (1973). *CA* **79** (1973) 95626.

339 F. S. Clark and L. W. Banister, *US Pat.* 3 751 367 (1973); *CA* **79** (1973) 116840.

340 H. F. Askew, C. J. Harrington and T. J. P. Bridgewater, *Ger. Offen.* 2 531 086 (1976); *CA* **85** (1976) 65537.

341 Hoechst A. G., *Res. Discl.* **189** (1980) 23; *CA* **92** (1980) 183272.

342 K. Von Werner and W. Knoblauch, *Ger. Offen.* 2 926 269 (1981); *CA* **94** (1981) 194799.

343 K. Nobusuke and H. Sugiyama, *Japan Pat.* 73–43 792 (1973); *CA* **81** (1974) 12534.

344 Y. Hori, *Japan Pat.* 70–30 921 (1970); *CA* **75** (1971) 9339.

345 Y. Hori, *Japan Pat.* 72–25 245 (1972); *CA* **78** (1973) 46954.

346 S. Ikeda, Y. Yuzawa and I. Takahashi, *Japan Pat.* 74–106 916 (1974); *CA* **84** (1976) 48589.

347 N. Kito and H. Sugiyama, *Japan Pat.* 73–43 792 (1973); *CA*; **81** (1974) 125347.

348 Imperial Chem. Ind., *Neth. Appl.* 6 407 917 (1965); *CA* **62** (1965) 16084.

349 J. E. Helbig, D. O. Nelsen, R. E. Pennington and I. J. Satterfield, *US Pat.* 3 232 704 (1966); *CA* **64** (1966) 10803.

350 G. C. Monroe and L. W. Patton, *Belg. Pat.* 666 793 (1966); *CA* **65** (1966) 5371.

351 Shell Int. Res. Maatschapij N. V., *Neth. Appl.* 6 612 501 (1967); *CA* **67** (1967) 53655.

352 M. A. McMahon and A. Matthew, *US Pat.* 3 410 913 (1968); *CA* **70** (1969) 46827.

353 J. Alagy, L. Asselineau and C. Busson, *French Pat.* 1 568 753 (1969); *CA* **72** (1970) 110890.

354 P. G. Field, Ger. Offen. 2 253 963 (1973); *CA* **79** (1973) 79456.

355 C. M. King and M. T. Musser, *US Pat.* 3 864 380 (1975); *CA* **82** (1975) 155421.

356 E. V. Lebedev, I. V. Vasilenko and V. T. Sklyar, *Dopovidi Akad. Nauk Ukr. SSR* **B37** (1975) 717; *CA* **83** (1975) 178203.

357 I. V. Vasilenko, P. L. Klimenko and E. V. Lebedev, *Neftepererab. Neftekhim. (Kiev)* **13** (1975) 136.

358 I. V. Vasilenko, E. V. Lebedev and P. L. Klimenko, *Neftekhimiya* **15** (1975) 286; *CA* **83** (1975) 42647.

359 P. L. Won, V. A. Itskovich and V. M. Potekhin, *Neftekhimiya* **18** (1978) 755; *CA* **90** (1979) 103479.

360 P. L. Won, V. A. Itskovich and V. M. Potekhin, *Neftekhimiya* **20** (1980) 568; *CA* **93** (1980) 238444.

361 V. A. Itskovich, P. L. Won, T. G. Maslyanskaya, L. P. Tsudikova and V. V. Fokin, *Neftekhimiya* **18** (1978) 603; *CA* **89** (1978) 196672.

362 V. A. Itskovich, V. M. Potekhin and L. P. Tsudikova, *Zh. Prikl. Khim.* **51** (1978) 2583.

363 V. A. Itskovich, V. V. Fokin, T. G. Maslyanskaya, L. P. Tsudikova, P. L. Won and Ts. Ralchevski, *Okislenie Organ. Soedin. Zhidk. Faze* **1978** 130; *CA* **91** (1979) 38593.

364 T. G. Maslyanskaya, V. A. Itskovich, V. M. Potekhin and L. V. Shumkova, *Zh. Prikl. Khim.* **52** (1979) 220.

365 T. G. Maslyanskaya, V. A. Itskovich, V. M. Potekhin, V. A. Proskuryanov and E. N. Dyatlova, *Zh. Prikl. Khim.* **52** (1979) 228; *CA* **90** (1979) 167771.

366 R. A. Sheldon and J. A. Van Doorn, *J. Catal.* **34** (1974) 242.

367 G. Caprara, L. Rivolta and C. Bonacossa, *Ital. Pat.* 932 322 (1972); *CA* **86** (1977) 43214.

368 L. C. Fetterly, *US Pat.* 3 391 213 (1968); *CA* **69** (1968) 51520.

369 R. C. Wade, *US Pat.* 3 536 741 (1970); *CA* **74** (1971) 53990.

370 P. F. Wolf, J. E. McKeon and D. W. Cannell, *J. Org. Chem.* **40** (1975) 1875.

371 D. W. Young, *US Pat.* 3 081 321 (1963); *CA* **60** (1964) 2833.

372 G. F. Pollnow and D. T. Haworth, *US Pat.* 3 284 408 (1966); *CA* **66** (1967) 19198.

373 M. S. Matta, *Nucl. Sci. Abstr.* **22** (1968) 22898.

374 T. Mieda and H. Imai, *Japan Pat.* 71–35 375 (1971); *CA* **76** (1972) 73058.

375 T. Mieda and H. Imai, *Japan Pat.* 70–35 916 (1970); *CA* **74** (1971) 64668.

376 R. C. Wade, *US Pat.* 3 502 703 (1970); *CA* **72** (1970) 121007.

377 T. Shima, T. Urasaki and I. Oka, *Polymer Prepr. Am. Chem. Soc.* **13** (1972) 578; *CA* **80** (1974) 37681.

378 P. G. S. Fields, *Brit. Pat.* 1 413 475 (1975); *CA* **84** (1976) 73672.

379 H. Nakai, *Japan Pat.* 73–20 422 (1973); *CA* **80** (1974) 60624.

380 H. Yamana and H. Nakai, *Japan Pat.* 73–20 422 (1973); *CA* **80** (1974) 60623.

381 A. K. Hoffmann and W. M. Thomas, *US Pat.* 2 934 526 (1960); *CA* **1960**, 17327.

382 T. Shima, *Japan Pat.* 67–8947 (1967); *CA* **68** (1968) 30984.

383 H. Steadly, *US Pat.* 3 391 123 (1968); *CA* **69** (1968) 37028.

384 H. Juenger and F. Weissenfels, *Ger. Offen.* 2 214 821 (1973); *CA* **80** (1974) 84153.

385 I. I. Kolodkina, E. I. Kichuzhkina, E. A. Ivanova and A. M. Yurkevich, *USSR Pat.* 406 841 (1973); *CA* **83** (1975) 98199.

386 I. I. Kolodkina, A. M. Yurkevich and E. F. Panarin, *Vysokomol. Soedin.* **A18** (1976) 490.

387 I. I. Kolodkina, D. G. Valkovskii, E. I. Pichuzhkina, E. A. Ivanova, S. V. Rogozhin and A. M. Yurkevich, *Vysokomol. Soedin.* **A18** (1976) 47.

388 R. F. K. Meredith, *Ger. Offen.* 2 452 407 (1975); *CA* **83** (1975) 116357.
389 G. Wulff, A. Sarhan, J. Gimpel and E. Lohmar, *Chem. Ber.* **107** (1974) 3364.
390 E. S. Lopata, and S. R. Riccitiello, *J. Appl. Polym. Sci.* **21** (1977) 91.
391 S. J. Groszos, *US Pat* 3 029 210 (1962); *CA* **57** (1962) 2409.
392 M. A. Kasem and H. R. Richards, *Ind. Engng Chem. Prod. Res. Develop.* **11** (1972) 114.
393 R. Leipins, S. Gilbert, F. Tibbetts and J. Kearney, *J. Appl. Polym. Sci.* **17** (1973) 2523.
394 S. H. Roth and J. Green, *US Pat.* 3 843 596 (1974); *CA* **82** (1975) 45329.
395 A. Yamada and K. Kimura, *Japan Kokai* 77–72 735 (1977); *CA* **87** (1977) 152934.
396 R. C. Wade, *Ger. Offen.* 2 021 060 (1971); *CA* **74** (1971) 127016.
397 S. Ikeda, Y. Yuzawa and I. Takahashi, *Japan Pat.* 75–10 813 (1975); *CA* **85** (1976) 24256.
398 H. Zenchi, *Japan Pat.* 76–07 133 (1976); *CA* **85** (1976) 128457.
399 P. Goeuriot, F. Thevenot and J. H. Driver, *Thin Solid Films* **78** (1981) 67; *CA* **95** (1981) 118999.
400 J. D. Banhurst and D. K. Shumway, *US Pat.* 4 110 426 (1978); *CA* **90** (1979) 76418.
401 F. Cauyolle and G. Bergal, *Compt. Rend. C. R. Acad. Sci. Paris* **228** (1949) 1249, 1516; **230** (1950) 1101; **231** (1950) 1550; **232** (1951) 109.
402 K. Torssell, *Physiol. Plantarum* **9** (1956) 652.
403 J. C. Gage, *Brit. J. Ind. Med.* **27** (1970) 1; *CA* **73** (1970) 12650.
404 A. B. Borkovec, J. A. Settepani and G. C. La Breque, *J. Econ. Entomol.* **62** (1969) 1472.
405 J. A. Settepani, M. M. Crystal and A. B. Borkovec, *J. Econ. Entomol.* **62** (1969) 375; *CA* **70** (1969) 105408.
406 D. M. Uppdegraff, *J. Infect. Diseases* **114** (1964) 304.
407 J. L. Delarue and R. L. Fallard, *Belg. Pat.* 640 031 (1964); *CA* **63** (1965) 631.
408 M. A. A. F. de C. T. Carrondo and A. C. Skapski, *Acta Crystallogr.* **B34** (1978) 3551.
409 E. Koberstein, H. G. Bachmann, H. Gebauer, G. Köhler, E. Lakatos and G. Nonnenmacher, *Z. Anorg. Allg. Chem.* **374** (1970) 125.
410 U. Sommer and G. Heller, *J. Inorg. Nucl. Chem.* **34** (1972) 2713.
411 K. F. Jahr, G. Heller and U. Sommer, *J. Inorg. Nucl. Chem.* **30** (1968) 2544.
412 G. Heller and J. Zambelli, *Z. Naturforsch.* **32b** (1977) 1393.
413 M. Maneva, *CA* **88** (1978) 145404.
414 N. T. Toroptseva, *Izv. Vyssh. Uchebn. Zaved., Khim. Khim. Teknol.* **19** (1976) 1470; *CA* **86** (1977) 35711.
415 N. T. Toroptseva, N. E. Khomutov and T. A. Lgalova, *Zh. Prikl. Khim.* **49** (1976) 2392.
416 V. N. Rassokhina, Yu. S. Plyshevskii, Z. I. Shishkina, I. A. Kantseva and I. A. Leont'eva, *Tr. Uralsk. N. I. Khim. In-ta* **1976** 63; *CA* **88** (1978) 82989.
417 Yu. S. Plyshevskii, G. M. Smirnova and V. N. Rassokhina, *USSR Pat.* 406 797 (1973); *CA* **81** (1974) 172385.
418 M. S. Dobrolyubova and A. B. Tsentsiper, *Izv. Akad. Nauk SSSR, Ser. Khim.,* **1974**, 1218; *CA* **81** (1974) 98825.
419 L. G. Kochkina, N. N. Loginova, Yu. A. Panshin, O. B. Khachaturyan and N. E. Khomutov, *Zh. Prikl. Khim.* **51** (1978) 1250.
420 G. Heller and D. Marquard, *Z. Naturforsch.* **33b** (1978) 159.

5

Boron–Sulphur and Boron–Selenium Heterocycles

W. SIEBERT

University of Heidelberg, Federal Republic of Germany

5.1 Introduction

Cyclic boron–sulphur compounds were first obtained by Stock[1] when boron halides BX_3 (X = Cl, Br) were reacted with monosulphane H_2S. The products, however, were not recognized as rings, but were thought to be adducts of the boron halides with B_2S_3. The cyclic nature of the compounds was elucidated by Wiberg[2] and Goubeau,[3] who both contributed much to the development of boron chemistry.

Cyclic boron–sulphur compounds that have been synthesized in the past dozen years are described in this chapter. The field has been reviewed by Muetterties, Haiduc, Siebert, Sommer, Nöth and Krebs.[B 1 – B 7]

The combination of borylene moieties (R—B:) with sulphur atoms can lead to a variety of boron–sulphur ring systems:

(1) (2) (3) (4) (5) (6)

THE CHEMISTRY OF INORGANIC HOMO-
AND HETEROCYCLES VOL 1 ISBN 0-12-655775-6

Nineteen ring types are possible if only four-, five- and six-membered rings are considered, of which 1,3,2,4-dithiadiboretane (1), 1,3,5,2,4,6-trithiatriborinane (2), 1,2,4,3,5-trithiadiborolane (3), 1,3,2,4,5-dithiatriborolane (5), and 1,4,2,3,5,6-dithiatetraborinane (6) have been prepared. The reactions of R—BX$_2$ and H$_2$S$_2$[4] as well as of C$_6$H$_5$—BBr$_2$ and sulphur[5] were expected to yield 1,2,4,5,3,6-tetrathiadiborinane (4); however, in both cases the 1,2,4,3,5-trithiadiborolane (3) was formed. The tetrathiadiborinane (4) and its isomer 1,2,3,5,4,6-tetrathiadiborinane can be isolated as adducts with N-containing donor molecules.

Substitution of the silyl group in a 1,3,2,3,5-dithiasiladiborolane derivative by RB(SMe)$_2$ resulted in the formation of the dithiatriborolane (5).[6]

Attempts to prepare the heterocycle RBS$_5$ from (C$_5$H$_5$)$_2$TiS$_5$ and RBX$_2$ resulted in the formation of (3) and sulphur. Since some of the sulphur was obtained as cycloheptasulphur S$_7$, it is likely that two molecules of the intermediate RBS$_5$ had formed (3) and the resulting sulphur fragments S$_3$ and S$_4$ had recombined to yield S$_7$.

In addition to the boron-substituted B$_2$S$_2$, B$_2$S$_3$, B$_3$S$_3$, B$_3$S$_2$, and B$_4$S$_2$ rings, a very interesting macrocyclic molecule B$_8$S$_{16}$ has been reported.[7]

A variety of compounds are known in which other heteroatoms are incorporated into the B–S framework. Thus, combining the building blocks

with other atoms or groups containing carbon, silicon or nitrogen leads to new ring systems. Only carbon-free rings are considered in this book, and those containing nitrogen and silicon are discussed elsewhere in this volume.

5.2 Rings with three-coordinate boron

5.2.1 The B$_8$S$_{16}$ molecule

The discovery of boron–sulphur species in the mass spectra of boron-sulphides was the first indication of the possible existence of neutral oligomeric boron–sulphur compounds. Gilles et al.[8] detected (BS$_2$)$_n$$^+$ ions and concluded that the BS$_2$ oligomers B$_6$S$_{12}$, B$_7$S$_{14}$, B$_8$S$_{16}$, B$_9$S$_{18}$, and B$_{10}$S$_{20}$ are in equilibrium with a 0.5–1.0% solution of sulphur in B$_2$S$_3$. The high intensity of the B$_8$S$_{16}$$^+$ peak suggests that the reaction between B$_2$S$_3$ and sulphur leads to a volatile B$_8$S$_{16}$ compound. This is in agreement with the fact that sulphur-rich boron sulphides are more volatile than pure B$_2$S$_3$.[9]

A polymeric boron–sulphur compound of the composition $(B_2S_4)_n$ was obtained from the reaction of 3,5-dibromo-1,2,4,3,5-trithiadiborolane (**3a**) with monosulphane (equation 1). The initially formed thioboric acid (**3b**) easily loses H_2S to yield a polymeric product $(B_2S_4)_n$,[10] in which the trithiadiborolane rings are probably linked together by sulphur bridges.[11]

$$Br-B\overset{S-S}{\underset{S}{\diagup}}B-Br \xrightarrow[-2HBr]{2H_2S} HS-B\overset{S-S}{\underset{S}{\diagup}}B-SH \xrightarrow{-H_2S} \frac{1}{n}\left[S-B\overset{S-S}{\underset{S}{\diagup}}B-\right]_n \quad (1)$$

(**3a**)   (**3b**)

The compound B_8S_{16} (**7**) has been obtained by the following methods:[12]

   (a) by heating a B_2S_3/S mixture (1 : 1.5) to 300 °C/10^{-4} bar in a quartz tube (two-zone oven with a sharp temperature profile, 300/100 °C);

   (b) as shown in equation (2), by treating 3,5-dibromo-1,2,4,3,5-trithiadiborolane with dilute solutions of the H_2S generator trithiocarbonic acid $S{=}C(SH)_2$ in CS_2 (6% yield):

$$4\ Br-B\overset{S-S}{\underset{S}{\diagup}}B-Br \xrightarrow[\substack{-4CS_2\\-8HBr}]{4\ SC(SH)_2} \quad (2)$$

(**3a**)

(**7**)

The colourless needles that were obtained decompose above 115 °C and are extremely sensitive towards water.

Solid boron disulphide is made up of planar B_8S_{16} molecules (D_{4h} symmetry). Each boron atom forms three boron–sulphur bonds, B–SS 1.822 Å, B–SB (ring) 1.789 Å, B–SB (bridge) 1.823 Å. The average length of the B–S bond in B_8S_{16} (1.811 Å) is similar to the average length in B_2S_3 (1.808 Å). The difference in the B–S bonds within the trithiadiborolane rings is a reflection of the different (p–p)π bonds in the molecule (the transannular S···S distance in B_8S_{16} is 4.667 Å). In the mass spectrum the strongest peaks are of the molecular ion and of the $B_2S_3^+$ fragments.[12]

5.2.2 1,3,2,4-Dithiadiboretanes $(RBS)_2$

Wiberg[13] reported the formation of the dimeric thioboric acid $(HS{\cdot}BS)_2$ (**1**), R = SH, m.p. 120 °C, from H_2S and BBr_3. However, attempts to repeat the synthesis of this compound were unsuccessful.[14] Distillation of $(EtBS)_3$ yielded the dimer $(EtBS)_2$ (1), R = Et.[15] The reaction of the borane adduct $H_3B.NEt_3$ with H_2S at high temperatures led to the diethylamino derivative of 1,3,2,4-dithiadiboretane, $(Et_2N—BS)_2$ (1), R = NEt_2, a thermally stable compound, which does not show any tendency to form the corresponding trimer.[16]

The 1,3,2,4-dithiadiboretanes have not yet been structurally characterized. The only available evidence of a B_2S_2 ring stems from an X-ray diffraction study of diboron trisulphide.[12] A structural characterization of B_2S_3 was made possible when Krebs[12] obtained suitable crystals by heating $Ag_3B_5S_9$. It was found that planar six-membered B_3S_3 rings and four-membered B_2S_2 rings are connected by sulphur bridges to form almost-planar two-dimensional layers. Each B_3S_3 ring is linked to one B_2S_2 and two B_3S_3 rings. Although the B—S bonds of both rings do not exhibit large differences (average B—S 1.808 Å), the BSB bond angle in the B_2S_2 ring is only 76°, while in the B_3S_3 ring it is 110°. The transannular B $\cdots$ B (2.237 Å) and S $\cdots$ S distances (2.878 Å) are remarkably short and are a reflection of ring strain. The building blocks of polymeric B_2S_3 (B_2S_2 and B_3S_3 rings) are found in the monomeric boron–sulphur compounds.

5.2.3   1,3,5,2,4,6-Trithiatriborinanes $(RBS)_3$

Different names, such as borsulfole,[13] borthiine and cyclotriborthiane are used in the literature for the six-membered B_3S_3 ring. The chemistry of $(RBS)_3$ was explored by Wiberg[13] and his results have been fully reviewed by Haiduc.[B 2]

Synthesis. The classic synthesis of $(RBS)_3$ compounds[1,13] is based upon the reaction of boron trihalides or organoboron dihalides with H_2S as a source of sulphur. Sodium sulphide Na_2S,[17] di-t-butylsulphane Bu_2^tS,[18] tetramethylcyclodisilthiane $(Me_2SiS)_2$,[19] hexamethylcyclotrisilthiane $(Me_2SiS)_3$,[19,20] and mercury(II) sulphide HgS[20] have been employed in the synthesis of $(RBS)_3$, by using the reactions of these reagents with organoboron dihalides RBX_2.

Redox reactions between iodoboranes XBI_2 (X = I or organic group) and elemental sulphur result in the formation of B–S ring compounds[10,21] and iodine:

$$2\ X—BI_2 + \tfrac{3}{8}S \xrightarrow[-2I_2]{25\,°C} X—B\overset{\displaystyle S—S}{\underset{\displaystyle S}{\diagup\diagdown}}B—X \xrightarrow[120\,°C]{XBI_2} (X—B—S)_3 + I_2 \qquad (3)$$

(3)

In the first step 1,2,4,3,5-trithiadiborolanes (3) are obtained.[4] At higher temperatures an additional iodoborane molecule attacks the S–S bridge in (3) and a formal insertion of borylene X–B: into the S–S bond yields the six-membered ring system (RBS)$_3$. Replacement of the halogen atoms (X = Br, I) in (XBS)$_3$ by HS groups leads to metathioboric acid (HS·BS)$_3$, of which several metal salts $M_3B_3S_6$ exist.[22] These ternary compounds are obtained in the high-temperature reactions of the boron sulphides with metal sulphides or of boron, H_2S and M_2S:

$$3\ K_2S + 6\ B + 9\ H_2S \longrightarrow 2\ K_3B_3S_6 \tag{4}$$

X-ray diffraction studies of these extremely air-sensitive salts have confirmed the planar structure of the B_3S_3 rings.[23]

Structure and bonding. The structure of (RBS)$_3$ was derived from Raman data[24] and molecular-weight determinations[2] and was proved later by X-ray structure analysis. The crystal structure[25] of the tribromo derivative (BrBS)$_3$ was redetermined.[26] The average values of the bond distances and the ring angles are B–S 1.807 Å, B–Br 1.895 Å and BSB 109.2°, SBS 130.7°. The compound (BrBS)$_3$ consists of an almost-planar six-membered ring, similar to that of (HSBS)$_3$,[14] in which the important parameters are B–S 1.803(5) Å, B–SH 1.816(6) Å and SBS 130.0(4)°, BSB 109.9(1)°. The length of the B–S bond in the ring is slightly less than that of the exocyclic B–S bond. However, the exocyclic B–S bond is considerably shorter in the isotypic compounds (NaSBS)$_3$ and (KSBS)$_3$ than in (HSBS)$_3$ (1.816(6) Å). This is a reflection of a higher degree of π-bonding in the B–S$^-$ group.

Physical properties. The compounds of the formula (RBS)$_3$ (where R = Cl, Br, I, SH, Bu^t, Bu^n, Bu^i, Ph, NMe$_2$, NPh$_2$) are all colourless. Their thermal stability depends on the substituent R at the boron atom. For R = Cl, Br or I the stability increases with increasing molecular weight of the halogen: for example (ClBS)$_3$ decomposes at room-temperature into BCl_3 and B_2S_3. Above 80 °C, (BrBS)$_3$ rapidly yields polymeric B_2S_3 and BBr_3, whereas (IBS)$_3$ is stable and sublimes at 200 °C.[11] However, (IBS)$_3$ is light-sensitive. Molecular-weight determinations of (MeBS)$_3$ indicate that in solution the trimeric nature of the compound diminishes and that oligomeric or polymeric products (MeBS)$_n$ ($n > 3$) are formed. On the other hand, the distillation of (EtSBS)$_3$ causes a ring contraction, to yield the corresponding 1,3,2,4-dithiadiboretane.

Chemical properties. Because of the high Lewis acid nature of the 1,3,5,2,4,6-trithiatriborinane system and the weakness of the B–S bond, the ring is easily cleaved by nucleophilic reagents such as water, alcohols,

amines and fluorides to yield H_2S or metal sulphides and the corresponding B–O, B–N or B–F compounds. In compounds with sterically or electronically active R groups, the sensitivity towards hydrolysis is reduced.

The low thermal stability of $(BrBS)_3$ can be used to synthesize the 1,2,4,3,5-trithiadiborolane ring. In the presence of sulphur the decomposition of $(BrBS)_3$ to B_2S_3 and BBr_3 is suppressed, since sulphur is incorporated into the B–S framework.[29] The result is the formation of the thermally stable 1,2,4,3,5-trithiadiborolane ring with a disulphane bridge (3), R = Br. This change from a six-membered B_3S_3 to a five-membered B_2S_3 ring is also observed when boron sulphide $(B_2S_3)_n$ and sulphur are heated together. The product $(BS_2)_n$ is a new B–S phase, in which 1,2,4,3,5-trithiadiborolane rings are linked together by sulphane bridges.[22]

The high thermal stability of $(IBS)_3$ with respect to B_2S_3 and BI_3 on the one hand and its reactivity on the other hand makes $(IBS)_3$ an ideal source for the I—B≡S fragment. This group is essential for the synthesis of 1,2,5-thiadiborolene rings (p. 148).[75-78]

5.2.4   1,2,4,3,5-Trithiadiborolanes

Synthesis. High yields of 3,5-dibromo-1,2,4,3,5-trithiadiborolane (**3a**) are obtained from H_2S_2 or H_2S_n ($n \sim 6$, a mixture of polysulphanes) and BBr_3 in CS_2.[29,30] Similarly, the dimethyl and diphenyl derivatives of (**3**), R = Me, Ph, are formed from $MeBBr_2$ or $PhBCl_2$ and polysulphanes[30] or Na_2S_2.[17] Iodoboranes are easily oxidized by compounds containing S–S groups.[10] Thus the reaction of BI_3 or $PhBI_2$ with elemental sulphur at room temperature leads to the 1,2,4,3,5-trithiadiborolane ring (3), R = I or Ph, and elemental iodine:[4,30]

$$2\ RBI_2 + \tfrac{3}{8} S_8 \longrightarrow R_2B_2S_3 + 2\ I_2 \tag{5}$$

Cleavage of the C–S bond in di-t-butyldisulphane by an organodiiodoborane (equation 6) is another example of the high tendency of formation[18] of the B_2S_3 ring, which is also formed when $(BrBS)_3$ and sulphur are heated (equation 7):

$$2\ RBI_2 + 2\ Me_3C—S—S—CMe_3 \longrightarrow R_2B_2S_3 + 4\ Me_3CI + \tfrac{1}{8}S_8 \tag{6}$$

$$2\ (BrBS)_3 + \tfrac{3}{8}S_8 \longrightarrow 3\ Br_2B_2S_3 \tag{7}$$

Thermal rearrangement of diboryldisulphanes[28] also yields the B_2S_3 ring (equation 8). Oxidation of trimethylsilylbis(dimethylamino)borane by sulphur (equation 9) leads to $Me_3Si—S_n—B(NMe_2)_2$, from which $(Me_3Si)_2S$ is eliminated. The resulting $(Me_2N)_2B—S_n—B(NMe_2)_2$ loses $B(NMe_2)_3$ and

$(Me_2N)_2B_2S_3$ is formed.[31] An analogous reaction is observed with the corresponding tin compound.[32]

$$2 R_2B\text{—}S\text{—}S\text{—}BR_2 \longrightarrow R_2B_2S_3 + 2 BR_3 + \tfrac{1}{8} S_8 \qquad (8)$$

$$4 Me_3Si\text{—}B(NMe_2)_2 + \tfrac{5}{8} S_8 \longrightarrow (Me_2N)_2B_2S_3$$
$$+ 2 (Me_3Si)_2S + 2 B(NMe_2)_3 \qquad (9)$$

Structure and bonding. For the 1,2,4,3,5-trithiadiborolane ring (**3**) a planar B_2S_3 skeleton was proposed[29,30] and subsequently proved by electron-diffraction studies[33] on $Cl_2B_2S_3$ and $Me_2B_2S_3$. Nonplanarity was found for $Ph_2B_2S_3$ by an X-ray diffraction study;[22] the unique sulphur is located 0.12 Å above the plane passing through the BS_2B atoms. The planes of the phenyl rings are rotated by *c.* 20° with respect to the plane of the B_2S_3 ring. Important molecular parameters are summarized in Table 5.1. CNDO/2-calculations indicate for $R_2B_2S_3$ the presence of some $(B\text{–}S)\pi$ bond character.[34,35]

Table 5.1   Bond lengths and angles in $R_2B_2S_3$.

R	B–S	S–S	B–R (Å)	BSB	SBS	BSS
Me	1.803	2.076	1.569	101.6°	117.7°	101.5°
Cl	1.790	2.067	1.758	96.7°	121.8°	99.8°
Ph	1.807	2.062	1.55			

Physical properties. Trithiadiborolanes exhibit high thermal stability, which suggests that the B_2S_3 skeleton possesses favourable geometric and electronic properties. The thermal transformation of the boron sulphide $(B_2S_3)_n$ by sulphur into $(BS_2)_n$ is evidence of a change in ring size from B_3S_3 and B_2S_2 in $(B_2S_3)_n$ to B_2S_3 in $(B_2S_4)_n$. The ring contraction of $(BrBS)_3$ with sulphur to $Br_2B_2S_3$ is additional proof for the high thermal stability of the B_2S_3 framework. Most of the compounds (R = Cl, Br, alkyl, aryl, SMe, NMe_2) are stable up to 200 °C. At elevated temperatures $I_2B_2S_3$ decomposes to yield elemental iodine and a boron–sulphur polymer because of its highly reactive B–I functions. Although a large number of derivatives are known, $(RO)_2B_2S_3$ and $F_2B_2S_3$ are obviously unstable.[30]

Pure 1,2,4,3,5-trithiadiborolane $H_2B_2S_3$ does not exist in the liquid phase since it readily polymerizes.[36]

Chemical properties. The reactivity of trithiadiborolanes is determined by the weakness of the endocyclic B–S bond and the exocyclic B–R bond. Several types of reaction have been reported for trithiadiborolanes: (*a*) the formation of donor–acceptor compounds between Lewis bases and the sp^2-hybridized boron atoms of $R_2B_2S_3$; (*b*) the substitution of the exocyclic substituent R; (c) the substitution of the entire B–R group by B–X; (d) the controlled replacement of the sulphane or disulphane bridge by amines or hydrazines; and (e) reactions of $R_2B_2S_3$ with alkenes, alkynes, isocyanates and carbodiimides, which lead to cyclic products with ring carbon atoms and are beyond the scope of this book.

Proton-active agents such as H_2O, ROH, RSH and RNH_2 can cause complete break-up of the B_2S_3 framework, yielding RBX_2, H_2S and H_2S_2; the latter rapidly decomposing to H_2S and sulphur.

Replacement of the substituent. Only minute quantities of $Cl_2B_2S_3$ are obtained from BCl_3 and H_2S_n ($n = 2$–6) because of the reduced reactivity of the B–Cl bond compared with that of the B–Br bond. The replacement of Br in $Br_2B_2S_3$ is achieved with HCl and affords $Cl_2B_2S_3$ in 70% yield. Attempts to synthesize the fluoro derivative $F_2B_2S_3$ from $Br_2B_2S_3$ and AsF_3 or SbF_3 have failed.[30] Instead, the B_2S_3 ring is ruptured and BF_3 and As_2S_3 are formed.

Attempts to synthesize 1,2,4,3,5-trithiadiborolane $H_2B_2S_3$, the parent compound of this class, from $Br_2B_2S_3$ and $NaBH_4$, $LiBH_4$ or $LiAlH_4$ were unsuccessful.[30] The reaction with tetrapropyldiborane (equation 10) resulted in exchange of the alkyl group. The action of $H_3B.THF$, tetrapropyldiborane and 9-borabicyclo-[3.3.1]-nonane (9-BBN) on a series of 3,5-substituted 1,2,4,3,5-trithiadiborolanes was followed by NMR spectroscopy in order to find the optimum conditions for the formation of $H_2B_2S_3$.[36] The compounds $(Me_2N)_2B_2S_3$ and 9-BBN proved to be the best reagents for the preparation of $H_2B_2S_3$ in solution (equation 11). The volatile compound polymerizes readily. As a result, monomeric $H_2B_2S_3$ does not exist in the liquid phase.[56]

$$Br_2B_2S_3 + (Pr_2BH)_2 \longrightarrow Pr_2B_2S_3 + 2(PrBrBH) \qquad (10)$$

$$(Me_2N)_2B_2S_3 + 2\,(C_8H_{14}BH)_2 \longrightarrow$$
$$H_2B_2S_3 + (C_8H_{14})BH.Me_2NBC_8H_{14} \qquad (11)$$

The formation of adducts with trimethylamine, $H_2B_2S_3.NMe_3$ and $H_2B_2S_3.2NMe_3$, stabilizes the trithiadiborolane. The adducts slowly lose trimethylamine.

The replacement of bromine in $Br_2B_2S_3$ by R_2NH,[30] $RNHLi$[42] and R_2NLi[42] leads to the amino-substituted derivatives $(R_2N)_2B_2S_3$ (3), R = NR_2, in good yield.

Attempts to prepare $(RO)_2B_2S_3$ from $Br_2B_2S_3$ and ROH or $B(OR)_3$ have failed. The substitution of Br occurs with EtSH to yield $(EtS)_2B_2S_3$. However, an excess of mercaptan cleaves the ring.

Compound (3a) reacts with AgNCS to yield the yellow compound

(SCN)$_2$B$_2$S$_3$[42] Organo-substituted compounds R$_2$B$_2$S$_3$ are usually obtained from RBX$_2$ and H$_2$S$_n$ (n = 2–6). Their synthesis is also possible from reactions of Br$_2$B$_2$S$_3$ with organolithium reagents.[37]

The mechanism of the exchange reactions with 1,2,4,3,5-trithiadiborolane has been extensively studied by Nöth.[37] [10]B labelling has shown that an exchange of substituents between 1,2,4,3,5-trithiadiborolanes X$_2$B$_2$S$_3$ and Y$_2$B$_2$S$_3$ can proceed by two routes.[38] A transfer of X and Y from one ring to the other occurs by an exo-process. Mixed trithiadiborolanes (XB)(YB)S$_3$, in which the ring boron atoms have not been exchanged, are the result. In an endo-process the boryl groups B–X and B–Y are exchanged. The rapid exo-reaction is accompanied by a slower endo-reaction, which leads to a statistical distribution of the B isotopes. Alkyl trithiadiborolanes only undergo an endo-process, since a lone pair of electrons at the alkyl group is not available.

Small alkyl groups (R = Me) in (R$_2$N)$_2$B$_2$S$_3$ allow an exocyclic exchange[39] with BBr$_3$ to take place. The bulky diisopropylamino group prevents an electrophilic attack by BBr$_3$ at the N atom and results in an endocyclic exchange. Coordination of BBr$_3$ at one of the sulphur atoms of the bridge has been observed.

Substitution of the S and S$_2$ bridges. Primary amines react with 3,5-dimethyl-1,2,4,3,5-trithiadiborolane Me$_2$B$_2$S$_3$ by opening up the ring. This is followed by a cyclocondensation, to yield 1,2,4,3,5-dithiazadiborolidines[40,41] and 1,3,5,2,4,6-thiadiazatriborinanes. The treatment of 3,5-dihalogeno-1,2,4,3,5-trithiadiborolanes with trimethylsilyl- and pentafluorophenyl-N-sulphinylamine also leads to the S$_2$B$_2$N heterocycle.[42] With hydrazines, the S$_2$ bridge is replaced and 1,3,4,2,5-thiadiazadiborolidines[43,44,49] are formed. These rings are discussed in Chapter 2 (Boron–Nitrogen Heterocycles).

The S_2 bridge in trithiadiborolanes may be replaced by alkynes and alkenes, to yield 1,3,2-dithiaborolenes and 1,3,2-dithiaborolanes,[45] but these carbon-containing systems are not discussed further here.

Formation of donor–acceptor compounds. The unsubstituted compound $H_2B_2S_3$ (3), R = H, does not exist as a monomeric species in the liquid phase. In solution the monomer is in equilibrium with its dimer (8), which forms nonvolatile and insoluble oligomers:

$$2 H \text{—} \underset{(3)}{\overset{S\text{—}S}{B\diagdown_S\diagup B}} \text{—} H \; \rightleftharpoons \; (8) \longrightarrow \; [(HB)_2S_3]_n \quad (12)$$

The treatment of a solution of $H_2B_2S_3$ (3), R = H, in hexane with trimethylamine yields a colourless adduct, (9) which exhibits two doublets in the ^{11}B NMR spectrum (61.7 ppm (J(B–H) = 152 $\pm$ 4 Hz) and 8.5 ppm (J(B–H) = 130 $\pm$ 4 Hz) and a singlet in the ^1H NMR spectrum (δ = 1.67). The IR spectrum supports the proposed structure: ν(B–H) = 2550 cm^{-1} for the three-coordinate and ν(B–H) = 2430 (br) cm^{-1} for the four-coordinate boron–hydrogen group. The adduct decomposes at 70 °C.[36]

(9)   (10)

An excess of trimethylamine and (3), R = H, in toluene leads to a 1:2 adduct (10). On the basis of NMR ($\delta^{11}B$ = 8.9, J(B–H) = 120 Hz) and IR data (ν(B–H) = 2383 cm^{-1} for a four-coordinate B–H group) both boron atoms are tetracoordinated. The adduct slowly loses trimethylamine.

3,5-Bis(diisopropylamino)-1,2,4,3,5-trithiadiborolane forms a BBr_3 adduct. The equilibrium constant K_{290} is 95, and thermolysis of the adduct yields $Br_2B_2S_3$ and $(Me_2CH)_2N\text{—}BBr_2$.[39]

5.2.5 Tetrathiadiborinanes

Two isomeric six-membered B_2S_4 heterocycles have been reported. In both compounds the B_2S_4 ring is stabilized by N-bases at both of the boron atoms in the ring. The change from trigonal (sp^2) to tetrahedral boron (sp^3) causes a decrease in the SBS angle, and thus deviations from the favourable 1,2,4,3,5-trithiadiborolane structure are possible.

1,2,4,5-Tetrathia-3,6-diborinane. The nucleophilic attack of amines (primary, secondary) on trithiadiborolanes results in the rupture of the heterocycle, and may lead to several products such as aminomercaptoborane, "aminoperthioborane" and bis(aminoboryl)disulphane:[41]

$$ \text{R-B} \overset{\text{S-S}}{\underset{\text{S}}{\diagup}} \text{B-R} \xrightarrow{\text{amines}} \text{R-B} \overset{\text{SH}}{\underset{\text{NR}_2}{\diagup}} , \quad \underset{\substack{\\ \text{N} \\ \diagup \diagdown \\ \text{H} \quad \text{R}}}{\text{R}} \overset{\text{S}}{\underset{\diagdown}{\text{B}}} \overset{\diagup}{\underset{\text{S}}{\diagdown}} \overset{\text{R}}{\underset{\substack{\\ \text{N} \\ \diagup \diagdown \\ \text{H}}}{\text{B}}} \text{R} \quad + \cdots \quad (13) $$

The "aminoperthioboranes" are only slightly soluble and slightly volatile. The dimer is observed in the mass spectrum. The ^{11}B NMR shift of $\delta = 11$ (R = Me) is strong evidence of four-coordinate boron atoms. A strong IR absorption in the 3120–3000 cm^{-1} region is indicative of N—H groups. These data support the formation[46] of a diadduct (11) of the unknown $R_2B_2S_4$ ring system (4):

$$ \underset{\text{Me}}{\overset{\text{HR}_2\text{N}}{}} \overset{\text{S-S}}{\underset{\text{S-S}}{\diagup \diagdown}} \overset{\text{Me}}{\underset{\text{NR}_2\text{H}}{}} $$

(11)

1,2,3,5-Tetrathia-4,6-diborinane. This ring (12) is not known as a monocyclic species, but it has been identified in a cage structure (13) as an adduct of the 4,6-dibromo derivative with 1,2,3-trimethyl-1,3,2-diazaborolidine. Thus, when impure and aged 3,5-dibromo-trithiadiborolane $Br_2B_2S_3$ reacted with 1,2,3-trimethyl-1,3,2-diazaborolidine (equation 14), a crystalline compound was isolated. An X-ray structure analysis showed the material to be the cage adduct (13).[47]

$$ \underset{\text{S}}{\overset{\text{R}}{\underset{\diagdown}{\text{B}}}} \overset{\text{S}}{\underset{\text{S}}{\diagup \diagdown}} \overset{\text{R}}{\underset{\diagup}{\text{B}}} $$

(12)

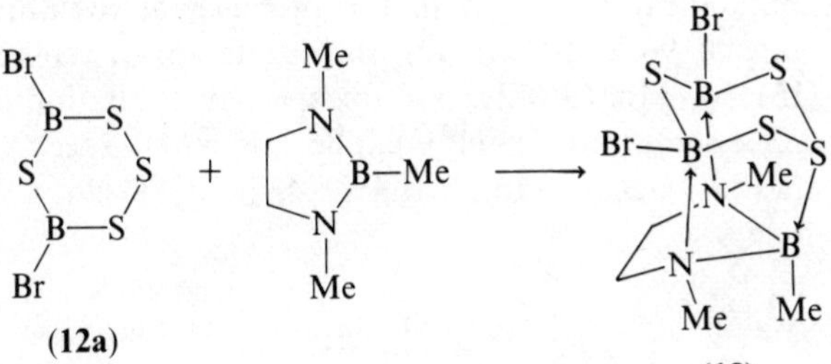

(14)

The B–S bonds in the $Br_2B_2S_4$ ring are in the 1.86–1.91 Å range. The intermolecular B–S bond is 2.04 Å and the S–S bonds (2.03, 2.04 Å) are shorter than those in S_8 (2.059 Å). Attempts to prepare the adduct from dibromotrithiadiborolane, sulphur and 1,3,2-diazaborolidine have failed.[47] Most likely, the heterocycle (12a) was already present in the dibromotrithia-diborolane prepared from BBr_3 and H_2S_n ($n \sim 6$). The $Br_2B_2S_3$ distilled from the reaction mixture is a yellow liquid, whereas pure $Br_2B_2S_3$ is only slightly yellow.[30] The colour of the crude product probably stems from heterocycles with more than three sulphur atoms in the ring.

The reaction of Cp_2TiS_5 with RBX_2 should lead to the sulphur-rich heterocycle 1,2,3,4,5,6-pentathiaborinane[27,40] (14), but besides Cp_2TiX_2 and sulphur the only boron–sulphur ring that was isolated was the 1,2,4,3,5-trithiadiborolane (3):

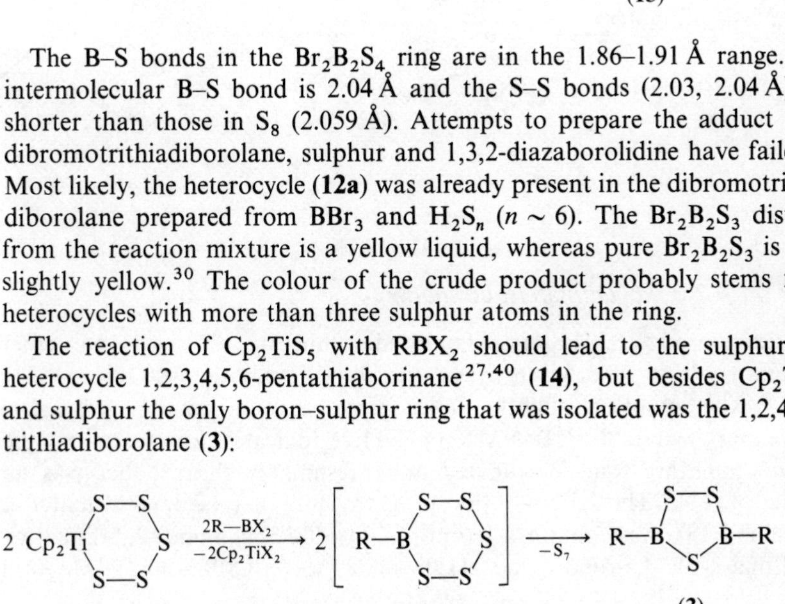

(15)

It is presumably formed from the intermediate, which supplies the R—B—S and R—BS$_2$ fragments. The discovery of small amounts of S_7, probably formed from the S_4 and S_3 fragments that result when two molecules of R—BS$_5$ are converted into (3), support the intermediate formation of a BS$_5$ heterocycle.

5.2.6 1,3,2,4,5-Dithiatriborolane

The preferential formation of 1,2,4,3,5-trithiadiborolane $R_2B_2S_3$ in various reactions is undoubtedly a result of the favourable geometry and electronic interactions in the B_2S_3 framework. In a B_3S_2 ring the electronic situation should be less favourable, unless the electron-deficient character of the boron atoms is reduced by electron-donating groups.[6] The reaction of (15) with diethylaminodibromoborane in the presence of pyridine yields (16) (65% yield, b.p. 94–96 °C/10^{-3} Torr). In an attempt to replace the Me_2N groups in (15) by $MeB(SMe)_2$, an unexpected formation of (17) was observed[6] (58% yield, m.p. 73–74 °C). The ^{11}B NMR spectra of (16) and (17) each shows two signals ((16): δ = 46.0, 42 (sh)), (17): δ = 72.0 and 43.5 (1 : 2 ratio)).

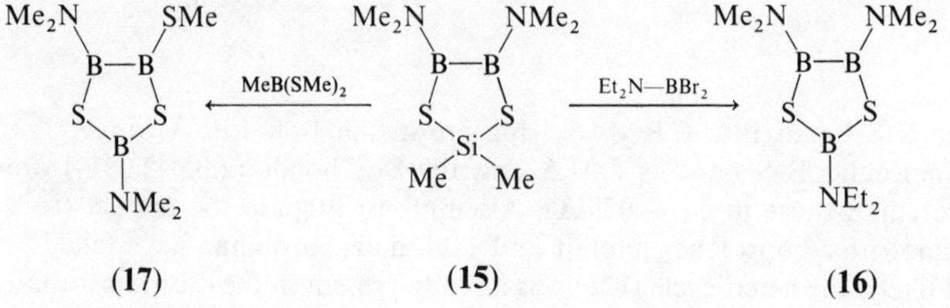

(17) (15) (16) (16)

5.2.7 1,4,2,3,5,6-Dithiatetraborinane

The reaction of tetrakis(dimethylamino)diborane(4), HCl and H_2S in ether (equation 17) is reported to yield the dithiatetraborinane (18) in 90% yield.[48] The colourless, slightly air-sensitive compound can be sublimed *in vacuo*, m.p. = 112 °C. ^{11}B NMR (δ = 11.2) indicates, however, that the boron atoms are four-coordinated and presumably the product was an adduct of (18).[6] The 2,3,5,6-tetrakis(dimethylamino)-1,4,2,3,5,6-dithiatetra-borinane (18) was obtained from 4,5-bis(dimethylamino)-2,2-dimethyl-1,3-dithia-2-sila-4,5-diborolane (19) and $Me_2N(Cl)B$—$B(Cl)NMe_2$ in 77% yield.[6]

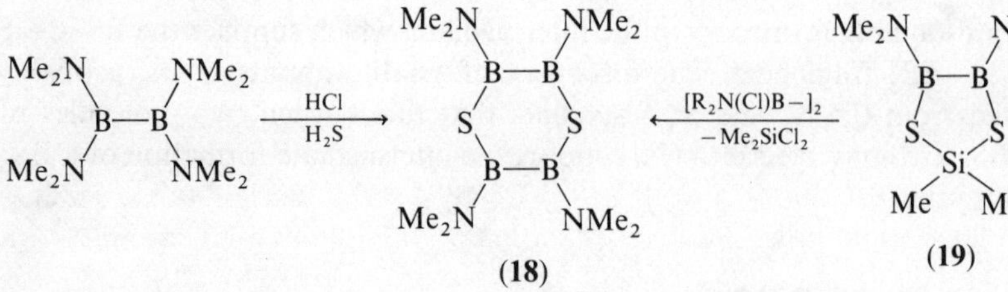

(18) (19) (17)

The constitution of (18) was proved by an X-ray structure analysis,[6] which showed that the molecule is nonplanar ($\sim D_{2d}$ symmetry). The C_2N and B_2S planes are almost coplanar, allowing an optimal (B–N)π interaction. The interplanar angle between the two C_2NB planes is 64.5°, and that between the two BSB planes is 32°, which results in considerable twisting of the molecule. The B–S bonds are quite long (1.84–1.87 Å; B–B 1.70 Å, B–N 1.37, 1.38 Å).

5.2.8 Thiasilaborolanes

Both 1,3,2,4,5-dithiasiladiborolane (**19**) and 1,3,4,5,2-dithiadisilaborolane (**20**) are known[6] and are discussed in Chapter 11:

$$R_2N-B-B-NR_2$$

(structure 19: five-membered ring with two S and Si(R$_2$))

$$R_2Si-SiR_2$$

(structure 20: five-membered ring with two S and B(Ph))

(**19**)   (**20**)

5.3 Rings with four-coordinate boron

Boron–sulphur ring compounds with sp³ hybridized boron atoms can be divided into several classes: (*a*) compounds with intermolecular donor–acceptor interactions, sulphur atoms not involved; (*b*) compounds with intra- or intermolecular donor–acceptor interactions with involvement of sulphur atoms; (*c*) metal thioborates with four boron sulphur bonds. A fourth class of compounds with tetracoordinate boron atoms is found in transition-metal complexes of boron–sulphur rings. The three substituents of the boron atoms, however, remain in the trigonal plane.

The electrophilic character of three-coordinate boron in boron–sulphur rings stems from the empty $2p_z$ orbital of the boron atom, which is used for bonding. The donor atom supplies an electron pair, and with the formation of the new bond a change of the hybridization from sp² to sp³ occurs. Compounds with a high degree of (B–X)π bonds possess only weak Lewis acidities.

The ¹¹B NMR spectra of boron compounds give information on the hybridization of the boron atoms.[50] Usually, the ¹¹B signals are shifted about 20–60 ppm to higher field, when the boron atoms change from sp² to sp³.

5.3.1 Intermolecular donor–acceptor compounds

In principle, each boron–sulphur ring described in this chapter should be capable of forming donor–acceptor compounds with Lewis bases such as NH_3, amines and phosphines.

A number of adducts of 1,3,5,2,4,6-trithiatriborinanes (2) have been reported;[2] however, no structural data are available. 1,2,4,3,5-trithiadiborolanes (3) form adducts with bases. They also can react with the Lewis acid BBr_3 to yield adducts.

A six-membered $(BSP)_2$ ring compound (21) is thought to be formed by dimerization of Me_2BSPPh_2:[51]

$$
\begin{array}{c}
Me_2 \\
B \\
Ph_2P^+ \diagup\;\overline{}\;\diagdown S \\
S \diagdown \;\diagup ^+PPh_2 \\
B \\
Me_2
\end{array}
$$

(21)

The reaction of B_2H_6 with H_2S results in the formation of a polymer $(HBS)_n$ and hydrogen. Presumably the adduct $H_3B.SH_2$ that is formed in the first step loses hydrogen to yield H_2B—SH, which through intermolecular interactions forms $(H_2B$—$SH)_n$ and finally $(HBS)_n$. The replacement of a hydrogen in H_2S by an alkyl group increases the Lewis-base character of the sulphur atom. On the other hand an exchange of H for halogen in B_2H_6 increases the Lewis acidity of the borane. Thus one expects favourable intermolecular donor–acceptor interactions to take place.

5.3.2 Methylthiodiborane and alkylthioboranes

$$
\begin{array}{ccc}
\begin{array}{c}
H\diagdown\;\;\diagup H \\
B \\
Me—S^+\;\;\diagup\overline{}\diagdown\;H \\
B \\
H\diagup\;\;\diagdown H
\end{array}
&
\begin{array}{c}
R \\
H_2\overline{B}—{}^+S \\
R—S^+\;\;{}^-BH_2 \\
H_2\overline{B}—{}^+S \\
R
\end{array}
&
\begin{array}{c}
Me_2 \\
H_2\overline{B}—{}^+N \\
Me—S^+\;\;{}^-BH_2 \\
H_2\overline{B}—{}^+N \\
Me_2
\end{array}
\end{array}
$$

(22)   (23)   (24)

Methanethiol reacts with B_2H_6 to yield the cyclic boron–sulphur compound (22) with a B–S–B and B–H–B bridge.[52] This diborane derivative slowly rearranges at room temperature to B_2H_6 and $(MeS$—$BH_2)_n$. On heating to 100 °C $(MeS$—$BH_2)_n$ is mainly depolymerized to trimeric $(MeS$—$BH_2)_3$ (23), R = Me. The ethyl and n-butyl analogues (R = Et or Bu^n) are also trimers $(RS$—$BH_2)_3$.[53,54] The reaction of MeSH, Me_2NH and $Me_2NH.BH_3$ yields the cyclic trimer (24) containing a single sulphur–boron donor–acceptor bond.[55]

5.3.3   Alkylthiodihalogenoboranes and alkylthio(halogeno)-alkylboranes

Alkylthiodihalogenoboranes (25) can be obtained by several routes. Reactions between boron trihalides and alkylthiols[56,57] lead to RS—BX_2 (equation 18); these compounds are also formed from tris(alkylthio)boranes and boron trihalides (equation 19), as well as from dialkyldisulphanes and boron triiodide (equation 20):[58]

$$BX_3 + RS—H \longrightarrow RS—BX_2 + HX \qquad (18)$$

$$BX_3 + 2(RS)_3B \longrightarrow 3\ RS—BX_2 \qquad (19)$$

$$2\ BI_3 + RS—SR \longrightarrow 2\ RS—BI_2 + I_2 \qquad (20)$$

The methylthiodihalogenoboranes (MeS—$BX_2)_n$ are cyclic oligomers.[57] Ebullioscopic molecular-weight determinations have shown that MeS—BBr_2 is 80% dimeric. This result is an indication that in solution an equilibrium exists between the monomer and the trimer.[59] Cryoscopic molecular-weight measurements[60] with MeS—BI_2 in EtBr have clearly demonstrated the trimeric nature of the compound. This was also confirmed by the mass spectrum.[61] The ^{1}H NMR results on MeS—BI_2, freshly prepared from $2BI_3$ and B(SMe)$_3$, showed the presence of a considerable amount of the monomer (δ = 2.73,s) which upon heating forms the trimer (δ = 3.1 broad).[58] Above 200 °C, MeI is eliminated from (MeS—$BI_2)_3$ to yield (IBS)$_3$. The crystal structures of MeS—BX_2 (X = Cl, Br, I) were determined,[59] and it was found that all three compounds are isostructural and form trimers. The six-membered rings are in the chair conformation.

(25), X = Cl, Br or I

The boron atoms are tetrahedrally coordinated and one methyl group is in an axial position (which reduces the point-group symmetry from C_{3v} to C_s). The bond lengths are B–Cl = 1.80 Å, B–Br = 1.96 Å, B–S = 1.95 Å, C–S = 1.84 Å.

Although the organo(halogeno)organothioboranes R(X)B—SR' have three different substituents, they are remarkably stable with respect to symmetrization to yield BR_3, BX_3 and B(SR')$_3$. The stability increases as X changes from Cl to I. The compounds $R(X)B$—SR' (R' = Me, Et; X = Cl, Br, I; R = Pr^n, Bu^n, C_5H_{11}, Ph) can be distilled.[61-64] However, it is not known if these compounds form trimers. For $[MeS$—$Me(I)]_3$, m.p. 99–101 °C, several ^1H NMR signals were observed[61] (δ = 2.61, 2.80, 2.83 (S–$C\underline{H}_3$), 1.18, 1.31, 1.43 (B–$C\underline{H}_3$)), which is an indication of the existence of structural isomers. In the mass spectrum the $(M–I)^+$ ion was found.

5.4   Metal thioborates

The ternary compounds $M_3^IB_3S_6$, i.e. M^IBS_2, are derivatives of trimeric metathioboric acid $(HSBS)_3$. In $Pb_2B_2S_5$[22,23] and $Ag_3B_5S_9$[22,23] the boron atoms are four-coordinated to sulphur atoms. Tetrahedral BS_4 units form a tetrameric adamantane-like anion $B_4S_{10}^{8-}$ (26):

(26)

Deep-red $Ag_3B_5S_9$ contains $B_{10}S_{20}$ units, formed from ten BS_4 tetrahedra. The average B–S bond lengths in $Pb_2B_2S_5$ and $Ag_3B_5S_9$ are 1.92 and 1.91 Å respectively.[22,23] Both compounds are stable towards air and water, since the boron atoms are four-coordinated.

An unusual structure is found for the perthioborate $TlBS_3$,[22] obtained from Tl_2S, B and S at 800 °C. The polymeric chain-like $(BS_3^-)_n$ anion is built from BS_4 tetrahedra (B–S 1.92 Å), in which adjacent borons are bridged by one S and one S_2 group. The structure can be described as a polyspirocyclic system containing nonplanar B_2S_3 rings. The relationship to polymeric $(BS_2)_n = [(B_2S_3)S]_{n/2}$ is given by nucleophilic addition of the anion S_2^{2-} to the sp^2 boron atoms of adjacent 1,2,5-trithiadiborolane rings:

$$(BS_2)_n + \tfrac{1}{2}n\, S_2^{2-} \longrightarrow (BS_3^-)_n \tag{21}$$

5.5 Boron–selenium rings with three-coordinate boron

5.5.1 1,3,5,2,4,6-Triselenatriborinane

The boron halides BX_3 (X = Cl, Br, I) react with H_2Se in boiling CS_2, C_6H_6 or C_6H_{12} to give polymeric products $(XBSe)_n$ in low yields (7–20%).[65,66] When BI_3 and $(Me_3Si)_2Se$ are reacted at room temperature, a quantitative formation of $(I—B—Se)_n$ occurs.[67,75] In contrast with boron trihalides, the reaction of phenyldibromoborane and hexamethyldisilylselenane leads to the 1,3,5,2,4,6-triselenatriborinane ring (27):

$$3 \ PhBBr_2 + 3(Me_3Si)_2Se \longrightarrow \underset{(27)}{\overset{\displaystyle Ph}{Ph—B}} \qquad (22)$$

The slightly yellow compound melts at 240 °C and polymerizes slowly in benzene. The compound is sensitive towards moisture and reacts with sulphur to give $(PhBS)_3$ and selenium.

5.5.2 1,2,4,3,5-Triselenadiborolane

The B_2Se_3 ring was first obtained[69] when tributylborane and selenium were heated at 220–250 °C (R = Bu^n):

$$2 \ BR_3 + 3 \ Se \overset{\Delta}{\longrightarrow} \underset{(28)}{R—B} \quad B—R + 2 \ H_2 + 4 \ C_4H_8 \qquad (23)$$

This unusual reaction has been studied[70] to elucidate the formation of (28). It was found that diboryldiselenanes[28] rearrange at temperature above 130 °C to yield (28) (see equation 24). Under these conditions, tributylborane decomposes to C_4H_8 and Bu_2BH, which reacts with selenium to give $R_2B—Se—Se—BR_2$.

$$2 \ R_2B—Se—Se—BR_2 \overset{\Delta}{\longrightarrow} R_2B_2Se_3 + 2 \ BR_3 + Se \qquad (24)$$

Redox reactions summarized in equation (25) between iodoboranes and red selenium lead to (28) in good yields:[71,72]

$$2\ \text{R—BI}_2 + \tfrac{3}{8}\ \text{Se}_8 \longrightarrow \text{R}_2\text{B}_2\text{Se}_3 + 2\ \text{I}_2 \tag{25}$$
$$\text{R} = \text{I, Me, Ph}$$

Furthermore, the ring system (28) is obtained from the reactions:[28,71,72]

$$(\text{RBH}_2)_2 + \tfrac{3}{8}\ \text{Se}_8 \longrightarrow \text{R}_2\text{B}_2\text{Se}_3 + 2\ \text{H}_2 \tag{26}$$

$$2\ \text{Cp}_2\text{TiSe}_5 + \text{R—BX}_2 \longrightarrow \text{R}_2\text{B}_2\text{Se}_3 + 7\ \text{Se} + 2\ \text{Cp}_2\text{TiX}_2 \tag{27}$$

An unusual rearrangement of ArSe—BBr$_2$[73] and of ArSe—BI$_2$[74] leads to the B$_2$Se$_3$ ring:

$$3\text{PhSeBI}_2 \longrightarrow \text{Ph}_2\text{B}_2\text{Se}_3 + \text{PhBI}_2 + \text{I}_2 \tag{28}$$

The yellow to orange coloured compounds (28), R = alkyl, aryl, exhibit high thermal stability. However, the compounds are easily destroyed by oxygen, to yield (RBO)$_3$ and red selenium, as well as by proton-active reagents such as H$_2$O, ROH, and RNH$_2$. Treatment with sulphur replaces the selenium in R$_2$B$_2$Se$_3$:

$$\text{R}_2\text{B}_2\text{Se}_3 + \tfrac{3}{8}\ \text{S}_8 \longrightarrow \text{R}_2\text{B}_2\text{S}_3 + \tfrac{3}{8}\ \text{Se}_8 \tag{29}$$

The structure of the ring is most likely analogous to that of the 1,3,4,3,5-trithiadiborolane ring.

5.5.3 1,3,4,2,5-Selenadiazadiborolidine and 1,2,4,3,5-Diselenaazadiborolidine

Reaction of (29) with N,N'-dimethylhydrazine yields the selenadiazaborolidine ring (30), R = Me, Pr:

(29)   (30)

The monomeric compounds are slightly yellow, and are easily hydrolysed to RB(OH)$_2$, H$_2$Se and dimethylhydrazine.[70]

The action of aniline on 3,5-dipropyltriselenadiborolane results in the replacement of the selena bridge, whereby yellow 4-phenyl-3,5-dipropyl-1,2,4,3,5-diselenaazadiborolidine is formed.[70]

5.6   Boron–selenium rings with four-coordinate boron

5.6.1   Dimeric and trimeric alkylselenodihalogenoboranes $(RSeBX_2)_{2,3}$

The reaction of RSeH, RSeNa or $B(SeR)_3$ and BBr_3 leads to $(RSe-BBr_2)_n$, which, according to cryoscopic measurements, are dimeric for R = Me. The ^1H NMR spectra in CS_2 of $RSe-BBr_2$ (R = Me, Et) indicate a monomer–dimer equilibrium.[73]

Redox reactions between RBI_2 (R = I, Me) and Me_2Se_2 lead to MeSe— $BI(R)$ (R = I, Me), which according to cryoscopic and mass-spectroscopic studies are trimers.[74] When $MeSeBI_2$ is obtained from BI_3 and $B(SeMe)_3$ under mild conditions, the monomer $MeSeBI_2$ can be detected in the ^1H NMR spectrum (δ = 2.63, in CS_2), which on heating irreversibly forms the trimer $(MeSe-BI_2)_3$. Because of the strong Se→B donor–acceptor bonds in $[MeSe-B(R)I]_3$ (R = I, Me), the boron atoms cannot be attacked by ROH, H_2O and amines. However, they react with RSe—SeR to yield the corresponding selenoboranes and iodine.

Most likely, the trimers have structures analogous to those of the corresponding $(MeS-BX_2)_3$ compounds.

5.6.2   μ-Methylselenadiborane

$K[CH_3Se(BH_3)_2]$, obtained from diborane and CH_3SeK, yields on treatment with iodine μ-$CH_3SeB_2H_5$ (31). From methylselenodiborane, diborane is then eliminated and polymeric $(CH_3Se-BH_2)_n$ is formed:[53]

(31)

Bibliography

B 1   E. L. Muetterties, *The Chemistry of Boron and Its Compounds* (Wiley, New York 1967), p. 647.

B 2   I. Haiduc *The Chemistry of Inorganic Ring Systems* (Wiley-Interscience, London, 1970).

B 3   W. Siebert, *Chem. Ztg* **98** (1974) 478.

B 4   W. Siebert and H. Sommer, Boron compounds with sulfur, in *Gmelin Handbuch Anorganische Chemie, Ergänzungswerk zur 8. Aufl. Boron Compounds* 3 (Springer, Berlin, 1975).

B 5   W. Siebert, in *Houben–Weyl: Methoden der Organischen Chemie*, Vol. XIII 13a, edited by R. Köster (Thieme, Stuttgart, 1982)

B 6   H. Nöth and B. Wrackmeyer, Nuclear magnetic resonance spectroscopy of boron compounds, in *NMR-Basic Principles and Progress*, Vol. 14, edited by P. Diehl, E. Fluck and R. Kosfeld (Springer, Berlin, 1978).

B 7   B. Krebs, *Angew. Chem.* **95** (1983), 113

References

1   A. Stock and F. Blix, *Ber. Deut. Chem. Ges.* **34** (1901) 3039.

2   E. Wiberg and W. Sturm, *Z. Naturforsch.* **8b** (1953) 529; *Angew. Chem.* **67** (1955) 483.

3   J. Goubeau and H. Keller, *Z. Anorg. Allg. Chem.* **272** (1953) 303.

4   M. Schmidt and W. Siebert, *Angew. Chem.* **76** (1964) 687, **78** (1966) 607.

5   M. F. Lappert and B. Prokai, *J. Chem. Soc.* A **1967**, 129.

6   H. Nöth, H. Fußstetter, H. Pommerening and T. Taeger, *Chem. Ber.* **113** (1980) 342.

7   B. Krebs and H.-U. Hürter, *Angew. Chem.* **92** (1980) 479.

8   F. T. Greene and P. W. Gilles, *J. Am. Chem. Soc.* **84** (1962) 3598; **86** (1964) 3964; J. G. Edwards and P. W. Gilles, *Adv. Chem. Ser.* **72** (1968) 211.

9   H. Chen and P. W. Gilles, *J. Am. Chem. Soc.* **92** (1970) 2309.

10   W. Siebert, Habilitationsschrift, Universität Würzburg (1971).

11   W. Siebert and K. Sommer, in *Gmelin Handbuch der Anorganischen Chemie, Ergänzungswerk zur 8. Aufl. Bd. 19. Borverbindungen*, Part 3, p. 16 (Springer, Berlin, 1975).

12   H. Diercks and B. Krebs, *Angew. Chem.* **89** (1977) 327; *Angew. Chem. Int. Ed. Engl.* **16** (1977) 313.

13   E. Wiberg and H. Sturm, *Angew. Chem.* **67** (1955) 483.

14   W. Schwarz, H. D. Hausen, H. Hess, M. Mandt, W. Schmelzer and B. Krebs, *Acta Crystallogr.* **B29** (1973) 2029.

15   W. Wiberg and H. Sturm, *Z. Naturforsch.* **10b** (1955) 114.

16   J. A. Forstner and E. L. Muetterties, *Inorg. Chem.* **5** (1966) 164.

17   M. Schmidt and F. Rittig, *Z. Naturforsch.* **25b** (1970) 1062.

18   M. Schmidt and F. Rittig, *Z. Anorg. Allg. Chem.* **394** (1972) 152.

19   E. W. Abel, D. A. Armitage and R. P. Bush, *J. Chem. Soc.* A **1965**, 3045.

20   R. H. Cragg, M. F. Lappert and B. P. Tilley, *J. Chem. Soc.* A **1967**, 947.

21   W. Siebert, *Chem. Ztg* **98** (1974) 478.

22   B. Krebs, *Angew. Chem.* **95** (1983) 113.

23   B. Krebs and H. Diercks, *Acta Crystallogr.* **A31** (1975) 566.

24   J. Goubeau, cited in ref. 2

25   U. V. Zvonkova, *Kristallografiya* **3** (1958) 569.

26   W. Schwarz, H. D. Hansen and H. Hess, *Z. Naturforsch.* **29b** (1974) 596.

27   F. Riegel, PhD dissertation, Universität Würzburg (1973).

28   F. Riegel and W. Siebert, *Z. Naturforsch.* **29b** (1974) 719.

29   M. Schmidt and W. Siebert, *Z. Anorg. Allg. Chem.* **345** (1966) 87.

30   M. Schmidt and W. Siebert, *Chem. Ber.* **102** (1969) 2763.

31   W. Biffar, H. Nöth and R. Schwerthöffer, *Liebigs Ann. Chem.* **1981**, 2067.

32   H. Nöth and R. Schwerthöffer, *Chem. Ber.* **114** (1981) 3056.

33   (a) H. M. Seip, R. Seip and W. Siebert, *Acta Chem. Scand.* **27** (1973) 15.
      (b) A. Almenningen, H. M. Seip and P. Vassbotn, *Acta Chem. Scand.* **27** (1973) 21.

34   O. Gropen and P. Vassbotn, *Acta Chem. Scand.* **27** (1973) 3079.

35   J. Kroner, D. Nölle, H. Nöth and W. Winterstein, *Chem. Ber.* **108** (1975) 3807.

36 H. Nöth and R. Staudigl, *Z. Anorg. Allg. Chem.* **481** (1981) 41.
37 H. Nöth and T. Taeger, *Z. Naturforsch.* **34b** (1979) 135.
38 H. Nöth, R. Staudigl and R. Brückner, *Chem. Ber.* **114** (1981) 1871.
39 H. Nöth and R. Staudigl, *Chem. Ber.* **115** (1982) 3011.
40 See ref. 11, pp. 42, 43.
41 D. Nölle, H. Nöth and T. Taeger, *Chem. Ber.* **110** (1977) 1643.
42 A. Meller and C. Habben, *Monatsh. Chem.* **113** (1982) 139.
43 See ref. 11, pp. 42, 43, and ref. 27.
44 D. Nölle and H. Nöth, *Z. Naturforsch.* **27b** (1972) 1425.
45 C. Habben, W. Maringgele and A. Meller, *Z. Naturforsch.* **37b** (1982) 43.
46 T. Taeger, PhD dissertation, Universität München, (1977).
47 H. Nöth and R. Staudigl, *Chem. Ber.* **115** (1982) 813.
48 S. C. Malhotra, *Inorg. Chem.* **3** (1964) 862.
49 H. Nöth and R. Ullmann, *Chem. Ber.* **108** (1975) 3125.
50 H. Nöth and B. Wrackmeyer, Nuclear magnetic resonance spectroscopy of boron compounds, in *NMR—Basic Principles and Progress*, Vol. 14 edited by P. Diehl, E. Fluck and R. Kosfeld (Springer, Berlin, 1978).
51 H. Vahrenkamp, *J. Organomet. Chem.* **28** (1971) 167.
52 A. B. Burg and R. I. Wagner, *J. Am. Chem. Soc.* **76** (1954) 3307.
53 E. L. Muetterties, N. E. Miller, K. J. Packer and H. C. Miller, *Inorg. Chem.* **3** (1964) 870.
54 B. M. Mikhailov, T. A. Shchegoleva, E. M. Shashkava and V. D. Sheludyakov, *Bull. Acad. Sci. USSR, Div. Chem. Sci.* **1962**, 1143.
55 A. B. Burg, *Inorg. Chem.* **11** (1972) 2283.
56 E. Wiberg and W. Sütterlin, *Z. Anorg. Allg. Chem.* **202** (1931) 37.
57 J. Goubeau and H. W. Wittmeier, *Z. Anorg. Allg. Chem.* **270** (1952) 16.
58 W. Siebert, F. R. Rittig and M. Schmidt, *J. Organomet. Chem.* **22** (1970) 511.
59 S. Pollitz, Fr. Zettler, D. Forst and H. Hess, *Z. Naturforsch.* **31b** (1976) 897.
60 E. G. Höfling, PhD dissertation, Tech. Hochschule Stuttgart (1959).
61 W. Siebert and A. Ospici, *Chem. Ber.* **105** (1972) 454.
62 W. Siebert and M. Füller, unpublished results.
63 B. M. Mikhailov and T. A. Shchegoleva, *Izv. Akad. Nauk SSSR, Otd. Khim. Nauk* **1956**, 508.
64 B. M. Mikhailov, T. A. Shchegoleva, *Izv. Akad. Nauk SSSR, Otd. Khim. Nauk* **1957**, 1080.
65 J. Cueilleron and R. Hillel, *Bull. Soc. Chim. Fr.* **1968**, 3635.
66 S. Gurrieri, *Bull. Sedute Accad. Gioenia Sci. Nat. Catania* **72** (1960) 667.
67 E. Kiewert, PhD dissertation, Universität Würzburg (1970).
68 M. Schmidt and E. Kiewert, *Z. Naturforsch.* **26b** (1971) 613.
69 B. M. Mikhailov, T. A. Shchegoleva, *Izv. Akad. Nauk SSSR, Otd. Khim. Nauk* **1959**, 356.
70 W. Siebert and F. Riegel, *Chem. Ber.* **106** (1973) 1012.
71 M. Schmidt, W. Siebert and E. Gast, *Z. Naturforsch.* **22b** (1967) 557.
72 M. Schmidt, W. Siebert and F. R. Rittig, *Chem. Ber.* **101** (1968) 281.
73 M. Schmidt and H. D. Block, *Z. Anorg. Allg. Chem.* **377** (1970) 305; H. D. Block, PhD dissertation, Universität Würzburg, (1968).
74 W. Siebert and A. Ospici, *Chem. Ber.* **105** (1972) 454.
75 W. Siebert and B. Asgarouladi, *Z. Naturforsch.* **30b** (1975) 647.
76 W. Siebert, R. Full, Th. Renk and A. Ospici, *Z. Anorg. Allg. Chem.* **418** (1975) 273.
77 B. Asgarouladi, R. Full, K. J. Schaper and W. Siebert, *Chem. Ber.* **107** (1974) 34.
78 W. Siebert, R. Full, J. Edwin and K. Kinberger, *Chem. Ber.* **111** (1978) 823.

6

Aluminium–Nitrogen Rings and Cages

M. CESARI and S. CUCINELLA

ENI, R & D Divisions, Milano, Italy

6.1 Introduction

Many results on the synthesis, structural characterization and applications of aluminium–nitrogen ring and cage compounds have been accumulated in recent years. The synthesis of aluminium–nitrogen compounds has progressed considerably through the discovery of new and more direct reaction paths. New compounds have been obtained and characterized, mainly by NMR and mass spectrometry. At the same time, the application of Al–N ring and cage compounds in organic synthesis and in polymerization catalysis has been expanded.

Before 1970, only one cage-type molecule, the "cubane" tetramer (PhAlNPh)$_4$ had been structurally analysed by diffraction methods.[70,113] Since then, many compounds have been isolated and their crystal structure resolved. Some of them exhibit very complex molecular architecture.

In addition to the section on aluminium–nitrogen compounds published in Haiduc's 1970 monograph,[B 3] these compounds have been reviewed as classes of hydrido-[B 2] and alkylaluminium[B 5] derivatives. Selected aspects of Al–N hydridoaluminium cages have also been reported.[B 1] More recently, a comprehensive review has been published.[B 4]

6.2 Synthesis

6.2.1 Aluminium–nitrogen ring compounds

This group comprises cyclic dimers and trimers of alkylamido or dialkyl-amidoaluminium derivatives (1) and (2) respectively and dimers of alkyl-imidobis(dihydroaluminium) compounds (3):

$$R_{3-m}Al(NHR')_m \qquad R_{3-m}Al(NR'_2)_m \qquad (H_2Al)_2NR'$$

$$\textbf{(1)} \qquad\qquad \textbf{(2)} \qquad\qquad \textbf{(3)}$$

where R = H, alkyl group, halogen; R' = alkyl; m = 1, 2 or 3. The tendency of these compounds to associate through Al–N–Al bonds is responsible for the formation of four- and six-membered $(AlN)_n$ rings:

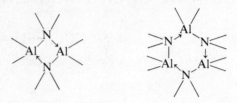

With a few exceptions, these compounds are solid at room temperature. The known derivatives are shown in Table 6.1 where the association degrees and the preparation methods discussed below are given.

Table 6.1 Association degrees and preparation methods for cyclic Al–N compounds.

Compound	Association degree n	Preparation method (reaction no.)	M.p. (°C)	Refs.
$H_2Al-NMe_2$	3	(10)		45, 65, 94
$H_2Al-NEt_2$	2	(10)	30	65
$H_2Al-NPr^i_2$	3	(8)		40
$H_2Al-N(CH_2)_2$	3	(8)		98
$H_2Al-N(CH_2)_3$	3	(8)		98
$H_2Al-N(CH_2)_4$	3	(8), (10)	90	65, 98
$H_2Al-N(CH_2)_5$	av. 2.4	(8), (10)	60	65, 98
$Me_2Al-NMe_2$	2	(8)		51, 71
$Me_2Al-N(CH_2)_2$	3	(8)	193	99, 12
$Me_2Al-N(CH_2)_3$	av. 2.5	(8)	114–116	99
$Me_2Al-N(CH_2)_4$	2	(8)	55–58	99
$Me_2Al-N(CH_2)_5$	2	(8)	70	95, 99
$Me_2Al-NPh_2$	2			102
$Me_2Al-NMePh$, *cis* and *trans*	2	(8)		104
$Me_2Al-N(Et)C_2H_4NMe_2$	1–2	(8)		17
$Me_2Al-N(Me)SiMe_3$	2	(8)	58–61	91
$Me_2Al-N=CMe_2$	2	(8)		93, 105
$Me_2Al-N=C(NMe_2)_2$	2	(9)	98	96
$Et_2Al-N(CH_2)_2$	3	(8)		99
$Et_2Al-N(CH_2)_5$	2	(8)		95
$Et_2Al-NPh_2$	2	(8)		57
$Et_2Al-N=C(NMe_2)_2$	av. 2.2	(8)		96

Table 6.1 (continued)

Compound	Association degree n	Preparation method (reaction no.)	M.p. (°C)	Refs.
$Bu^i_2Al—N(CH_2)_2$	2	(8)		99
$Bu^i_2Al—N(CH_2)_5$	2	(8)	65	95
$(C_3H_5)_2Al—N(CH_2)_2$	3	(8)		72
$HAl(NMe_2)_2$	av. 2.4	(10), (11)		65
$HAl(NEt_2)_2$	av. 2.4	(10)		65
$HAl[N(CH_2)_4]_2$	2	(10)		65
$HAl[N(CH_2)_5]_2$	av. 2.2	(10)	100	65
$Al(NMe_2)_3$	av. 2.2		88, 89	45, 65
$H_2Al—NHBu^t$	2	(1)		77
$D_2Al—NHBu^t$	2	(1)		77
$Me_2Al—NHMe$, *trans*	3	(1)	109–111	3
cis	3	(1)	113–114	3
$Me_2Al—NHEt$	3	(1)		3
$Et_2Al—NHMe$	3			3
$Me_2Al—NHPr^i$	2	(1)	52–54	5
$Et_2Al—NHPr^i$	2–3	(1)		5
$Al(NHPr^i)_3$	2	(3)		31
$(Me_3SiCH_2)_2Al—NHMe$	2	(2)		76
$EtClAl—NMe_2$	2	(8)		50
$EtBrAl—NMe_2$	2	(8)	123–124	48
$EtIAl—NMe_2$	2	(8)		49
$HClAl—NHBu^t$	2	(8)		77
$EtClAl—NHBu^t$	2	(8)		20
$EtBrAl—NHBu^t$	2	(8)	83–92	20
$EtIAl—NHBu^t$	2	(8)	96–115	20
$Cl_2Al—NMe_2$	2	(12)–(14)	238	1, 15, 45
$Cl_2Al—N=CBu^i_2$	2	(14)	124–126	97
$Cl_2Al—N=CPh_2$	2	(14)	199–200	97
$Cl_2Al—N=C(NMe_2)_2$	2	(14)		96
$Br_2Al—NMe_2$	2	(13)		2
$Br_2Al—N=CPh_2$	2	(14)	246–249	97
$I_2Al—NMe_2$	2	(13)		2
$Cl_2Al—NHPr^i$	2			32
$Cl_2Al—NHBu^t$	2	(8)		77
$ClAl(NMe_2)_2$	2			45
$(BH_4)_2Al—NMe_2$	2		156	62

New alkylamido dihydrido- and dialkylaluminium compounds (**5**) have been synthesized by the following reaction ($R = H$, D, $R' = Bu^t$;[77] $R = Me$, $R' = Me$, Et;[3,47] $R = Me$, Et, $R' = Pr^i$ [5]):

$$AlR_3 + R'NH_2 \longrightarrow R_3Al.R'NH_2 \xrightarrow[-RH]{} \tfrac{1}{n}[R_2Al—NHR']_n \qquad (1)$$
$$\qquad\qquad\qquad\quad (4) \qquad\qquad\qquad (5)$$

If trialkylaluminium is the starting material then the intermediate complex (**4**) can be isolated at room temperature.[3,5] The unusual reaction (2) leads to the formation of compound (6),[76] and dimeric $Al(NHPr^i)_3$ was formed in reaction (3):[31]

$$(Me_3SiCH_2)_3Al + MeNH_2 \longrightarrow \tfrac{1}{2}[(Me_3SiCH_2)_2Al\!-\!NHMe]_2 + SiMe_4$$
$$(6) \tag{2}$$

$$(Pr^iNHAl\!-\!NPr^i)_4 + 4\,Pr^iNH_2 \longrightarrow 2\,[Al(NHPr^i)_3]_2 \tag{3}$$

t-Butylamidoaluminium halides (8) ($m = 1$ or 2) have been derived from (5) by substitution of the hydride hydrogen with chlorine[77] (reaction 4) and by decomposition of the complex (7), obtained directly from Et_2AlCl and amine, or according to reaction (5) (where $R' = H, D; X = Cl, Br$):[20]

$$(H_2Al\!-\!NHBu^t)_2 + m\,HgCl_2 \longrightarrow (H_{2-m}Cl_mAl\!-\!NHBu^t)_2 + mH_2 + mHg$$
$$(8) \tag{4}$$

$$AlEt_3 + Bu^tNR'_3X \xrightarrow[-EtR']{} Et_2XAl.Bu^tNR'_2 \xrightarrow[-EtR']{} \tfrac{1}{2}\,(EtXAl\!-\!NR'Bu^t)_2 \tag{5}$$
$$\qquad\qquad\qquad\qquad (7) \qquad\qquad\qquad (8)$$

Alkylimidobis(hydridoaluminium) derivatives (3) were reported to be very unstable.[B2] The use of dimethylaminopropylamines in reactions (6) and (7) ($R' = (CH_2)_3$ and CH_2CHMe) afforded stable dimers:[92]

$$2\,AlH_3 + Me_2N(CH_2)_3NH_2 \xrightarrow[-2\,H_2]{THF} \tfrac{1}{2}\,[(H_2Al)_2N(CH_2)_3NMe_2]_2.THF \tag{6}$$

$$2\,NaAlH_4 + Me_2N\!-\!R'\!-\!NH_2.HCl \xrightarrow[\substack{-4\,H_2 \\ -2\,NaCl}]{THF} \tfrac{1}{2}\,[(H_2Al)_2N\!-\!R'\!-\!NMe_2]_2 \tag{7}$$

The reaction (8) of trialkylaluminium (or AlH_3) trimethylamine adduct with dialkylamines has been extended to novel amidoaluminium derivatives of secondary cyclic amines and 1,1,3,3-tetramethylguanidine. The intermediate complexes (9) can be isolated. The individual compounds thus prepared can be identified with the aid of Table 6.1:[12,17,72,91,95,96,98,99]

$$AlR_3 + m\,R'_2NH \longrightarrow R_3Al.NHR'_2 \xrightarrow[-m\,RH]{} \tfrac{1}{n}\,[R_{3-m}Al(NR'_2)_m]_n \tag{8}$$
$$\qquad\qquad\qquad\quad (9) \qquad\qquad\qquad\quad (2)$$

The formation of $(Me_2Al\!-\!N\!\!=\!\!CMe_2)_2$ (10) by thermolysis[93] of $K[Al_2Me_6SCN]$ has been described. Compound (10) has also been obtained from N-chloro-2-propanimine with trimethylaluminium:[105]

$$AlMe_3 + Me_2C\!\!=\!\!NCl \longrightarrow \tfrac{1}{2}\,(Me_2Al\!-\!N\!\!=\!\!CMe_2)_2 + MeCl \tag{9}$$

Considerable progress has been achieved with the direct synthesis of dialkyl-amidohydridoaluminium compounds from aluminium powder and amine, under hydrogen pressure (3000–4000 p.s.i.) at relatively high temperature (100–150 °C) in benzene (reaction 10) ($m = 1$ or 2; $R_2' = Me_2$, Et_2, $(CH_2)_4$, $(CH_2)_5$; $n = 2$ or 3). The predominant formation of bis(dialkylamido)-hydridoaluminium ($m = 2$) is observed in every case, even when using an excess of aluminium with respect to the amine. Only low yields of mono-(dialkylamido)dihydridoaluminium have been observed when the reactions were run at lower temperature.[7,65]

$$Al + \tfrac{3}{2} H_2 + m \; R_2'NH \longrightarrow \tfrac{1}{n} [H_{3-m}Al(NR_2')_m]_n + m \; H_2 \qquad (10)$$

In addition, bis(dimethylamido)hydridoaluminium is the only product of the reaction of aluminium and tris(dimethylamido)aluminium under hydrogen pressure:[65]

$$\tfrac{1}{2} [Al(NMe_2)_3]_2 + \tfrac{1}{2} Al + \tfrac{3}{4} H_2 \longrightarrow \tfrac{3}{4} [HAl(NMe_2)_2]_2 \qquad (11)$$

The dimer $(EtRAl—NMe_2)_2$ ($R = Cl,^{50}$ Br^{48} and I^{49}) has been obtained by thermal decomposition of diethylaluminium–dimethylamine adducts, according to reaction (8). Amidoaluminium halides have also been obtained by reacting amidodihydrido or dialkylaluminium derivatives with HCl (reaction 12),[45] HgX_2 ($X = Cl,^{1,45}$ Br^2 and I^2) (reaction 13) or $AlCl_3,^{59}$ or by reaction of aluminium halides ($X = Cl$, Br) with alkylamidolithium (reaction 14)[45,96,97]:

$$(H_2Al—NMe_2)_2 + 4HCl \longrightarrow (Cl_2Al—NMe_2)_2 + 4 \; H_2 \qquad (12)$$

$$(R_2Al—NMe_2)_2 + 2 \; HgX_2 \longrightarrow (X_2Al—NMe_2)_2 + 2 \; R—R + 2 \; Hg \qquad (13)$$

$$AlX_3 + LiNR_2' \longrightarrow \tfrac{1}{n} (X_2Al—NR_2')_n + LiX \qquad (14)$$

On adding diborane to a solution of $(H_2Al—NMe_2)_3$ in diethyl ether or toluene $[(BH_4)_2Al—NMe_2]_2$ is formed.[62]

6.2.2 Aluminium–nitrogen cage compounds

This group includes a large variety of polyalkylimidopolyaluminium compounds, derived from primary amines, and a small number of derivatives from secondary amines.[79,85,101] All these compounds are solid; they are listed in Table 6.2.

Polyalkylimidopolyaluminium compounds derived from primary amines can be considered to result by the combination of (RAl—NR')mers with themselves or with (–RHAl—NHR'–)mers, in which R = alkyl, hydrogen or halogen. In addition, the complexation of LiH or AlH₃, or the partial substitution of aluminium atoms with Ca or Mg atoms, results in the formation of novel structures.

The molecular structures of Al–N cages range from the simple Al_4N_4 cube skeleton to much more complex spatial arrangements, as described in the following section and illustrated there by the structures (23)–(35).

Table 6.2　Aluminium–nitrogen cage compounds.

Compound	Preparation method (reaction no.)	Refs.
$(HAlNPr^i)_4$	(23)	21, 29, 39
$(HAlNBu^s)_4$	(23)	21, 29
$(HAlNBu^t)_4$	(15), (18), (23), (24)	21, 27, 28, 29, 31, 42, 77, 78
$(DAlNBu^t)_4$	(15)	78
$(HAlNC_6H_{11})_4$	(23)	29
$[HAlNCH(Me)Ph]_4$	(15)	6, 21, 22
$(MeAlNPr^i)_4$	(15)	21, 33, 39
$(EtAlNPr^i)_4$	(15)	21, 33
$(PhAlNPh)_4$	(15)	70, 113
$(HAlNEt)_n$ ($n = 6, 7, 8$ mixture)	(15), (24)	27, 31
$(HAlNPr^n)_n$ ($n = 6, 7, 8$ mixture)	(15), (17)	27, 42
$(HAlNPr^n)_6$	(16), (17), (24)	21, 27, 31, 38
$(HAlNPr^i)_6$	(15), (16), (19)–(21) (23)–(25), (33)	24, 21, 27–29, 33, 42, 109, 77
$(HAlNBu^n)_n$ ($n = 6, 7, 8$ mixture)	(15), (22)–(24)	27, 29, 31, 42
$(HAlNBu^s)_6$	(15), (23), (24)	27, 29, 31
$(HAlNBu^i)_6$	(15), (23)	21, 27, 29
$(HAlNC_6H_{11})_6$	(15), (23), (24)	27, 29, 31
$[HAlNCH(Me)Ph]_6$	(15)	6, 22
$[HAlNCH(Me)Ph]_6 \cdot \frac{1}{2}C_6H_{14}$		41
$[HAlN(CH_2)_3NMe_2]_6$	(15), (23), (24)	43
$(MeAlNPr^i)_6$		33
$(HAlNPr^n)_7$	(16)	27
$(MeAlNMe)_7$	(15)	4, 54, 55
$(EtAlNMe)_7$		55
$(HAlNEt)_8$	(16)	77
$(HAlNPr^n)_8$	(16), (17), (24)	21, 27, 28, 31
$(MeAlNMe)_8$	(15)	3, 4
$(HAlNPr^i)_{10}$	(16)	27
$(HAlNEt)_{15}$	(16)	77
$(HAlNEt)_{16}$	(16)	77
$(ClAlNPr^i)_4$	(30), (31)	21, 32, 77
$(ClAlNBu^t)_4$	(30)	21, 32, 77
$(BrAlNBu^t)_4$		77
$(ClAlNPr^n)_n$ ($n = 6, 8, 10$ mixture)		32
$(ClAlNPr^i)_6$		32, 37
$(Pr^iHNAlNPr^i)_4$	(32), (34)	31

Table 6.2 (continued)

Compound	Preparation method (reaction no.)	Refs.
$(Bu^nHNAlNBu^n)_6$	(32)	31
$[(HAlNPr^i)_{6-n}(MeAlNPr^i)_n]$	(26)	33
(n = 1–6 mixture)		
$[(MeAlNPr^i)(MeAlNPr^i)_5]$	(26)	33
$[(Me_{0.83}H_{0.17}AlNPr^i)_6(MeAlNPr^i)_6]$	(26)	37
$[(HAlNPr^i)_{6-n}(EtAlNPr^i)_n]$	(26)	33
(n = 1–4 mixture)		
$[(HAlNPr^i)_{4-n}(ClAlNPr^i)_n]$		78
(n = 1–4 mixture)		
$[(HAlNBu^t)_{4-n}(ClAlNBu^t)_n]$ (n = 1, 2 mixture)		32
(n = 1–4 mixture)		78
$[(HAlNBu^t)(ClAlNBu^t)_3]$		77
$[(HAlNBu^t)(BrAlNBu^t)_3]$		77
$[(HAlNBu^t)_{4-n}(IAlNBu^t)_n]$ (n = 2, 3 mixture)		77
$[(HAlNBu^t)_2(ClAlNBu^t)_2]$		21
$[(HAlNBu^t)_3(ClAlNBu^t)]$		21
$[(HAlNPr^i)_{6-n}(ClAlNPr^i)_n]$ (n = 1–6 mixture)		32
(n = 0–4 mixture)		78
$[(HAlNEt)_{8-n}(ClAlNEt)_n]$ (n = 0–8 mixture)		77
$[(HAlNEt)_4(ClAlNEt)_4]$		77
$[(HAlNBu^t)_3(H_2AlNBu^t)]$	(35)	21, 27
$[Al_3H_3(NBu^t)_2(NHBu^t)_2]$	(15)	77
$[\{HAlNCH(Me)Ph\}_3\{H_2AlNHCH(Me)Ph\}]$	(35)	21, 22
$[\{HAlNCH(Me)CH_2NMe_2\}_{4-n}\{H_2AlNCH-$		
$(Me)CH_2NMe_2\}]$ n = 1, 2	(36), (37)	43
$[\{HAlN(CH_2)_3OMe\}_4\{H_2AlNH(CH_2)_3OMe\}_2]$	(35)	43
$[\{HAlN(CH_2)_3NMe_2\}_3\{H_2AlNH(CH_2)_3NMe_2\}_3]$	(35)	43
$[(HAlNEt)_6(H_2AlNHEt)_2]$	(35)	77
$[(HAlNPr^i)_2(H_2AlNHPr^i)_3]$		83
$[(HAlNPr^i)_2(H_2AlNHPr^i)_2\{HAlNCH(Me)CH_2NMe_2\}]$	(38)	86
$[(MeAlNMe)_6(Me_2AlNHMe)_2]$	(35)	4
$[(ClAlNEt)_4(Cl_2AlNHEt)_2]$		77
$(HAlNPr^i)_6.AlH_3$		84
$[HAlN(CH_2)_3NMe_2]_6.2LiH$	(38)	43, 86
$H(HAlNPr^i)_5AlH_2.LiH.Et_2O$		25
$[(THF)MgNBu^t(HAlNBu^t)_3]$	(39)	34, 36
$[(3THF)MgNBu^t(HAlNBu^t)_3]$	(40)	34
$[(3THF)CaNBu^t(HAlNBu^t)_3].THF$	(39)	36
$[(2THF)CaNBu^t(HAlNBu^t)_3]$	(39)	34
$[(RAl(NR_2')_2NR'']_2$		
$R' = R'' = Me; R = H, Cl, Br, I$	(41)	79
$R' = Me; R'' = Pr^i; R = H$	(41)	79
$[\{ClAl(NMe_2)\}_2NMe]_2$		101
$HAl[\{EtN(CH_2)_2)NEt\}AlH_2]_2$		85
$(H_{0.85}Cl_{0.15})Al[\{EtN(CH_2)_2NEt\}AlH(H_{0.7}Cl_{0.3})]_2$		85

Prismo-poly-μ_3-alkylimidopoly(hydrido- or alkylaluminium), (RAl—NR′)$_n$ (12) have been obtained by reactions (15) and (16), in various solvents.[3,4,6,22,27,33,42,43,77] The nature of (12) is strongly influenced by the steric hindrance of R′. If R is an alkyl group, the intermediate product (11) (reaction 15) is stable at room temperature and completion of the reaction requires higher temperatures. If R is H the intermediate (11) is stable only at subambient temperatures; only $(H_2Al—NHBu^t)_2$ and $(D_2Al—NHBu^t)_2$ have been reported to be stable at room temperature.[77] Individual compounds prepared by reactions (15) and (16) can be traced with the aid of Table 6.2.

$$AlR_3 + R′NH_2 \xrightarrow[-RH]{} \tfrac{1}{m}(R_2Al—NHR′)_m \xrightarrow[-RH]{} \tfrac{1}{n}(RAlNR′)_n \qquad (15)$$
$$\quad\quad\quad\quad\quad\quad\quad\quad\quad\quad (11) \quad\quad\quad\quad\quad (12)$$

$$LiAlH_4 + RNH_3Cl \longrightarrow \tfrac{1}{n}(HAlNR)_n + 3\,H_2 + LiCl \qquad (16)$$
$$\quad\quad\quad\quad\quad\quad\quad\quad (13)$$

The tetramer $(PhAlNPh)_4$ has been produced in the reaction of $AlPh_3$ with aniline.[70] The formation of an insoluble product from $AlH_3.NMe_3$ and methylamine has been cited.[77]

Reaction (15) for AlH_3 and a number of amines in THF has been followed conductimetrically,[30,42] and some intermediates, for example $(H_2Al)_2NR$ and $Al(NR—AlH_2)_3$, have been recognized in the case of Pr^iNH_2 and Bu^nNH_2.[30] Starting from Pr^iNH_2, the intermediate $(H_2Al—NHPr^i)_2$ has been isolated.[77]

A minor route to $(HAlNPr)_n$ (where n ranges from 6 to 9) is represented by:[27]

$$AlH_3.NMe_3 + EtCN \longrightarrow \tfrac{1}{n}(HAl—NPr)_n + NMe_3 \qquad (17)$$

The tetramer $(HAlNBu^t)_4$ (14) has been obtained by reaction of $(H_2Al—NPr^i_2)$ with Bu^tNH_2 (reaction 18), and the hexamer $(HAlNPr^i)_6$ (15) by reactions (19) and (20):[31]

$$\tfrac{1}{2}(H_2Al—NPr^i_2)_2 + Bu^tNH_2 \longrightarrow \tfrac{1}{4}(HAlNBu^t)_4 + Pr^i_2NH + H_2 \qquad (18)$$
$$\quad\quad\quad\quad\quad\quad\quad\quad\quad\quad\quad\quad (14)$$

$$[Al(NHPr^i)_3]_2 + 4\,AlH_3.NMe_3 \longrightarrow (HAlNPr^i)_6 + 6\,H_2 + 4\,NMe_3 \qquad (19)$$

$$3\,(Pr^iHN—AlNPr^i)_4 + 12\,AlH_3.NMe_3 \longrightarrow$$
$$4(HAlNPr^i)_6 + 12\,H_2 + 12NMe_3 \qquad (20)$$
$$(15)$$

The formation of (13) by reactions (21) and (22) has been studied conductimetrically in THF:[42]

$$3 \text{ HAl(NHPr}^i)_2 + 3 \text{ AlH}_3 \longrightarrow (\text{HAlNPr}^i)_6 + 6 \text{ H}_2 \qquad (21)$$

$$2 \text{ Al(NHBu)}_3 + 4 \text{ AlH}_3 \longrightarrow \tfrac{6}{n} (\text{HAlNBu})_n + 6 \text{ H}_2 \qquad (22)$$

Recent progress in the field has brought some more economical synthetic routes. Thus reaction (23) allows the synthesis of (13) ($R = Pr^i$, Bu^n, Bu^i, Bu^s and Bu^t; $R = C_6H_{11}{}^{29}$) and saves hydride hydrogen, since MH can be recovered and used for the preparation of the starting MAlH_4 (M = Li, Na). In addition to the nature of R, the solvent has some influence upon the product, by determining different values for n.[29]

$$\text{MAlH}_4 + \text{RNH}_2 \longrightarrow \tfrac{1}{n} (\text{HAlNR})_n + \text{MH} + 2 \text{ H}_2 \qquad (23)$$
$$(13)$$

The direct route, reaction (24), has allowed the synthesis of several (13) ($n = 4$–10; R = Et,[31] Pr,[28,31] Pr^i,[28,31,109] $(CH_2)_3NMe_2$,[43] Bu^n,[31] Bu^s,[31] Bu^t,[28,31] C_6H_{11}[31]) from aluminium powder and primary amines under hydrogen pressure and heating. For example, the synthesis of $(\text{HAlNPr}^i)_6$ (15) occurs at 175–180 °C and at c. 2400 p.s.i. of H_2 in diethyl ether.[31] The reaction occurs in the presence of activators such as Na, NaAlH_4, (13) itself and others, used in small ($\leqslant 5 \text{ mol}\%$ with respect to the amine)[28,31,43] or larger[109] amounts.

$$\text{Al} + \text{RNH}_2 \longrightarrow \tfrac{1}{n} (\text{HAlNR})_n + \tfrac{1}{2} \text{ H}_2 \qquad (24)$$

In addition, $(\text{HAlNPr}^i)_6$ (15) is the only product of the following reaction, which occurs under hydrogen pressure:[31]

$$3 (\text{Pr}^i\text{HN—AlNPr}^i)_4 + 12 \text{ Al} + 6 \text{ H}_2 \longrightarrow 4 (\text{HAlNPr}^i)_6 \qquad (25)$$

The stability of (15) has allowed the substitution of hydride hydrogens with alkyl groups according to the following reaction ($z = xy$, R = Me, Et).[33]

$$(\text{HAlNPr}^i)_6 + y \text{ AlR}_3 \longrightarrow (\text{HAlNPr}^i)_{6-z}(\text{RAlNPr}^i)_z + y \text{ R}_{3-x}\text{AlH}_x \qquad (26)$$

Depending on the value of y, reaction (26) has yielded products with a different degree of substitution. With an excess of trialkylaluminium $(\text{HAlNPr}^i)(\text{MeAlPr}^i)_5$, $(\text{MeAlNPr}^i)_6$ and $(\text{HAlNPr}^i)_2(\text{EtAlNPr}^i)_4$ have been obtained.

The exchange of hydride hydrogens in (13) with halogen atoms can involve either simple substitution, according to reactions (27)–(29), or the substitution reaction is accompanied by the cleavage of Al–N bonds with subsequent rearrangement of the molecular structure:

$$(\text{HAlNR})_n + n \text{ HCl} \xrightarrow{\text{Et}_2\text{O}} (\text{ClAlNR})_n + n \text{ H}_2 \qquad (27)$$

$$(HAlNR)_n + \tfrac{1}{2}n \ HgX_2 \longrightarrow (XAlNR)_n + \tfrac{1}{2}n \ Hg + \tfrac{1}{2}n \ H_2 \qquad (28)$$

$$(HAlNR)_n + nTiCl_4 \longrightarrow (ClAlNR)_n + nTiCl_3 + \tfrac{1}{2}n \ H_2 \qquad (29)$$

The tetramer $(HAlNBu^t)_4$ **(14)** has been transformed into isostructural, partially and completely halogenated compounds, by reactions with HCl,[32] $HgCl_2$,[32,77] $HgBr_2$[77] and $TiCl_4$.[32] Attempts to obtain $(IAlNBu^t)_4$ by reaction of **(14)** with HgI_2 has yielded a mixture of $[(HAlNBu^t)_{4-n}(IAlN-Bu^t)_n]$ (n = 2 and 3).[77] The preparation of $(ClAlNPr^i)_6$ **(16)** from $(HAlNPr^i)_6$ **(15)** was successful using $TiCl_4$; attempts to obtain **(16)** by reactions (27) and (28) gave a mixture of partially and completely chlorinated hexamers, beside other chlorinated cage and cyclic Al–N compounds.[32] The partial chlorination of $(HAlNEt)_8$ by $HgCl_2$ afforded $(HAlNEt)_4(ClAlNEt)_4$, but decomposition products, for example $(ClAlNEt)_4(Cl_2AlNHEt)_2$, were formed on attempted complete substitution.[77] A mixture of $(ClAlNPr)_n$, with n = 6, 8 and 10, was obtained from $(HAlNPr)_8$ with $TiCl_4$.[32] Redistribution of Al-substituents between $(HAlNBu^t)_4$ and $(ClAlNPr^i)_4$ occurs without molecular rearrangements.[78] Alternative routes to $(ClAlNR)_4$ (R = Pr^i, Bu^t) are illustrated by the following reactions:[77]

$$\tfrac{1}{2} (ClHAlNHR)_2 \longrightarrow \tfrac{1}{4} (ClAlNR)_4 + H_2 \qquad (30)$$

$$H_2AlCl.NEt_3 + Pr^iNH_2 \xrightarrow[-NEt_3]{} \tfrac{1}{4} (ClAlNPr^i)_4 + 2 \ H_2 \qquad (31)$$

Cage compounds $(RHN{-}AlNR)_n$ **(17)** have been obtained by direct synthesis under hydrogen pressure and at high temperature (R = Pr^i, n = 4; R = Bu^n, n = 6).[31]

$$Al + 2 \ RNH_2 \longrightarrow \tfrac{1}{n} (RHN{-}AlNR)_n + \tfrac{3}{2} H_2 \qquad (32)$$

The compounds **(17)** have been shown to be intermediates in the direct synthesis (reaction 24) of $(HAlNR)_n$. They form $(HAlNR)_n$ in reactions (25) and (33) (R = Pr^i, m = 4, n = 6; R = Bu^n, m = 6, n = 6, 7 and 8):[31]

$$\tfrac{1}{m} (RHN{-}AlNR)_m + AlH_3.NMe_3 \longrightarrow \tfrac{2}{n} (HAlNR)_n + \tfrac{3}{2} H_2 + NMe_3 \qquad (33)$$

The reverse reaction has also been reported:[31]

$$2 (HAlNPr^i)_6 + 12 \ Pr^iNH_2 \longrightarrow 3 (Pr^iHN{-}AlNPr^i)_4 + 12 \ H_2 \qquad (34)$$

Compounds $(RAlNR')_{n-m}(R_2AlNHR')_m$ **(18)** in which n is an even number, have been shown to be precursors of $(RAlNR')_n$ **(12)**. In fact, **(18)** is converted to **(12)** when heated.[4,27,43,77] Some of these compounds have been obtained according to reactions (35) (R = H or Me; R' = Me, Et, Bu^t, $(CH_2)_3OMe$, $(CH_2)_3NMe_2$),[4,6,22,27,43,77] (36) (R = $CHMeCH_2NMe_2$)[43] and (37) (R = $CHMeCH_2NMe_2$, $(CH_2)_3NMe_2$):[43]

$$n \, R_3Al.B + n \, R'NH_2 \xrightarrow[-nB]{} (RAlNR')_{n-m}(R_2AlNHR')_m + (2n-m) \, RH$$
$$(35)$$

$$4 \, LiAlH_4 + 4 \, RNH_2 \longrightarrow (HAlNR)_2(H_2AlNHR)_2 + 6 \, H_2 + 4 \, LiH \quad (36)$$

$$n \, Al + n \, RNH_2 + \tfrac{1}{2}m \, H_2 \longrightarrow (HAlNR)_{n-m}(H_2AlNHR)_m + (\tfrac{1}{2}n - m) \, H_2$$
$$(37)$$

The compound $H_3Al_3(NBu^t)_2(NHBu^t)_2$ is formed in reaction (15), but using a ratio $RNH_2 : AlH_3 = 4 : 1$. It is believed to be an intermediate in the formation of $(HAlNBu^t)_4$.[77]

The formation of Al–N cages is complex and unpredictable. Thus $(HAlNPr^i)_3(H_2AlNHPr^i)_2$ has been separated (as a minor component)[83] from the reaction products between $AlCl_3$ with NaH and isopropylamine (molar ratio amine : $AlCl_3 = 0.93$). By using a smaller amount of amine,[14] the same reaction produces $(HAlNPr^i)_6.AlH_3$[84] and $H(HAlNPr^i)_5AlH_2$. $LiH.Et_2O$ when NaH is replaced by LiH.

The hexamer $[HAlN(CH_2)_3NMe_2]_6.2LiH$ has been obtained according to reaction (38),[43] whereas $(HAlNPr^i)_2(H_2AlNHPr^i)_2HAlNCH(Me)CH_2$ NMe_2 was obtained from the reaction of isopropylamine, 2-dimethylamino-isopropylamine and AlH_3:[86]

$$6 \, LiAlH_4 + 6 \, RNH_2 \longrightarrow (HAlNR)_6.2LiH + 4 \, LiH + 12 \, H_2 \quad (38)$$

Mixed imidoderivatives, containing –HAl—NR– and –Ca(THF)—NR– or –Mg(THF)—NR– units, have been prepared by the following reaction (M = Ca, $x = 3$, $y = 2$ or 3; M = Mg, $x = 5$, $y = 1$):[26,34]

$$M(AlH_4)_2.x \, THF + AlH_3.NMe_3 + 4 \, Bu^tNH_2 \longrightarrow$$
$$(HAlNBu^t)_3 \, (yTHF)MNBu^t + (x-y) \, THF + 8 \, H_2 + NMe_3 \quad (39)$$

A mixed derivative has also been obtained by direct synthesis in THF under a hydrogen pressure of 2700 p.s.i. at 100 °C:[34]

$$3 \, Al + Mg + 4 \, Bu^tNH_2 \xrightarrow{THF} (HAlNBu^t)_3 \, (3THF)MgNBu^t + \tfrac{5}{2} \, H_2 \quad (40)$$

Aluminium–nitrogen compounds (19) with the adamantane structure[101] have been obtained[79] from the following reaction (R = H, Cl, Br, I, R' = R'' = Me; R = H, R' = Me, R'' = Pr^i):

$$2 \, RHAl—NR'_2 + R''NH_2 \longrightarrow \tfrac{1}{2} \, [(RAlNR'_2)_2NR'']_2 + 2 \, H_2 \quad (41)$$
$$(19)$$

The hydride hydrogen in (19) (R = H) can be substituted with halogens, in reactions with mercury(II) halides.[79]

The compound HAl $[\{EtN(CH_2)_2NEt\}AlH_2]_2$ has been prepared by reaction of $AlH_3.Et_2O$ with N,N'-diethylethylenediamine.[85] A partially chlorinated analogue was formed by the same method, but starting from $AlH_{3-x}Cl_x$.[85]

6.3 Structure and bonding

In recent years a considerable number of X-ray structure determinations of aluminium–nitrogen rings and cages have been performed. This section will cover only the most significant results. Some of the structures are illustrated schematically by (23)–(35)

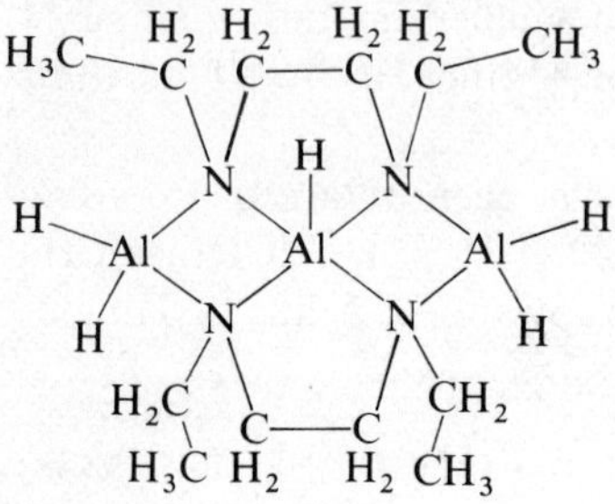

(23)

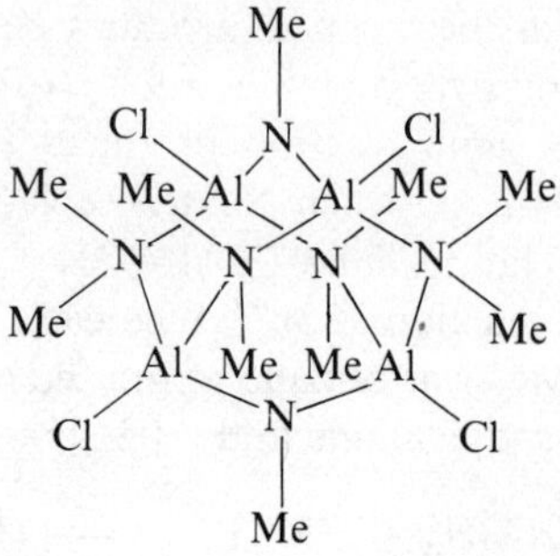

(24)

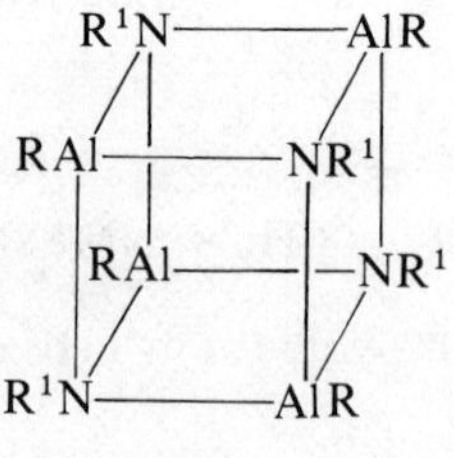

(25)

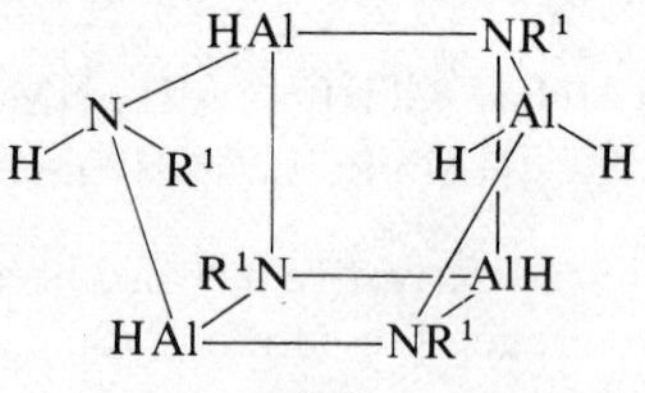

(26)

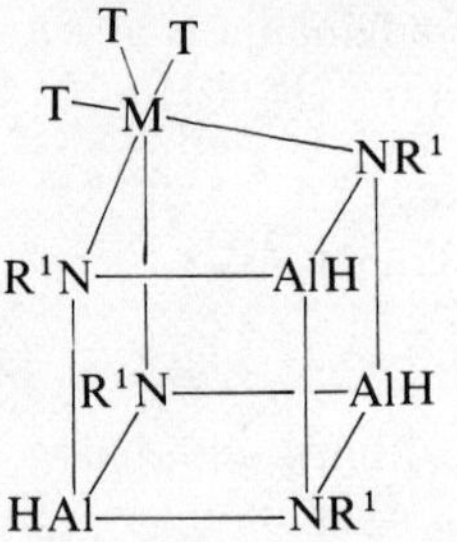

(27)

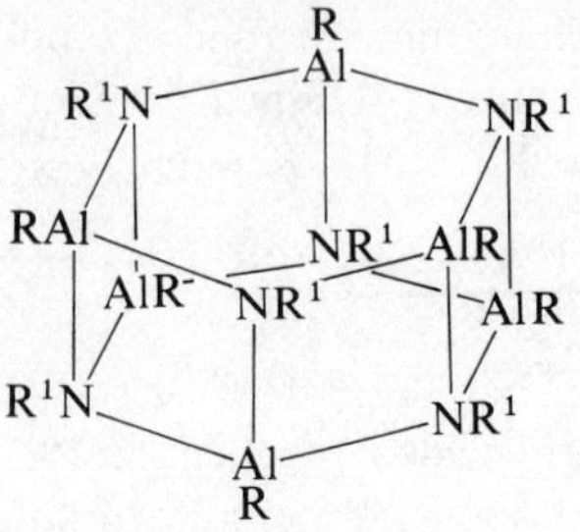

(28)

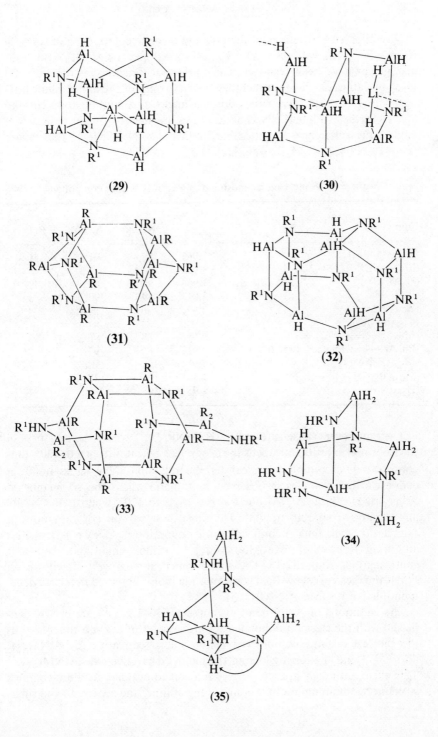

(29)

(30)

(31)

(32)

(33)

(34)

(35)

Several crystal structures of dimeric and trimeric compounds of formula $(R_2AlNR'_2)_n$ (2) with $n = 2,3$ (R = alkyl, halogen or H; R' = alkyl or alkylidene) have been resolved. Their principal structural parameters are given in Table 6.3. The four-membered Al_2N_2 rings of the dimers show little departure from planarity, with endocyclic angles at Al and N atoms close to 90°. Nevertheless, most AlNAl bond angles are slightly greater than NAlN angles; the difference is noticeable in $(Me_2Al—N=CMe_2)_2$ (20)[93] and $(Ph_2Al—N=CPh.C_6H_4Br)_2.2C_6H_6$ (21)[69]

Table 6.3 Bond lengths and bond angles in AlN ring compounds.

Compound	Al–N (Å)	NAlN	AlNAl	Refs.
$(Me_2AlNMe_2)_2$	1.963	88.4°	91.6°	51
$(Me_2AlNHPr^i)_2$, cis	1.952	87.15°	91.85°	5
$(H_2AlNPr^i_2)_2$	1.966	91.5°	88.5°	40
$(I_2AlNMe_2)_2$	1.943	89.1°	90.9°	2
$(Br_2AlNMe_2)_2$	1.946	90.9°	89.1°	2
$(Cl_2AlNMe_2)_2$	1.910	88.6°	91.4°	1
$(Me_2Al—N=CMe_2)_2$	1.926	83.7°	96.3°	93
$(Ph_2Al—N=CPh.C_6H_4Br)_2.2C_6H_6$	1.917	82.9°	97.1°	69
$(Me_2AlNHMe)_3$	1.953	102.1°	122.3°	47
$[Me_2AlN(CH_2)_2]_3$	1.925	101.7°	120.8°	12
$(H_2AlNMe_2)_3$	1.93	108°	115°	94

The only reverse case is found in $(H_2AlNPr^i_2)_2,^{40}$ where the bulkiness of the isopropyl substituents opens the exocyclic angle at nitrogen to 120°, thus considerably exceeding the ideal tetrahedral value. The same trend of endocyclic bond angles is observed in trimers, where the six-membered Al_3N_3 rings exhibit AlNAl bond angles close to 120°, whereas the NAlN angles range from 100° to 108°. The rings show various conformations: a cyclohexane-type (chair) conformation is found in $(H_2AlNMe_2)_3$;[94] the cis and trans isomers of $(Me_2AlNHMe)_3$[47,71] exhibit chair and skew-boat conformations respectively. The stereoisomers interconvert slowly on an NMR time scale. A skew-boat conformation is also observed in the ethylene-imino derivative $[Me_2AlN(CH_2)_2]_3.$[12]

The variations of Al–N bond distances (1.910–1.966 Å) are clearly connected with the electron-donor–acceptor character of the substituents; thus the shortest distance is observed in the chloro derivative $(Cl_2AlNMe_2)_2$ (22),[1,15,71] and the longest one in the methyl derivative $(Me_2AlNMe_2)_2.$[51] This effect, although present, is very difficult to evaluate in more complex structures containing multiple connections of four- and six-membered rings.

The shortest Al $\cdots$ Al nonbonded distance is found in (22) (2.66 Å), the shortest N $\cdots$ N distance in both alkylidene derivatives (20) and (21) (mean 2.87 Å). The structures of $HAl[\{EtN(CH_2)_2NEt\}AlH_2]_2$ and of its isostructural partially chlorinated analogue have been reported.[85] The molecule of the hydride derivative is built up of two AlH_2 group and one AlH group, connected through two diethylethylenediamido groups (23). The compound contains two nearly planar Al_2N_2 rings and a central Al atom in a distorted trigonal bipyramidal coordination with an axial NAlN bond angle of 154.3° and mean Al–N distances 2.000(5) Å (axial) and 1.922(2) Å (equatorial).

The crystal structure of the cage molecule $[\{ClAl(NMe_2)_2\}_2NMe]_2$ has been reported;[101] it is a representative of several compounds with adamantane structures of type (24).[79] The molecule displays four tetrahedrally coordinated N atoms, but two other N atoms show a trigonal planar coordination (CNAl angle 118.7°); mean Al–N distances are 1.92(2) Å (to four-coordinate N) and 1.79(3) Å (to three-coordinate N).[101]

Table 6.4 Bond lengths and bond angles in AlN cage compounds.

Compound	Al–N (Å)	NAlN	AlNAl	Refs.
$(MeAlNPr^i)_4$	1.923	89.6°	90.4°	39
$(HAlNPr^i)_4$	1.913	89.9°	90.1°	39
$(PhAlNPh)_4$	1.915	89.8°	90.2°	70
$(HAlNPr^i)_6$	1.898^a	116.4^{ob}	123.2^{ob}	24
	1.956^c	91.4^{od}	88.5^{od}	
$(HAlNPr^n)_6$	1.890^a	115.2^{ob}	124.3^{ob}	38
	1.959^c	91.2^{od}	88.6^{od}	
$(HAlNCHMe,Ph)_6$	1.890^a	116.1^{ob}	123.6^{ob}	41
	1.981^c	91.2^{od}	88.7^{od}	
$(MeAlNPr^i)_6$	1.917^a	115.7^{ob}	123.9^{ob}	37
	1.964^c	91.4^{od}	88.55^{od}	
$(ClAlNPr^i)_6$	1.906^a	122.0^{ob}	122.0^{ob}	37
	1.955^c	117.1^{od}	88.1^{od}	
$(MeAlNMe)_7$	1.91	110.0^{ob}	120^{ob}	54, 55
		90^{od}	89^{od}	
$(HAlNPr^n)_8$	1.916	114.1^{ob}	120.8^{ob}	38
		91.1^{od}	88.8^{od}	

^{a} In the six-membered Al_3N_3 ring (bond length).
^{b} Bond angle in the six-membered ring.
^{c} Transverse Al–N bond (connecting two Al_3N_3 rings).
^{d} In the four-membered Al_2N_2 ring (bond angle).

Several crystal structures of prismo-poly-μ_3-alkylimidopoly(hydrido or alkyl)aluminium (12) have been determined (see Table 6.4). Compounds of general formula $(RAlNR')_n$ with $n = 4, 6, 7$ and 8 display a "regular closed cage" skeleton, with four-coordinate Al and N atoms. The tetramers (12) ($n = 4$) have a cubane skeleton of nearly planar Al_2N_2 rings (structure 25). The structures of compounds with R = H, Me, R' = Pr^i [39] and R = R' = Ph [70] have been determined. ^1H NMR studies[21] strongly support the formation of tetrameric "open cages" (26) of composition $[(HAlNR)_3$-$(H_2AlNHR)]$ as precursors[27,43] of $(HAlNR)_4$. They are relatively stable at room temperature; the stability is enhanced by bulkiness of R', and solid (23) with R' = Bu^t,[27] $CH(Ph)Et$[6,22] and $CH(Me)CH_2NMe_2$[43] have been separated.

Tetrameric compounds in which one Al atom is substituted by an alkaline earth metal (M) have been structurally elucidated.[37] In $[(HAlNBu^t)_3$ $(THF.MgNBu^t)]$ and in $[(HAlNBu^t)_3(THF)_3.CaNBu^t)]$ the Mg and Ca atoms reach tetrahedral and octahedral distorted coordination respectively by complexing one or three THF molecules (structure 27).

The structures of several hexamers (12) ($n = 6$) have been determined; [24,37,38,41] the Al_6N_6 framework is a hexagonal prismatic cage, containing four- and six-membered rings (28). The hydrido derivative $(HAlNPr^i)_6$ (15) has two conformers in the solid state, differing by the rotation of an isopropyl group.[24] In this compound, as well as in other analogues listed in Table 6.4, there are significant differences between the Al–N bond distances in the Al_3N_3 and Al_2N_2 rings of the cage.

The structure of the adduct $(HAlNPr^i)_6.AlH_3$ (29)[84] may be derived from the structure of a regular closed hexamer (28) by breaking two adjacent Al–N bonds; the complexed Al atoms achieve five-coordination (trigonal bipyramidal). The adduct $H(HAlNPr^i)_5AlH_2LiH.Et_2O$ can be described as a pseudohexameric cage, consisting of a five-membered fragment Al–N–Al–N–Al, crosslinked to a six-membered Al_3N_3 ring (30).[25]

The structure of the adduct $[HAlN(CH_2)_3NMe_2]_6.2LiH$ may be considered as deriving from (28), where two opposite Al–N transverse bonds are broken.[86]

Heptameric cage molecules, represented by two compounds $(MeAlNMe)_7$ and $(EtAlNMe)_7$ have been reported,[54,55] and the former was structurally elucidated. The structure (31) of the Al_7N_7 unit can be derived by connecting two cubane skeletons and by removing an Al and an N atom.

The compound $(HAlNPr^i)_8$ contains an Al_8N_8 skeleton, which may be formally derived by combining a hexameric cage with a square Al_2N_2 ring (32). An alternation of longer and shorter Al–N bonds is observed along the hexagonal rings.[38]

An octameric cage Al_8N_8, containing two amido–alane bridges[4] is $[(Me_2AlNHMe)_2(MeAlNMe)_6]$ (33), a precursor of (28) (R = R' = Me) and $(MeAlNMe)_8$.

Pentameric cages in the series $(RAlNR')_n$ do not exist, but an Al_5N_5 framework has been found in the compound $[(HAlNPr^i)_2(H_2AlNPr^i)_3]$ **(34)**; the structure consists of a six-membered ring fragment $(HAlNPr^i)_2$ $(H_2AlNHPr^i)$ in a skew-boat conformation,[83] crossed on both sides by a bridging unit $-(HAlNHPr^i)-$; the Al–N bond distances range from 1.901 to 1.985 Å. The compound[86] $[(HAlNPr^i)_2(H_2AlNPr^i)_2$ $\{HAlNCH(Me)CH_2-NMe_2\}]$ can also be regarded as a pentameric derivative. In **(35)** $R^1 = Pr^i$ the curved arrow represents the $-CH(Me)CH_2NMe_2$ group.

6.4 Spectra

Only few articles[21,27,65,77] have reviewed the most significant results obtained by NMR (^1H, ^{13}C, ^{27}Al, ^{14}N) and mass spectrometry, which are by far the most widely used techniques for the identification and characterization of these compounds.

Hydride hydrogen resonances occur in the range 4–5 ppm,[21,27,65] which is abnormally low for a main-group metal hydride; in many cases these cannot be observed because of the extensive broadening caused by the relatively large electric quadrupole moment associated with the ^{27}Al nucleus.[65] Sometimes a broad signal at about 0.3 ppm is reported for the N–H group protons, as in the case of the *cis* and *trans* isomers of $(Me_2AlNHMe)_3$.[3] The ^1H and ^{13}C NMR spectra of $[H_{3-m}Al(NR_2)_m]_n$ ($m = 1, 2; n = 2, 3$), $(HAlNR)_4$ and $(HAlNR)_6$ systems support the geometric equivalence of N atoms,[21,65] in agreement with X-ray data. The octameric cage structure $(HAlNR)_8$ shows two environments for N atoms; however, a unique series of signals was found for $R = Pr^n$,[21,27] in contrast to two series for the $R = Et$ derivative.[77]

The ^{27}Al and ^{14}N NMR spectra give similar information. Chemical-shift values of 130 ppm for ^{27}Al and -220 ppm for ^{14}N are typical for the four-coordinate Al and N atoms.[77]

The influence of polar solvents has been studied for $(HAlNPr^i)_n$ with $n = 4$ and 6.[21] The change in the chemical shifts depends on the electron-donor effect of the solvent coordinated to Al atoms.

In open cage molecules $[(HAlNR)_{n-m}(H_2AlNHR)_m]$ with $n = 4$, $m = 1$;[21,27,43] $n = 6$, $m = 2$ and 3,[43] the magnetic equivalence of N-alkyl groups is destroyed, giving rise to complex spectra which are difficult to assign unambiguously. The same effect is observed by partial substitution of hydride hydrogens with alkyl groups or halogens.[21,32,33,77]

In infrared spectra[27] Al–H stretching frequencies occur mostly in the range 1820–1875 cm^{-1}. A shift to lower frequencies has been observed in the partially halogenated derivatives $[(HAlNBu^t)(ClAlNBu^t)_3]$ (1931 cm^{-1}) and $[(HAlNBu^t)_2(IAlNBu^t)_2]$ (1875 cm^{-1}).[77]

No detailed report on the mass-spectral fragmentation of Al–N ring or cage compounds exists. The main use of mass spectra is confined to the determination of the association degree in vapour phase.

The mass spectrum of $(Me_2AlNHMe)_3$ shows no parent ion, and the strongest peaks correspond to loss of methyl groups from the trimeric and dimeric ions.[3] The mass spectra of $R_2AlN(CH_2)_5$ (R = Me, Et, Bu^i)[95] and $Me_2AlN(CH_2)_n$ (n = 2, 3, 4 and 5)[99] show loss of a methyl group or of an imino group from the dimeric and trimeric parent ions. The compounds $(Cl_2AlNMe_2)_2$[45] and $[Et_2AlN{=}C(NMe_2)_2]_2$[96] maintain their dimeric nature in the vapour phase. The mass spectrum of $(Cl_2AlN{=}CPh_2)_2$ has also been reported.[97]

The mass spectra of closed cage compounds $(RAlNR')_n$, where R = H[27] and Cl[32,77], are dominated by the fragments originating from the fission of the C–C linkages adjacent to one nitrogen of the cage.

6.5 Chemical properties and reactions

6.5.1 Chemical stability

The synthesis and handling of AlN ring and cage compounds require an inert atmosphere of anhydrous nitrogen or argon. With very few exceptions[33] these compounds are air-sensitive. Water causes rapid hydrolysis to alumina, amine and, in the case of hydrides and alkylaluminium derivatives, hydrogen or alkane.

Most of these compounds are soluble in ethereal solvents, in aromatic hydrocarbons and, depending on the bulkiness of the alkyl groups, also in aliphatic hydrocarbons.

As a rule, the Al–N–Al bridges of AlN rings are stable in hydrocarbon solutions and some association is retained in ether.[B2] However, Me_2AlNPh_2, which is a dimer, is cleaved by NMe_3, Et_2O and Me_2S with formation of adducts.[90]

6.5.2 Reactions

Reactions occurring with preservation of the AlN skeleton have been reported in Section 6.2. This section includes reactions that imply ring or skeleton cleavage.

The cyclic oligomers of $Al(NMe_2)_3$ (36)[63] and $HAl(NMe_2)_2$[64] react with excess diborane to form $(Me_2NBH_2)_2$, μ-$Me_2NB_2H_5$ (37), $Al(BH_4)_3.Et_2O$, $[Me_2NAl(BH_4)_2]_2$ (38) and two new compounds $H_2B(NMe_2)_2Al(BH_4)_2$ and μ-$(Me_2N)_2B_3H_7$. Diborane adds to H_2AlNMe_2 in Et_2O or toluene to form (38). Excess diborane causes slow formation of (37) and aluminium borohydride. Compound (38) is formed in THF; it reacts with excess borane to produce (37) and $Bu^nOAl(BH_4)_2$.[62] Compound (36) reacts with $Fe(CO)_5$[88] and $Ni(CO)_4$[87] to form the carbene complexes $[(CO)_4FeC-(NMe_2)OAl(NMe_2)_2]_2$ and $[(CO)_3NiC(NMe_2)OAl(NMe_2)_2]_2$ respectively. Dimeric Br_2AlNMe_2 reacts with $Fe_3(CO)_{12}$ to yield $[(CO)_3(Me_2NAl)-FeBr_2]_2$.[89] The product of CS_2 insertion into $Me_2AlNPhR$[103] is $Me_2AlS_2C—NPhR$ (R = Me, Ph).

6.6 Uses

Some of the compounds involved with the following uses have not been characterized at molecular level; however, on the basis of general knowledge of AlN chemistry, their cyclic or cage structure is very likely.

Aldehydes have been prepared by the reduction of carboxylic acids,[74] esters[75] and N,N-disubstituted carboxamides[73] using $HAl(NR_2)_2$ (39). H_2AlNR_2 and (39) are regioselective reagents for conjugate reduction of enones.[9] The hexamer $(HAlNPr^i)_6$ (15) reduces aldehydes and ketones to alcohols in good yields, while carboxylic acids and their derivatives are reduced in low yields.[67] Compound (15) has also been found suitable for the selective reduction of steroid dicarbonyl compounds.[81,82] Chiral N-methyl-N-phenethylamidodihydroaluminium and chiral prismo-poly-μ_3-phenethylimidopoly(hydridoaluminium) (40) promote the asymmetric reduction of ketones[46] and racemic phosphine oxides.[23] Compounds (40) are also active in the reduction of sulphur-containing functional groups such as sulphoxides, sulphilimines and sulphoximides;[19] among the three molecular species of (40), namely $(HAlNR)_6$, $(HAlNR)_4$ and $[(HAlNR)_3-(H_2AlNHR)]$, only the last one has been found active in the selective resolution of racemic sulphoxides.[6]

Derivatives (39) in the presence of catalytic amounts of transition metal compounds, for example $(C_5H_5)_2TiCl_2$, are hydrometallation systems for olefins;[8,11] the product can be hydrolysed to give alkanes or can be reacted with carbonyl compounds to form tertiary amines.[11] The compounds H_2AlNMe_2, $HAl(NMe_2)_2$ and $(HAlNR)_n$ (13), in conjunction with cobalt and nickel derivatives, are homogeneous systems for the hydrogenation of mono- and diolefins.[44] Depending on the type of ligands, $R_2AlNR'_2$ (41), have been found to react with benzophenone in different ways. For R = Cl the main reaction is 1,2-addition of Al–N to the C=O bond; for R = Et C=O reduction is promoted.[60] Esters have yielded carboxamides when reacted with Me_2AlNR_2.[16,68] The treatment of epoxides with Et_2AlNR_2 gives β-aminoalcohols,[80] whereas they are converted to allylic alcohols when treated with $Et_2AlNR'R''$ from hindered amines.[106] Et_2AlNMe_2 (42) reacts with β-propiolactone to give selectively $Et_2AlOCH_2CH_2CONMe_2$.[52] Acid anhydrides $Y(CO)_2O$ react with one mole of (42) to give Me_2NCO_2-YCO_2AlEt_2 and with two moles of the same compound to form the diamide.[52]

Five types of reaction have been found to occur between Me_2AlNMe-$SiMe_3$ and ketones, esters and amides, depending on the structure of the carbonyl compounds.[91] Addition of the Al–N bond present in Et_2AlNR_2 (R = Me, Et) to the $C\equiv N$ bond of benzonitrile[53,59] and p-substituted benzonitriles[53] yields diethylaluminium benzamidine derivatives; the addition is suppressed for sterically hindered diphenylamido- and dicyclohexyl-amidodiethylaluminium.[57] Acetonitrile and benzyl cyanide react with (42) to give crystalline addition products.[53] Compound (42) has been added to ketenimines $Me_2C=C=NAr$ (Ar = Ph, p-tolyl) and the hydrolysis of the adduct has yielded the amidine $Me_2CHC(=NAr)NArC(NMe_2)=CMe_2$.[102]

The compounds $R_2AlN=CR'R''$ react with ammonia or primary amines to yield $(R_2AlNH_2)_2$ and the Schiff bases $R'R''C=NR$.[18] Compounds of the type $Cl_nEt_{2-n}AlN=CHR$ react by addition of the AlN bond to the $C\equiv N$ group of aromatic nitriles to give $Cl_nEt_{2-n}AlN=C(Ph)$—$N=CHR$.[56] Compound (15) reacts with caprolactam to yield aluminium caprolactamate.[66] $Et_2AlN=CHPh$ has been treated with a suspension of naphthalene and potassium in THF to yield $PhCHKNKAlEt_2$.[58]

Compounds of formula Ph_2AlNR_2 decompose thermally to give hydrocarbons, olefins and a residue of empirical formula $(AlNPh)_x$, which can be hydrolysed to primary amines.[10] Dialkylamidoaluminium compounds decompose at 300–500 °C and 30–40 Torr to give aluminium nitride and other products.[100]

The use of AlN compounds as polymerization initiators is very limited. Et_2AlNEt_2 has been found to be an active catalyst for the copolymerization of carbon dioxide and epoxycyclohexane.[107] The copolymerization of methyl-α-phenylacrylate with $Bu_2^iAlNPh_2$ has given a syndiotactic polymer in low yield.[108] The compounds $Et_2AlNHPr^i$ and Et_2AlNMe_2 polymerize acetaldehyde and butyraldehyde to crystalline polymers.[5,61]

The AlN compounds can be used as components of catalytic polymerization systems in conjunction with transition-metal compounds, for the polymerization of α-olefins and conjugated dienes. Thus R_2AlNR_2' have been used as cocatalysts in the polymerization of propylene,[112] the derivatives of hindered amines being more active. $Al(NMe_2)_3$ is a component of a vanadium catalytic system for ethylene–butadiene copolymers.[35] Liquid polymers have been obtained from higher α-olefins by using polyalkylimido-polyhydridoaluminium (PIA)–$TiCl_4$.[111] The application of PIA–titanium-chloride systems has afforded an industrial development in polyethylene[110] and 1,4-*cis*-polyisoprene[13] processes. In particular, the advances in catalysts based on $(HAlNR)_n$–$TiCl_4$ for the polymerization of isoprene to high *cis*-polymer have been reported in detail[B 1,13], with emphasis on the influence of the chemical and/or structural modifications of $(HAlNR)_n$ on the catalytic activity.

Bibliography

B 1   S. Cucinella, *Chim. Ind. (Milan)* **59** (1977) 696.
B 2   S. Cucinella, A. Mazzei and W. Marconi, *Inorg. Chim. Acta Rev.* **4** (1970) 51.
B 3   I. Haiduc, *The Chemistry of Inorganic Ring Systems* (Wiley-Interscience, London, 1970).
B 4   M. F. Lappert, P. P. Power, A. R. Sanger and R. C. Srivastava, *Metal and Metalloid Amides* (Ellis Horwood/Wiley, New York, 1980).
B 5   T. Mole and E. A. Jeffery, *Organoaluminium Compounds* (Elsevier, Amsterdam 1972).

References

1   A. Ahmed, W. Schwarz and H. Hess, *Acta Crystallogr.* **B33** (1977) 3574.
2   A. Ahmed, W. Schwarz and H. Hess, *Z. Naturforsch.* **33b** (1978) 43.
3   K. J. Alford, K. Gosling and J. D. Smith, *J. Chem. Soc., Dalton Trans.* **1972**, 2203.
4   S. Amirkhalili, P. B. Hitchcock and J. D. Smith, *J. Chem. Soc., Dalton Trans.* **1979**, 1206.
5   S. Amirkhalili, P. B. Hitchcock, A. D. Jenkins, J. Z. Nyathi and J. D. Smith, *J. Chem. Soc., Dalton Trans.* **1981**, 377.
6   R. Annunziata, G. Borgogno, F. Montanari, S. Quici and S. Cucinella, *J. Chem. Soc., Perkin Trans.* **1981**, 113.
7   E. C. Ashby and R. A. Kovar, *J. Organomet. Chem.* **22** (1970) C34.
8   E. C. Ashby and A. S. Noding, *Tetrahedron Lett.* **1977**, 4579.
9   E. C. Ashby and J. J. Lin, *Tetrahedron Lett.* **1976**, 3865.
10   E. C. Ashby and G. F. Willard, *J. Org. Chem.* **43** (1978) 4750.
11   E. C. Ashby and S. A. Noding, *J. Org. Chem.* **44** (1979) 4364.
12   J. L. Atwood and G. D. Stucky, *J. Am. Chem. Soc.* **92** (1970) 285.
13   A. Balducci, M. Bruzzone, S. Cucinella and A. Mazzei, *Rubber Chem. Technol.* **48** (1975) 736.
14   A. Balducci, Unpublished data (1975).
15   T. C. Bartke, A. Haaland and D. P. Novak, *Acta Chem. Scand.* **A29** (1975) 273.
16   A. Basha, M. Lipton and S. M. Weinreb, *Tetrahedron Lett.* **1977**, 4171.
17   O. T. Beachley and K. C. Racette, *Inorg. Chem.* **14** (1975) 2534.
18   B. Bogdanovic and S. Konstantinovic, *Ann.* **738** (1970) 202.
19   G. Borgogno, F. Montanari, S. Quici and S. Cucinella, Unpublished data (1981).
20   R. E. Bowen and K. Gosling, *J. Chem. Soc., Dalton Trans.* **1974**, 964.
21   C. Busetto, M. Cesari, S. Cucinella and T. Salvatori, *J. Organomet. Chem.* **132** (1977) 339.
22   C. Busetto, S. Cucinella and T. Salvatori, *Inorg. Chim. Acta* **26** (1978) L51.
23   E. Cernia, G. M. Giongo, F. Marcati, W. Marconi and N. Palladino, *Inorg. Chim. Acta* **11** (1974) 195.

24 M. Cesari, G. Perego, G. Del Piero, S. Cucinella and E. Cernia, *J. Organomet. Chem.* **78** (1974) 203.

25 M. Cesari, G. Perego, G. Del Piero, M. Corbellini and A. Immirzi, *J. Organomet. Chem.* **87** (1975) 43.

26 S. Cucinella, G. Dozzi and A. Mazzei, *J. Organomet. Chem.* **63** (1973) 17.

27 S. Cucinella, T. Salvatori, C. Busetto, G. Perego and A. Mazzei, *J. Organomet. Chem.* **78** (1974) 185.

28 S. Cucinella, A. Mazzei and G. Dozzi, *J. Organomet. Chem.* **84** (1975) C19.

29 S. Cucinella, G. Dozzi, A. Mazzei and T. Salvatori, *J. Organomet. Chem.* **90** (1975) 257.

30 S. Cucinella, G. Dozzi, M. Bruzzone and A. Mazzei, *Inorg. Chim. Acta* **13** (1975) 73.

31 S. Cucinella, G. Dozzi, C. Busetto and A. Mazzei, *J. Organomet. Chem.* **113** (1976) 233.

32 S. Cucinella, T. Salvatori, C. Busetto and A. Mazzei, *J. Organomet. Chem.* **108** (1976) 13.

33 S. Cucinella, T. Salvatori, C. Busetto and M. Cesari, *J. Organomet. Chem.* **121** (1976) 137.

34 S. Cucinella, G. Dozzi, G. Perego and A. Mazzei, *J. Organomet. Chem.* **137** (1977) 257.

35 S. Cucinella, A. Mazzei and A. De Chirico, *Eur. Polym. J.* **12** (1977) 65.

36 G. Del Piero, M. Cesari, S. Cucinella and A. Mazzei, *J. Organomet. Chem.* **137** (1977) 265.

37 G. Del Piero, G. Perego, S. Cucinella, M. Cesari and A. Mazzei, *J. Organomet. Chem.* **136** (1977) 13.

38 G. Del Piero, M. Cesari, G. Perego, S. Cucinella and E. Cernia, *J. Organomet. Chem.* **129** (1977) 289.

39 G. Del Piero, M. Cesari, G. Dozzi and A. Mazzei, *J. Organomet. Chem.* **129** (1977) 281.

40 G. Del Piero, Unpublished data (1978).

41 G. Del Piero, S. Cucinella and M. Cesari, *J. Organomet. Chem.* **173** (1979) 263.

42 G. Dozzi, S. Cucinella, A. Mazzei and T. Salvatori, *Inorg. Chim. Acta* **15** (1975) 179.

43 G. Dozzi, C. Busetto, T. Salvatori and S. Cucinella, *J. Organomet. Chem.* **192** (1980) 17.

44 G. Dozzi, S. Cucinella and A. Mazzei, *J. Organomet. Chem.* **164** (1979) 1.

45 R. Ehrlich, *Inorg. Chem.* **9** (1970) 146.

46 G. M. Giongo, F. Di Gregorio, N. Palladino and W. Marconi, *Tetrahedron Lett.* **1973**, 3195.

47 K. Gosling, G. M. McLaughlin, G. A. Sim and J. D. Smith, *J. Chem. Soc., Chem. Commun.* **1970** 1617.

48 K. Gosling, A. L. Bhuiyan and K. R. Mooney, *Inorg. Nucl. Chem. Lett.* **7** (1971) 913.

49 K. Gosling and A. L. Bhuiyan, *Inorg. Nucl. Chem. Lett.* **8** (1972) 329.

50 K. Gosling and R. E. Bowen, *J. Chem. Soc., Dalton Trans.* **1974**, 1961.

51 H. Hess, A. Hinderer and S. Steinhauser, *Z. Anorg. Allg. Chem.* **377** (1970) 1.

52 T. Hirabayashi, K. Itoh, S. Sakai and Y. Ishii, *J. Organomet. Chem.* **25** (1970) 33.

53 T. Hirabayashi, K. Itoh, S. Sakai and Y. Ishii, *J. Organomet. Chem.* **21** (1970) 273.

54 P. B. Hitchcock, G. M. McLaughlin, J. D. Smith and K. M. Thomas, *J. Chem. Soc., Chem. Commun.* **1973**, 934.

55 P. B. Hitchcock, J. D. Smith and K. M. Thomas, *J. Chem. Soc., Dalton Trans.* **1976**, 1433.

56 H. Hoberg and J. Barluenga-Mur, *Ann.* **751** (1971) 86.

57 H. Hoberg and J. Barluenga-Mur, *Ann.* **748** (1971) 163.

58 H. Hoberg and U. Griebsch, *Synthesis* **1976**, 830.

59 H. Hoberg and J. Barluenga-Mur, *Ann.* **733** (1970) 141.

60 H. Hoberg and I. Tkatchenko, *Ann.* **751** (1971) 77.

61 A. D. Jenkins, J. Z. Nyathi and J. D. Smith, *Eur. Polym. J.* **18** (1982) 149.

62 P. C. Keller, *Inorg. Chem.* **11** (1972) 256.

63 P. C. Keller, *J. Am. Chem. Soc.* **94** (1972) 4020.

64 P. C. Keller, *J. Am. Chem. Soc.* **96** (1974) 3073.
65 R. A. Kovar and E. C. Ashby, *Inorg. Chem.* **10** (1971) 893.
66 O. Kriz and B. Casensky, *J. Organomet. Chem.* **161** (1978) 273.
67 O. Kriz, T. Stuchlik and B. Casensky, *Z. Chem.* **17** (1977) 18.
68 M. F. Lipton, A. Basha and S. M. Weinreb, *Org. Synth.* **59** (1980) 49.
69 W. S. McDonald, *Acta Crystallogr.* **B25** (1969) 1385.
70 T. R. R. McDonald and W. S. McDonald, *Acta Crystallogr.* **B28** (1972) 1619.
71 G. M. McLaughlin, G. Sim and J. D. Smith, *J. Chem. Soc., Dalton Trans.* **1972**, 2197.
72 J. Muller, K. Margiolis and K. Dehnicke, *J. Organomet. Chem.* **46** (1972) 219.
73 M. Muraki and T. Mukaiyama, *Chem. Lett.* **1975**, 875.
74 M. Muraki and T. Mukaiyama, *Chem. Lett.* **1974**, 1447.
75 M. Muraki and T. Mukaiyama, *Chem. Lett.* **1975**, 215.
76 J. Z. Nyathi, J. M. Ressner and J. D. Smith, *J. Organomet. Chem.* **70** (1974) 35.
77 H. Nöth and P. Wolfgardt, *Z. Naturforsch.* **31b** (1976) 697.
78 H. Nöth and P. Wolfgardt, *Z. Naturforsch.* **31b** (1976) 1201.
79 H. Nöth and P. Wolfgardt, *Z. Naturforsch.* **31b** (1976) 1447.
80 L. E. Overman and L. A. Flippin, *Tetrahedron Lett.* **1981**, 195.
81 M. Paglialunga Paradisi, G. Pagani Zecchini and A. Romeo, *Tetrahedron Lett.* **1977**, 2369.
82 M. Paglialunga Paradisi and G. Pagani Zecchini, *Tetrahedron* **37** (1981) 971.
83 G. Perego, G. Del Piero, M. Cesari, A. Zazzetta and G. Dozzi, *J. Organomet. Chem.* **87** (1975) 53.
84 G. Perego, M. Cesari, G. Del Piero, A. Balducci and E. Cernia, *J. Organomet. Chem.* **87** (1975) 33.
85 G. Perego, G. Del Piero, M. Corbellini and M. Bruzzone, *J. Organomet. Chem.* **136** (1977) 301.
86 G. Perego and G. Dozzi, *J. Organomet. Chem.* **205** (1981) 21.
87 W. Petz, *J. Organomet. Chem.* **55** (1973) C42.
88 W. Petz and G. Schmid, *Angew. Chem.* **84** (1972) 997.
89 W. Petz and G. Schmid, *J. Organomet. Chem.* **35** (1972) 321.
90 J. E. Rie and J. P. Oliver, *J. Organomet. Chem.* **80** (1974) 219.
91 T. Sakakibara, T. Hirabayashi and Y. Ishii, *J. Organomet. Chem.* **46** (1972) 231.
92 T. Salvatori, G. Dozzi and S. Cucinella, *Inorg. Chim. Acta* **39** (1980) 263.
93 S. K. Seale and J. L. Atwood, *J. Organomet. Chem.* **73** (1974) 27.
94 K. N. Semenenko, E. B. Lobkovskii and A. L. Dorosinskii, *Zh. Strukt. Khim.* **13** (1972) 696.
95 B. Sen and G. L. White, *J. Inorg. Nucl. Chem.* **35** (1973) 2207.
96 R. Snaith, K. Wade and B. K. Wyatt, *J. Chem. Soc. A* **1970**, 380.
97 R. Snaith, C. Summerford, K. Wade and B. K. Wyatt, *J. Chem. Soc. A* **1970**, 2635.
98 A. Storr, B. S. Thomas and A. D. Penland, *J. Chem. Soc. A* **1972**, 326.
99 A. Storr and B. S. Thomas, *J. Chem. Soc. A* **1971**, 3850.
100 Y. Takahashi, K. Yamashita, S. Motojima and K. Sugiyama, *Surface Sci.* **86** (1979) 238 (*CA.* **92** (1980) 33161).
101 U. Thewalt and I. Kawada, *Chem Ber.* **103** (1970) 2754.
102 K. Urata, K. Itoh and Y. Ishii, *J. Organomet. Chem.* **66** (1974) 229.
103 K. Wakatsuki, Y. Takeda and T. Tanaka, *Inorg. Nucl. Chem. Lett.* **10** (1974) 383.
104 K. Wakatsuki and T. Tanaka, *Bull. Chem. Soc. Jpn* **48** (1975) 1475.
105 F. Weller and K. Dehnicke, *Chem. Ber.* **110** (1977) 3935.
106 A. Yasuda, S. Tanaka, K. Oshima, H. Yamamoto and H. Nozaki, *J. Am. Chem. Soc.* **96** (1974) 6513.
107 Y. Yoshida and S. Inoue, *Polymer J.* **12** (1980) 763.

108 H. Yuki, K. Hatada, T. Niimoni, M. Hashimoto and J. Ohshima, *Polymer J.* **2** (1971) 629.
109 B. Casensky, J. Machacek and T. Hanslik, *Ger. Offen* 2 112 665.
110 M. Corbellini and A. Balducci, *UK Patents* 1 508 046, 1 508 048.
111 P. Girotti, R. Tesei and T. Floris *UK Patent* 1 468 723.
112 A. W. Langer, *US. Patents* 4 094 818; 4 224 181; 4 224 182
113 T. R. R. McDonald and W. S. McDonald, *Proc. Chem. Soc.* **1962**, 366.

7

Silicon Homocycles (Cyclopolysilanes) and Related Heterocycles[*]

EDWIN HENGGE and KARL HASSLER

Institute of Inorganic Chemistry, Technical University Graz, Austria

7.1 Homocyclic silanes

Cyclosilane rings of various sizes, containing three, four, five, six and up to 35 silicon atoms have been reported, together with a range of spirocyclic and polycyclic (cage) species. The first four members of the series are illustrated by formulae (1)–(4):

$$(1) \qquad (2) \qquad (3) \qquad (4)$$

Introduction of heteroatoms into polysilane rings results in a variety of heterocyclic polysilanes, which are treated in the last part of this chapter.

Review articles concerned with these compounds are listed in the bibliography.[B 1 – B 11]

7.1.1 Cyclotrisilanes

Little is known about rings with three silicon atoms. Although the photolysis of dodecamethylcyclohexasilane Si_6Me_{12} yields smaller rings, for example cyclotetra- and pentasilanes, no three-membered cyclosilane is

[*] The article represents the research work up to 1982. Since the article was written, many new results in this field have been published.

obtained.[1] A three-membered ring (1a) has been postulated as intermediate in the liquid-phase pyrolysis of neat methoxytris(trimethylsilyl)silane (R = Me):[2]

$$(R_3Si)_3SiOR \xrightarrow{250°} (R_3Si)_2Si: \longrightarrow (R_3Si)_2Si{=}Si(SiR_3)_2$$

$$
\begin{array}{ccc}
& \text{addition} & \text{dimerization} \\
(R_3Si)_2Si{-}Si(SiR_3)_2 & \xrightarrow[\substack{(R_3Si)_2Si:}]{\text{insertion}} & (R_3Si)_2Si{-}Si(SiR_3)_2 \\
\underset{(SiR_3)_2}{Si} & & (R_3Si)_2Si{-}Si(SiR_3)_2 \\
\textbf{(1a)} & & \textbf{(2a)}
\end{array}
\tag{1}
$$

The final stable product is the four-membered ring compound (2a).

The only stable three-membered cyclosilane known so far was found in the reaction of bis(2,6-dimethylphenyl)dichlorosilane with lithium naphthalenide (R = $Me_2C_6H_3$):

$$3\ R_2SiCl_2 \xrightarrow{\text{Li naphthalenide}} Si_3R_6 \tag{2}$$

Hexakis(2,6-dimethylphenyl)cyclotrisilane (1), R = $Me_2C_6H_3$, is formed in 10% yield. The ^{1}H NMR spectrum is temperature-dependent, indicating hindered rotation of the aromatic rings about the Si–C bonds.[3]

An X-ray diffraction study shows that the three Si atoms form an isosceles triangle with bond angles SiSiSi of 58.7°–60.7°. One Si–Si bond is 5 pm shorter than the two others: 237.5 and 242.5 pm.[3]

7.1.2 Cyclotetrasilanes

Perphenylated cyclotetrasilane. The four-membered cyclosilane Si_4Ph_8 is one of the best-known cyclosilanes. It was first prepared, together with other cyclic polysilanes, in 1921 by Kipping[4] from diphenyldichlorosilane by reaction with sodium, and a radical disilane structure was postulated. The correct ring structure was established by Gilman.[5] The original preparation has been modified several times[5,5a,6] (best yield c. 50%), and other preparative routes have been described. Thus diphenylsilane Ph_2SiH_2 reacts with bis(t-butyl)mercury $HgBu_2^t$ to give the tetrasilane,[7] and the compound can also be produced electrochemically from diphenyldichlorosilane in dimethoxyethane with tetrabutylammonium perchlorate as supporting electrolyte.[8] A second electrochemical synthesis involves 1,2-dichlorotetraphenyldisilane and 1,4-dichlorooctaphenyltetrasilane $Cl(SiPh_2)_nCl$ (n = 2 and 4) as starting materials.[9]

Octaphenylcyclotetrasilane is a white crystalline powder (m.p. 321–323 °C), which is not very soluble in benzene and other organic solvents. Mass spectra (70 eV) show a strong parent molecular ion ($m/e = 728$) with isotope satellites. The fragmentation scheme shows preferred cleavage of an Si–Si bond, and one of the most intense peaks is $Si_3Ph_5^+$, formed by abstraction of $SiPh_3$.[10] The crystal structure[11] exhibits a nonplanar ring with a fold angle of c. 12°. The Si-Si distances are 237 and 238 pm.

Infrared spectra have been measured,[6,12] and, in contrast with the perphenylated five- and six-membered analogues, the cyclotetrasilane exhibits only one strong absorption at c. 330 cm^{-1} in the range below 400 cm^{-1}. Raman spectra have also been discussed.[13]

The UV spectrum shows bands at 270, 234 and 205 nm.[6,14] The ^{29}Si NMR spectrum shows only one singlet at -20.93 ppm (relative to external TMS).[15] The ^{13}C NMR data[16] indicate great flexibility of the ring system and semiempirical force-field calculations have been used to interpret the relative energies of the possible conformations of the Si_4 ring.

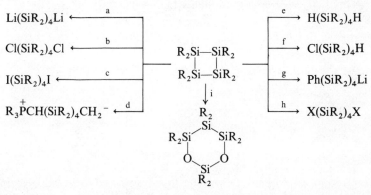

Figure 7.1 Cleavage reactions of octaphenylcyclotetrasilane (R = Ph). Reagents: a = Li metal;[5] b = Cl_2,[23] $HgCl_2$, $SnCl_2$,[21] PCl_5,[23] or $Cl_2HCCHCl_2$;[5] c = I_2;[26] d = R_3P—CH_2;[25] e = H_2(Pt);[24] f = Bu^tCl;[22] g = PhLi;[19] h = HX;[20] i = $PhNO_2$.[4,17,18]

The main product of pyrolysis of Si_4Ph_8 is triphenylsilane.[6] The chemical reactivity of Si_4Ph_8 is relatively high in comparison with other perphenylated cyclosilanes, as a result of the high strain in the four-membered ring. Ring cleavage is possible with a range of compounds, and these reactions are summarized in Fig. 7.1. An important cleavage takes place with lithium in THF and the product, 1,4-dilithio-octaphenyltetrasilane $Li(SiPh_2)_4Li$, can be isolated.[27] This highly reactive compound is the precursor of several homo- and heterocyclic polysilanes. Some of the known reactions are summarized in Fig. 7.2. The yields are in general low and the cyclotetrasilane is often reformed; side reactions lead to polymeric materials.

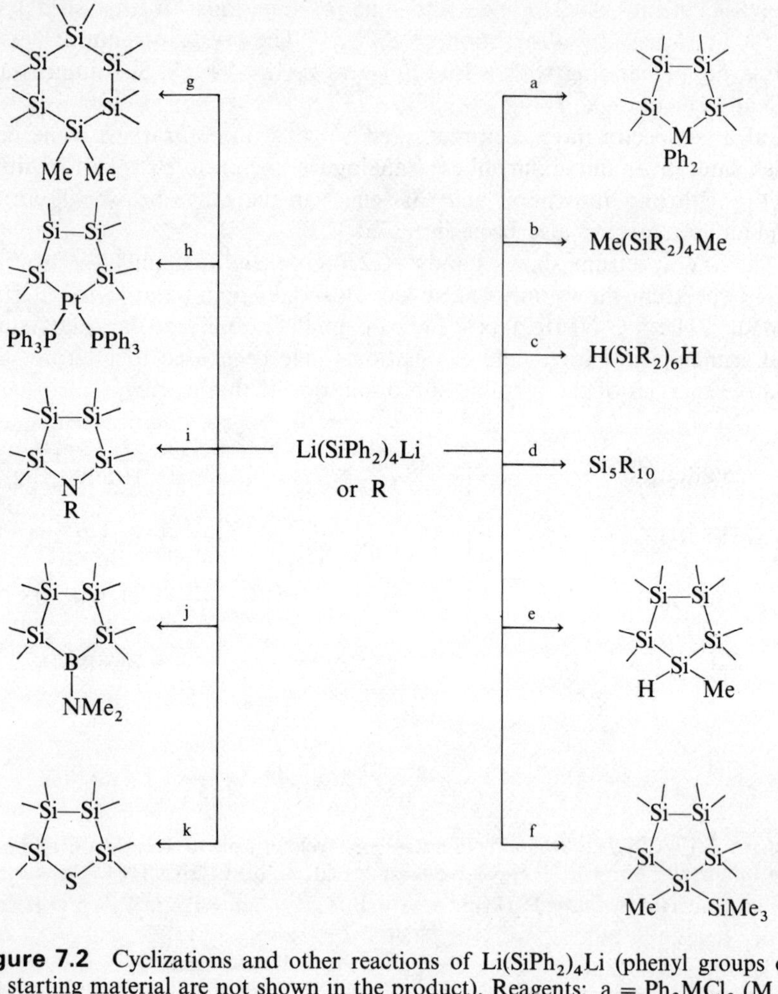

Figure 7.2 Cyclizations and other reactions of $Li(SiPh_2)_4Li$ (phenyl groups of the starting material are not shown in the product). Reagents: a = Ph_2MCl_2 (M = Ge, Sn, Si);[35] b = $(MeO)_3PO$;[36] c = Ph_2SiHCl;[36,39] d = K_2PtCl_6 or $PhCl$;[28] e = $MeSiHCl_2$;[32] f = $Cl_2MeSiSiMe_3$;[37,38] g = $Cl(SiMe_2)_2Cl$;[32] h = $(PPh_3)_2Pt(C_2H_4)$;[33] i = $RNCl_2$;[27] j = $Cl_2B—NMe_2$;[27,34] k = SCl_2.[35]

In some ring-closure reactions, for example $Li(SiPh_2)_4Li$ with K_2PtCl_6, the main product is Si_5Ph_{10} instead of the expected tetrasilane,[28] showing that thermodynamically the five-membered ring compound is more stable.

With dichlorodiphenylmethane, no carbon-containing ring system could be obtained from $Li(SiPh_2)_4Li$.[29]

Octamethylcyclotetrasilane. The permethylated cyclotetrasilane Si_4Me_8 was found together with the corresponding pentamer during photolysis (2537 Å) of dodecamethylcyclohexasilane.[30,31] The reaction proceeds with generation of dimethylsilylene species, as was shown by trapping experiments.[40] The cyclotetrasilane was formed only in very small yield and was purified by chromatography. Recently, a new route has been found for preparing larger amounts. The starting materials are the cyclic Si_4Cl_8 or Si_4Br_8 derivatives, and with mild methylating agents, such as dimethylzinc, octamethylcyclotetrasilane is formed in 40% yield.[41]

The tetramer Si_4Me_8 is a fairly air-sensitive white crystalline compound, which is stable at room temperature under nitrogen for several days. It is sublimable (40 °C, 0.001 Torr) and decomposes at about 70 °C. The IR and Raman spectra have been measured and assigned,[41,42] and, in this connection, the deuterated derivative $Si_4(CD_3)_8$ was also prepared. Normal-coordinate analysis shows that the Si–Si stretching force constant (139 N/m) is substantially lower than that in the corresponding five- and six-membered rings.[42] The ^{29}Si NMR shift (-27.72 ppm, relative to TMS) has been compared with data for other cyclosilanes.[43] Other NMR data (e.g. $^{29}Si-^1H$ coupling constants) and ^{13}C NMR spectra have also been reported.[41]

An X-ray structure determination[44] shows that the Si_4 ring is planar, with Si–Si 236 pm. This compound and octakis(trimethylsilyl)cyclotetrasilane are the only planar Si_4 ring compounds.

Halogenated cyclotetrasilanes. Octachloro-, octabromo- and octaiodocyclotetrasilanes Si_4X_8 (X = Cl, Br, I) can be prepared from Si_4Ph_8 by using the known cleavage of phenyl–silicon bonds with HX in the presence of AlX_3. Although with HX alone there is only incomplete substitution of the phenyl groups,[45,46] in the presence of AlX_3 as a catalyst, complete substitution is possible.[47] This reaction is possible without ring cleavage using especially mild conditions.[43,48,49]

The perhalogenated cyclotetrasilanes are yellow crystalline compounds, very sensitive to moisture. Si_4Cl_8 is decomposed at *c.* 100 °C, and the stability order is $Si_4Cl_8 < Si_4Br_8 < Si_4I_8$.

A complete normal-coordinate analysis was carried out using measured IR and Raman frequencies. Selection rules indicate that the Si_4 ring is nonplanar and the Si–Si stretching force constants are lower than in the corresponding five- and six-membered analogues. The Si–halogen force constants appear to be somewhat higher.[49,50] Empirical force-field calculations with an MMI program and using experimental data for similar compounds show that a puckered ring conformation with D_{2h} symmetry is more stable than the planar D_{4h} conformation.[51]

NMR spectral data (^{29}Si) are known only for Si_4Cl_8 and Si_4Br_8. The chemical shifts are markedly different from those of the five- and six-membered analogues, as a consequence of the high ring strain.[43]

Other cyclotetrasilanes. In addition to the compounds described above, derivatives with different alkyl and aryl groups have been characterized. Using Et_2SiCl_2, it is possible to prepare Si_4Et_8, as a colourless liquid, which reacts with a large number of organic and inorganic compounds.[52] Octakis(p-tolyl)cyclotetrasilane has been known for a long time,[53] but cleavage reactions with $Si_4(p$-tolyl)$_8$ are much easier than with the phenyl analogue, and more than one Si–Si bond is usually cleaved.[54]

The recently prepared octakis(trimethylsilyl)cyclotetrasilane $Si_4(SiMe_3)_8$ (**2a**) is formed in the liquid-phase pyrolysis of neat methoxytris(trimethylsilyl)silane, cited above (reaction 1). The X-ray structure determination indicates a planar four-membered ring structure, with a Si–Si distance of 240 pm.[2]

t-Butylmethyldichlorosilane reacts with Na/K alloy in THF to produce 1,2,3,4-tetra-t-butyl-1,2,3,4-tetramethylcyclotetrasilane.[55] In this reaction, two of the four possible isomers, (**2b**) and (**2c**) are formed in a 2 : 1 ratio, but only traces of the third isomer (**2d**) were observed.[57]

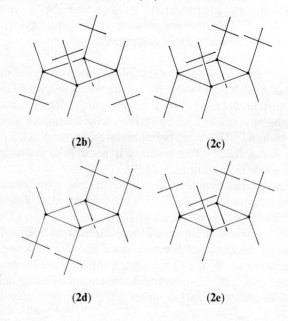

(**2b**)   (**2c**)

(**2d**)   (**2e**)

The most-favoured isomer (**2b**) was isolated by fractional crystallization from acetone. The third isomer (**2d**) can also be formed by photolysing mixtures of (**2b**) and (**2c**). A reinvestigation of this system has shown that all the four isomers can be isolated and characterized by NMR spectroscopy.[60]

A crystal structure determination[56] for isomer (**2b**) showed that the four-membered ring is not planar and the fold angle (36.8°) is unusually large. The bond lengths are Si–Si 237.7 pm and Si–C(Me) 189.3 pm, Si–C(Bu^t) 191.8 pm; the SiSiSi angle (86.99°) is very small.

The strained four-membered ring of $(SiMeBu^t)_4$, further distorted by the four Bu^t groups, gives rise to extreme electronic properties, such as the long-wavelength UV transition at 300 nm ($\varepsilon = 290$)[55,57] and a very low first vertical ionization potential of 7.82 eV.[59]

The compounds (**2b**)–(**2e**) are very stable towards oxidation,[58] and the lack of reactivity towards a variety of reagents is due to very effective shielding by the t-butyl groups. Oxidation of all four isomers with *m*-chloroperbenzoic acid, however, leads to $(Bu^tMeSi)_4O_n$, with $n = 1, 2$ and 4.[61]

Chlorodemethylation is possible with hydrogen chloride in the presence of $AlCl_3$ yielding mainly the mono- and dichloro-substitution products. This reaction seems to be strongly stereospecific, as shown by the retention of configuration of (**2b**) on chlorodemethylation.[58]

7.1.3 Cyclopentasilanes

Perphenylated cyclopentasilanes. Decaphenylcyclopentasilane Si_5Ph_{10} is the most stable cyclopentasilane. It was originally prepared by Kipping[4] by reaction of Ph_2SiCl_2 with sodium, but the postulated structure was wrong; the correct cyclic pentameric structure was suggested by Gilman.[12] Today the best preparation is the reaction of Ph_2SiCl_2 with lithium in THF, which gives a yield of 70–80%. The cyclopentasilane is also found in small quantities in a ring-expansion reaction between $Li(SiPh_2)_4Li$ and Ph_2SiCl_2.[35] A better yield is possible if $Li(SiPh_2)_nLi$ ($n = 4$ or 5) reacts with K_2PtCl_6.[28]

The pentamer Si_5Ph_{10} is an air-stable crystalline white solid, soluble in benzene and other organic solvents. It has a high melting point (474 °C). The ^{29}Si NMR chemical shift ($\delta = -34.41$ ppm) is very similar to that for the related cyclohexasilane.[43] Infrared and Raman spectra have been measured,[12,28] and the most intense Raman band at 517 cm^{-1} can be assigned to a ring pulsation, strongly coupled with v_sSi–Ph.[13] The mass spectrum shows the parent peak.[10] UV spectra show absorptions at 251 and 205 nm, with a shoulder at 230 nm.[6,14]

The main product of thermal decomposition is triphenylsilane.[6] A crystal structure determination shows a puckered ring in a form intermediate between C_s and C_2 symmetry. The Si–Si bond lengths vary between 237.1 and 241.3 pm, and SiSiSi angles between 102.7° and 106.7°.[65]

Many reactions have been carried out with Si_5Ph_{10}. The ring can be cleaved by a variety of reagents, and some of these reactions are summarized in Fig. 7.3.

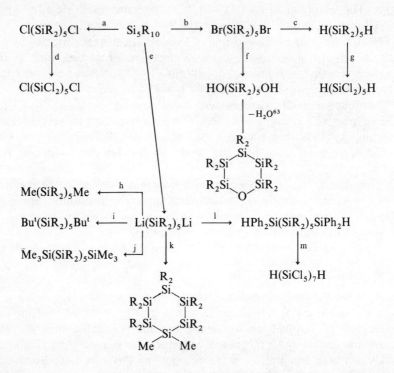

Figure 7.3 Ring-opening reactions of Si_5Ph_{10} (R = Ph) and further conversions of the linear products. Reagents: a = PCl_5 or Cl_2;[23] b = Br_2;[12] c = $LiAlH_4$;[12] d = $HCl/AlCl_3$;[64] e = Li metal;[12] f = H_2O;[12] g = $HCl/AlCl_3$;[62] h = $(MeO)_3PO$ or $MeMgX$;[12] i = $(Bu^tO)_3PO$;[12] j = Me_3SiCl;[12] k = Me_2SiCl_2;[12] l = $ClSiPh_2H$;[12] m = $HCl/AlCl_3$.[62]

In contrast with the cyclotetrasilane, Si_5Ph_{10} does not react with either hydrogen[24] or hydrogen halides.[20] Also, no cyclic products are observed in the reaction of the dilithiopentasilane $Li(SiPh_2)_5Li$ with dichloromethanes.[29]

In some reactions the phenyl groups are substituted without ring cleavage. Thus Si_5Ph_{10} reacts with HI in a sealed tube to form benzene and

$Si_5Ph_5I_5$, in which each silicon carries one phenyl group and one iodine.[46] From this compound, several new derivatives $Si_5Ph_5X_5$, with X = Me, Bu, OMe or H, have been prepared.[66] Hydrolysis leads to a yellow polymer in which Si_5 rings are connected through oxygen bridges.

^{1}H NMR investigation of $Si_5Ph_5H_5$ shows two singlets in the ratio 2 : 3,[67] and this can be explained by the configuration shown in (3a). The proton shift depends on the number of phenyl groups. The monohydride Si_5Ph_9H can be synthesized from $Li(SiPh_2)_4Li$ and $PhSiHCl_2$.[67]

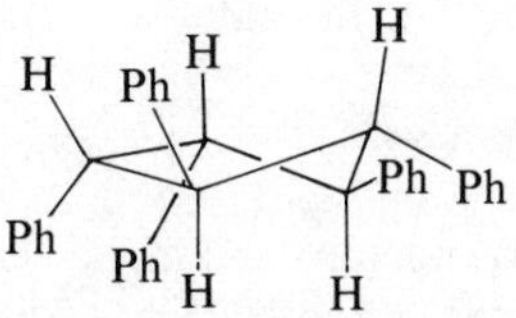

(3a)

Permethylated cyclopentasilanes. The pentamer Si_5Me_{10} was first isolated from the reaction of dimethyldichlorosilane with sodium–potassium alloy.[68,69] It was separated from other ring products $(SiMe_2)_n$, with $n \geqslant 6$, by gas chromatography. The highest yield, c. 10% (15–20% of all cyclic material), is achieved under special reaction conditions.[69] There is a kinetic preference for the formation of the five- (and seven-) membered ring, but the six-membered ring compound is thermodynamically preferred. Si_5Me_{10} is also observed in small quantities when Si_6Me_{12} or larger ring compounds are refluxed in a dilute solution of the naphthalene radical anion in THF,[69] or in the reaction of linear permethylpolysilanes, for example $Ph_3Si(SiMe_2)_n$—$SiPh_3$ with triphenylsilyllithium.[70] Recently, pure Si_5Me_{10} (free of other oligomers) was obtained by methylation of Si_5Cl_{10} with $ZnMe_2$ or $AlMe_3$.[41]

Aluminium chloride catalyses skeletal transformations of methylcyclosilanes, and Si_6Me_{12} undergoes a ring contraction to give silyl-substituted cyclopentasilanes (R = Me):[71,72]

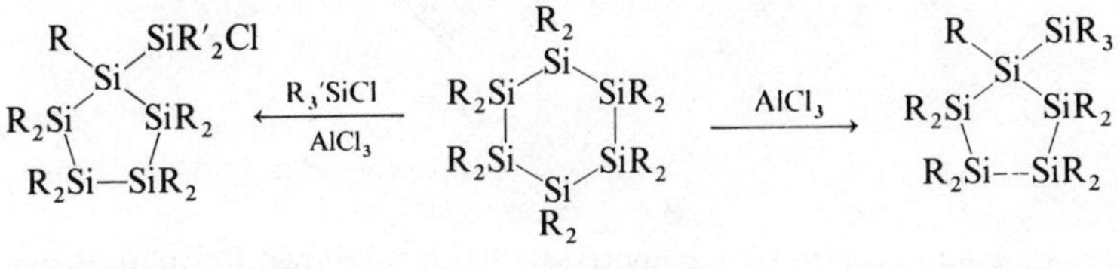

(5)

(3b)   (4)   (3c)

A range of $Si_5Me_9(SiR_2'X)$ derivatives, with $X = SiMe_2Ph$, OEt, Cl, and Me^{73} was prepared from $Si_5Me_9(SiMe_2Cl)$.

With a six-membered ring carrying a Me_3Si side-chain, the $AlCl_3$-catalysed reaction yields (pentamethyldisilyl)nonamethylcyclopentasilane, which undergoes a further rearrangement to 1,1-bis(trimethylsilyl)octamethylcyclopentasilane (R = Me):[31,72]

$$\qquad\qquad\qquad\text{(3d)}\qquad\qquad\text{(3e)}\qquad\qquad\text{(6)}$$

A similar skeletal rearrangement can be carried out photochemically.[30,31,40] For example, the hexamer $(SiMe_2)_6$ rearranges on irradiation (2537 Å) to Si_5Me_{10} among other rings and polymers. The photolysis proceeds via generation of dimethylsilylene, and in the presence of HCl the silane Me_2SiHCl is formed.[31] Similarities between the photolysis and mass spectrometry of organocyclopolysilanes have also been discussed.[74]

Infrared and Raman spectra of Si_5Me_{10} and $Si_5(CD_3)_{10}$ have been reported and a normal-coordinate analysis has been performed. ^{1}H and ^{13}C NMR data, mass spectra, UV maxima and IR–Raman data were compared for $(SiMe_2)_n$ (n = 5–19) and discussed.[75]

The crystal structure of Si_5Me_{10} has not yet been determined. One of the difficulties may be the solid-state transition to a plastic crystalline phase at 234 K; molecular motion in the brittle plastic phase was studied by NMR relaxation methods.[76]

Other cyclopentasilane derivatives. Two iron-containing derivatives (5) and (6) are formed in the reaction of a mixture of $Si_5Me_9(SiMe_2Cl)$ and $Si_5Me_8(Cl)(SiMe_2Cl)$ with $Na^+[CpFe(CO)_2]^-$. Their crystal structures were determined.[77]

$$\qquad\qquad\qquad\text{(5)}\qquad\qquad\qquad\qquad\qquad\text{(6)}$$

In (5) the Si_5 ring exhibits C_s symmetry with four Si atoms coplanar and the fifth silicon out of plane. The average Si–Si bond length is 235.6 pm.

In (**6**) the Si_5 ring is in a distorted C_s conformation with an average bond length of 232.6 pm.[77]

Cyclopentasilane derivatives with different substituents (**3f**) are formed in the cocondensation of Me_2SiCl_2 and $RR'SiCl_2$ with an alkali metal (RR' = MePh, Ph_2, $MeBu^t$ and Bu^t_2):[78]

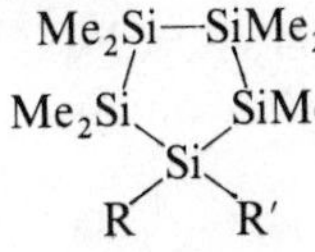
Me2Si—SiMe2
Me2Si
SiMe2
Si
R
R′

(**3f**)

The main product of the reaction of Et_2SiCl_2 with lithium is Si_5Et_{10}, but other ring compounds are also formed. However, when potassium is used the yield of the five-membered ring is raised to 91%.[79]

Because of steric hindrance it is difficult to obtain deca(t-butyl)cyclopentasilane. The compound is not formed when di(t-butyl)dichlorosilane reacts with an alkali metal, but a yield of 1.4% was obtained using potassium biphenylide.[80]

The cyclopentasilanes Si_5R_{10} (**3**), with R = Et, Pr, Bu, Me_2CHCH_2, were prepared in 10–72% yields from R_2SiCl_2 and lithium metal.[81,82] $MeBu^tSiCl_2$ reacts with Li to form both cyclotetra- and cyclopentasilanes. All four possible isomers of $(MeBu^tSi)_5$ have been identified.[60]

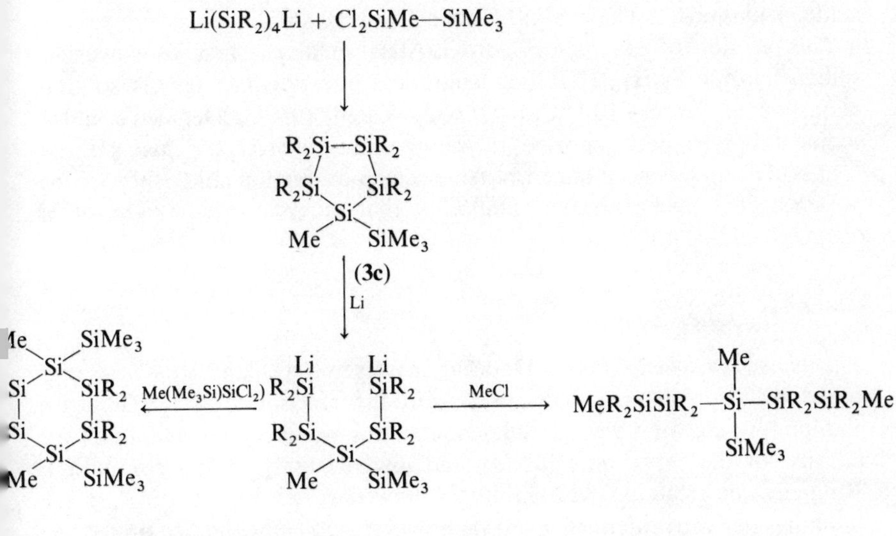
Li(SiR2)4Li + Cl2SiMe—SiMe3
R2Si—SiR2
R2Si
SiR2
Si
Me
SiMe3
(3c)
Li
Me(Me3Si)SiCl2
MeCl
MeR2SiSiR2—Si—SiR2SiR2Me
SiMe3

Figure 7.4 Reactions of five-membered cyclosilanes (R = Ph).

Ring-expansion reactions of $Li(SiPh_2)_4Li$ with Ph_2SiCl_2 gives the per-phenylated cyclopentasilane; other homo- and heterocyclic systems can also be formed.[35] With 1,1-dichlorotetramethyldisilane (Fig. 7.4) a five-membered ring with a silicon-containing side-chain (3c) is formed in small yield. This ring can be specifically cleaved by lithium, and the cleavage product can be used as the starting material to form new chain and ring compounds, as shown in Fig. 7.4 (R = Ph).[37,38]

Deca(p-tolyl)cyclopentasilane was obtained from octa(p-tolyl)cyclotetra-silane on reaction with Li. The dilithio compound formed reacts with K_2PtCl_4 to give the five-membered ring.[28]

Halogenated cyclopentasilanes and Si_5H_{10}. The perphenylated pentamer Si_5Ph_{10} reacts with HBr in a sealed tube to give Si_5Br_{10} and benzene.[83] Analytical and spectroscopic investigations confirm the cyclic structure.[84] If an aluminium halide is present as a catalyst, this compound can be obtained using a normal bottle-flask technique; additionally, Si_5Cl_{10} and Si_5I_{10} can be synthesized.[47,49]

All three known perhalogenated cyclopentasilanes (Si_5F_{10} is not yet known) are crystalline compounds, very sensitive to moisture, but stable under nitrogen. The best solvent for their recrystallization is ^1H-perfluoro-octane.[85]

The crystal structures of Si_5Br_{10} and Si_5I_{10} show a nonplanar Si_5 ring.[86] Infrared and Raman spectra have been reported and Si–Si stretching force constants have been determined[87] in good agreement with empirical force-field calculations.[51] The ^{29}Si NMR spectra show singlet peaks.[43,49]

The reaction of decahalides with $LiAlH_4$ yields the first known cyclic silicon hydride Si_5H_{10}.[84] It is a colourless liquid, which ignites sponta-neously in air (m.p. $-10.5\,°C$, b.p. (extrapolated) $195\,°C$). Detailed spectro-scopic data have been reported, including those for Si_5D_{10}.[88] Also ^1H and ^{29}Si NMR data were obtained.[84] An electron-diffraction study showed the presence of a puckered ring, similar to that of cyclopentane. The Si–Si distance is $234.2\,pm$.[89]

7.1.4 Cyclohexasilanes

Phenylated cyclohexasilanes. Dodecaphenylcyclohexasilane was first pre-pared by Kipping[4] in small yield from the reaction of Ph_2SiCl_2 with sodium. An Si_6 ring was postulated, but this was very difficult to prove because of the very low solubility and low reactivity of this compound. Ring-cleavage reaction with lithium followed by characterization of the dimethyl derivative derived from it, however, indicates the Si_6 ring struc-ture.[122] More convincing was the mass-spectroscopic evidence,[10,90] but vibrational spectroscopic results are ambiguous because of strong coupling between the phenyl modes and the Si–Si modes.[13]

On treatment with hydrogen bromide, it is possible to split off six phenyl groups without cleavage of the ring,[13] to form $Si_6Ph_6Br_6$; with $LiAlH_4$ the hexabromide can be reduced to $Si_6Ph_6H_6$.[13]

A crystal structure determination of $Si_6Ph_{12}.7C_6H_6$ adduct (obtained from benzene solution), shows that the Si_6 ring has a symmetrical chair conformation. The mean Si–Si bond distance is 239.4 pm.[91] The compound melts with decomposition at $c.$ 500 °C, but differential thermal analysis also shows a small endothermic affect at 370 °C and a strong exothermic effect at 410 °C.[6] The compound shows a UV absorption shoulder at 248 nm.[14]

Permethylated cyclohexasilanes. Dodecamethylcyclohexasilane Si_6Me_{12} (**4**), R = Me, was first prepared by Burkhard[92] in very low yield. Later, the yield of the reaction of Me_2SiCl_2 with an alkali metal was increased to 84%.[93] It was pointed out that Si_6Me_{12} is the most thermodynamically stable ring compound in the $(SiMe_2)_n$ series.[69] Equilibria among the cyclic compounds with $n = 5, 6$ and 7 lead to the following thermodynamic values for the redistribution reactions:[94]

$$n = 5\text{–}6: \quad \Delta H = -18 \text{ kcal/mol}, \quad \Delta S = -20 \text{ cal deg}^{-1} \text{ mol}^{-1},$$

$$n = 7\text{–}6: \quad \Delta H = -3 \text{ kcal/mol}, \quad \Delta S = +33 \text{ cal deg}^{-1} \text{ mol}^{-1}.$$

Most of the hexamer Si_6Me_{12} seems to be produced by depolymerization of the polymer $(SiMe_2)_x$, which is the major initial product in the reaction of Me_2SiCl_2 with an alkali metal.[69,75]

Other preparative routes are known. With an α,ω-bis(triphenylsilyl)poly-dimethylsilane starting material, the end groups can be split off in the presence of a catalytic amount of Ph_3SiLi, and cyclization then takes place:[70]

$$Ph_3Si(SiMe_2)_nSiPh_3 \xrightarrow{Ph_3SiLi} (SiMe_2)_n + Si_2Ph_6 \qquad (7)$$

$$n = 5, 6$$

This type of reaction also occurs with α,ω-diphenylpolydimethylsilanes:[31,95]

$$Ph(SiMe_2)_nPh \xrightarrow[THF]{Ph_3SiLi} Si_6Me_{12} + Ph(SiMe_2)_nPh \qquad (8)$$

$$n = 3\text{–}5 \qquad\qquad\qquad n = 2\text{–}5$$

Ring expansion from Si_5Me_{10} to Si_6Me_{12} is catalysed by Ph_3SiLi,[70] and treatment of 1,2-dimethoxytetramethyldisilane with NaOMe also leads to Si_6Me_{12} and $Me_2Si(OMe)_2$.[96] The use of ultrasonic waves in the reaction of

Me_2SiCl_2 and lithium leads to a good yield of Si_6Me_{12} together with formation of the five-membered ring.[97] An alternative method starts with Si_6Cl_{12}, which can be methylated with $ZnMe_2$ in very good yields.[41]

Dodecamethylcyclohexasilane is a white air-stable crystalline solid (m.p. 526 K), soluble in most organic solvents. The Si_6 ring exhibits a chair conformation; the Si–Si distance is 234 pm.[98] There is a solid-state transition to a plastic crystalline phase at 350 K, and molecular motions in the brittle and plastic phases were investigated by proton NMR relaxation methods.[76]

UV absorption spectra, together with those of other permethylated rings, and ^{13}C, ^1H[75] and ^{29}Si NMR[43] data have been measured and interpreted. The IR and Raman frequencies have been assigned for Si_6Me_{12} and $Si_6(CD_3)_{12}$. A normal-coordinate analysis shows an Si–Si stretching force constant of 160 N/m, which is low in comparison with other cyclohexasilane derivatives.[42]

The stability of Si_6Me_{12} to oxidation was investigated. Only small amounts of siloxanes are formed with O_2 or H_2O_2, but cyclic polysiloxanes are found on oxidation with CrO_3, $K_2Cr_2O_7$ or $KMnO_4$ in solution.[99] Reaction of Si_6Me_{12} with HCl/Bu^tCl yields a mixture of linear $Cl(SiMe_2)_nCl(H)$ species, with $n = 1$–6.[22]

Chlorine[100] and bromine[101] react with Si_6Me_{12} with ring cleavage, and the product is a mixture of α,ω-dihalopolydimethylsilanes $Cl(SiMe_2)_nCl$, with $n = 2$–6. With tetracyanoethylene, Si_6Me_{12} forms a charge-transfer complex.[102,103]

The hexamer Si_6Me_{12} can be photolysed by irradiation with UV light (2537 Å) to give $(SiMe_2)_n$, with $n = 4$, 5, and polymers.[30,31,40] The photolysis proceeds with generation of dimethylsilylene,[40] as proved by trapping experiments with Et_2MeSiH. Attempts have been made to measure the kinetics of gas-phase insertion reactions of dimethylsilylene, generated by photolysis of Si_6Me_{12},[104] and irradiation in the presence of hydrogen chloride leads to a high yield of Me_2SiHCl.[105] Irradiation of Si_6Me_{12} in either rigid hydrocarbon glasses at 77 K or an argon matrix yields, in addition to Si_5Me_{10}, a yellow species, which is probably dimethylsilylene.[106]

Aluminium chloride skeletal rearrangements[31,71,73,107,108] take place with Si_6Me_{12} to form both five-membered rings and bicyclic compounds. The compound can be chlorodemethylated without change of ring size by passing dry hydrogen chloride through a solution of Si_6Me_{12} to form $Si_6Me_{11}Cl$. This monochlorinated cyclohexasilane then undergoes a ring contraction on treatment with aluminium chloride to form $Si_5Me_9(SiMe_2Cl)$.

Halogenated cyclohexasilanes and Si_6H_{12}. All the phenyl groups from Si_6Ph_{12} can be split off with HX/AlX_3 systems to give Si_6X_{12} (X = Br and Cl).[47] The corresponding iodide was prepared later,[49] but Si_6F_{12} is still unknown. The perhalogenated cyclohexasilanes are moisture-sensitive sublimable crystalline solids; their IR and Raman spectra have been assigned, and a normal-coordinate analysis suggests that the Si–Si and Si–X force constants are very similar to those of the cyclopentasilanes.[87,109]

The relative energies of possible conformations of the perhalogenated cyclohexasilanes have been estimated using empirical force-field calculations.[51] The results show the lowest energy for the chair conformation.

The ^{29}Si NMR data are very similar to those of cyclopentasilanes. The spectra are singlets.[43]

From Si_6Cl_{12} and $LiAlH_4$ a second cyclic silicon hydride Si_6H_{12} has been prepared[110] as a colourless liquid. It ignites spontaneously in air, and slow decomposition occurs on irradiation. Detailed thermodynamic data are given in ref. 111. The ^1H and ^{29}Si NMR spectra have been reported.[110] Infrared and Raman data[112] are in agreement with a nonplanar ring in a chair conformation, and a normal-coordinate analysis has been performed. Empirical force-field calculations show the lowest energy for the chair conformation, but energy differences between the conformations are low.[113] As a result, an electron-diffraction analysis[114] at 130° shows the presence of both the twist and boat forms in addition to the chair conformation. The Si–Si distance is 234.2 pm and the SiSiSi angle 110.3°.

Other derivatives. Treatment of Et_2SiCl_2 with an alkali metal gives the cyclic compounds $(SiEt_2)_n$, with n = 5–8, which can be separated by HPLC. The main products are rings with n = 5 and 7.[79] Condensation of Me_2SiCl_2 and $RR'SiCl_2$ (in a ratio 5 : 1) with an alkali metal provides five- and six-membered ring cyclosilanes $RR'Si(SiMe_2)_4$ and $RR'Si(SiMe_2)_5$, with $RR' = Ph_2$, $MeBu^t$, Bu_2^t and MePh.[78]

Six-membered ring compounds with silicon-containing side-chains are formed in the reactions shown in Fig. 7.4 above, which involve skeletal rearrangements.[38,108] A cyclohexasilane $Si_6Me_4Ph_8$ has been synthesized according to the following reaction:[32]

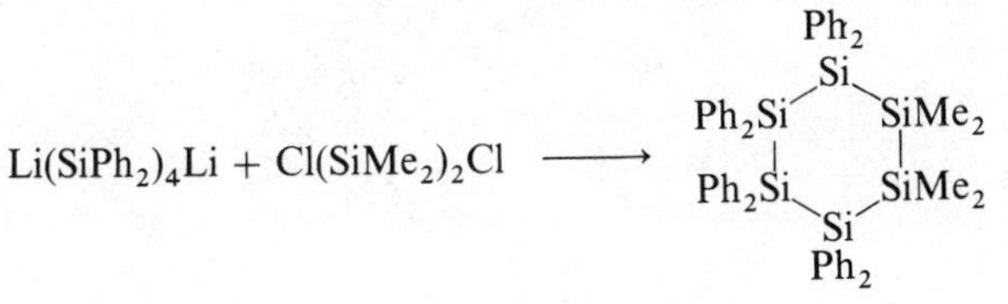

$$Li(SiPh_2)_4Li + Cl(SiMe_2)_2Cl \longrightarrow \begin{array}{c} Ph_2 \\ Si \\ Ph_2Si \diagup \quad \diagdown SiMe_2 \\ | \qquad\qquad | \\ Ph_2Si \diagdown \quad \diagup SiMe_2 \\ Si \\ Ph_2 \end{array} \qquad (9)$$

7.1.5 Cyclosilanes $(SiR_2)_n$ with $n > 6$

Compounds with homocyclic systems $(SiR_2)_n$ with $n > 6$ are rare. Some have already been mentioned above. The cycloheptasilane $(SiMe_2)_7$ is obtained in $c.$ 2% yield in the usual synthesis from Me_2SiCl_2 with Na/K alloy,[69] but using Li in THF at or below 0 °C better yields of larger rings are obtained, for example 7.2% for $n = 7$, 15.8% for $n = 8$ and 3.3% for $n = 9$.[115] With HPLC even larger rings were found and identified by mass spectrometry; in this way rings up to $n = 35$ are observable. A synthesis using Me_2SiCl_2 with Na/K alloy under special reaction conditions was used for their preparation, and UV, IR, Raman, mass and NMR data are listed for most of these compounds.[75,116]

In addition to the permethylated cyclosilanes, perethylcyclosilanes $(SiEt_2)_n$, with $n \leqslant 8$, have been described.[79]

7.1.6 Polycyclic systems

The first cage polysilane to be reported was tetradecamethylbicyclo[2.2.2]-octasilane Si_8Me_{14},[117] but six other polycyclic polysilanes were found later.[118] These were prepared by treating mixtures of $MeSiCl_3$ and Me_2SiCl_2 with Na/K alloy in THF in the presence of naphthalene. The structures of Si_7Me_{12} (**8**) and Si_8Me_{14} (**9**) were assigned following NMR investigations, and an X-ray structure determination on Si_9Me_{16} showed the presence of the bicyclic skeleton (**10**).[119] Probable structures for other observed compounds are as follows[118] (positions of Si atoms only are shown):

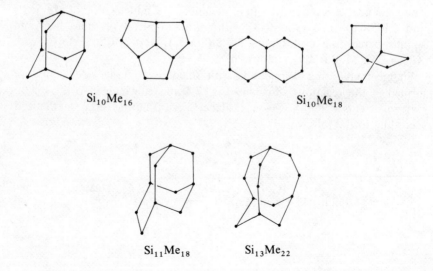

$Si_{10}Me_{16}$ $Si_{10}Me_{18}$

$Si_{11}Me_{18}$ $Si_{13}Me_{22}$

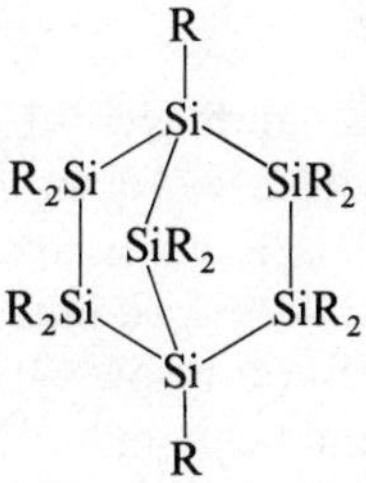

$$Si_7Me_{12}$$
(8)

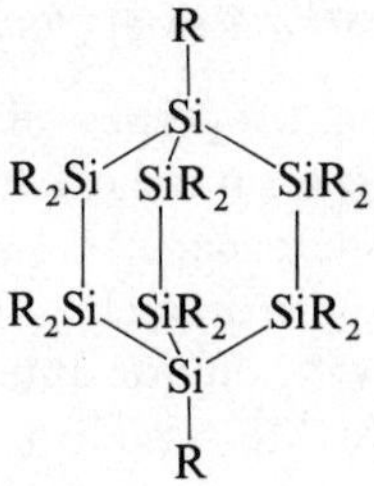

$$Si_8Me_{14}$$
(9)

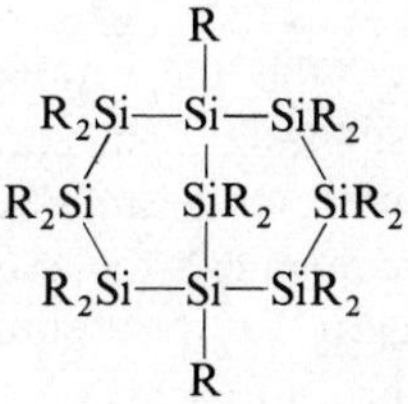

$$Si_9Me_{16}$$
(10)

The bicyclic compounds Si_8Me_{14} **(9)** and $Si_{10}Me_{18}$ **(12)** were found to isomerize as follows on heating with $AlCl_3$ in refluxing benzene (R = Me):[108]

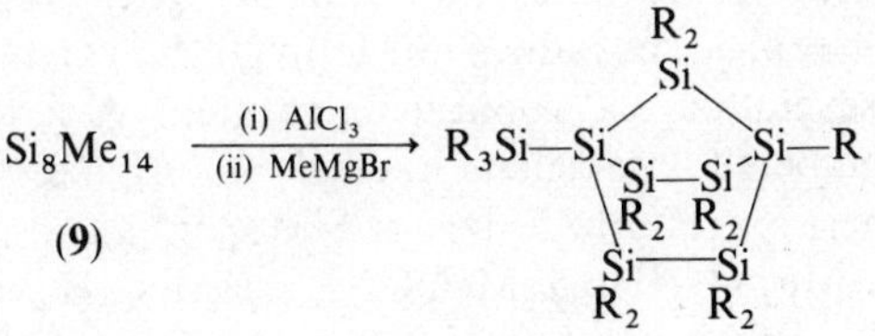

(11)

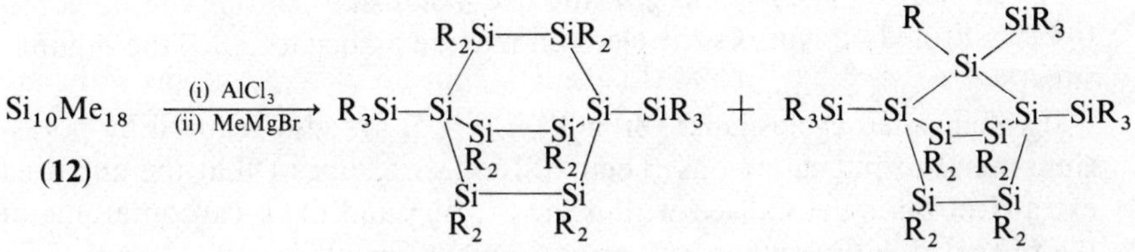

(13)   **(14)**

A further bicyclic system **(15)**, R = Me, is produced by the reaction of nonamethylchlorocyclopentasilane Si_5Me_9Cl with Na/K alloy, and this compound undergoes a skeletal rearrangement with $AlCl_3$ to give a mixture containing compounds **(13)** and **(14)**:[108]

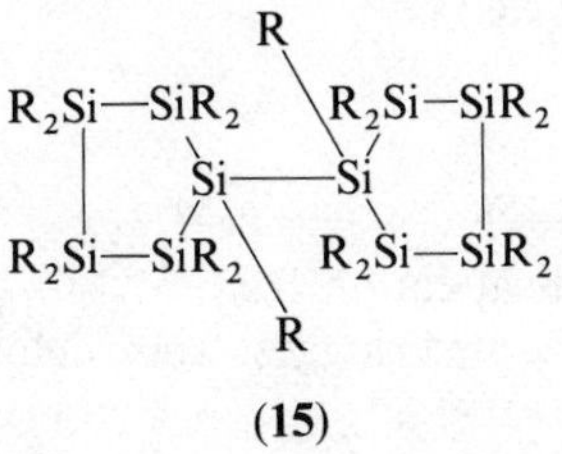

(15)

7.1.7   Radical anions and cations of cyclosilanes

One of the most outstanding properties of alkyl- and aryl-substituted cyclosilanes is their tendency to form radical anions. This was demonstrated first with dodecamethylcyclohexasilane,[120] using Na/K alloy or by an electrochemical reduction at approximately $-100\,°C$. The dark blue radical anion $[Si_6Me_{12}]^{\doteq}$ is stable only in solution at temperatures below $-50\,°C$. It was shown that Si_5Me_{10} and Si_4Me_8 also form such radical anions, but this was not possible with Si_7Me_{14}. Surprisingly, chemical reduction of Si_7Me_{14} gives only $[Si_5Me_{10}]^{\doteq}$. Detailed ESR investigations of $[Si_6Me_{12}]^{\doteq}$ and $[Si_5Me_{10}]^{\doteq}$ show that the unpaired electron is completely delocalized over the entire cyclosilane ring. The order of relative reducibility is:[121,122]

$$(Me_2Si)_5 > benzene > (Me_2Si)_6 > (Me_2Si)_7$$

It is also possible to form radical anions from Si_6Et_{12}, $Si_6Me_6Et_6$,[123] and $(SiMeBu^t)_n$ with $n = 4, 5$. The radical $[SiMeBu^t]_n^{\doteq}$ shows remarkable stability and persists for several days at room temperature.[60] A series of substituted five- and six-membered cyclosilane radical anions, i.e. $[Si_5Me_9X]^{\doteq}$ and $[Si_6Me_{11}X]^{\doteq}$, where $X = H$, Ph, benzyl or $SiMe_3$,[124] and $[Si_5Me_9X]^{\doteq}$, where $X = $ biphenyl, $p\text{-}Me_3SiC_6H_4$, $p\text{-}MeOC_6H_4$, $p\text{-}MeC_6H_4$ and $[Si_5Me_9Si\text{-}Me_2X]^{\doteq}$, where $X = $ Ph, $p\text{-}Me_3SiC_6H_4$ and $p\text{-}Me_2NC_6H_4$,[125] were investigated in detail by ESR and NMR spectroscopy. In cyclopentasilanes the unpaired electron is delocalized in the five-membered Si_5 ring, but in some aryl-substituted derivatives the electron may be associated with the organic ring.

Perphenylated cyclosilanes $(SiPh_2)_n$, $n = 4, 5$, are also reduced by potassium metal to radical anions. Their ESR spectra suggest that the unpaired extra electron is delocalized within the Si_n ring and the α-carbon atoms of the phenyl ring. Possibly, these radical anions are planar.[126]

It is also possible to produce radical cations by treating the permethylated cyclosilanes shown below (methyl groups omitted) with an $AlCl_3$–CH_2Cl_2 mixture, whose redox potential is sufficient to oxidize compounds with a first ionization potential below 8 eV:

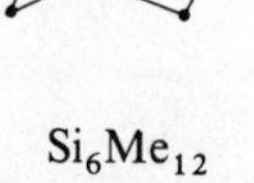

Si_6Me_{12} 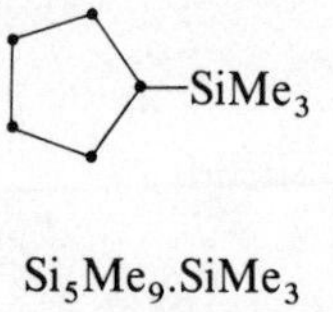 $Si_5Me_9.SiMe_3$ 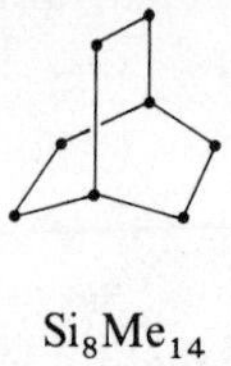 Si_8Me_{14} 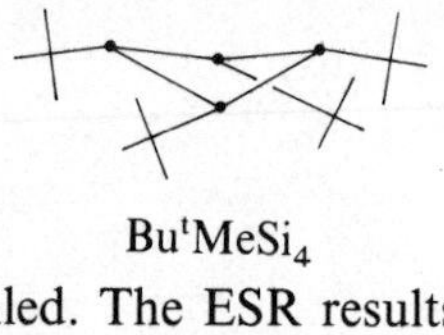 Bu^tMeSi_4

Attempts to obtain a radical cation from Si_5Me_{10} failed. The ESR results on the radical cations formed suggest spin delocalization into the Si–Si σ-system.[127]

7.2 Heterocyclic silanes

In this section heterocyclic systems are reviewed that contain at least one Si–Si bond.

7.2.1 Heterocyclic silanes containing Group III elements

No cyclic structures containing Si–Al, Si–Ga, Si–In or Si–Tl bonds are known, but a Si_4B ring (16), R = Ph, R' = NMe_2, is formed from $Li(SiPh_2)_4Li$ and Cl_2B—NMe_2:[27,34]

$$
\begin{array}{c}
R_2Si\!-\!SiR_2 \\
R_2Si\diagdown\quad\diagup SiR_2 \\
B \\
| \\
R'
\end{array}
$$

(16)

Attempts to prepare $(SiPh_2)_4BPh$ using $PhBCl_2$ were not, however, successful and it is supposed that substituents increasing the electron density at boron are necessary to stabilize the Si–B bonds.

7.2.2 Heterocyclic silanes containing Group IV elements

From the reaction of mixtures of Me_2SiCl_2 and Me_2GeCl_2 with lithium, a series of compounds of the general formula $Me_{10}Si_xGe_{5-x}$ (where $x = 0$–5), $Me_{12}Si_yGe_{6-y}$ (where $y = 0$–6) and $Me_{14}Si_zGe_{7-z}$ (where $z = 0$–7) are obtained.[128] The distribution of the products (established by gas chromatography) is similar to that observed in the preparation of the $(SiMe_2)_n$ series, in that the six-membered ring products are predominant. Small amounts of eight-membered ring compounds were also detected. Two six-membered ring compounds (17) and (18), R = Me, have been prepared in pure form.[128]

An alternative route to Ge- and Sn-containing heterocycles (19) and (20), R = Ph, uses the reactions of $Li(SiPh_2)_4Li$ with Ph_2GeCl_2 and Ph_2SnCl_2 respectively.[35]

$$
\begin{array}{cccc}
\begin{array}{c}
R_2 \\
Ge \\
R_2Si\diagup\;\diagdown SiR_2 \\
| \qquad\quad | \\
R_2Si\diagdown\;\diagup SiR_2 \\
Ge \\
R_2
\end{array}
&
\begin{array}{c}
R_2 \\
Si \\
R_2Si\diagup\;\diagdown SiR_2 \\
| \qquad\quad | \\
R_2Si\diagdown\;\diagup SiR_2 \\
Ge \\
R_2
\end{array}
&
\begin{array}{c}
R_2Si\!-\!SiR_2 \\
R_2Si\diagdown\quad\diagup SiR_2 \\
Ge \\
R_2
\end{array}
&
\begin{array}{c}
R_2Si\!-\!SiR_2 \\
R_2Si\diagdown\quad\diagup SiR_2 \\
Sn \\
R_2
\end{array}
\\
(17) & (18) & (19) & (20)
\end{array}
$$

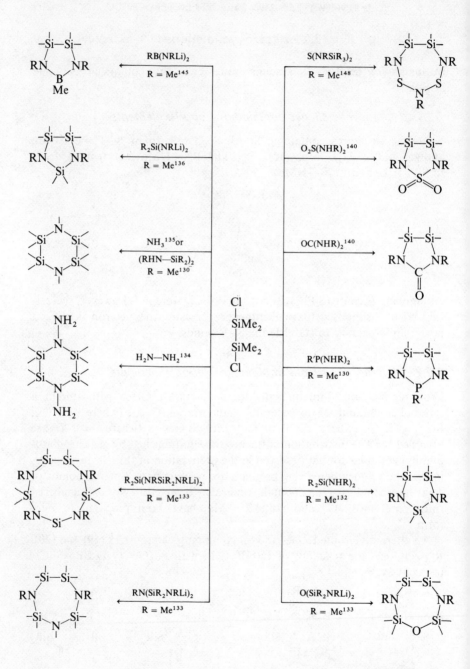

Figure 7.5 Preparation of Si–Si heterocycles from $ClMe_2Si$—$SiMe_2Cl$ (the substituents on the silicon atoms are not shown).

No Si_4Pb ring is formed if Ph_2PbCl_2 is used, because the dichloroplumbane is reduced to elemental lead.

Several spiroheterocyclosilanes, containing germanium, titanium and zirconium[153−160] are known; the first type is discussed in Chapter 8.

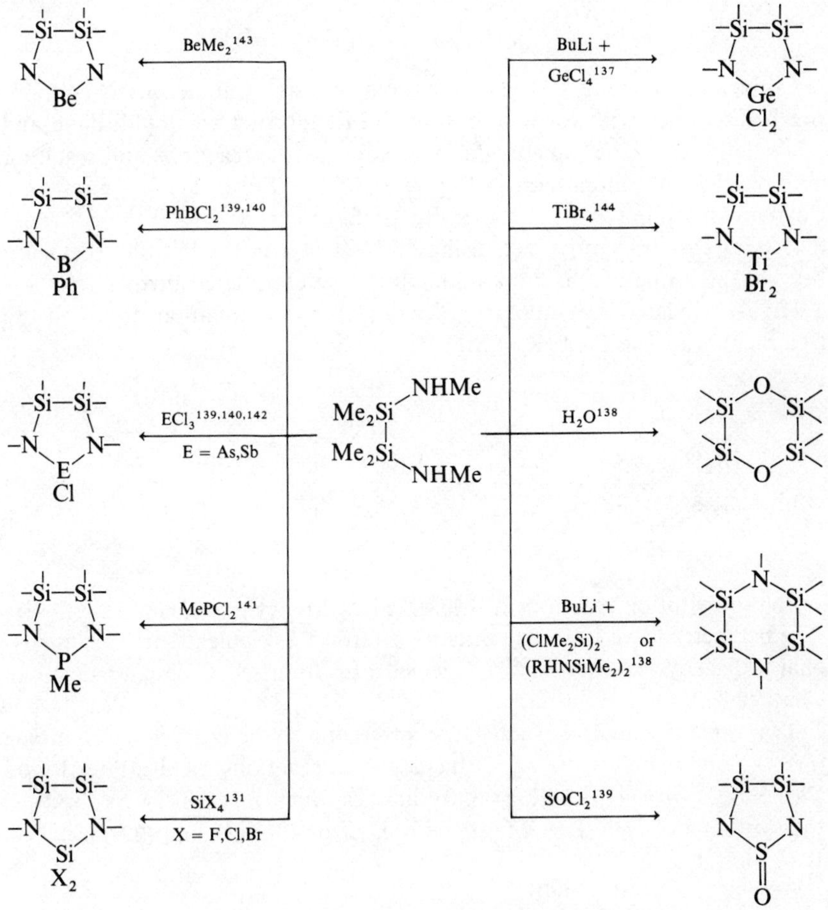

Figure 7.6 Preparation of Si–Si heterocycles from $[(MeNH)Me_2Si_2]$.

7.2.3 Heterocyclic silanes containing Group V elements

A wide variety of nitrogen-containing heterocyclic silanes, with from three to nine ring atoms, are known, including polycyclic structures. The only known three-membered ring system Si_2N is formed in a pyrolysis reaction:[129] (R = Me)

$$R_3Si(N\!\!-\!\!SiR_2)_2NLi \xrightarrow[150-195\,°C]{100-110\,atm} \begin{matrix} R_2Si\!\!-\!\!SiR_2 \\ \diagdown N \diagup \\ | \\ SiR_3 \end{matrix} \qquad (10)$$

Convenient starting materials for the synthesis of numerous N-containing heterocycles with Si–Si bonds are 1,2-dichlorotetramethyldisilane and 1,2-bis(alkylamino)tetramethyldisilane. Some of the reactions and resulting heterocycles are summarized in Figs. 7.5 and 7.6. For a review see ref. 147. Carbon-containing heterocycles with Si–Si bonds,[140,146,168,177,178] (e.g. $Si_2N_2C_2$) can be synthesized using $LiRN(CH_2)_2NRLi$,[146] but these are beyond the scope of this book and will not be considered further.

Highly silylated structures are obtained by metallation followed by treatment with Me_3SiCl (R = Me):[151]

$$(11)$$

For substitution reactions of $Cl_2Me_4Si_2N_2Me_2$ with amines see ref. 131.

Other heterocyclic silanes containing Group V elements are disilaphospha(III)diazacyclopentanes,[149,150] which are discussed in greater detail in Chapter 8.

Two of the possible routes for preparing N-heterocycles containing $(SiR_2)_n$ fragments with $n > 2$ include the reactions of $I(SiPh_2)_4I$ and $Cl(SiR_2)_nCl$ (n = 4–6) with primary amines. Such reactions were used for the synthesis of (21), R = Me, R' = Me, Et or Ph,[27,35] and (22), n = 2–4, R = Me:[152]

(21)   (22)

Phosphorus-containing heterocycles (23)–(24) are obtained from either Si–Li or P–Li compounds, by treating them with phosphorus or silicon halides respectively.

R$_2$Si—SiR$_2$
R$_2$Si   SiR$_2$
       P
       R′
(23)

R$_2$Si   SiR$_2$
R$_2$Si   SiR$_2$
       P
       R′
(24)

R$_2$ Si   SiR$_2$
R$_2$Si   SiR$_2$
       P
       R′
(25)

R$_2$Si—SiR$_2$
R$_2$Si   SiR$_2$
       P—SiR$_2$
       R′
(26)

R$_2$Si—PR′
R$_2$Si—PR′
(27)

R′P—PR′
R$_2$Si   SiR$_2$
       Si
       R$_2$
(28)

R$_2$Si   PR′
R$_2$Si   PR′
       Si
       R$_2$
(29)

An example of the first type is the formation of Ph$_8$Si$_4$PPh **(23)**, R = Ph, from Li(SiPh$_2$)$_4$Li and PhPCl$_2$,[35] while reactions of α,ω-dichloropermethyl-polysilanes with PhPLi$_2$,[161] MePLi$_2$[152] and LiPhP—PPhLi[162] yield the compounds **(24)**–**(26)**, R = Me, R′ = Me or Ph, and **(27)**–**(29)**, R = Me, R′ = Ph. Spectral data, including ^{1}H, ^{29}Si, ^{13}C and ^{31}P NMR parameters, have been reported for these compounds and for (PhP)$_2$(SiMe$_2$)$_3$ and (PhP)$_2$(SiMe$_2$)$_4$.[162]

The crystal structures of (PhP)$_2$(SiMe$_2$)$_3$[162] and PhP(SiMe$_2$)$_2$PPh[163] have been determined. The five-membered P$_2$Si$_3$ ring is puckered and exists in the *trans* form, while the six-membered P$_2$Si$_4$ ring has a chair conformation.

7.2.4 Heterocyclic silanes containing Group VI elements

The oldest known heterocycles are those containing oxygen as the hetero-element, namely Si$_4$Ph$_8$O and Si$_4$Ph$_8$O$_2$, prepared by Kipping.[4,18] Octa-phenylcyclooxatetrasilane Si$_4$Ph$_8$O can also be prepared by hydrolysing the diacetoxy derivative AcO(SiPh$_2$)$_4$OAc.[164] Other Si$_4$O rings, containing p-tolyl[53] and benzyl[165] substituents were also described early. The compound Si$_4$Ph$_8$O$_2$ can be prepared by condensation of the disilane diol HO—SiPh$_2$SiPh$_2$—OH, which confirms structure **(31)**.[17] Recently, the preparation of the six-membered ring compound Ph$_{10}$Si$_5$O **(33)**, R = Ph, was reported,[63] using the hydrolysis of Cl(SiPh$_2$)$_5$Cl (prepared from Si$_5$Ph$_{10}$ and PCl$_5$). An X-ray investigation indicates a chair conformation for the Si$_5$O ring.[166]

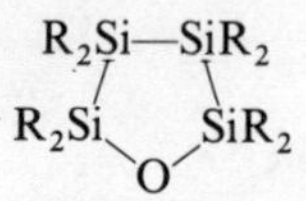

(30) 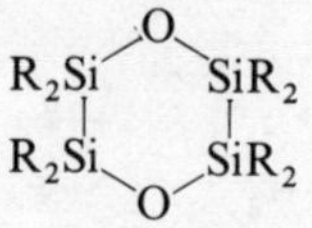   (31) 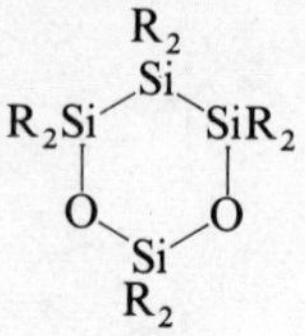   (32) 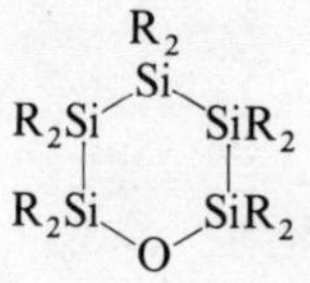   (33)

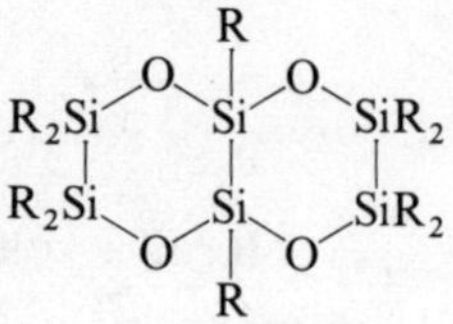

(34) 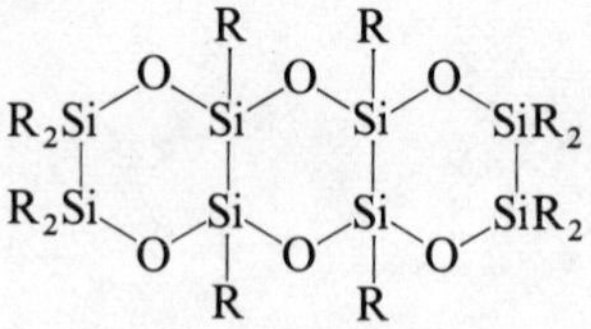   (35)

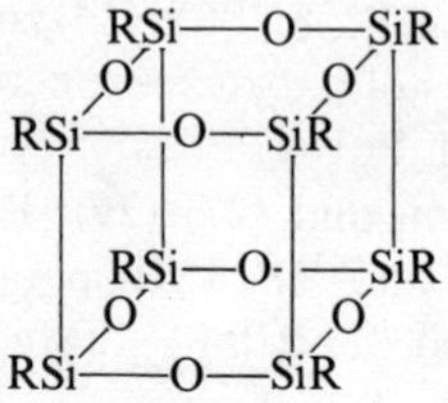

(36)

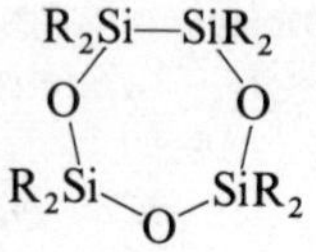

(37) 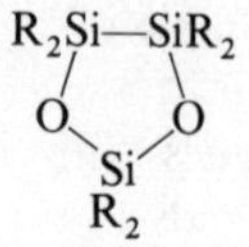   (38)

A new route to oxygen-containing heterocycles was suggested in relation to the oxidation of tetramesityldisilene, which contains a very reactive Si=Si bond (Mes = mesityl):[167]

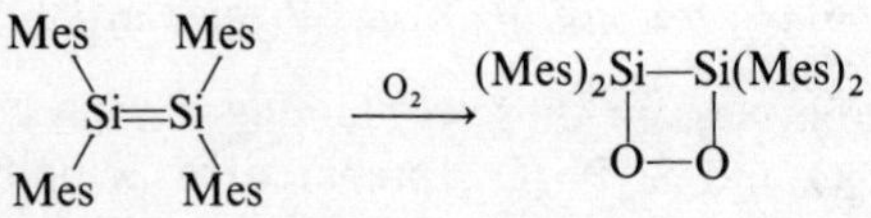

$$\begin{array}{c} \text{Mes} \quad \text{Mes} \\ \text{Si}=\text{Si} \\ \text{Mes} \quad \text{Mes} \end{array} \xrightarrow{\text{O}_2} \begin{array}{c} (\text{Mes})_2\text{Si}-\text{Si}(\text{Mes})_2 \\ \text{O}-\text{O} \end{array}$$ (12)

A six-membered Si_4O_2 ring derivative $O(Si_2Me_4)_2O$ (31), R = Me, was isolated from the hydrolysis of either 1,2-dichloro- or 1,2-difluorotetramethyldisilane.[169] Its structure was determined by X-ray analysis,[170] showing a chair conformation for the ring.

Cyclic and cage-like methylsilicon telomers (31), R = Me, and (34)–(38), R = Me, have been obtained[171] by acid hydrolysis or cohydrolysis of $(EtO)_2MeSi-SiMe(OEt)_2$ and $Me_2Si(OMe)_2$. Crystal and molecular structures have been determined for (35) and (36), R = Me.[172,173]

A new route to cyclic siloxanes containing Si—Si units is via silylene insertion into an Si—O bond (R = Me):[174]

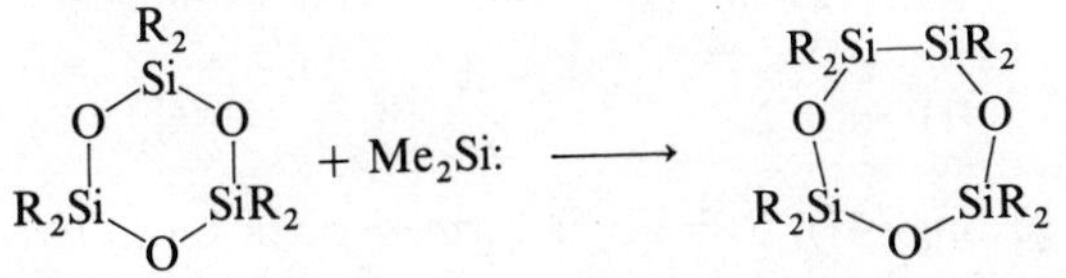

$$\text{(13)}$$

Similarly, carbon-containing C—Si—Si—O rings are formed by analogous insertions.[176,177]

Heterocycles containing oxygen are also formed when various $(SiMeBu^t)_4$ isomers react with m-chloroperbenzoic acid, to give the oxygen-insertion products $(SiMeBu^t)_4O_n$, with $n = 1, 2$ and 4.[61]

Rings containing both oxygen and sulphur(VI) are obtained by treating α,ω-dichloropermethylpolysilanes with sulphuric acid. Thus the rings (39) and (40), R = Me, have been prepared:[175]

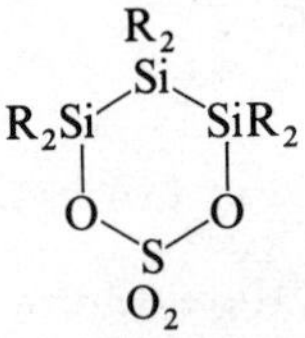

(39) 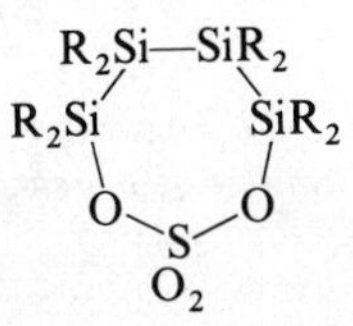 (40)

A relatively small number of heterocycles containing sulphur or selenium (41)–(48) exist:

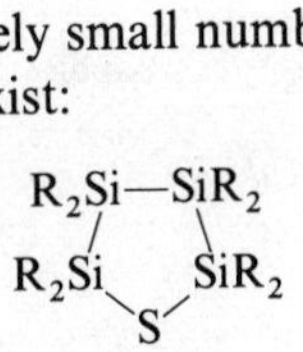

(41) 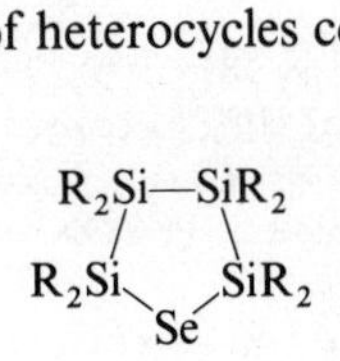 (42) 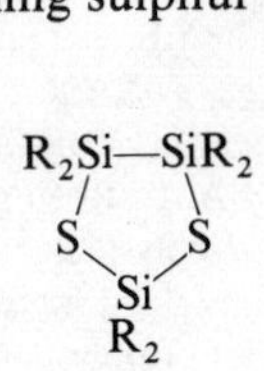 (43)

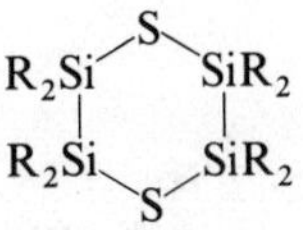

(44) 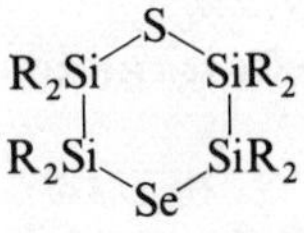 (45) 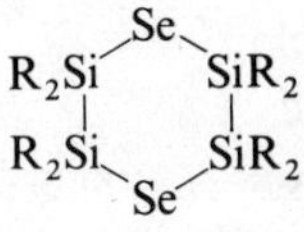 (46)

$$
\begin{array}{cc}
\underset{\substack{\text{Si} \\ R_2\text{Si} \quad \text{SiR}_2 \\ R_2\text{Si} \quad \text{SiR}_2 \\ \text{S}}}{R_2} & \underset{\substack{\text{Si} \\ R_2\text{Si} \quad \text{SiR}_2 \\ R_2\text{Si} \quad \text{SiR}_2 \\ \text{Se}}}{R_2} \\
(47) & (48)
\end{array}
$$

A five-membered ring (41), R = Ph, is formed when $Li(SiPh_2)_4Li$ is treated with SCl_2.[35] Similar permethylated ring systems (41) and (42), R = Me, are obtained when octamethylcyclotetrasilane Si_4Me_8 reacts wih elemental sulphur or selenium.[178] The five-membered ring compound $Me_6Si_3S_2$ (43) has also been prepared.[179]

Six-membered rings (44), R = Me, can be prepared by reacting 1,2-dichlorotetramethyldisilane with H_2S.[180]

Ring systems with two heteroatoms (44)–(48) result (R = Me) when Si_4Me_8 is treated with stoichiometric amounts of sulphur or selenium.[181] Sulphur and selenium also react with Si_5Me_{10} to form monothia- and monoselenacyclosilanes (47) and (48), R = Me, respectively.[182]

Bibliography

B 1   H. Gilman and G. L. Schwebke, *Adv. Organomet. Chem.* **1** (1964) 90 (cyclic polysilanes).

B 2   I. Haiduc, *Chemistry of Inorganic Ring Systems*, Part 1 (Wiley-Interscience, London, 1970), p. 52 (cyclopolysilanes).

B 3   M. Kumada, *Intra-Sci. Chem. Rep.* **7** (1973) 121 (skeletal transformations of polysilanes).

B 4   R. West, *Ann. NY Acad. Sci* **239** (1974) 1 (aromatic properties of cyclopolysilanes).

B 5   M. Kumada, *J. Organomet. Chem.* **100** (1975) 127 (skeletal transformations of polysilanes).

B 6   R. West and E. Carberry, *Science* **189** (1975) 179 (permethylcyclopolysilanes).

B 7   E. Hengge, *Top. Curr. Chem.* **51** (1976) 1 (polysilanes).

B 8   E. Hengge, in *Homoatomic Rings, Chains, Macromolecules of Main Group Elements*, edited by A. L. Rheingold (Academic Press, New York, 1977), p. 235 (cyclopolysilanes and linear polysilanes).

B 9   M. Ishikawa and M. Kumada, *Rev. Silicon, Germanium, Tin, Lead Compds* **4** (1979) 7 (photolysis of polysilanes).

B 10   M. Ishikawa and M. Kumada, *Adv. Organomet. Chem.* **19** (1981) 51 (photolysis of cyclopolysilanes).

B 11   R. West, *Pure Appl. Chem.* **54** (1982) 1041 (cyclic polysilanes).

References

1   M. Ishikawa and M. Kumada, *Adv. Organomet. Chem.* **19** (1981) 51.

2   Yue-Shen Chen and P.O. Gaspar, *Organometallics* **1** (1982) 1410.

3   S. Masamune, Y. Hanzawa, S. Murakami and J. F. Blount, *J. Am. Chem. Soc.* **104** (1982) 1150.

4   F. S. Kipping and J. E. Sands *J. Chem. Soc.* **119** (1921) 830, 848.
5   A. Jarvie, H. S. Winkler, D. J. Peterson and H. Gilman, *J. Am. Chem. Soc.* **83** (1961) 1921.
5a H. Gilman, D. J. Peterson, R. A. Tomasi and R. L. Harrell, *J. Organomet. Chem.* **4** (1965) 167.
6   M. M'hirsi and M. Brini, *Bull. Soc. Chim. Fr.* **1968**, 1509.
7   U. Blaukat and W. P. Neumann, *J. Organomet. Chem.* **63** (1973) 27.
8   E. Hengge and G. Litscher, *Monatsh. Chem.* **109** (1978) 1217.
9   E. Hengge and H. Firgo, *J. Organomet. Chem.* **212** (1981) 155.
10 Th. Kinstle, I. Haiduc and H. Gilman, *Inorg. Chim. Acta* **3** (1969) 373.
11 L. Párkányi, K. Sasvári and J. Barta, *Acta Crystallogr.* **B34** (1978) 883.
12 H. Gilman and G. Schwebke, *J. Am. Chem. Soc.* **86** (1964) 2693.
12a H. Gilman and G. Schwebke, *J. Organomet. Chem.* **3** (1965) 382
13 E. Hengge and F. Lunzer, *Monatsh. Chem.* **107** (1976) 371.
14 H. Gilman and W. A. Atwell, *J. Organomet. Chem.* **4** (1965) 176.
15 E. Hengge, H. Söllradl and H. Schmölzer, unpublished results.
16 H. Hönig and K. Hassler, *Monatsh. Chem.* **113** (1982) 285.
17 A. Jarvie, H. S. Winkler and H. Gilman, *J. Org. Chem.* **27** (1962) 614.
18 F. S. Kipping, *J. Chem. Soc.* **123** (1923) 2590.
19 A. Jarvie and H. Gilman *J. Org. Chem.* **26** (1961) 1999.
20 H. Gilman, D. R. Chapman and G. L. Schwebke, *J. Organomet. Chem.* **14** (1968) 267.
21 H. Gilman and A. Jarvie, *Chem. Ind. (London)* **1960** 965.
22 H. Gilman and D. R. Chapman *Chem. Ind. (London)* **1965** 1788.
23 H. Gilman and D. R. Chapman, *J. Organomet. Chem.* **8** (1967) 451.
24 H. Gilman and R. A. Tomasi, *J. Organomet. Chem.* **9** (1967) 223.
25 H. Gilman and R. A. Tomasi, *J. Org. Chem.* **27** (1962) 3647.
26 F. S. Kipping, *J. Chem. Soc.* **123** (1923) 2598.
27 E. Hengge and D. Wolfer, *J. Organomet. Chem.* **66** (1974) 413.
28 M. E. Lemanski and E. P. Schram, *Inorg. Chem.* **15** (1976) 2515.
29 P. H. Sen, Th. B. Brennan and H. Gilman, *Indian Chem. Soc.* **51** (1974) 561.
30 M. Ishikawa and M. Kumada, *J. Chem. Soc., Chem. Commun.* **1970**, 612.
31 M. Kumada, *J. Organomet. Chem.* **100** (1975) 127.
32 E. Hengge, R. Sommer, *Monatsh. Chem.* **108** (1977) 1413.
33 M. F. Lemanski, E. P. Schram, *Inorg. Chem.* **15** (1976) 1489.
34 E. Hengge and D. Wolfer, *Angew. Chem.* **85** (1973) 304.
35 E. Hengge and U. Brychcy, *Monatsh. Chem.* **97** (1966) 1309.
36 H. Gilman, R. Harrell, K. J. Chang and S. Cottis, *J. Organomet. Chem.* **2** (1964) 434.
37 E. Hengge and G. Kollmann, *J. Organomet. Chem.* **92** (1975) C 43.
38 E. Hengge and G. Kollmann, *Monatsh. Chem.* **109** (1978) 477.
39 H. J. S. Winkler and H. Gilman, *J. Org. Chem.* **27** (1962) 254.
40 M. Ishikawa and M. Kumada, *J. Organomet. Chem.* **42** (1972) 325.
41 E. Hengge, H. G. Schuster and W. Peter, *J. Organomet. Chem.* **186** (1980) C 45.
42 K. Hassler, *Spectrochim. Acta* **37A** (1981) 541.
43 D. Kovar, K. Utvary and E. Hengge, *Monatsh. Chem.* **110** (1979) 1295.
44 Ch. Kratky, H. G. Schuster and E. Hengge, *J. Organomet. Chem.* **247** (1983) 253.
45 E. Hengge, G. Bauer and H. Marketz, *Z. Anorg. Allg. Chem.* **394** (1972) 93.
46 E. Hengge and H. Marketz, *Monatsh. Chem.* **100** (1969) 890.
47 E. Hengge and D. Kovar, *J. Organomet. Chem.* **125** (1977) C 29.
48 E. Hengge and D. Kovar, *Z. Anorg. Allg. Chem.* **458** (1979) 163.
49 E. Hengge and D. Kovar, *Angew. Chem.* **93** (1981) 698.
50 K. Hassler, E. Hengge and D. Kovar, *J. Mol. Struct.* **66** (1980) 25.
51 H. Hönig and K. Hassler, *Monatsh. Chem.* **113** (1982) 129.

52 C. W. Carlson and R. West, *Organometallics* **2** (1983) 1792, 1801.
53 A. R. Steele and F. S. Kipping, *J. Chem. Soc.* **1929**, 2545.
54 M. F. Lemanski and E. P. Schramm, *Inorg. Chem.* **15** (1976) 2515.
55 M. Biernbaum and R. West, *J. Organomet. Chem.* **77** (1974) C 13.
56 C. J. Hurt, J. C. Calabrese and R. West, *J. Organomet. Chem.* **91** (1975) 273.
57 M. Biernbaum and R. West, *J. Organomet. Chem.* **131** (1977) 179.
58 M. Biernbaum and R. West, *J. Organomet. Chem.* **131** (1977) 189.
59 T. B. Block, M. Biernbaum and R. West, *J. Organomet. Chem.* **131** (1977) 199.
60 B. J. Helmer and R. West, *Organometallics* **1** (1982) 1458.
61 B. J. Helmer and R. West, *Organometallics* **1** (1982) 1463.
62 E. Hengge and G. Miklau, *Z. Allg. Anorg. Chem.* **508** (1984) 33.
63 E. Hengge and H. Stüger, *Monatsh. Chem.* **111** (1980) 1043.
64 H. C. Marsmann, W. Raml and E. Hengge, *Z. Naturforsch.* **35b** (1980) 35.
65 L. Párkányi, K. Sasvári, J. P. Declerc and G. Germain, *Acta Crystallogr.* **B34** (1978) 3678.
66 E. Hengge and H. Marketz, *Monatsh. Chem.* **101** (1970) 528.
67 E. Hengge, D. Kovar and H. P. Söllradl, *Monatsh. Chem.* **110** (1979) 805.
68 E. Carberry and R. West, *J. Organomet. Chem.* **6** (1966) 582.
69 E. Carberry and R. West, *J. Am. Chem. Soc.* **91** (1969) 5440.
70 M. Kumada, M. Ishikawa, S. Sakamoto and S. Maeda, *J. Organomet. Chem.* **17** (1969) 223.
71 M. Ishikawa and M. Kumada, *J. Chem. Soc., Chem. Commun.* **1969**, 567.
72 M. Ishikawa and M. Kumada, *Organometallics* **1** (1982) 317.
73 M. Ishikawa and M. Kumada, *Synth. Inorg. Metalorg. Chem.* **1** (1971) 229.
74 C. Lageot, J. C. Maire, M. Ishikawa and M. Kumada, *J. Organomet. Chem.* **57** (1973) C 39.
75 L. F. Brough and R. West, *J. Am. Chem. Soc.* **103** (1981) 3049.
76 D. W. Larsen, B. A. Soltz, F. E. Stary and R. West, *J. Phys. Chem.* **84** (1980) 1340.
77 T. Drahnak, R. West and J. C. Calabrese, *J. Organomet. Chem.* **198** (1980) 55.
78 B. J. Helmer and R. West, *J. Organomet. Chem.* **236** (1982) 21.
79 C. W. Carlson, K. Matsumura and R. West, *J. Organomet. Chem.* **194** (1980) C 5.
80 G. R. Husk, R. Wexler and B. M. Kilcullen, *J. Organomet. Chem.* **29** (1971) C 49.
81 Chisso Corp. Japan, P. 81, 123993 (1981) (*CA* **96** (1982) 69201).
82 H. Watanabe, T. Muraoka, Y. Kohara and Y. Nagai, *Chem. Lett.* **1980**, 735.
83 E. Hengge and G. Bauer, *Angew. Chem.* **85** (1973) 304.
84 E. Hengge and G. Bauer, *Monatsh. Chem.* **106** (1975) 503.
85 E. Hengge and W. Veigl, unpublished results.
86 C. Kratky, E. Hengge, W. Stüger and A. L. Rheingold, *Acta Crystallogr.* **C41** (1985) 824.
87 K. Hassler, D. Kovar and E. Hengge, *Spectrochim. Acta* **34A** (1978) 1199.
88 F. Höfler, G. Bauer and E. Hengge, *Spectrochim. Acta* **32** (1976) 1435.
89 Z. Smith, H. M. Seip, E. Hengge and G. Bauer, *Acta Chem. Scand.* **A30** (1976) 697.
90 K. Kühlein and W. P. Neumann, *J. Organomet. Chem.* **14** (1968) 317.
91 M. Dräger and K. G. Walker, *Z. Anorg. Allg. Chem.* **479** (1981) 65.
92 C. A. Burkhard, *J. Am. Chem. Soc.* **71** (1949) 963.
93 R. West, L. Brough and W. Wojnowski, *Inorg. Synth.* **19** (1979) 265.
94 L. F. Brough and R. West, *J. Organomet. Chem.* **194** (1980) 139.
95 M. Kumada, S. Sakamoto and M. Ishikawa, *J. Organomet. Chem.* **17** (1969) 231.
96 H. Watanabe, K. Higuchi, M. Kobayaski, T. Kitahara and Y. Nagai, *J. Chem. Soc., Chem. Commun.* **1977**, 704; Jap. Kokai 7924874 (*CA* **91** (1979) 140984).
97 P. Boudjouk and B. H. Han, *Tetrahedron Lett.* **22** (1981) 3813.

97   P. Boudjouk and B. H. Han, *Tetrahedron Lett.* **22** (1981) 3813.
98   H. L. Carrell and J. Donohue, *Acta Crystallogr.* **B28** (1972) 1566.
100 W. Wojnowski, C. J. Hurt and R. West, *J. Organomet. Chem.* **124** (1977) 271.
101 C. J. Hurt, W. Wojnowski and R. West, *J. Organomet. Chem.* **140** (1977) 133.
102 V. Traven and R. West, *J. Am. Chem. Soc.* **95** (1973) 6824.
103 H. Bock and W. Ensslin, *Angew. Chem. Int. Ed. Engl.* **10** (1971) 404.
104 M. T. I. Davidson and A. N. Ostak, *J. Organomet. Chem.* **206** (1981) 149.
105 M. Ishikawa and M. Kumada, *J. Chem. Soc., Chem. Commun.* **1971**, 507.
106 T. J. Drahnak, J. Michl and R. West, *J. Am. Chem. Soc.* **101** (1979) 5427.
107 M. Ishikawa and M. Kumada, *Synth. Inorg. Metalorg. Chem.* **1** (1971) 191.
108 M. Ishikawa, M. Watanabe, J. Iyoda and M. Kumada, *Organometallics* **1** (1982) 317.
109 E. Hengge and K. Hassler, unpublished results.
110 E. Hengge and D. Kovar, *Angew. Chem.* **89** (1977) 417.
111 E. Hengge and D. Kovar, *Z. Anorg. Allg. Chem.* **459** (1979) 123.
112 K. Hassler, E. Hengge and D. Kovar, *Spectrochim. Acta* **34A** (1978) 1193.
113 J. P. Hummel, J. Stackhouse and K. Mislow, *Tetrahedron* **33** (1977) 1925.
114 Z. Smith, A. Almenningen, E. Hengge and D. Kovar, *J. Am. Chem. Soc.* **104** (1982) 4362.
115 K. Matsumura, L. F. Brough and R. West, *J. Chem. Soc., Chem. Commun.* **1978**, 1092.
116 L. F. Brough, K. Matsumura and R. West, *Angew. Chem.* **91** (1979) 1022; *Angew. Chem. Int. Ed. Engl.* **18** (1979) 955.
117 A. Indriksons and R. West, *J. Am. Chem. Soc.* **92** (1970) 6704.
118 R. West and A. Indriksons, *J. Am. Chem. Soc.* **94** (1972) 6110.
119 W. Stollings and J. Donohue, *Inorg. Chem.* **15** (1976) 524.
120 G. R. Husk and R. West, *J. Am. Chem. Soc.* **87** (1965) 3993.
121 E. Carberry, R. West and G. E. Glass, *J. Am. Chem. Soc.* **91** (1969) 5446.
122 R. West and E. Carberry, *Science* **189** (1975) 179.
123 V. V. Bukhtiyarov, S. P. Solodovnikov, O. M. Nefedov and V. I. Shiryaev, *Izv. Akad. Nauk SSSR, Ser. Khim.* **5** (1968) 1012.
124 R. West and E. S. Kean, *J. Organomet. Chem.* **96** (1975) 323.
125 A. C. Buchanan III and R. West, *J. Organomet. Chem.* **172** (1979) 273.
126 M. Kira, H. Bock and E. Hengge, *J. Organomet. Chem.* **164** (1979) 277.
127 H. Bock, W. Kaim, M. Kira and R. West, *J. Am. Chem. Soc.* **101** (1979) 7667.
128 E. Carberry and B. Danlock, *J. Organomet. Chem.* **22** (1970) C 43.
129 W. Fink, *Helv. Chim. Acta* **46** (1963) 720.
130 O. J. Scherer, W. Glässel and R. Thalacker, *J. Organomet. Chem.* **70** (1974) 61.
131 U. Wannagat and M. Schlingmann, *Z. Anorg. Allg. Chem.* **419** (1976) 48.
132 U. Wannagat and O. Brandstätter; *Monatsh. Chem.* **97** (1966) 1352.
133 U. Wannagat and S. Meier, *Z. Anorg. Allg. Chem.* **392** (1972) 179.
134 J. W. Conolly, B. F. Stockton, C. M. Emerick and E. A. Bernheim, *Inorg. Chem.* **9** (1970) 93.
135 U. Wanngat and O. Brandstätter, *Angew. Chem.* **75** (1963) 345.
136 U. Wannagat, *Angew. Chem.* **76** (1964) 234.
137 U. Wannagat, R. Seifert and M. Schlingmann, *Z. Anorg. Allg. Chem.* **439** (1978) 83.
138 U. Wannagat, H. Autzen and M. Schlingmann, *Z. Anorg. Allg. Chem.* **419** (1976) 41.
139 U. Wannagat and M. Schlingmann, *Abh. Braunschw. Wiss. Ges.*, XXIV, 79 (1973/74).
140 U. Wannagat, M. Schlingmann and H. Autzen, *Chem. Ztg* **98** (1974) 372.
141 U. Wannagat, M. Schlingmann and H. Autzen, *Chem. Ztg.* **98** (1974) 111.
142 U. Wannagat and M. Schlingmann, *Z. Anorg. Allg. Chem.* **424** (1976) 87.
143 D. J. Brauer, H. Bürger, H. H. Moretto, U. Wannagat and K. Wiegel, *J. Organomet. Chem.* **170** (1979) 161.

144   H. Bürger and K. Wiegel, *J. Organomet. Chem.* **124** (1977) 279.
145   I. Geisler and H. Nöth, *J. Chem. Soc., Chem. Commun.* **1969**, 775.
146   U. Wannagat and G. Eisele, *Monatsh. Chem.* **109** (1978) 1059.
147   U. Wannagat, *Chem. Ztg.* **97** (1973) 105; U. Wannagat, M. Schlingmann and H. Autzen, *Z. Naturforsch.* **31b** (1975) 621.
148   O. J. Scherer and G. Wolmershauser, *Z. Naturforsch.* **29b** (1974) 5.
149   H. Autzen and U. Wannagat, *Z. Anorg. Allg. Chem.* **420** (1976) 139.
150   U. Wannagat and H. Autzen, *Z. Anorg. Allg. Chem.* **420** (1976) 132.
151   U. Wannagat, T. Blumenthal, G. Eisele, A. Koenig and R. Schächter: *Z. Naturforsch.* **36b** (1981) 1479.
152   T. H. Newman, R. West and R. T. Oakley. *J. Organomet. Chem.* **197** (1980) 159.
153   M. Schlingmann and U. Wannagat, *Z. Anorg. Allg. Chem.* **419** (1976) 115.
154   M. Schlingmann and U. Wannagat, *Chem. Ztg.* **98** (1974) 457.
155   H. Bürger, K. Wiegel, U. Thewalt and D. Schomburg, *J. Organomet. Chem.* **87** (1975) 301.
156   H. Bürger, M. Schlingmann and G. Pawelke, *Z. Anorg. Allg. Chem.* **419** (1976) 121.
157   M. Schlingmann and U. Wannagat, *Z. Anorg. Allg. Chem.* **429** (1977) 74.
158   M. Schlingmann and U. Wannagat, *Z. Anorg. Allg. Chem.* **419** (1976) 108.
159   U. Wannagat, R. Seifert and M. Schlingmann, *Z. Naturforsch.* **32b** (1974) 869.
160   U. Wannagat and R. Seifert, *Z. Anorg. Allg. Chem.* **439** (1978) 90.
161   R. T. Oakley, D. A. Stanisklawski and R. West, *J. Organomet. Chem.* **157** (1978) 389.
162   T. H. Newman, J. C. Calabrese, R. T. Oakley, D. A. Stanislawski and R. West, *J. Organomet. Chem.* **225** (1982) 211.
163   A. W. Cordes, P. F. Schubert and R. T. Oakley, *Can J. Chem.* **57** (1979) 174.
164   H. Gilman and W. H. Atwell, *J. Org. Chem.* **28** (1963) 2905.
165   A. R. Steele and F. S. Kipping, *J. Chem. Soc.* **1928**, 1431.
166   L. Párkányi, E. Hengge and H. Stüger, *J. Organomet. Chem.* **251** (1983) 167.
167   R. West, M. J. Fink and J. Michl, *Science* **214** (1981) 1343.
168   P. Boudjouk, B. H. Han and K. R. Anderson, *J. Am. Chem. Soc.* **104** (1982) 4992.
169   M. Kumada, M. Yamaguchi, Y. Yamamoto, J. Nakajima and K. Shiina, *J. Org. Chem.* **21** (1956) 1264.
170   T. Takano, N. Kasai and M. Kakudo, *Bull. Chem. Soc. Jpn* **36** (1963) 585.
171   M. Kumada, M. Ishikawa and B. Murai, *Kogyo Kagaku Zasshi* **66** (1963) 637 (*CA* **59** (1963) 15303).
172   T. Higuchi and A. Shimada, *Bull. Chem. Soc. Jpn* **39** (1966) 1316.
173   T. Higuchi and A. Shimada, *Bull. Chem. Soc. Jpn* **40** (1967) 752.
174   I. A. McLure and J. F. Neville, *J. Chem. Thermodynamics* **14** (1982) 385.
175   H. S. D. Soysa, H. Okinoshima and W. M. P. Weber, *J. Organomet. Chem.* **133** (1977) C17.
176   M. Kumada and S. Maeda, *Inorg. Chim. Acta* **1** (1967) 105.
177   H. Okinoshima and W. M. P. Weber, *J. Organomet. Chem.* **150** (1978) C25.
178   D. Seyferth, D. C. Annarelli and D. P. Duncan, *Organometallics* **1** (1982) 1288.
179   E. Hengge and H. G. Schuster, *J. Organomet. Chem.* **231** (1982) C17.
180   H. Nöth, H. Fusstetter, H. Pommerening and T. Taeger, *Chem. Ber.* **113** (1980) 342.
181   U. Wannagat and O. Brandstätter, *Monatsh. Chem.* **94** (1963) 1090.
182   E. Hengge and H. G. Schuster; *J. Organomet. Chem.* **240** (1982) C65.
183   M. Wojnowska, W. Wojnowski and R. West, *J. Organomet. Chem.* **199** (1980) C1.

8

Silicon–Nitrogen Heterocycles

U. KLINGEBIEL

University of Göttingen, Federal Republic of Germany

8.1 Introduction

The field of silicon–nitrogen rings has attracted much attention in recent years and has been reviewed several times.[B1–B6] This chapter deals with silicon–nitrogen heterocycles and heterocyclosilazanes, i.e. rings formed by partial replacement of Si or N atoms by other elements, reported after mid-1969. References published before this date have been fully covered in ref. B 1.

There is a great diversity of inorganic ring systems containing silicon and nitrogen; these can be classified as follows.

(a) Four- (1), six- (2) and eight- (3) membered heterocycles, formed by regular alternation of silicon and nitrogen atoms, i.e. cyclosilazanes:

(1) (2) (3)

(b) Silicon–nitrogen heterocycles, (4)–(13), containing bonds between like atoms, which appear as products of insertion of silicon atoms into nitrogen homocycles, or of nitrogen heteroatoms into silicon homocycles:

THE CHEMISTRY OF INORGANIC HOMO-
AND HETEROCYCLES VOL 1 ISBN 0-12-655775-6

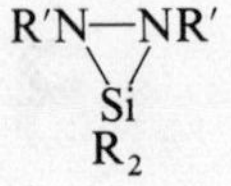

(4)

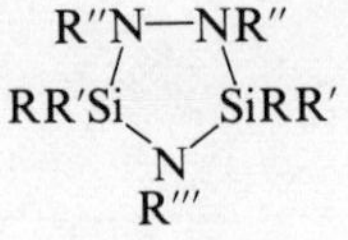

(5)

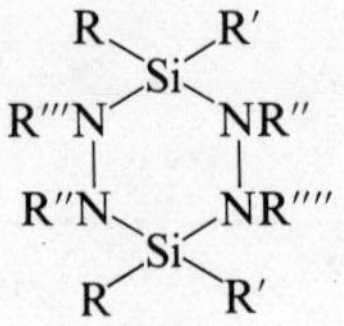

(6)

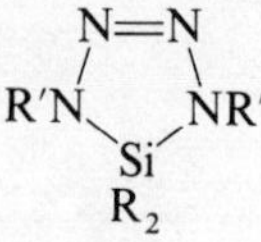

(7)

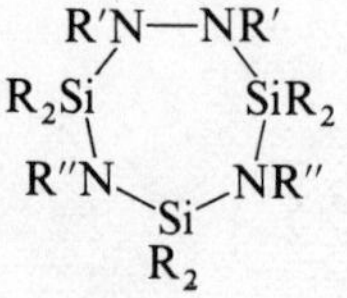

(8)

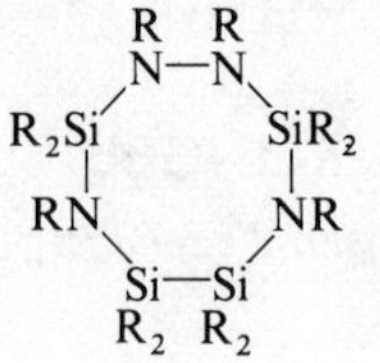

(9)

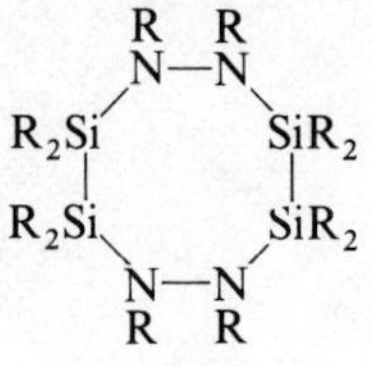

(10)

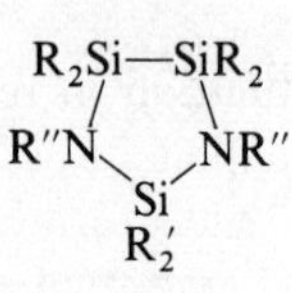

(11)

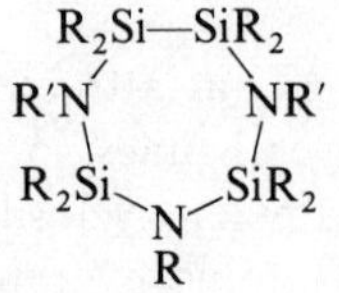

(12)

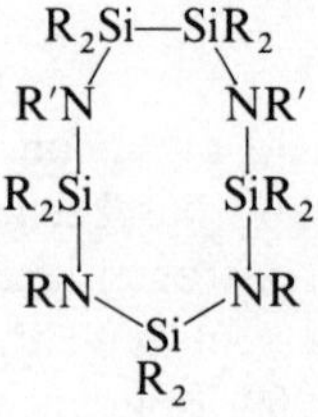

(13)

(c) Heterocycles resulting from partial replacement of silicon with main-group elements such as germanium, tin, lead, phosphorus, arsenic, antimony (boron derivatives have been treated in Chapter 2). The main types are illustrated by formulae (14)–(20):

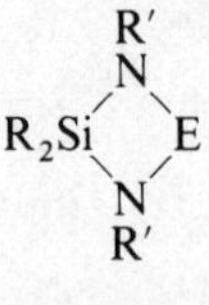

(14)

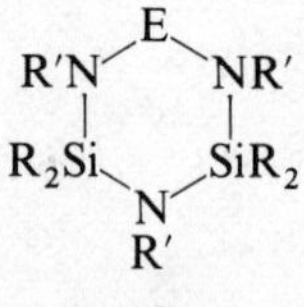

(15)

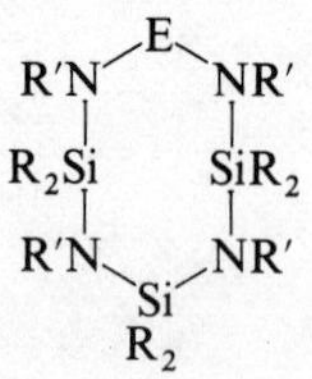

(16)

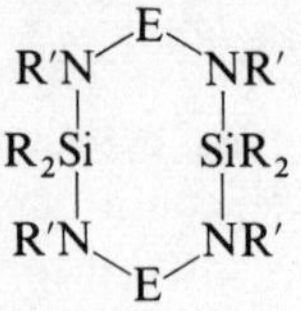

(17)

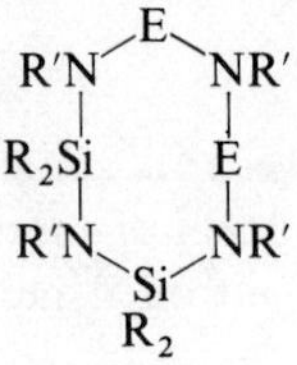

(18)

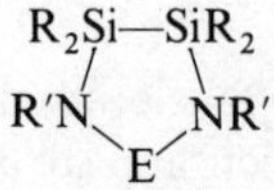

(19)

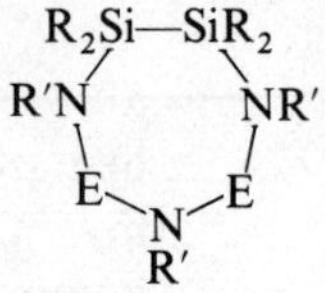

(20)

(*d*) Another class of cyclic silicon–nitrogen compounds contain oxygen (cyclosilazoxanes) or sulphur (cyclosilthiazanes); these result by partial replacement of nitrogen by another electronegative element (oxygen, sulphur) and are illustrated by (**21**)–(**37**):

(**21**)   (**22**)   (**23**)

(**24**)   (**25**)   (**26**)   (**27**)

(**28**)   (**29**)   (**30**)   (**31**)

(**32**)   (**33**)   (**34**)   (**35**)

(36)   (37)

(e) Some silicon–nitrogen heterocycles contain four different elements in the ring; these can be regarded as resulting from partial replacement of both silicon and nitrogen in a parent cyclosilazane, and are illustrated by (38)–(40).

(38)   (39)   (40)

For all five classes (a)–(e) of silicon–nitrogen heterocycles cited, in addition to monocyclic systems, various polycyclic structures are known. These will be presented at the appropriate place in the text.

8.2 Cyclosilazanes

8.2.1 Cyclodisilazanes

The ammonolysis of organodichlorosilanes[1] and the condensation of aminosilanes[2] are classical preparative methods for four- to eight-membered cyclosilazanes. These methods have been used for the synthesis of a large number of new derivatives. New routes have been found particularly for the preparation of four-membered cyclodisilazanes.

Cyclodisilazanes with different substituents $R'R''Si_2N_2R'''R^{iv}$ were synthesized in the stepwise reactions of a dichlorosilane R_2SiCl_2 with two different primary amines, followed by metallation with Bu^nLi and treatment with a second organodichlorosilane $R'R''SiCl_2$ (R = Me, Ph; R' = Me, Cl; R'' = Me, Et, Vi, Cl; R''' = CH_2Ph, Cy; R^{iv} = Et, Pr^i, Cy). The N-lithium route to cyclodisilazanes was also used for N-fluoroalkylsubstituted rings $[Me_2SiN\{CMe(CF_3)_2\}]_2$ (1), R = Me, R' = $CMe(CF_3)_2$,[4,5] which can be prepared from $Me_2Si[NLi\{CMe(CF_3)_2\}]_2$ and Me_2SiCl_2. Another cyclodisilazane $(Me_2SiNC_6F_5)_2$ has been prepared similarly, from $C_6F_5NLi_2$ and Me_2SiCl_2.[5a]

An interesting reaction for the formation of cyclodisilazanes is the cleavage of the Si—N bond by organodihalogenosilanes. In this way sulphamoyl-substituted cyclodisilazanes $[R_2SiNSO_2R']_2$ (1), R = Me, Ph; R′ = NMe_2 and 1,3-morpholino, have been obtained from R_2SiCl_2 and $R'SO_2N(SiMe_3)_2$.[6] Similar reactions were used for the preparation of N,N'-bis(trialkylphosphino)cyclodisilazanes:[7a,b]

$$2\,R_3P{=}N{-}SiR_3 + 2\,SiF_4 \xrightarrow{-2\,R_3SiF} R_3P{=}N\overset{+}{\underset{}{}}\,\,\overset{+}{}N{=}PR_3 \quad (1)$$

with the F_3Si / SiF_3 bridging ring as drawn.

$$(R_3P{=}N{-})_2SiMe_2 + Me_2SiX_2 \longrightarrow \left[R_3P{=}\overset{+}{N}\,\,\overset{+}{}N{=}PR_3 \right]^{2+} 2X^- \quad (2)$$

$$R = Me,\ Et;\ X = Cl,\ Br,\ I$$

The crystal structure of N-(trifluorosilyl)trimethylphosphinimine demonstrates that the molecule is dimeric, $[F_3Si{-}N{-}PMe_3]_2$, with a planar four-membered Si_2N_2 ring containing trigonal bipyramidal silicons and trigonal nitrogens.[7a] This is a rare example of a cyclodisilazane containing five-coordinate silicon.

The ammonolysis of bis(triorganosiloxy)dichlorosilanes $(R_3SiO)_2SiCl_2$ (R = Me, Et) is reported to produce N,N'-bis(triorganosiloxy)cyclodisilazanes (1), R = R′ = $OSiMe_3$ and $OSiEt_3$, R″ = H, among other products.[8] A similar ring was also obtained by heating bis(triethylsiloxy)diaminosilane, $(Et_3SiO)_2Si(NH_2)_2$, at 280–340 °C for 32 h.[8]

N-Substituted aminosilanes $Si(NHR)_4$ (R = Bu, Ph) are cyclized by thermal condensation and cocondensation with $Me_2Si(NHPh)_2$ to give N-substituted cyclodisilazanes (e.g. (1), R = R′ = BuNH, R″ = Bu) and organopolycyclodisilazanes.[9,10]

Other N-substituted cyclodisilazanes (e.g. $[(Me_3SiNH)_2SiNH]_2$ (1), R = R′ = Me_3SiNH, R″ = H) are formed by the ammonolysis of bis-(trimethylsilyl)aminotrichlorosilane $(Me_3Si)_2N{-}SiCl_3$ after the molecular rearrangement of the initial product $[(Me_3Si)_2N(NH_2)SiNH]_2$ (1), R = $(SiMe_3)_2N$, R′ = NH_2, R″ = H, by heating to 200 °C.[11]

N-Trimethylsilyl-substituted cyclodisilazanes (1), R = Me, Ph, R′ = Ph, R″ = $SiMe_3$, were obtained[11,12] by reaction of $MePhSiCl_2$ with

$(Me_3Si)_2NH$ in the presence of pyridine, and by pyrolysis of Ph_2Si-$(NHSiMe_3)_2$.

Ring-closure reactions occur on treating silylamines, $R_2Si(NHSiR_3)_2$ with organosilanes (e.g. $PhSiH_3$) and on heating $-SiH$—NH— compounds such as HR_2Si—NH—SiR_2H (R = Me) in the presence of R_3SiOK.[13,14] The products include dimers (1), R = R' = Me, R" = $SiMe_3$ or $SiHMe_2$, and the trimer (2), R = R' = Me, R" = $SiHMe_2$, and $N(SiHMe_2)_3$.

Some new preparative methods for cyclodisilazanes use the [2 + 2] cyclo-addition of intermediate silicon–nitrogen doubly bonded species.[15-28] Thus gas-phase pyrolysis of 1,1-disubstituted silacyclobutanes with an N-phenylimine affords roughly comparable yields of olefinic products and the corresponding cyclodisilazane:[15]

The thermolysis of 1,2,3,-triaza-4-sila-l-cyclopentene led to the formation, among other products, of a cyclodisilazane (R = Me):[16]

Numerous cyclodisilazanes (1), R = F, Me, Bu^t, Ph; R′ = Me, Ph, N(SiMe$_3$)$_2$, Bu^t; R″ = Pr^i, Bu^t, Ph, C$_6$H$_4$Me-4, etc., are obtained in the reactions of aminofluorosilanes RR′FSi—NHR″ with lithium organics or lithium amides, via rather stable lithium aminofluorosilanes RR′FSi—NR″Li, after LiF elimination:[17-30]

$$\underset{\underset{F}{|}}{RR'Si}-NR''H \xrightarrow{\text{LiR}} \underset{\underset{F}{|}\ \underset{Li}{|}}{RR'Si}-NR'' \xrightarrow[-\text{LiF}]{} RR'Si{=}NR'' \longrightarrow (1) \qquad (3)$$

N-Lithiated N-organo-N′,N′-bis(trimethylsilyl)aminofluoroorganosilanes RFSi[N(SiMe$_3$)$_2$(NR′Li) cyclize by LiF elimination with simultaneous 1,3-migration of a trimethylsilyl group from one nitrogen atom to another to form a cyclodisilazane (1), R = Me, Ph; R′ = N(SiMe$_3$)Alk with Alk = Me, Pr^i, Bu^t, Ph; R″ = SiMe$_3$:[26]

Cross-dimerization is observed on LiF elimination from two similar lithio-aminosilanes FMe$_2$Si—N(C$_6$H$_2$Me$_3$)Li and MePhFSi—N(C$_6$H$_4$Me)Li to form the cyclodisilazane Me$_3$PhSi$_2$N$_2$(C$_6$H$_4$Me)(C$_6$H$_2$Me$_3$).[27]

Under suitable experimental conditions, the reactions of organosilicon halides MeSiCl$_3$ with the N-lithiated silylamines LiNBu^t(SiMe$_3$) lead to the formation of Si$_2$N$_2$ derivatives (1), R = Me, R′ = Cl, R″ = Bu^t.[28]

Cyclodisilazanes with different substituents are formed by stepwise reactions of aminofluorosilanes with another fluorosilane or lithiated amine. Ring closure also occurs in reactions with lithiated amines and fluorosilanes:[21,29]

$$R_2Si\underset{NR''H}{\overset{NR'Li}{<}} + R'''R^{iv}SiF_2 \longrightarrow \underset{\underset{R''NH}{|}}{R_2Si}-NR'-\underset{\underset{F}{|}}{SiR'''R^{iv}} \tag{5}$$

$$\downarrow +BuLi_1-BuH_1-LiF$$

$$\underset{\underset{F}{|}}{R_2Si}-NR'-\underset{\underset{F}{|}}{SiR'''R^{iv}} + 2NHR''Li \longrightarrow R_2Si\underset{\underset{R''}{N}}{\overset{\overset{R'}{N}}{<}}\underset{R^{iv}}{\overset{R'''}{>}}Si + H_2NR + 2LiF \tag{6}$$

The reactivity of monoaminofluorosilanes with butyllithium is reported to decrease with increasing size of the substituents. Reactions observed for aminofluorosilanes include [2 + 2] cycloaddition cited above, ring closure with C–H bond cleavage or migration of a methanide ion.[18,20,22,24,30,31] The products obtained in reactions of butyllithium with (t-butylamino)-(organotrimethylsilyl)aminofluorosilanes are unsymmetrically substituted cyclodisilazanes, whose formation can be attributed to a nucleophilic 1,3-rearrangement at silicon (R = Me, Vi, Bu^s, Bu^t; R' = Bu^t or SiMe$_3$):[18,20,22,24,30]

$$RFSi\underset{NR'SiMe_3}{\overset{NHBu^t}{<}} \overset{BuLi}{\longrightarrow} \underset{+}{R\underset{NR'SiMe_3}{\overset{NBu^t}{>}}Si} \overset{\sim Me^-}{\longrightarrow} \underset{Me}{\overset{R}{>}}Si\underset{\underset{R'}{N}}{\overset{\overset{Bu^t}{N}}{<}}SiMe_2 \tag{7}$$

Two diastereoisomeric cyclodisilazanes are formed by LiF elimination from lithiated 1-[bis(trimethylsilyl)amino]-1-fluoro-3,3,3-trimethyl-1-[(trimethylsilyl)amino]disiloxane,[31] $(Me_3Si)_2N$—$SiF(OSiMe_3)$—$NLi(SiMe_3)$. Attempts to prepare $N(SiMe_2NHR)_3$ from $N(SiMe_2Cl)_3$ and RNH_2 (R = Me, Et, Pr^i) afforded only cyclodisilazanes with aminosilane side groups, $Me_4Si_2N_2R(SiMe_2NHR)$.[32]

8.2.2 Cyclotri- and cyclotetrasilazanes

A widely used method for the preparation of cyclotri- and cyclotetrasilazanes is that by which the cyclosilazanes were obtained for the first time,[1] namely by the reaction of ammonia or primary amines with dihalosilanes. Many new cyclotri- and cyclotetrasilazanes (2) and (3)[33–36] containing alkyl, aryl, vinyl, vinyloxyethyl or trifluoropropyl groups on the silicon atom were synthesized by this reaction, including those with R = Me, Et, R' = C_2H_4Ph, $CH_2CHMePh$, $CH_2CHMeCOOMe$, $(CH_2)_3CN$;[33] R = Me, Ph, R' = Pr^i, C_6H_{11}, OBu^t;[34] R = Me, R' = Ph, $(CH_2)_2CF_3$;[35] R = Me, Et, R' = $(CH_2)_2OCH$=CH_2[36] (R" = H in all cases).

The reaction of diaminodiphenylsilane $Ph_2Si(NH_2)_2$ with dichloromethyl-vinylsilane $MeViSiCl_2$ leads to the formation of three six-membered rings $MeViPh_4Si_3N_3H_3$, $Me_2Vi_2Ph_2Si_3N_3H_3$ and $(MeViSiNH)_3$.[37]

Dimethyldiphenylsilane $SiMe_2Ph_2$ and dimethyldibenzylsilane $SiMe_2$-$(CH_2Ph)_2$ react in liquid ammonia with KNH_2 to give H_2N—$SiMe_2$—NH—$SiMe_2$—NHK; exclusively C_6H_5 and $PhCH_2$ groups are split off and substituted by NH_2. The reaction of the potassium compound thus formed with an equivalent amount of NH_4Cl gives a mixture of $(Me_2SiNH)_3$ and $(Me_2SiNH)_4$.[38] A cyclotrisilazane $Me_6Si_3N_3H_2(CH_2Ph)$ is obtained in the reaction of benzylamine with an aminosilane $Me_2Si(NH\text{-}SiMe_2H)_2$.[13,14]

Another method often used for the synthesis of cyclosilazanes, especially with different substituents, is the intermolecular reaction of diaminodisila-zanes with organohalogenosilanes after metallation or in the presence of triethylamine ($R = Me$, $R' = Me$, Et, Pr, Bu, $R'' = Me$, Pr^i, Bu, $R''' = Me$, Ph, Cl, $R^{iv} = Me$, Cl):[39,40,41]

$$RN\begin{matrix} SiMe_2-NHR' \\ \\ SiMe_2-NHR'' \end{matrix} \xrightarrow[\text{(ii) } R'''R^{iv}SiCl_2]{\text{(i) } 2BuLi \text{ or } NEt_3} RN\begin{matrix} \overset{Me_2}{Si}-\overset{R'}{N} \\ \diagup \qquad \diagdown \\ \qquad \qquad SiR'''R^{iv} \\ \diagdown \qquad \diagup \\ \underset{Me_2}{Si}-\underset{R''}{N} \end{matrix} \qquad (8)$$

The condensation of dichloro-organodisilazanes $R'N(SiMe_2Cl)_2$ with organoaminosilanes $Me_2Si(NHR)_2$ was used to prepare other unsymmetrical cyclotrisilazanes, $Me_6Si_3N_3R_2R'$ ($R = Et$, Pr, Bu, $R' = Me$, Et).[42]

The seven-membered Si_3N_4 chain of 1,5-bis(dialkylamino) octamethyltri-siladiazanes $R_2Si(NR\text{—}SiR_2\text{—}NHR')_2$ can undergo ring closure, with elimination of amine $R'NH_2$, to form cyclotrisilazanes with varied substituents in position 2, $R_6Si_3N_3R_2R'$ ($R = Me$, $R' = Et$, Pr); the same starting compound, after metallation with BuLi and treatment with $RR''SiCl_2$, gives cyclotetrasilazanes with varied substituents in positions 1, 2 and 8, namely $R_7R''Si_4N_4R_2R_2'$ ($R = Me$, $R' = Pr$, $R'' = Me$, Vi).[43a,b]

A new route for the preparation of nonamethylcyclotrisilazane $(Me_2SiNMe)_3$ (**2**), $R = R' = R'' = Me$, in high yields was found in the reaction of Me_2SiCl_2 with $MeNH_2$ and catalytic amounts of $(NH_4)_2SO_4$.[44]

The stepwise formation of octamethylcyclotetrasilazane (**3**), $R = R' = Me$, $R'' = H$, was studied by the ammonolysis of 1,3-dichlorotetra-methyldisilazane, $HN(SiMe_2Cl)_2$.[45] Some eight-membered ring compounds, i.e. deca-,[46] undeca-[47] and dodecamethylcyclotetrasilazane[48] $Me_8Si_4N_4Me_nH_{4-n}$ ($n = 2$, 3 and 4) were prepared for the first time by various routes.

The reaction products formed by thermal transformation of bis(methyl-aminodimethylsilyl)tetramethylcyclodisilazane $(Me_2Si—NSiMe_2NHMe)_2$[49] were found to be a mixture of N-methylcyclotrisilazanes.[46]

Cyclotetrasilazanes $Me_4PhRR'R''Si_4N_4H_4$, with a specific arrangement of the organic substituents in the ring, have been obtained by the hetero-functional condensation of tetraorganodiaminodisilazanes $H_2NSi-MePh—NH—SiR(Me)NH_2$ with tetraorganodichlorodisilazanes $ClR'Si-(Me)—NH—Si(Me)R''Cl$ $(R = R' = R'' = Me,$ Ph; $R' = R'' = Vi;$ $R = R'' = Me;$ $R = R' = Ph, R'' = Me)$.[50] Some of the products are mixtures of isomers. Chloro- and fluorosubstituted cyclotrisilazanes $(RXSiNR')_3$, with R = Me, Et, Ph, R' = Me, Et, X = F, Cl, were obtained by elimination of halogenotrimethylsilane from compounds of the compo-sition $RSiX_2(NR'SiMe_3)$.[28]

An unsymmetrical fluoro-substituted cyclotetrasilazane $R_4Ph_2F_2Si_4N_4R_4$ has been obtained in the reaction of a silylaminofluorosilane $R_2Si(NR—SiPhF_2)_2$ with the dilithiated diaminosilane $R_2Si(NRLi)_2$ (R = Me).[51]

Hexachlorocyclotrisilazane $(Cl_2SiNH)_3$ (2), R = R' = Cl, R'' = H, has been isolated by partial ammonolysis of silicon tetrachloride in diethyl ether at below $-60\,°C$.[52]

The formation of six-membered rings is observed in the reaction of hexamethyldisilazane $(Me_3Si)_2NH$ with dichlorodimethylsilane Me_2SiCl_2 in the presence of pyridine (to yield (2), R = R' = Me, R'' = H), and by elimination of Me_3SiCl from (diethylchloro)trimethyldisilazane $Me_3SiNH-SiEt_2Cl$ under similar conditions (to yield (2), R = R' = Et, R'' = H).[53]

The reactive intermediate monomer $R'R''Si=NR'''$ formed from organo-silyl azides $R'R''R'''Si—N_3$, inserts into the Si—N bond of a cyclodisilazane $(R_2SiNR)_2$ to form the six-membered ring-expanded products $R_4R'R''Si_3N_3R_2R'''$ (R = Me, R' = R'' = Me, R''' = Ph; R' = R''' = Me, R'' = Ph; R' = R'' = R''' = Ph).[54]

8.2.3 Polycyclic silazanes

There are three types of compound containing two or more cyclosilazane units: (a) *spirocyclic silazanes*, in which the rings are connected by sharing a silicon atom; (b) *fused cyclosilazanes*, in which the rings are attached through two or more common atoms (both silicon and nitrogen); (c) *bridge-linked cyclosilazanes*, in which the rings are connected through silicon or nitrogen bridges, or in some cases through oxygen, boron and other bridging groups.

The preparative methods for polycyclic silazanes are similar to those for the single rings:

(a) halogeno- and aminofunctional silazanes are possible precursors to spirane systems;

(b) the thermal condensation of aminosilanes leads to the formation of polycyclic and polymeric silazanes;

(c) some bicyclic silazanes are prepared by ring–ring interconversions in the Si–N system (see Section 8.2.7).

Spirocyclic silazanes. The following spirocyclic silazanes have been reported in the literature:

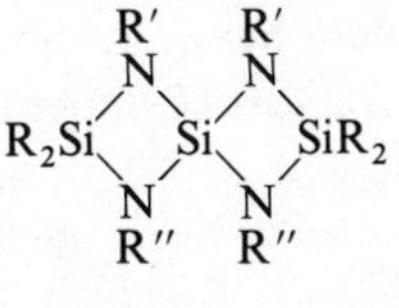

$$\underset{(41)}{\overset{\displaystyle R'\ \ \ R'}{\underset{\displaystyle R''\ \ R''}{N\ \ \ N}}}\quad R_2Si\diagdown Si\diagup SiR_2$$

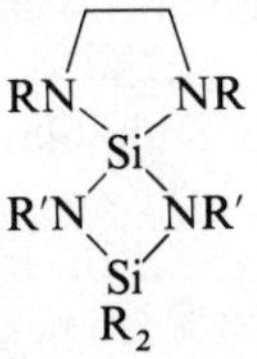

(41)   (42) 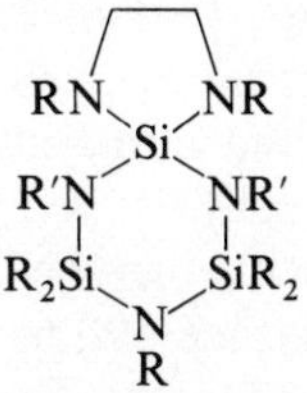   (43) 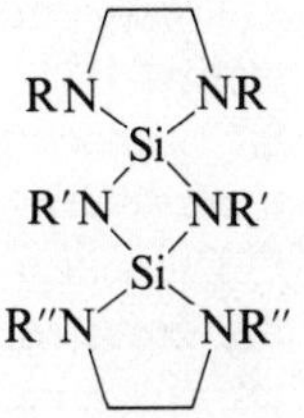   (44)

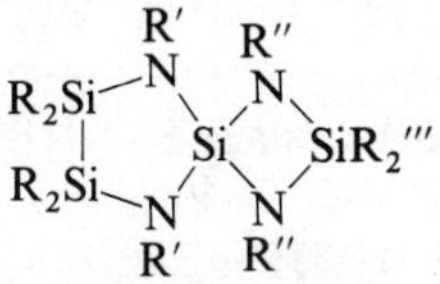

(45) 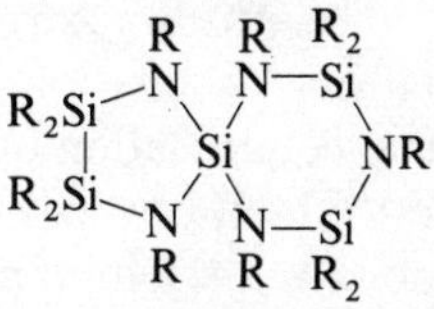   (46)

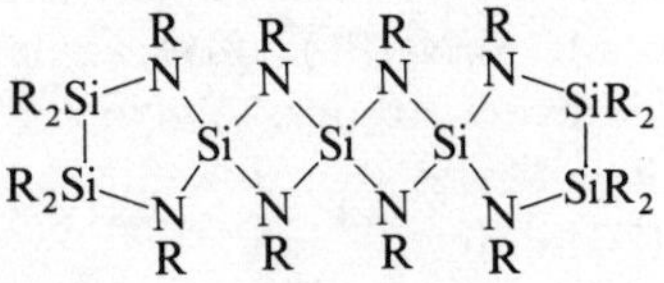

(47)

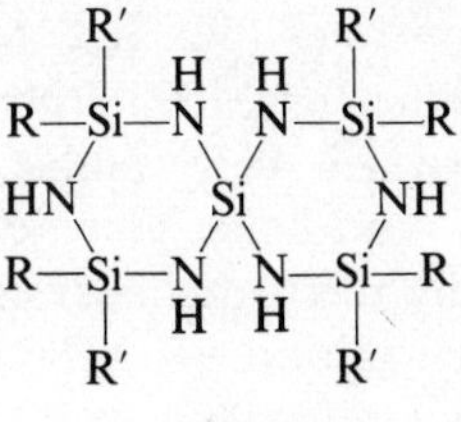

(48)

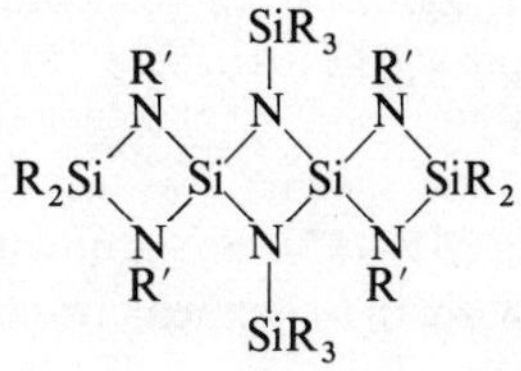

(49) 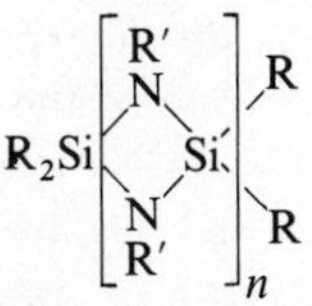   (50) 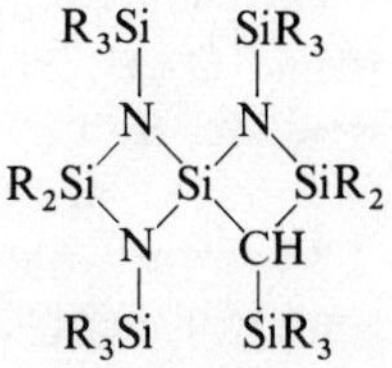   (51)

A spiro-bicyclic silazane (41), R = Ph, R' = CH$_2$Ph, R'' = Et, was formed from the dilithium derivative of bis(ethylbenzylamino)diphenyl-silane Ph$_2$Si(NEtLi)(NBzLi) and silicon tetrachloride.[3] Analogous ring systems (41), R = Ph, R' = R'' = Me, Et, Pr^i, are obtained from methyl-, ethyl- and isopropylamines.[55]

Di- and trispiro systems (42), (43) and (44), R = Me, SiMe$_3$, R' = Me, Et, Pr, R'' = Me, are prepared in reactions of N-lithiated bis(alkylamino)-1,3-diaza-2-silacyclopentanes with appropriate dichloro silicon derivatives.[56a,b]

Treatment of the dilithio compound of 4,4-bis(methylamino)hexa-methyl-1,2,4-cyclotrisila-3,5-diazane with R''$_2$SiCl$_2$, MeN(SiMe$_2$Cl)$_2$ and SiCl$_4$ gave spirodi- and tetracyclic compounds (45), (46) and (47) respectively (R = Me, R' = Et, R'' = Me, Ph).[57,58]

The dilithiated disilazane HN(SiMePhNHLi)$_2$ treated with SiCl$_4$ gave the expected spirosilazane (48) as a 28 : 48% isomeric mixture.[59] A dispiro tetrasilazane (49) was formed in a three-stage reaction sequence by treating R$_3$SiNHSiCl$_3$ with BuLi at −40 °C to give [Cl$_2$Si—NSiR$_3$]$_2$, followed by treatment with PhNHLi and then with BuLi and R$_2$SiCl$_2$ (R = Me, R' = Ph).[60] This dispiro structure was confirmed by an X-ray crystal structure analysis.[60]

The thermal condensation of tetrabutylaminosilane Si(NHBu)$_4$ and tetra-anilinosilane Si(NHPh)$_4$ led to the formation of four-membered spiro-cyclic compounds (41) and polymers (R = NHBu, R' = R'' = Bu, R = NHPh, R' = R'' = Ph).[10] The thermal condensation of tetraanilinosilane with dimethyldianilinosilane Me$_2$Si(NHPh)$_2$ at 360–430 °C gave spiro-cyclic compounds (50), R = Me, R' = Ph, n = 1–3.[9] The condensation of MeSi(NHPh)$_3$ or MePhSi(NHPh)$_2$ with Me$_2$Si(NHPh)$_2$ gave analogous products, which were soluble in aromatic hydrocarbons.[9] Polymers with an Si$_2$N$_2$ backbone unit (e.g. (50), R = Me$_3$SiNH, R' = SiMe$_3$) are obtained by thermal condensation of Si(NHSiMe$_3$)$_4$.[61]

A phenyl-substituted spirane (41), R = R' = R'' = Ph, is formed in the reaction of Ph$_2$SiH$_2$ with Si(NHPh)$_4$ in the presence of catalytic Me$_3$SiOK.[62]

Dropwise addition of SiCl$_4$ to NaN(SiMe$_3$)$_2$ at 200 °C gives the compound C$_{17}$H$_{49}$N$_3$Si$_7$, which has the unexpected spirocyclic structure (51), R = Me, with a carbon atom in one of the rings.[65]

Fused cyclosilazanes. The fused cyclosilazanes (52)–(57) are known; some will be discussed in Section 8.2.7 in relation to the chemical trans-formations of cyclosilazanes.

The reaction of N(SiMe$_2$Cl)$_3$ with NH$_3$ led to (Me$_2$SiNH)$_3$, oligomers, and the bicyclic compound of structure (52), R = Me.[63]

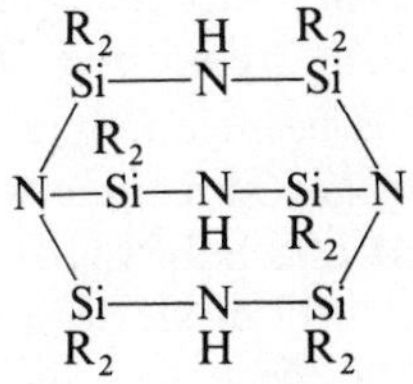

R₂ H R₂
Si—N—Si
R₂
N—Si—N—Si—N
H R₂
Si—N—Si
R₂ H R₂

(52)

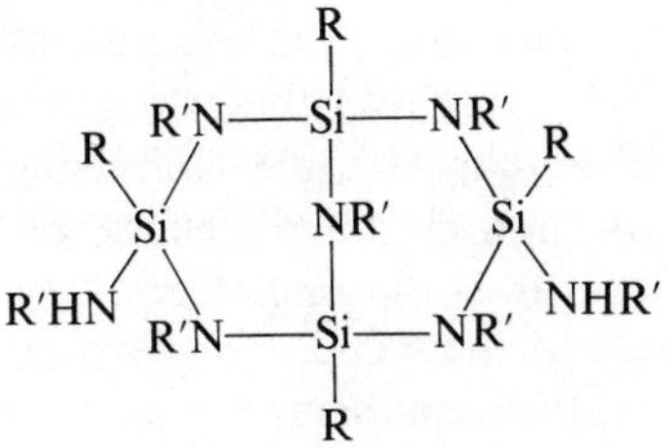

R
|
R'N—Si—NR'
R Si NR' Si R
R'HN N—Si—N NHR'
R'N R'N—Si—NR'
R

(53)

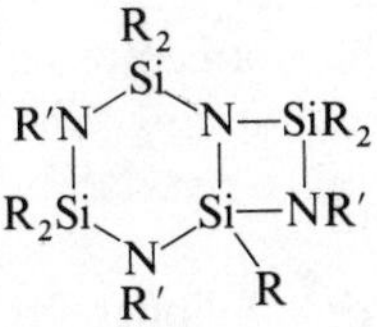

R₂
Si
R'N N—SiR₂
R₂Si Si—NR'
N
R' R

(54)

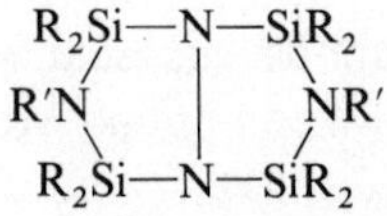

R₂Si—N—SiR₂
R'N NR'
R₂Si—N—SiR₂

(55)

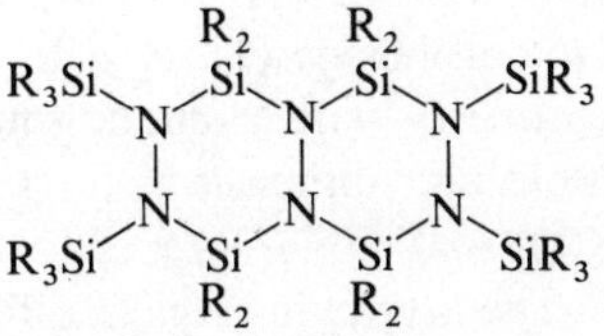

R₂ R₂
R₃Si Si Si SiR₃
N N N
R₃Si Si Si SiR₃
N N N
R₂ R₂

(56)

R''' R R'''
N—Si—N
R' Si Si R'
R'' N—Si—N R''
R''' R R'''

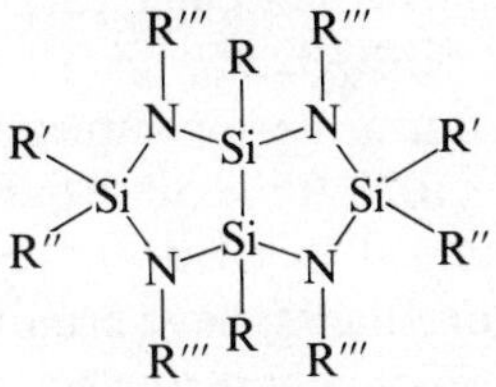

(57)

Aminocyclodisilazanes (**1**), R = alkyl, R' = NHR'', R'' = alkyl, were formed by heating RSi(NHR'')₃ to 210 °C and these underwent further condensation to form bicyclic compounds (**53**).[64]

The reaction of cyclosilazanes with metalorganic compounds, such as BuLi, is an interesting and useful preparative method for the synthesis of various polycyclic silazanes through ring–ring interconversions. These reactions and those leading to the formation of bridge-linked cyclosilazanes will be dealt with in Section 8.2.7.

8.2.4 Structure and bonding

Single crystal X-ray structure determinations have been carried out for several cyclosilazanes. The structures of cyclodisilazanes (**1**) have without exception a planar ring. The substituents on the ring nitrogens are coplanar with the Si_2N_2 ring and the nitrogen atom is sp^2 hybridized. The SiNSi angle is larger than 90°, the NSiN ring angle is correspondingly smaller. The inductive and mesomeric effects of the nitrogen substituents influence the bond lengths in the ring. Thus $-I$ and $-M$ effects of the substituents lead, because of the increase or decrease respectively of electron density at the nitrogen, to a lengthening, and $+I$ and $+M$ effects to a shortening of the Si–N bond.

The molecules of tetramethyl-N,N'-diphenylcyclodisilazane (1), $R = R' = Me$, $R'' = Ph$,[66] and tetramethyl-N,N'-bis(3,5-dimethylphenyl)-cyclodisilazane (1), $R = R' = Me$, $R'' = 3,5-Me_2C_6H_3$,[25] have a centre of symmetry. The aromatic rings are almost coplanar with the Si_2N_2 ring. The molecular dimensions for the second compound are Si–N 173.6 pm, NSiN 85.1° and SiNSi 94.9°.

The aromatic rings in difluoro-di-t-butyl-N,N'-bis(2,4,6-trimethylphenyl)-cyclodisilazane[25] (1), $R = F$, $R' = Bu^t$, $R'' = 2,4,6-Me_3C_6H_2$, are rotated out of the Si_2N_2 plane by 67° about the N–C bond. The consequence is a longer N–C bond distance. The molecular parameters are Si–N 173.2 and 172.6 pm, NSiN 87.1° and SiNSi 92.9°.

The crystal structures of two more complex cyclodisilazane derivatives (1), $R = Me$, $R' = N(SiMe_3)Bu^t$, $R'' = SiMe_3$ and (1), $R = Ph$, $R' = N(Si-Me_3)Pr^i$, $R'' = SiMe_3$, obtained as cited above by LiF elimination from lithiated organo(amino)fluorosilanes, have been determined. The structures proved that during the reaction a trimethylsilyl group migrated from one nitrogen atom to another, as shown in reaction (4) (p. 227). The Si–N bond lengths are in the range 173.2–175.8 pm.[26]

The crystal structures of two diastereoisomers of the cross-dimerisation product of the Si=N monomers, formed as cited above (by LiF elimination from a lithiated fluoroaminodisiloxane) have been determined.[31] The bond lengths and angles of both isomers are very similar. The molecules differ in the torsion angles of the side chains.

An interesting compound, in which the cyclodisilazane ring bears two cyclotrisiloxanyl substituents connected through –Me_2Si—O– groups at the nitrogen sites (58), $R = Me$, has been investigated structurally.[67,68] The endocyclic Si–N bonds are in the range 173.8–174.8 pm and the exocyclic Si–N bonds are shorter, 170.8 pm.

(58)

The molecular structure of hexamethylcyclotrisilazane, **(2)**, R = R′ = Me, R″ = H, has been investigated by electron diffraction in the gas phase.[69] The study yielded the following geometrical parameters: Si–N 172.8 pm, NSiN 108.4°, SiNSi 126.8°. The Si_3N_3 ring is puckered, but the deviation from planarity is relatively small.

A crystal structure determination of another trimer, hexachlorocyclotrisilazane **(2)**, R = R′ = Cl, R″ = H,[70] showed a planar ring and shorter Si–N bonds in this inorganic derivative, Si–N 168 pm, SiNSi 131°, NSiN 109°.

The Si_3N_3 ring in 2,4,6-trifluoro-1,3,5-trimethyl-2,4,6-triphenylcyclotrisilazane **(2)**, R = F, R′ = Ph, R″ = Me,[71] is virtually planar, and the molecule lies on a crystallographic threefold axis. The Si–N distance is 170.5 pm and the SiNSi and NSiN angles are 129.8° and 110.1° respectively.

In 2,2,4,4,6,6-hexamethyl-1,3,5-tris(trimethylsilyl)cyclotrisilazane **(2)**, R = R′ = Me, R″ = SiMe₃,[72] the Si_3N_3 six-membered ring is an almost ideal boat with an average Si–N bond length of 174.4 pm.

An unusual tetracyclic compound, incorporating a cyclotrisilazane ring in a trispiro structure also containing three cyclotrisiloxane rings **(59)**, R = Ph, has been investigated by X-ray diffraction:[73]

(59)

The central trisilazane ring is approximately planar and the Si–N bond length is 169 pm; NSiN 109.5°, SiNSi 129.5°.

An X-ray structural study of octamethyltetraphenylcyclotetrasilazane **(3)**, R = Me, R′ = Ph, R″ = Me,[74] shows that the molecule is situated at a centre of symmetry; the tetrasilazane ring has a general boat conformation and Si–N 172 pm, NSiN 109.5°, SiNSi 129.3°.

In the dodecamethyl derivative **(3)**, R = R′ = R″ = Me,[75] the crystal structure determination shows that the Si_4N_4 ring is nonplanar (boat conformation) with Si–N 173.5 pm and NSiN 110.7°, SiNSi 122.8°.

The structure of the spirocyclic compound **(51)** cited above was confirmed by X-ray diffraction, which established the presence of only one Si_2N_2 ring in the molecule, while the second is an Si_2NC ring.[70]

The molecule of the spirosilazane **(49)**, R = Me, R′ = Ph,[60] is centrosymmetric with four colinear silicon atoms. The molecular parameters are Si–N 172.7–174.2 pm, NSiN 85.3° and SiNSi 94.7°.

8.2.5 Spectra

Physical methods such as infrared or NMR spectroscopy are largely used for structure elucidation of cyclosilazanes. Infrared spectra are used for distinction of dimeric and trimeric rings.[B 1,7,76,77] Comparison of the infrared spectra of the cyclodisilazanes with the cyclotrisilazane showed that the vibration mode $v_{as}SiNSi$ depends on the ring size and on the substituents attached to silicon and nitrogen. A strong skeleton band at about 1035 cm^{-1} is characteristic of four-membered rings. The $v_{as}SiNSi$ stretching mode of cyclotrisilazanes appears at about 930 cm^{-1}.

A detailed Raman study of hexamethylcyclotrisilazane, together with new far-infrared data are discussed.[78] The vibrational spectra are well accounted for on the basis of a planar structure of D_{3h} symmetry.[B 1,78]

[15]N- and N-deuterium-substituted organocyclosilazanes have been synthesized (by reactions of organodichlorosilanes with $^{15}NH_3$ or ND_3) and their IR spectra analysed with the object of establishing the influence of the structure of the compounds on the position of the absorption bands corresponding to the Si–N and N–H bonds.[79] The IR spectra of isotopically labelled and nonlabelled compounds are identical, with the exception of the absorption bands corresponding to Si–N and N–H bonds, for which shifts to lower frequencies are observed.

Investigations of the UV spectra of cyclosilazanes show characteristic absorption bands in the range 185–220 nm.[69,80] This absorption can be assigned tentatively to an n → σ* transition of the unshared electron pair of the nitrogen atom.

NMR data of silicon compounds are listed in some reviews.[81,82] Some ^{1}H NMR spectra of substituted cyclotrisilazanes, for example the *trans* and *cis* isomers of (MePhSiNH)$_3$ have been discussed in terms of conformational analysis.[83] The experimental spectra are consistent with an equilibrium involving chair conformers. The barrier to chair–chair interconversion in hexamethylcyclotrisilazane was estimated at 14.6–16.7 kJ/mol.[83] The temperature dependence of ^{1}H NMR spectra of cyclotetrasilazanes provides valuable information relating to configuration of substituents.[84]

The effect of the substituents of cyclosilazanes on the ^{29}Si NMR chemical shifts has been studied and compared with the effect in siloxanes.[85]

The NMR spectrum of tridecamethyltricycloheptasilazane demonstrated a 3-chair cyclic system, containing 6 axial, 6 equatorial and one central Me–Si units.[86]

8.2.6 Other physical properties

Cyclosilazanes are thermally quite stable, colourless crystalline solids or liquids, soluble in organic solvents.[B 1] Below their melting points the rings often form waxy solids, which show characteristics of an ideal crystal and a liquid.[48,87]

Dipole moments and ionization constants have been determined for a number of cyclotrisilazanes.[88] The dipole moments suggest that cyclotrisilazanes have a nonplanar structure. The relatively low values of the ionization constants of cyclotrisilazanes are explained by high π interaction through the Si–N bonds in the Si_2N groups. The existence of only one ionization constant of cyclotrisilazanes is associated with configurational peculiarities.[88] Vapour pressure data for $(Me_2SiNH)_{3,4}$ are given in refs. 201 and 202.

8.2.7 Chemical properties and reactions

Reactions with preservation of ring size. The total substitution of chlorine atoms in $[Cl_2Si—NSiCl_3]_2$ (1), R = R' = Cl, R'' = $SiCl_3$ was achieved by MeLi, MeOH, EtOH or $MeNH_2$, whereas Me_2NH displaced only nine Cl atoms to give $(Me_2N)_3ClSi_2N_2\{Si(NMe_2)_3\}_2$, and Et_2NH displaced six Cl atoms to form $[(Et_2N)ClSi—N\{Si(NEt_2)_2Cl\}]_2$ (1), R = Cl, R' = NEt_2, R'' = $Si(NEt_2)_2Cl$.[89]

Total substitution of the Cl atoms of the six-membered ring $(Cl_2SiNH)_3$ to $(R_2SiNH)_3$ (R = NCO, EtO, Me_2N) occurred with AgNCO, EtOH and Me_2NH.[90]

Amino-, $Me_5(RR'N)Si_3N_3Me_3$ and silylamino- derivatives $Me_5\{(Me_3-Si)_2N\}Si_3N_3Me_3$ were obtained from chlorooctamethylcyclotrisilazane $ClMe_5Si_3N_3Me_3$ and methylamines RR'NH (R = Me, H, R' = Me) and $NaN(SiMe_3)_2$ respectively.[41] A nitrogen-bridged bicyclic compound (60) was obtained by reacting the same chlorocyclotrisilazane $ClMe_5Si_3N_3Me_3$ with an *N*-lithiated derivative $Me_6Si_3N_3Me_2Li$:[41]

(60)

Reactions with preservation of the ring size were also carried out with Na-octamethylcyclotrisilazane $Me_6Si_3N_3Me_2Na$ and monochloro-organosilanes ClSiMeRR' to give *N*-substituted silyl derivatives $Me_6Si_3N_3Me_2$ (SiMeRR') (R = Me or Cl).[91]

Numerous derivatives of cyclotri- and tetrasilazanes, for example $Me_6Si_3N_3H_2(SiRR'F)$ and $Me_8Si_4N_4H_3(SiF_3)$, were prepared in reactions of lithiated cyclosilazanes with fluorosilanes $RR'SiF_2$ and SiF_4

(R = H, F, Me, Bu^i, Ph; R' = F, Bu^i, Bu^s, Bu^t, Ph, N(SiMe$_3$)$_2$, NBu^t(SiMe$_3$)).[92-98] Disubstituted compounds Me$_6$Si$_3$N$_3$H(SiRR'F)$_2$ were obtained in the reaction of dilithium cyclotrisilazane Me$_6$Si$_3$N$_3$HLi$_2$ with fluorosilanes RR'SiF$_2$ (R = Me, Ph, F; R' = Me, Bu^s, Ph) or in reactions of lithiated fluorosilyl-substituted cyclosilazanes with fluorosilanes.[92,93,96-98]

Silyl-bridged cyclosilazanes (61), R = Me, R' = H, F, Me, Ph, R'' = H, SiMe$_2$F, SiMeF$_2$, SiPhF$_2$, etc., were synthesized in reactions of fluorosilyl-substituted and lithiated cyclosilazanes:[92-94]

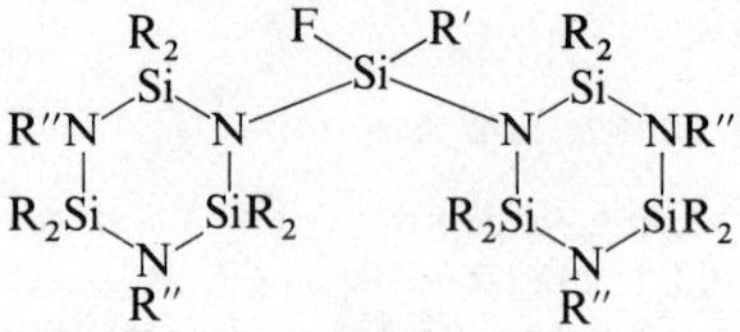

(61)

Analogous methods have been used for the preparation of tris(fluorosilyl),[95,98] trimethylsilyl(fluorosilyl)stannyl-, and boryl-substituted compounds Me$_6$Si$_3$N$_3$RR'R'', with R = H, SiMe$_2$F, SiBu^{t}F$_2$, BF$_2$; R' = H, SiF$_3$, SiMe$_3$, SiMe$_2$F, SiMeF$_2$, SiMeBu^{s}F, SiBu^{t}F$_2$, SiPhF$_2$; R'' = SiF$_3$, SiMe$_3$, SiPhF$_2$, SiBuPhF, BFN(SiMe$_3$)$_2$, BPhNMe$_2$, SnMe$_3$, in various combinations.[76,95,99-101]

Ring coupling occurred in the reaction of lithiated cyclotrisilazane with chloropentamethylborazine[95] and with BF$_3$.Et$_2$O,[101] to give the bicyclic compounds (62) and (63) (R = Me):

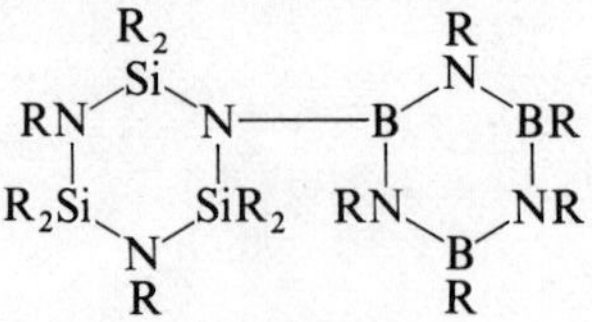

(62) 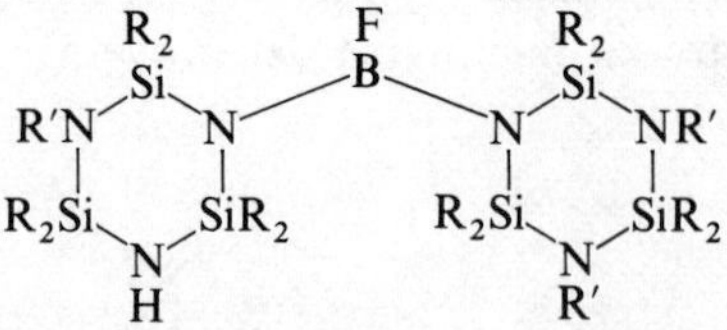   (63)

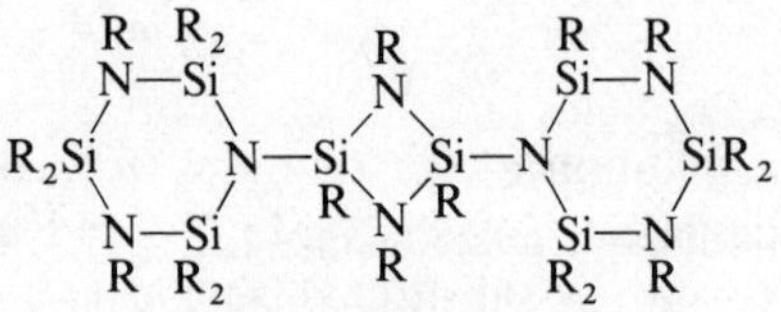

(64)

The reaction of a fluoro(methylamino)silyl-substituted cyclotrisilazane with BuLi led to a system (64), R = Me, consisting of two cyclotrisilazane units connected through a cyclodisilazane.[95] In this reaction the ring size was only partially preserved.

One of the compounds discussed, namely difluorosilanediyl-1,1'-bis(hexamethylcyclotrisilazane) (61), R = Me, R' = F, R" = H, has been investigated by X-ray diffraction.[102] It was found that the Si_3N_3 rings are far from planar. They do not correspond to any ideal conformation of six-membered rings, but approximate most closely the boat conformation.

The tetracyclic compound $[(CH_2)_3NSiOEt]_3$ (65), R = OEt, reacts with EtLi and PhLi to give substitution products (65), R = Et, Ph:[103]

(65)

When mixtures of the rings $(Me_2SiNR)_3$ (R = Me, Et) were heated at moderate temperatures in the absence of any catalyst, alkyl exchange was observed.[104]

Octamethylcyclotetrasilazane reacts with halides of the early transition series to give adducts of the type $(Me_2SiNH)_4.2MCl_3.2L$, where M = Ti, L = THF; M = V, L = THF, and M = Cr, L = NMe_3, and $(Me_2SiNH)_4.2MCl_4$, where M = Ti.[105] Lewis-base participation of the tetrameric Si_4N_4 ring is seen to involve only two of four available nitrogen centres in M–N bonding. The trimer $(Me_2SiNH)_3$ and $TiCl_4$ form $(Me_2SiNH)_3.TiCl_4$ as a neutral six-coordinate complex.[105]

Treatment of the lithiated ring $(Me_2SiNH)_4$ with $MePOCl_2$ in boiling toluene[106] gave a phosphorus-containing product to which a bicyclic structure was speculatively assigned.

Rearrangements to rings of other sizes. The reaction of cyclotri- and tetrasilazanes $(Me_2SiNH)_n$, $n = 3$ and 4, with anhydrous hydrogen halides or with dimethyldihalogenosilanes leads to ring contraction with formation of cyclodisilazanes $[Me_2SiN(SiMe_2X)]_2$ (1), R = R' = Me, R" = $SiMe_2X$, X = Br, I.[107] Analogous cyclodisilazanes (1), R = Me, R' = Et, $CH_2CH_2CF_3$, Ph, Vi, R" = SiMeR'Cl, have been prepared by heating mixtures of $(RMeSiNH)_3$ and $RMeSiCl_2$.[108]

The reaction of N,N'-dilithio- and N,N',N"-trilithiohexamethylcyclotrisilazane with Me_3SiCl depends very strongly on the experimental conditions,[109,110] and gives a variety of products:

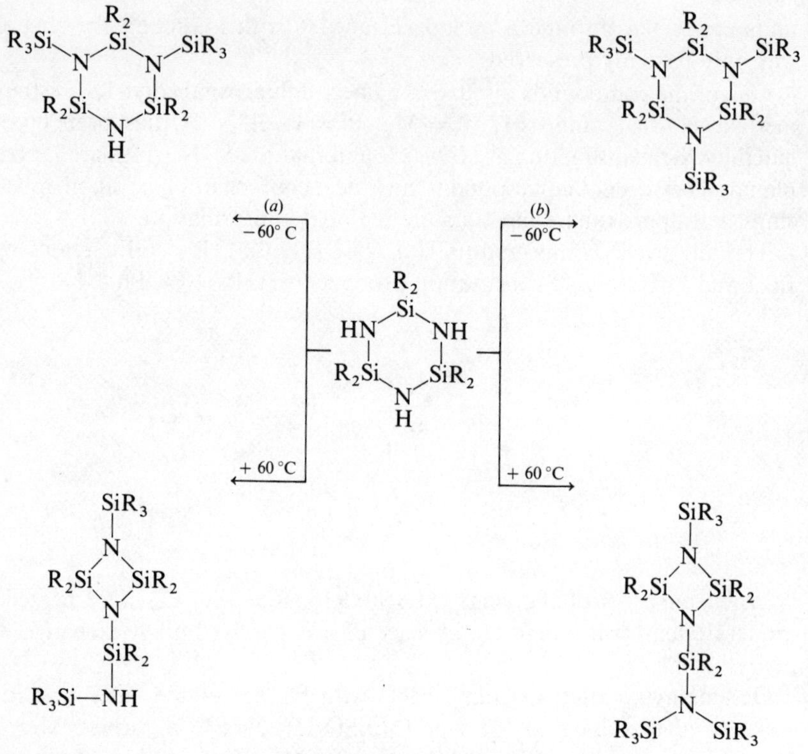

The reaction scheme shows cyclotrisilazane structures with reactions (a) and (b):

$$R_3Si-N(-SiR_2-)-N(-SiR_3)-Si(R_2)-N(-SiR_2-)... \text{(top left structure)}$$

Top left structure:

$$\begin{array}{c} R_2 \\ R_3Si-N \underset{Si}{\overset{}{}} N-SiR_3 \\ R_2Si \quad SiR_2 \\ N \\ H \end{array}$$

Top right structure:

$$\begin{array}{c} R_2 \\ R_3Si-N \underset{Si}{\overset{}{}} N-SiR_3 \\ R_2Si \quad SiR_2 \\ N \\ SiR_3 \end{array}$$

Center structure:

$$\begin{array}{c} R_2 \\ HN \underset{Si}{\overset{}{}} NH \\ R_2Si \quad SiR_2 \\ N \\ H \end{array}$$

(a) $-60°$ C (b) $-60°$C

$+60\,°C$

Bottom left structure (four-membered ring):

$$\begin{array}{c} SiR_3 \\ N \\ R_2Si \quad SiR_2 \\ N \\ SiR_2 \\ R_3Si-NH \end{array}$$

Bottom right structure (four-membered ring):

$$\begin{array}{c} SiR_3 \\ N \\ R_2Si \quad SiR_2 \\ N \\ SiR_2 \\ R_3Si-N-SiR_3 \end{array}$$

(a)   2 BuLi followed by 2 R_3SiCl
(b)   3 BuLi followed by 3 R_3SiCl

Silyl substitution is preferred at low temperatures in diglyme as solvent, but isomerization reactions are observed at higher temperatures. The isomerizations are reported to be base-catalysed and to take place after metallation.[109,110]

Ring contraction with formation of four-membered rings is also observed in the reaction of lithiated 1-fluorosilyl-5-trimethylsilylcyclotrisilazanes with compounds containing an acidic hydrogen atom.[111] Thus ring contraction is observed with MeOH, PhOH, $BuC \equiv CH$ and $PhC \equiv CH$.[77,111]

Simple substitution of the cyclotrisilazane competes with substitution and ring contraction to form the isomeric cyclodisilazane in reactions of fluorosilanes with lithiated cyclotrisilazanes.[76,77,96-100]

The reaction of a lithiated N-trimethylsilylcyclotrisilazane with Bu^tSiF_3 produced four compounds (66)–(69), R = Me, in a ratio of $1:1.5:1:0.2$; two of these are ring-contraction products:

$$\begin{array}{c} \text{SiR}_3 \\ | \\ \text{N} \\ \text{R}_2\text{Si} \quad \text{SiR}_2 \\ \text{HN} \quad \text{NLi} \\ \text{Si} \\ \text{R}_2 \end{array}$$

$+ \text{Bu}^t\text{SiF}_3$
$- \text{LiF}$

(66)   **(67)**   **(68)**   **(69)**

Ring-contraction reactions are used for the preparation of silyl-bridged cyclodisilazanes;[96,110] the cleavage of the Si–N bond of the six-membered ring depends on the nature of the substituted fluorosilyl groups (R = F, Me, R′ = F, Me or Bu^s, R″ = Me, Bu^s or Ph).

The structures of two tricyclic compounds thus obtained, consisting of three Si$_2$N$_2$ rings, bridged by SiMe$_2$ groups, were demonstrated by X-ray diffraction.[96]

Steric influence on ring contractions was studied in reactions of lithiated cyclotrisilazanes with bulky aminofluorosilanes of the type $RSiF_2$—NR'-($SiMe_3$) with R = F, Me, R′ = Bu^t or $SiMe_3$,[77,97] in which cyclodisilazane derivatives were formed.

The ring expansion of cyclodisilazanes to six-membered ring derivatives is also possible. Thus a lithiated cyclodisilazane equilibrates to a lithiated cyclotrisilazane, and a fused ring bicyclic derivative (54a) is also formed:[77]

(54a)

The molecular structure of the bicyclic compound (54a) was determined by X-ray diffraction.[112] The four-membered Si_2N_2 ring is almost planar. The dimensions are typical for Si_2N_2 rings, with NSiN angles less than and SiNSi greater than 90°. The six-membered ring is far from planar. The ring does not correspond to any of the ideal conformations.[112]

Several other ring contractions were observed in LiF elimination reactions from N-lithiated cyclotrisilazanes containing SiR_2F groups, leading to various compounds in which Si_3N_3 and Si_2N_2 rings are linked through SiNSiNSi or SiNSi bridges, or containing three Si_2N_2 rings, linked by SiR_2 groups,[98,100] but lack of space prevents their full presentation here. The reader is referred to the original articles.

Anionic rearrangements of cyclotetrasilazanes to form cyclotrisilazanes are observed in the preparation and methylation of eight-membered rings.[47,91] Thus on treating $RN(SiR_2$—$NRLi)_2$ with $HN(SiR_2Cl)_2$, or by reacting $R_8Si_4N_4R_2HLi$ with MeI, in addition to the expected Si_4N_4 ring derivative $R_8Si_4N_4R_3H$ (type **3**), an isomeric product containing the Si_3N_3 ring, namely $R_6Si_3N_3R_2(SiR_2NHR)$ (type **2**) (R = Me) was also obtained.[47,91]

Ring expansion of four-membered silylamino cyclodisilazanes to six-membered rings occurs with catalytic amounts of base, as shown by the following examples (R = Me, R' = H or $SiMe_3$):[32,63,77,109,110,113]

$$
\begin{array}{c}
\underset{|}{\overset{SiR_3}{N}} \\
R_2Si\diagdown{}_{\displaystyle N}\diagup SiR_2 \\
\underset{|}{} \\
SiR_2NHR'
\end{array}
\xrightarrow{\text{base}}
\begin{array}{c}
R_3Si-N\overset{\overset{R_2}{Si}}{\diagdown}NR' \\
R_2Si\diagdown{}_{\displaystyle \underset{|}{N}}\diagup SiR_2 \\
H
\end{array}
$$

$$
\begin{array}{c}
\underset{|}{\overset{H}{N}} \\
R_2Si\diagdown{}_{\displaystyle N}\diagup SiR_2 \\
\underset{|}{} \\
SiMe_3
\end{array}
\xrightarrow[\text{or acid}]{\text{base}}
\begin{array}{c}
\overset{SiR_3}{\underset{|}{N}} \\
R_2Si\diagup{}^{}\diagdown SiR_2 \\
HNNH \\
R_2Si\diagdown{}_{\displaystyle \underset{|}{N}}\diagup SiR_2 \\
SiR_3
\end{array}
+
\begin{array}{c}
HN\overset{\overset{R_2}{Si}}{\diagdown}NH \\
R_2Si\diagdown{}_{\displaystyle \underset{|}{N}}\diagup SiR_2 \\
SiR_3
\end{array}
+ R_2Si
\begin{array}{c}
\overset{NHSiR_3}{\underset{|}{\overset{SiR_2}{N}}} \\
\diagup{}^{}\diagdown SiR_2 \\
\underset{|}{} \\
\overset{SiR_2}{\underset{|}{NHSiR_3}}
\end{array}
$$

$$
\begin{array}{c}
\overset{NH_2}{\underset{|}{SiR_2}} \\
\underset{|}{N} \\
R_2Si\diagdown{}_{\displaystyle N}\diagup SiR_2 \\
\underset{|}{SiR_2} \\
NH_2
\end{array}
\xrightarrow{R_3SiO^-NMe_4^+}
\begin{array}{c}
HN\overset{\overset{R_2}{Si}}{\diagdown}NH \\
R_2Si\diagdown{}_{\displaystyle \underset{|}{N}}\diagup SiR_2 \\
\overset{SiR_2}{\underset{|}{NH_2}}
\end{array}
+ (R_2SiNH)_4
$$

Attempts to prepare $N(SiR_2NHR')_3$ from $N(SiR_2Cl)_3$ and $R'NH_2$ afforded only cyclodisilazanes $R_4Si_2N_2R'(SiR_2NHR')$; these compounds rearrange in the course of their reaction with LiBu to give cyclotrisilazane derivatives $R_6Si_3N_3R'_2Li$ (R = Me, R' = Me, Et, Pr^i). The ring expansion proceeds by an anionic mechanism and was not observed if only catalytic amounts of BuLi were added.[32]

The isomerization of a cyclodisilazane to a cyclic hydrazine derivative, containing a five-membered ring, is demonstrated in the reaction of the lithiated compound with trifluorosilane (R = Me, R' = Bu^t):[114]

$$R_2Si\overset{\overset{\displaystyle R'}{|}\underset{\displaystyle N}{}}{\underset{\overset{\displaystyle |}{N}}{}}SiR_2 \quad \xrightarrow[\text{(ii) PhSiF}_3]{\text{(i) LiBu}} \quad R_2Si\overset{\overset{\displaystyle R' \quad SiPhF_2}{N-N}}{\underset{\displaystyle \underset{R'}{N}}{}}SiR_2 \qquad (9)$$

$$\underset{\displaystyle R'}{\underset{\displaystyle |}{NH}}$$

Reactions with ring cleavage. Ring cleavage and formation of polymers occurred in the reaction of $(Cl_2SiNH)_3$ with $LiAlH_4$.[90] The cleavage of decachloro-N,N'-bis(silyl)cyclodisilazane $(Cl_2Si\!-\!NSiCl_3)_2$ (**1**), R = R' = Cl, R'' = $SiCl_3$, with anhydrous HCl leads to hexachlorodisilazane $(Cl_3Si)_2NH$.[89]

A convenient method for the synthesis of 1,3-dihalogenodisilazanes is the reaction of cyclosilazanes with hydrogen halides.[115] Thus the reaction of $(R_2SiNH)_3$ with HX in various molar ratios (1:9, 1:6 and 1:3) gave R_2SiX_2, $(XR_2Si)_2NH$ and $R_2Si(NHSiR_2X)_2$. The tetramer $(R_2SiNH)_4$ reacted with HX to produce $(XR_2Si)_2NH$ (R = Me or Ph, X = F, Cl, Br).

The redistribution between $(Me_2SiNEt)_3$ and Ph_2SiCl_2 in 1:3 molar ratio also causes ring cleavage to yield $ClSiMe_2\!-\!NEt\!-\!SiPh_2Cl$.[115]

The coalcoholysis of organocyclosilazanes with chloroorganosiloxanes is used for the preparation of organosilicon ethers.[116] Cyclic ethers of diorganosilanediols are formed in reactions with glycols, and the Si_3N_3 ring is fully cleaved in the process.[116]

Reactions of $(Me_2SiNH)_n$, $n = 3$ and 4, with stoichiometric amounts of KOH in MeOH gave $Me_2Si(OMe)_2$ and $Me_2Si(OMe)OK$, with liberation of ammonia. The cyclotrisilazane was much more reactive than the cyclotetrasilazane.[117] Treatment of $Me_8Si_4N_4H_3(SiMe_2NLi)$ with $MePOCl_2$ gave, in addition to a phosphorus-containing cyclotetrasilazane, the ring-cleavage product $H_2NSiMe_2NSiMeBuSiMe_2NHSiMe_2Bu$.[107]

Hexamethylcyclotrisilazane $(MeSiNH)_3$ reacts with H_2S to give mixed substitution products, by elimination of NH_4SH and formation of Si–S–Si bonds.[118] In the cothiolysis of dimethylcyclosilazanes with Me_2SiCl_2, a mixture of cyclic silthianes is formed[119] (see Chapter 11).

The cleavage of cyclosilazanes with boron halides leads to the formation of borazines, silaborazines and diborylamines.[101,120,121]

8.2.8 Uses

There have been relatively few investigations of the potential commercial uses of organocyclosilazanes.

A polysilazane was formed along with a tricyclotetrasilazane by heating a bicyclotetrasilazane to 325 °C for 200 h; the products are useful as coatings.[122]

Several attempts were made to prepare thermally stable silicon–nitrogen polymers from cyclosilazanes.[203-213] The SiN rings cannot be polymerized by ring-opening reactions similar to those of cyclosiloxanes, and generally only low-molecular-weight oligomers and condensation products were obtained. More successful was the synthesis of polymers by reacting various cyclosilazane derivatives with aromatic diols,[214-217] carborane diols[218] and siloxane diols.[219]

Cyclosilazanes were used in the surface treatment (for hydrophobization) of cellulose[220,221] and finely divided silica.[222] They were also used as components of catalytic compositions for the preparation of polyepoxides,[223] polymerization of α-olefins,[224] stabilizers for dicarboxylates, polyesters[225] and polypropylene,[226] inhibiting compositions[227] and vulcanizable mixtures.[228]

Organocyclosilazanes are useful reagents in organic synthesis, as divalent silylation agents for oximes[229,230] and hydroperoxides.[231] They can also serve as dehydration reagents for the synthesis of nitriles from carboxylic acids, amides and aldoximes.[232-235]

Hexamethylcyclotrisilazane is a starting material in the synthesis of some organosilicon compounds that exhibit spasmolytic and anticholesteremic properties.[236,237]

8.3 Silicon–nitrogen heterocycles with N–N bonds

8.3.1 Cyclosiladiazanes

The first three-membered ring (4), R = Bu^t, R' = SiMe$_3$ was obtained in the reaction of Bu$^t_2$SiF$_2$ with lithiated N,N'-bis(trimethylsilyl)hydrazine (Me$_3$Si)HN—N(SiMe$_3$)Li in several successive steps.[123]

8.3.2 Cyclodisilatriazanes

A survey of the synthesis, structure, properties and reactions of five-membered Si–N ring systems has been published.[124]

A cyclodisilatriazane (5), R = R' = R'' = R''' = Me, was constructed by reacting MeN(SiMe$_2$Cl)$_2$ with 1,2-dimethylhydrazine MeHN—NHMe or its dilithium derivative LiMeN—NMeLi.[125] The same product is formed by transamination of 1,3-bis(methylamino)disilazane MeN(SiMe$_2$—NHMe)$_2$ with 1,2-dimethylhydrazine hydrochloride,[125] or by amine elimination from 1,2-bis(aminodimethylsilyl)-1,2-dimethylhydrazine [(MeNH)SiMe$_2$-NMe–]$_2$.[126]

Another preparative method for five-membered rings (5) is based on the reaction of 1,2-bis(fluorosilyl)hydrazines $(FMe_2Si)R''N$—$NR''(SiMe_2F)$ with lithiated amines $LiNHR'''$ in a $1:2$ molar ratio ($R = R' = Me$, $R'' = Ph$ and $SiMe_3$, $R''' = Pr^i$, Bu^t, Ph or NMe_2),[127] or on the reaction of N-aminosilyl-N'-fluorosilylhydrazines $(R'''NH)Me_2Si$—$NR''NR''SiMe_2F$ with butyllithium.[114,128]

Analogous cyclizations occur in the reaction of bis(fluorosilyl)amines $(FRR'Si)(FRR''Si)NR'''$ and dilithiated hydrazines $LiHN$—$NR''Li$, or of monofluorosilyl-substituted hydrazines with butyllithium, to form unsymmetrically substituted cyclodisilatriazanes (5).[114] The substituents in these products are very different: F, Me or Bu^t at silicon sites and H, Me, Bu^t or mesityl at nitrogen sites.

The isomerization of a lithiated cyclodisilazane to a cyclodisilatriazane was cited above (equation 9).[114]

The isomeric N-fluorosilyl-N,N'-bis(trimethylsilyl)hydrazine (R_3Si)-$(FRR'Si)N$—$NHSiR_3$ and N'-fluorosilyl-N,N-bis(trimethylsilyl)hydrazine $(R_3Si)_2N$—$NH(SiRR'F)$, where $R = Me$, $R' = Me$, Ph, on treatment with BuLi, undergo ring-closure reactions to form (5), $R = Me$, $R' = Me$ or Ph, $R'' = SiMe_3$ and $R''' = N(SiMe_3)_2$. The structure of the compound with $R = Me$ and $R' = Ph$ has been determined by X-ray diffraction.[114] The Si_2N_3 ring is nonplanar. By-products in this reaction are the cyclodisilazane (1), $R = Me$, $R' = Me$ or Ph, $R'' = N(SiMe_3)_2$, and the cyclodisilatetrazane (6), $R = Me$, $R' = Me$ or Ph, $R'' = R''' = SiMe_3$.[114]

Hydrazine, or its dilithium derivative $LiHN$—$NHLi$, reacts with bis(halogenosilyl)amines $R'N(SiR_2X)_2$ ($X = Cl$, F) in a $1:2$ molar ratio to give a bicyclic compound (55), $R = Me$, $R' = Me$ or mesityl.[125,129] The structure of the mesityl derivative has been determined by X-ray diffraction.[129]

An interesting formation of a cyclodisilatriazane was found by the thermolysis and ring reorganization of a cyclosilatetrazene (7), $R = Me$:[130]

$$R_3Si—N\underset{\underset{R_2}{Si}}{\overset{N=N}{\diagdown}}N—N(SiR_3)_2 \quad \xrightarrow{-N_2} \quad R_2Si\overset{\overset{R}{N}—\overset{SiR_3}{N}}{\diagdown\underset{\underset{SiR_3}{N}}{\diagup}}SiR_2 \qquad (10)$$

(7)

8.3.3 Cyclodisilatetrazanes

Cyclo-3,6-disila-1,2,4,5-tetrazanes are six-membered rings (6) containing two N—N bonds.[B1] The fully methylated compound (6),

R = R' = R'' = R''' = Me, was prepared by the reaction of dimethylhydrazine MeHN—NHMe with Me_2SiCl_2.[126] The same product was obtained in the reaction of $ClMe_2SiSiMe_2Cl$ with LiHN—NHLi by cleavage of the Si–Si bond.[131] The structural changes accompanying the one-electron oxidation of this compound with $AlCl_3$–CH_2Cl_2 (to form a cation radical) were studied by ESR.[132]

Fluorosilylhydrazines $(FR_2Si)PhN-NH(SiR_3)$ react with butyllithium by elimination of LiF and formation of six-membered rings (6) R = R' = Me, R'' = Ph, R''' = $SiMe_3$.[114,128]

The bicyclic six-membered ring system (56) (p. 233) was obtained by the reaction of bis(N,N'-fluorodimethylsilyl-N,N'-trimethylsilyl)hydrazine $[(R_3Si)(FR_2Si)N-]_2$ with dilithiated hydrazine LiHN—NHLi.[114]

8.3.4 Cyclotrisilatetrazanes

Seven-membered silicon–nitrogen heterocyclic rings (8) were prepared from the dilithium derivatives of a bis(aminodimethylsilyl)-1,2-dimethylhydrazine $(LiR''N-SiR_2-NR'-)_2$, with Me_2SiCl_2 (R = R' = Me, R'' = Me, Pr).[126] There has been little investigation of these compounds, except for the usual characterization by spectroscopic techniques.

8.3.5 Cyclotetrasilatetrazanes

Two new isomers (9) and (10) of the known cyclotetrasilazane (3) were obtained by LiCl elimination. Thus a permethylated cyclo-3,5,6,8-tetrasila-1,2,4,7-tetrazane (9), R = Me, was formed in the reaction of the dilithiated bis(methylaminodimethylsilyl)-1,2-dimethylhydrazine $(LiRN-SiR_2-NR-)_2$ with ClR_2SiSiR_2Cl (R = Me).[126]

The isomeric cyclo-3,4,7,8-tetrasila-1,2,5,6-tetrazane (10), R = Me, was isolated from ClR_2SiSiR_2Cl and HRN—NHR (R = Me).[131]

8.4   Silicon–nitrogen heterocycles with an Si–Si bond

8.4.1 Cyclotrisiladiazanes

The five-membered ring system Si_2N_2Si (11) has been prepared from lithiated bis(alkylamino)disilanes $(LiR''N)R_2Si-SiR_2(NLiR'')$ with organodihalosilanes R'_2SiX_2 (X = F, Cl, Br).[133,134,135] Several derivatives with R = Me, Pr, R' = F, Cl, Br, Me and X = F, Cl, Br were isolated.

The chloro derivative (11), R' = Cl, is characterized by unusual stability in aqueous solution; however, aminolysis occurs quickly. Several spirocyclic compounds were prepared from Si_3N_2 ring derivatives:

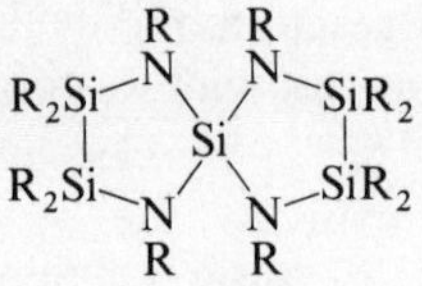

(70)

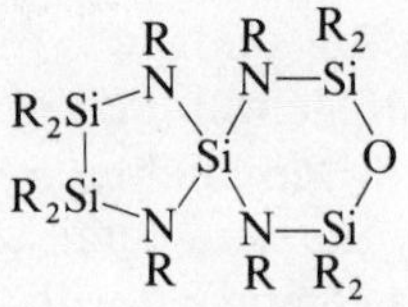

(71)

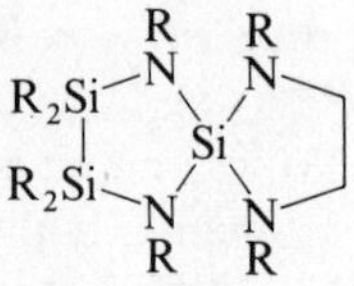

(72)

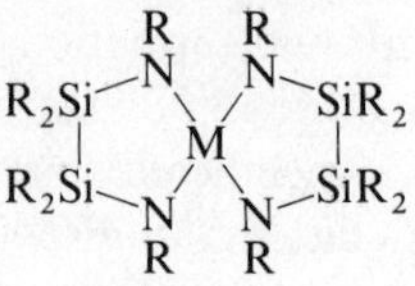

(73)

Metallation of bis(methylamino)-substituted cyclotrisiladiazane (11), R = R″ = Me, R′ = NHMe with butyllithium, followed by treatment with ClMe$_2$SiSiMe$_2$Cl, gave a spirobicyclic compound (70), R = Me.[57,136] Treatment of the dilithium compound of (11), R = R″ = Me, R′ = NHMe, with (ClMe$_2$Si)$_2$O gave another spirane (71), R = Me.[57] The same spirocyclic compound (71) was obtained by two other different routes.[58]

Other related spirocyclic compounds, containing an Si–Si bond[57,58] have already been cited above, (45)–(47).

A method for the synthesis of silicon–nitrogen–carbon spiro systems is the reaction of 2,2-diamino-1,3-diaza-2-silacyclopentane with BuLi and then with ClMe$_2$SiSiMe$_2$Cl to form (72), R = Me.[56]

The main features of the IR and Raman spectra of five-membered ring systems (11) were compared with those of the heteroatom analogues (19) and spirobicyclic derivatives (70) and (73) [(R = Me, E = GeCl$_2$, PMe, AsCl, SbCl, SO in (19) and M = Ge, Sn, Ti, Zr in (73)];[137] the variation of E and M produces a shift of nearly all skeletal vibrations, in comparison with the silicon rings (11) and (70).

Bicyclic compounds, namely bicyclo[3.3.0]tetrasila-1,3,5,7-tetrazanes (57), were prepared by reacting two moles of lithiated bis(alkylamino)silanes R′R″Si(NR‴Li)$_2$ with Cl$_2$RSi—SiRCl$_2$ (R = Me, R′ and R″ = Me or Ph, R‴ = Me, Et, CH$_2$Ph).[138] The structure of (57), R = R′ = R″ = Me, R‴ = CH$_2$Ph, has been determined by X-ray diffraction.[138]

8.4.2 Cyclotetrasilatriazanes

Cyclo-1,2,4,6-tetrasila-3,5,7-triazane (12) is a seven-membered ring containing an Si–Si bond. The synthesis of the permethylated derivative (12),

R = Me, is achieved by condensation of dilithiated 1,3-bis(methylamino)-disilazane $RN(SiR_2NRLi)_2$ with ClR_2SiSiR_2Cl.[139] This is the only representative of the class known so far.

8.4.3 Cyclopentasilatetrazanes

The nine-membered cyclo-1,2,4,6,8-pentasila-3,5,7,9-tetrazane **(13)**, R = Me, R' = Pr^n, was obtained from ClR_2SiSiR_2Cl and dilithiated 1,5-bis(propylamino)octamethyltrisiladiazane $R_2Si(NRSiR_2NR'Li)_2$.[139] No other derivative of this heterocycle is known.

8.5 Heterocyclosilazanes containing Group IV elements

Heterocyclosilazanes are inorganic ring compounds that result by replacing silicon in cyclosilazanes with other heteroatoms. The main types known have been cited above and are illustrated by **(14)–(20)**. Boron-containing heterocyclosilazanes are treated in Section 2.4.6.

8.5.1 Germanium-containing heterocyclosilazanes

This class includes monocyclic systems, corresponding to the types **(15)**, **(16)** and **(19)**, E = Ge, and several spirocyclic systems. An overall view of the germanium-containing heterocyclosilazanes is given by the structures **(74)–(84)**:

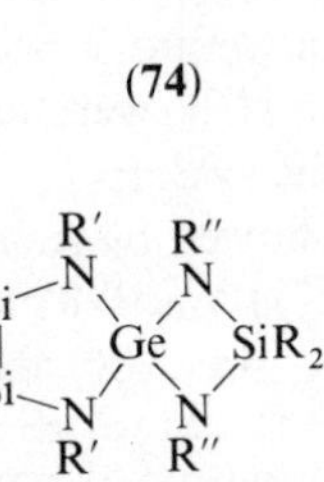

(74) 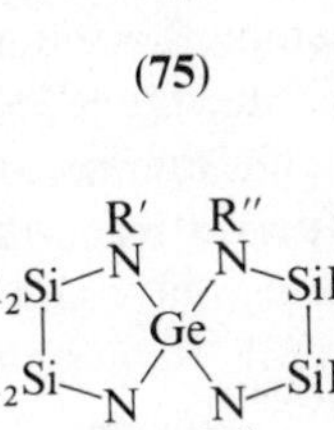   (75) 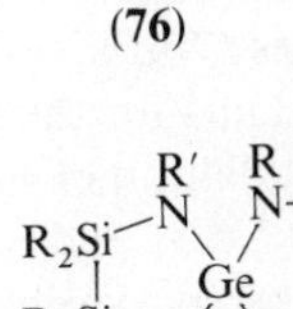   (76)

(77)   (78)   (79)

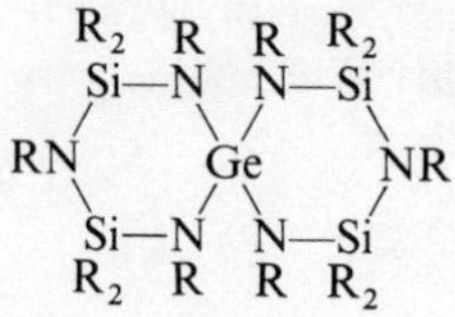

$$
\begin{array}{ccc}
\text{(80)} & \text{(81)} & \text{(82)}
\end{array}
$$

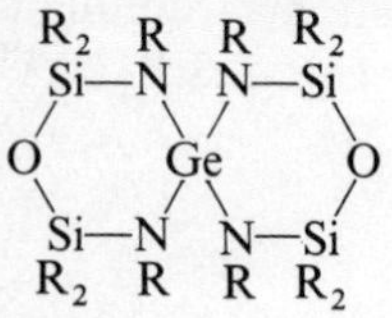

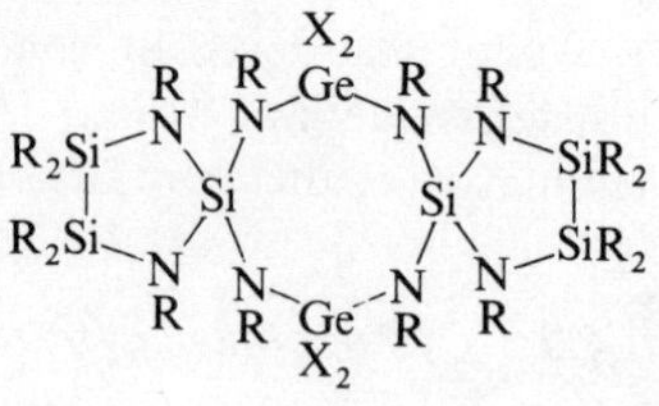

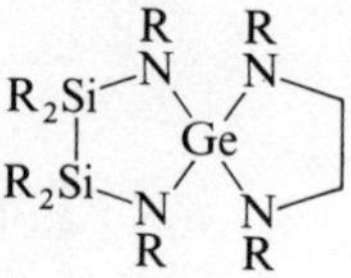

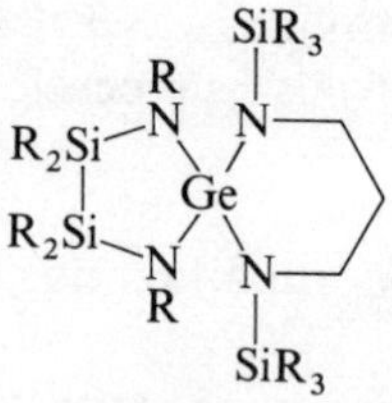

$$
\begin{array}{cc}
\text{(83)} & \text{(84)}
\end{array}
$$

Six-membered cyclo-1-germa-3,5-disila-2,4,6-triazanes (74), R = Me, R' = Cl, Bu, have been obtained by the reaction of diaminodisilazanes $RN(SiR_2NHR)_2$ (after metallation or in the presence of NEt_3) with $GeCl_4$.[140]

An eight-membered ring, cyclo-1-germa-3,5,7-trisila-2,4,6,8-tetrazane (75), R = Me, R' = Pr, has been similarly prepared from $R_2Si(NR-SiR_2NHR')_2$ and $GeCl_4$.[43b]

Five-membered-ring cyclo-1-germa-3,4-disila-2,5-diazanes (76), R = Me, R' = Me, Et, Pr, R'' = Cl, were obtained from $R'NHSiR_2SiR_2NHR'$ and $GeCl_4$ after preliminary metallation with BuLi or in the presence of NEt_3.[133,135,141] With a molar ratio of 2 : 1 the spirocyclic compounds (78) can be obtained.[136,141] The vibrational spectra of the permethylated compounds (76) and (78) have been studied.[137] Substitution reactions with the Si_2N_2Ge ring (76), R'' = Cl, were carried out at the germanium site with alcohols, amines and organometallic compounds.[141] The amino-substituted derivatives (76), R = Me, R' = Me, Et, R'' = NHMe, NHEt, were used as building units for the synthesis of a number of spirocyclic systems.[57,136,142] Thus metallation of the amino derivatives with BuLi, followed by treatment with Me_2SiCl_2 and $(ClMe_2Si-)_2$, gave the spiranes (77) and (78) respectively. The spirocyclic system (78), R = R' = R'' = Me, was also obtained by reacting $GeCl_4$ with $(LiMeNSiMe_2-)_2$.[143] The reaction of (76) R = Me, R' = Me, Et, R'' = NMeLi, with the dichlorodisiloxane $(ClMe_2Si)_2O$ yields the spirobicyclic system (79).[144]

Symmetrical six–six and five–eight–five spirocyclic compounds of germanium (80)–(82), R = Me, were obtained in reactions of $GeCl_4$ with dilithiated derivatives of $MeN(SiMe_2NHMe)_2$, $O(SiMe_2NHMe)_2$ and the five-membered-ring compound (76) R = R' = Me, R'' = NHMe, respectively.[144] Spirocyclic compounds containing a GeN_2C_2 ring (83) and (84) were obtained from (76), R = Me, R' = Me, Et R'' = Cl, and lithiated N,N'-dimethylethylenediamine or N,N'-bis(trimethylsilyl)propylenediamine.[143]

8.5.2. Tin-containing heterocyclosilazanes

The known Si–N–Sn ring systems are as follows:

(85)   (86)   (87)

(88)   (89)   (90)

(91)   (92)

Reactions of 1,3-bis(amino)disilazanes $RN(SiR_2NHR)_2$, after metallation or in the presence of NEt_3, with chloroderivatives of tin(IV), $R_2'SnCl_2$, led to novel six-membered Si_2SnN_3 ring systems (85), R = Me, R' = Cl, Bu, cyclo-1,3-disila-5-stanna-2,4,6-triazanes.[140]

An eight-membered $Si_2Sn_2N_4$ ring, cyclo-1,5-disila-3,7-distanna-2,4,6,8-tetrazane (86), R = Me, R′ = SO_2CF_3, has been formed in the reaction of $CF_3SO_2N(SnMe_3)_2$ with Me_2SiCl_2, by elimination of Me_3SnCl and $SnMe_4$.[145]

Bis(alkylamino)disilanes $(R′NHSiR_2-)_2$ (R = Me, R′ = Me, Et) react with R''_2SnCl_2 (R = Bu, Cl) to give five-membered cyclo-4,5-disila-2-stanna-1,3-diazanes (87), R = Me, R′ = Me, Et, R″ = Bu, Cl.[135,146] The symmetrical spirocyclic compound (88), R = R′ = Me, is a by-product in the reaction with $SnCl_4$.[146]

The most general method for preparing spirosilazanes (88)–(90) with tin in the centre is the reaction of dilithiated bis(amino)disilanes $(LiR′NSiR_2-)_2$ (R = Me, R′ = Me, Et),[57,136,146] diaminosilanes $R_2Si(NR′Li)_2$ (R = Me, R′ = Bu^t)[147] and diaminodisilazanes $R′N(SiR_2NR′H)_2$ (R = R′ = Me)[146] respectively with $SnCl_4$ in a molar ratio of 2 : 1.

A method for preparing unsymmetrical spirocyclic tin compounds (91) and (92) has been found in the reaction of the five-membered ring derivatives (87), R = Me, R′ = Et, R″ = Cl, with lithiated amino compounds $O(SiR_2NRLi)_2$ and $N,N′$-bis(trimethylsilyl)propylene-diamine respectively.[146]

8.5.3 Lead-containing heterocyclosilazanes

The only known derivative containing lead is a four membered ring SiN_2Pb, cyclo-2-plumba-4-sila-1,3-diazane (93), prepared by reacting the dilithiated compound $Me_2Si[NLiC(CF_3)_2Me]_2$ with Ph_2PbCl_2:[5]

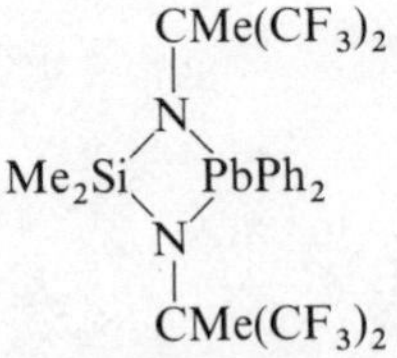

$$CMe(CF_3)_2$$
$$|$$
$$N$$
$$Me_2Si\diagup\diagdown PbPh_2$$
$$N$$
$$|$$
$$CMe(CF_3)_2$$

(93)

The scarcity of Si–N–Pb heterocycles is not surprising in view of the low stability of Pb–N bonds.

8.6 Heterocyclosilazanes containing Group V elements

Similar methods to those cited in the previous sections have been employed to some extent for generation of a variety of heterocycles containing phosphorus, arsenic or antimony in addition to silicon and nitrogen.

8.6.1 Phosphorus-containing heterocyclosilazanes

The known Si–N–P heterocycles are illustrated by (94)–(104). They represent the general types (14), (15), (16), (18) and (19), E = P. An additional source of diversity is the ability of phosphorus to occur in various oxidation and coordination states.

(94)   (95)   (96)   (97)

(98)   (99)   (100)

(101)   (102)   (103)   (104)

Several four-membered cyclo-2-phospha-4-sila-1,3-diazanes (94) have been prepared. Phosphorus halides undergo substitution reactions with metallated silylamides. Thus the bulky dilithiated di(t-butylamino)silane $R_2Si(NR'Li)_2$ reacted with Cl_2P—$NBu^t(SiMe_3)$ to give (94), R = Me, R′ = Bu^t, R″ = NBu$^t(SiMe_3)$.[148] The rotamers that result from hindered rotation about the P—N bond of the ring were isolated and characterized. Similar rotamers were obtained in reactions of chloro-substituted rings (94), R″ = Cl, with LiNR′SiMe$_3$ (R′ = Bu^t, SiMe$_3$).[149]

Lithiated aminofluorosilanes FR_2Si—NBu^tLi react with PCl_3 by elimination of LiCl and substitution. Depending on the bulkiness of the substituents and the reaction conditions, fluorosilanes are also cleaved and four-membered rings or aminoiminophosphanes are formed.[150] The four-membered ring derivative (94), $R = Me$, $R' = Bu^t$, $R'' = Cl$, was thus obtained:

$$2FR_2Si—NBu^tLi + PCl_3 \xrightarrow[-LiCl]{} ClP(NBu^t—SiR_2F)_2 \xrightarrow[-R_2SiF_2]{}$$
$$\xrightarrow[-R_2SiF_2]{} ClP(NBu^t)_2SiR_2 \qquad (11)$$

Aminoiminophosphanes react with halogenosilanes to form SiN_2P rings (94), $R_2 = Cl_2$, Br_2, ClMe, ClEt, ClPh, Me_2, $R' = SiMe_3$ or Bu^t and $R'' = X = Cl$, Br:[151-153]

$$(Me_3Si)_2N—P{=}N—R' \xrightarrow[-Me_3SiX]{+ R'R''SiX_2} R_2Si(NR')(NSiMe_3)PX \qquad (12)$$

A unique transformation was observed in the reaction of aminoiminothio-λ^5-phosphorane with $SnCl_4$, which gave an SiN_2P ring (94) coordinated to $SnCl_2$ ($R = Me$, $R' = Bu^t$):[154]

$$(13)$$

Trapping an aminoiminophosphorane R_3Si—$N{=}P$—$N(SiR_3)_2$ with a lithiated fluorosilylamine FR_2Si—$NR'Li$ led to the elimination of LiF, and a [2 + 2] cycloaddition product (94), $R = Me$, $R' = SiMe_3$ and mesityl, $R'' = N(SiR_3)_2$, was formed, presumably through the intermediacy of a silaimine $R_2Si{=}NR'$.[27]

Treatment of a bis(amino)-substituted cyclophosphazene with $MeSiHCl_2$ led to the formation of a spirocyclic system incorporating an SiN_2P ring ($R = CH_2CF_3$):[155]

$$(14)$$

The first six-membered Si_2PN_3 ring derivatives, cyclo-1,3-disila-5-phospha-2,4,6-triazanes (95), were obtained by the reactions of 1,3-bis(alkyl-amino)disilazanes $R'''N(SiR_2NR'H)_2$ with dichlorophosphines $R''PCl_2$, after metallation or in the presence of NEt_3 (R = R''' = Me, R' = Me and Et, R'' = Me, Et, Ph).[156,157]

Derivatives of $Si_2N_3P^V$ (96) and $[Si_2N_3P^V]^+$ (97) rings were formed in oxidation reaction of $Si_2N_3P^{III}$ (95). Thus reactions of (95) with elemental sulphur and Me_3SiN_3 resulted in oxidative addition of S and N—$SiMe_3$ groups to the phosphorus atom, with formation of (96), X = S, N—$SiMe_3$, while the reaction with CoI_2 gave a coordination compound with the ring (95) as ligand; the reactions with MeI and CS_2 gave 1:1 adducts with formation of the corresponding phosphonium compounds (97), R = R' = R''' = Me, R'' = Me, Ph.[157]

Eight-membered Si_3N_4P ring systems, cyclo-1-phospha-3,5,7-trisila-2,4,6,8-tetrazanes (98), were prepared by treatment of dilithiated Me_2Si-$(NMeSiMe_2NHR')_2$ with $R''PCl_2$ (R = Me, R' = Et, R'' = Et, Ph).[43a,b]

A $Si_3N_4P^V$ ring (99), R = Me, R' = H and $SiMe_2Bu$, R'' = Me, X = O, was obtained along with a bicyclic derivative in the reaction of a lithiated cyclotetrasilazane with $MeP(O)Cl_2$ in boiling toluene.[107]

The $Si_2N_4P_2$ ring present in cyclo-1,3-diphospha-5,7-disila-2,4,6,8-tetrazane (100), R = R' = Me, was formed from $RN(SiR_2NRH)_2$ with $RN(PR'Cl)_2$ in the presence of NEt_3.[157]

Five-membered cyclo-2-phospha-4,5-disila-1,3-diazanes (101) were prepared by reacting bis(alkylamino)disilanes, $(R'NHSiR_2-)_2$, with organo-phosphorus(III) halides, after metallation or in the presence of NEt_3 (R = Me, R' = Me, Et, Pr, R'' = Me, Bu^t).[133,158,159,160]

Cyclophosphadisiladiazanes (101), R = R' = Me, R'' = Bu^t or OMe, are also formed in the reaction of $R''P(NMeLi)_2$ with ClR_2Si—SiR_2Cl.[160]

The $Si_2N_2P^{III}$ ring compounds (101) were oxidized with O, S, Se, Me_3SiN_3 and $Ni(CO)_4$ to give 50–95% yields of the $Si_2N_2P^V$ ring derivatives (102), R = R'' = Me, R' = Me, Et or Pr, X = O, S, Se, $NSiMe_3$ and $Ni(CO)_3$.[135,161] Also, (101) was air-oxidized to give 75% yields of (102) R = R'' = Me, R' = Me, Et, Pr, X = O.[161]

The oxidation of (101) with methyl iodide and CS_2 gave phosphonium compounds (103), with R''' = Me, X = I^- and R''' = $C(S)S^-$.[159]

The formation of a cyclo-2-phospha(v)-4,5-disila-1,3-diazane (104), R = R' = Me, R'' = Bu^t, occurred in the reaction of $ClBu^tP(NMeSiMe_3)$ ($=NSiMe_3$) with $(ClSiMe_2)_2$, in which the Si–N bonds of the silylamino-iminophosphane were cleaved with cyclization.[160]

8.6.2 Arsenic- and antimony-containing heterocyclosilazanes

The known heterocycles of this class are as follows:

$$
\begin{array}{ccc}
\underset{\substack{R''N \diagdown \diagup NR'' \\ \underset{\underset{R'''}{|}}{As}}}{\overset{\overset{RR'}{Si}}{}} &
\underset{\substack{R_2Si \diagdown \diagup SiR_2 \\ \underset{\underset{R}{|}}{N}}}{\overset{\overset{R'}{As} \\ RN \diagdown \diagup NR}{}} &
\underset{\substack{R_2Si \diagdown \diagup SiR_2 \\ \underset{\underset{R}{|}}{N}}}{\overset{\overset{R'}{Sb} \\ RN \diagdown \diagup NR}{}} \\
(105) & (106) & (107)
\end{array}
$$

$$
\begin{array}{cc}
\underset{\substack{R'N \diagdown \diagup NR' \\ \underset{\underset{R''}{|}}{As}}}{\overset{R_2Si - SiR_2}{}} &
\underset{\substack{R'N \diagdown \diagup NR' \\ \underset{\underset{R''}{|}}{Sb}}}{\overset{R_2Si - SiR_2}{}} \\
(108) & (109)
\end{array}
$$

Cyclo-1-arsa-3-sila-2,4-diazanes (**105**) were prepared by treating a lith-iated aminofluorosilane $FPhBu^tSi$—NBu^tLi with AsF_3. Elimination of LiF and a difluorosilane, $PhBu^tSiF_2$, gives the four-membered SiN_2As ring (**105**), R = Ph, R″ = Bu^t, R‴ = F.[150]

The six-membered ring systems Si_2AsN_3 of cyclo-1-arsa-3,5-disila-2,4,6-triazanes (**106**) and Si_2SbN_3 of cyclo-1-stiba-3,5-disila-2,4,6-triazanes (**107**) were formed in the reaction of 1,3-bis(methylamino) pentamethyldisilazane, $RN(SiR_2NHR)_2$, after metallation or in the presence of NEt_3, with $AsCl_3$ and $SbCl_3$ respectively (R = Me, R′ = Cl).[140] The fully methylated deriva-tive (**106**), R = R′ = Me, was obtained by treatment with MeMgI.[140]

Five-membered Si_2N_2E rings (E = As, Sb) of cyclo-1-arsa-3,4-disila-2,5-diazanes (**108**) and cyclo-1-stiba-3,4-disila-2,5-diazanes (**109**) can be pre-pared in an analogous way to the Si_2N_2P ring system.[133,135,162] Thus 1,2-bis(alkylamino)disilanes $(R'NHSiR_2-)_2$ react with $AsCl_3$ and $SbCl_3$ respect-ively to give the arsenic (**108**) and antimony (**109**) ring systems (R = Me, R′ = Me, Et, Pr, R″ = Cl). The AsCl units of (**108**) can be substituted with $MeOH$, Me_2NH and MeMgI, leading to (**108**), R = Me, R′ = Me, Et, Pr, R″ = OMe, NMe_2, Me.[162.]

The condensation of $(Me_2SiNH)_4$ with $SbCl_3$ in refluxing toluene or at 300 °C in the absence of a solvent afforded several liquid and crystalline oligomeric products, some of which contained antimony, in addition to silicon and nitrogen. A tricyclic structure, consisting of fused Si_2SbN_3 rings, was speculatively suggested (but not proved) for one of the products.[163]

8.7   Cyclosilazoxanes

Cyclosilazoxanes are heterocycles formed by partial replacement of nitrogen of the cyclosilazane rings by oxygen.[B 1] Rings with four to twelve members belong to this class, and are illustrated by (21)–(35) (p. 223).

8.7.1   Cyclodisilazaoxanes

Two reports of four-membered Si_2NO ring systems have appeared.[31,164] Thus an NMR analysis of pentamethyldisiladioxazine showed that it exists in equilibrium with its tautomeric four-membered ring (21a), which predominates in polar solvents at low temperatures, but rearranges to the six-membered ring completely at 100 °C in benzonitrile (R = Me):[164]

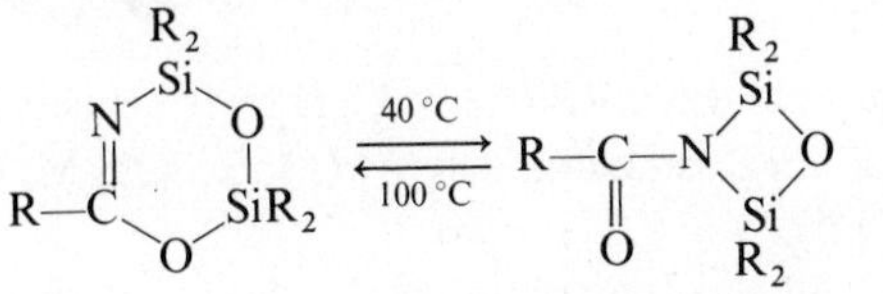

$$\text{(15)}$$

(21a)

A bis(isopropyltrimethylsilyl)amino-substituted Si_2NO ring (21b), R = Me, R′ = NPr(SiMe₃), R″ = Bu^t, has been prepared by cyclization of a 1-fluoro-3-aminodisiloxane with BuLi:[31]

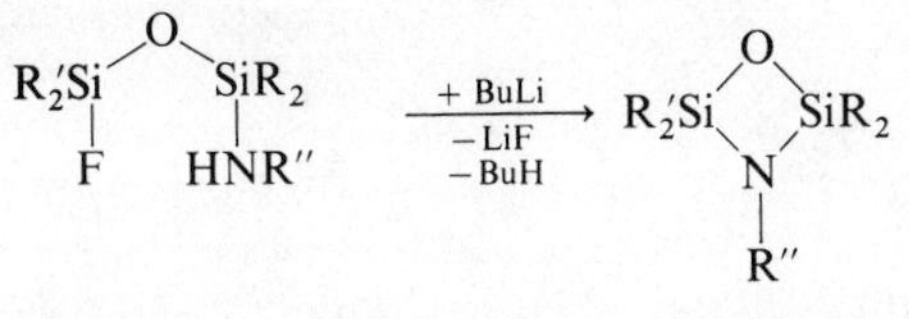

$$\text{(16)}$$

(21b)

8.7.2   Cyclotrisiladiazoxanes

Different methods have been used for the preparation of the six-membered Si_3N_2O ring system.

Treatment of 1,3-bis(methylamino)disiloxanes $O(SiMe_2NHMe)_2$ with dihalosilanes $RR'SiX_2$, after metallation or in the presence of NEt_3, leads to ring closure, with formation of six-membered rings (22), R = R‴ = Me, R′ = Me, R″ = Vi; R′ = R″ = Cl or Br.[165]

The use of a dipolar solvent (DMF) was necessary for the formation of a cyclic compound (22), R = R' = R'' = Me, R''' = Ph, in the reaction of aniline with $O(SiMe_2Cl)_2$; otherwise, linear oligomers were the only products.[166]

Cis-trans mixtures of cyclotrisiladiazoxane (22), R_2 = MePh, R' = Me, R'' = Cl, R''' = H, were obtained by cyclization of $PhMe[(H_2N)Si]_2O$ with Me_2SiCl_2 or $SiCl_4$ in triethylamine.[59]

The reaction of $(Me_2SiNH)_3$ with Me_2SiCl_2 in wet dioxane led to the formation of (22) and (23), R = R' = R'' = Me, R''' = H:[167]

$$(Me_2SiNH)_3 + Me_2SiCl_2 \xrightarrow[\text{dioxane}]{H_2O} R_6Si_3O_2(NH) + R_6Si_3O(NH)_2 \qquad (17)$$

$$(22) \qquad\qquad (23)$$

A single-step reaction of $MeRSiCl_2$ in C_6H_6 with excess NH_3 containing 0.5 mol H_2O gave cyclosilazoxanes[168] $(MeRSiO)_n(MeRSiNH)_{m-n}$ (m, n = 1–3, R = Me, Ph).

The reaction of 2,2-bis(4-hydroxyphenyl)propane with hexamethylcyclotrisiladiazoxane gave a yellow glassy polymer containing a Si_3N_2O ring in the repeating unit.[169]

Spirocyclic compounds (110)–(112) containing silazoxane rings have also been reported:

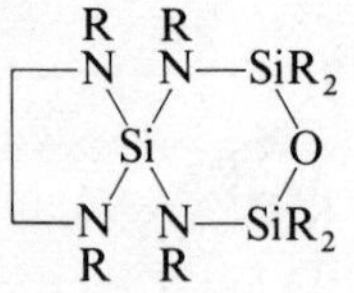

(110)

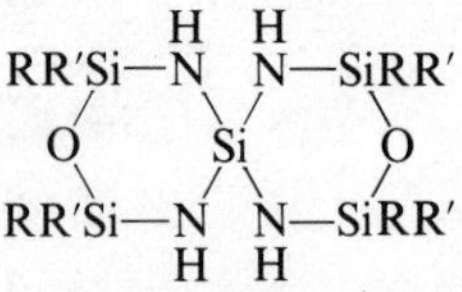

(111)

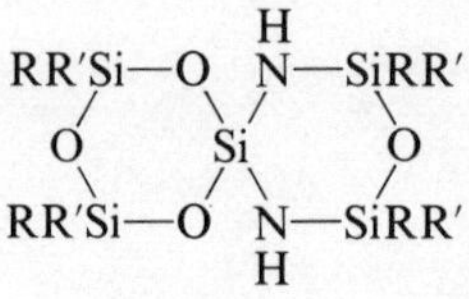

(112)

The spirocyclic compound (110), R = Me, was obtained in the reaction of dilithiated bis(methylamino)diazasilacyclopentane with $(ClSiMe_2)_2O$.[56b] The spirosilazoxane (111), R = Me, R' = Ph, was formed as an isomeric mixture by cyclization of the dilithium derivative $[MePh(NHLi)Si]_2O$ with $SiCl_4$.[59] The isomers of (112), R = Me, R' = Ph, were obtained similarly.[59]

8.7.3 Cyclotrisilazadioxanes

Cyclotrisilazadioxanes (23) can be prepared by treatment of silanediols $R_2Si(OH)_2$ with 1,3-bis(chlorosilyl)amines $RN(SiMe_2Cl)_2$ to form (23), R = Ph, R' = Me, R'' = H, Et.[170]

The reaction of $C_6F_5NLi_2$,[171] $PhNH_2$[166] or $(Me_3Si)_2NH$[172] with $Me_2Si(OSiMe_2Cl)_2$ was used to prepare the six-membered rings (23), R = R' = Me, R'' = H,[172] Ph[166] and C_6F_5.[171]

Hydrolysis of aminosilyl- and chlorosilyl-substituted cyclodisilazanes results in different products. If the disilazane $[Me_2SiN(SiMe_2Cl)]_2$ (1), R = Me, R' = Me, R'' = $SiMe_2Cl$, is added to an ether solution containing 2 mol H_2O and 1 mol NEt_3, the six-membered ring (23), R = R' = Me, R'' = $SiMe_2OH$, is obtained in 89% yield, and the expected four-membered-ring hydrolysis product (1), R = R' = Me, R'' = $SiMe_2OH$, in 10% yield; if the order of addition is reversed, a 98% yield of the four-membered ring is obtained.[173]

The mass spectra of the six-membered-ring derivatives (23) have been reported and analysed.[174,176] The crystal structure of 2,2,4,4,6,6-hexamethyl-5-hydroxydimethylsilylcyclotrisilazadioxane (23), R = R' = Me, R'' = $SiMe_2OH$, was demonstrated by X-ray diffraction.[175]

Some cyclotrisilazadioxanes (23), $R_2 = R_2' = Ph(NHR'')$, R'' = Me, Et, Pr, Cy, were obtained by heating the trisiloxane derivative $(R''NH)_2SiPhO-SiPh(NHR'')OSiPh(NHR'')_2$.[177]

The compound (23), $R_2 = R_2' = Ph(NHPh)$, was prepared by anilinolysis of the corresponding linear 1,3,5-triphenylpentachlorotrisiloxane.[177]

The cyclization of $ClSiMe_2(OSiMe_2)_nOSiMe_2N(SnMe_3)_2$ (n = 0–2) gave Me_3SnCl and the six-membered (23), R = R' = Me, R'' = $SnMe_3$, and eight-membered (25) and (27), R = Me, R' = $SnMe_3$.[178]

Bicyclic compounds (23), R = R' = Me, R'' = $P_3N_3F_5$ and $P_4N_4F_7$, were obtained from the reaction of linear $Cl(SiMe_2O)_nSiMe_2N(SnMe_3)_2$ (n = 1–3) with fluorophosphazenes $(F_2PN)_{3,4}$.[178]

8.7.4 Cyclotetrasilazoxanes

Cyclotetrasilazoxanes comprise four eight-membered rings (24)–(27), two of which are isomeric.[B 1]

With metallated 1,5-bis(alkylamino)trisiladioxanes, $R_2Si(OSiR_2NHR')_2$, organodichlorosilanes $R''R'''SiCl_2$ did not give the expected asymmetric, but rather the symmetric cyclotetrasiladiazadioxanes of type (25) (one SiR_2 replaced by $SiR''R'''$), i.e. $Me_6R''R'''Si_4N_2R_2'O_2$, with R' = Me, Et, R'' = Me, R''' = Me or Vi.[179]

The preparation of the symmetrical decamethylcyclotetrasiladiazadioxane

(25), R = R' = Me, occurred through several new routes, including thermal condensation of $O(SiR_2NHR)_2$ and metallation of $R_2Si(OSiR_2NHR)_2$ with BuLi, followed by treatment with R_2SiCl_2.[47] The permethylated cyclotetrasilatriazaoxane (24), R = R' = Me, could be isolated from dilithiated $O(SiMe_2NHMe)_2$ and $MeN(SiMe_2Cl)_2$.[47]

Hydrolysis is another frequently used method for the preparation of cyclosilazoxanes. Thus hydrolysis of the cyclodisilazane $Me_4Si_2N_2H$-$(SiMe_3)$ gave a bis(trimethylsilyl)-substituted eight-membered ring (25), R = Me, R' = $SiMe_3$.[113]

Cyclotetrasiladiazadioxanes are also formed by the homocondensation of aminosiloxanes. Heating $[(RHN)_2SiPh]_2O$ to 300 °C gave RNH_2 and the eight-membered-ring compound (25), R_2 = Ph(NHR'), R' = Me, Et, Pr.[177]

Trimethylstannyl-substituted eight-membered cyclosilazoxanes were formed by heating $Cl(SiMe_2)_nN(SnMe_3)_2$ (n = 1–3).[178]

Dilithiated pentafluoroaniline $C_6F_5NLi_2$ reacts with $O(SiMe_2Cl)_2$ to give an $Si_4O_2N_2$ ring derivative (25), R = Me, R' = C_6F_5.[171]

The structure of 2,4,6,8-tetramethyl-2,4,6,8-tetraphenylcyclotetrasiladiazadioxane (25), R_2 = MePh, R' = H, was determined by X-ray diffraction. The methyl groups in positions 2 and 4 are *trans* to those in positions 6 and 8. The eight-membered silazoxane ring has a chair conformation.

Hydrolysis of the substituents in cyclodisilazane derivatives occurs primarily with retention of the ring system when an HCl acceptor is present,[181,182] but 2–3% of a cyclotetrasila-1,3-diaza-5,7-dioxane (26), R = Ph, R' = H, was also obtained in the hydrolysis of $R_4Si_2N_2(SiR_2Cl)_2$ (1), R = R' = Ph, R'' = $SiPh_2Cl$.

Bis(hydroxydimethylsilyl)tetramethylcyclodisilazane (1), R = R' = Me, R'' = $SiMe_2OH$, undergoes a rearrangement with ring expansion to form an eight-membered cyclosilazoxane (25), R = Me, R' = H:[63]

$$\tag{18}$$

Treatment of $N(SiMe_2Cl)_3$ with $O(SiMe_2OH)_2$ gave a cyclotetrasilazatrioxane (27), R = Me, R' = $SiMe_2Cl$,[63] in which the remaining Cl can be substituted by reactions with ammonia and water.

The reaction of monomeric silaimines $R_2Si{=}NR$, generated by the gashase pyrolysis of the corresponding silylazides, with hexamethylcyclotrisitrisiloxane $(Me_2SiO)_3$ is an interesting method of preparing cyclotetrasilazatrioxanes (27):[183]

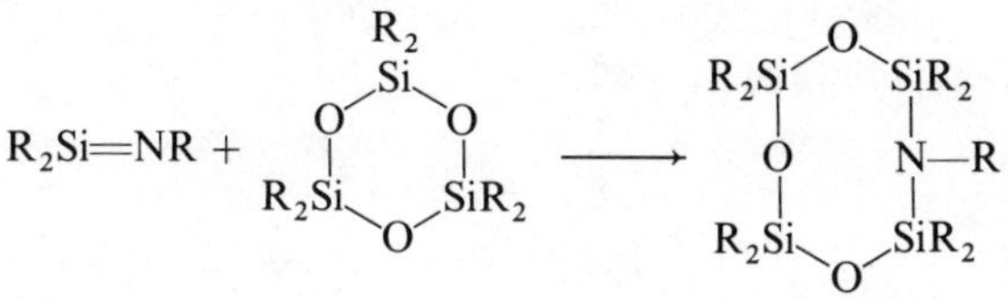

$$R_2Si{=}NR + \begin{array}{c} R_2 \\ Si \\ O \quad\quad O \\ R_2Si \quad\quad SiR_2 \\ O \end{array} \longrightarrow \begin{array}{c} R_2Si^{\diagup O}{}^{\diagdown}SiR_2 \\ O \qquad N{-}R \\ R_2Si^{\diagdown}{}_{O}{}^{\diagup}SiR_2 \end{array} \qquad (19)$$

(27)

8.7.5 Cyclopentasiladiazatrioxanes

A cyclopentasila-1,3-diaza-5,7,9-trioxane (28), R = Me, has been obtained in the rearrangement of an aminodimethylsilyl-substituted cyclotetrasilazatrioxane (27), R = Me, R' = $SiMe_2NH_2$, in the presence of tetramethylammonium silanolate.[63]

8.7.6 Cyclohexasiladiazatetroxanes

The only known member of this class has been obtained by reacting dilithiated pentafluoroaniline $C_6F_5NLi_2$ with $Me_2Si(OSiMe_2Cl)_2$; the product is dodecamethyl-N,N'-bis(pentafluorophenyl)cyclohexasiladiazatetroxane (29), R = Me, R' = C_6F_5.[171]

8.7.7 Rearrangements of cyclosilazoxanes to rings of other sizes

Cyclosilazoxanes having NH groups can be lithiated with butyllithium. Some of the lithium derivatives rearrange with a contraction of the ring and formation of one or more SiOLi side-chains. The derivatives, normal and rearranged, react with $ClSiMe_3$ to give the corresponding trimethylsilyl compounds.[184,185] A reduction in ring size is preferred at higher temperatures, and preservation of the ring size at lower temperatures. Some of these transformations are illustrated here (p. 262) (the methyl substituents at silicon are omitted). Thus, cyclotetrasila-1,5-diaza-3,7-dioxane (25), R = Me, R' = H) reacts with BuLi to give the monolithium derivative, and with two equivalents of BuLi to give the dilithium compound. The 1,3-bis(trimethylsilyl) derivative was obtained in the reaction with $ClSiMe_3$ only when the reaction was carried out at $-60\,°C$.[184] If the lithiated solution reached room temperature, the action of the chlorosilane produced a cyclodisilazane. A similar reduction in ring size occurred when a solution of the monolithiated derivative was allowed to reach room tem-

(a = ClSiMe₃)

perature, and a six-membered ring was the only product after trimethyl-silylation.

The ring-contraction reaction depends on ring geometry, and only substitution occurred in reactions of cyclotetrasilazadioxanes (23), R = R' = Me, R'' = H, and hexasila-1,7-diaza-3,5,9,11-tetraoxane (29), R = Me, R' = H, with BuLi and ClSiMe$_3$, following N-metallation and N—SiMe$_3$ coupling.[184]

An interesting substrate is cyclotrisiladiazoxane (22), R = R' = R'' = Me, R''' = H. Neither N-mono- nor dilithium derivatives undergo rearrangement by ring contraction, but the monolithium compound of the trimethylsilyl-substituted ring rearranges to a four-membered cyclodisilazane at room temperature (R = Me):[184]

$$
\begin{array}{c}
\underset{R_2Si\diagdown_{O}\diagup SiR_2}{\overset{\overset{\displaystyle R_2}{\underset{|}{Si}}}{HN\diagdown\diagup N—SiR_3}}
\end{array}
\xrightarrow[\text{(ii) ClSiR}_3]{\text{(i) BuLi}}
R_3Si—N\overset{\overset{\displaystyle R_2}{Si}}{\underset{\underset{\displaystyle R_2}{Si}}{}}N—SiR_2—O—SiR_3 \qquad (20)
$$

Infrared and Raman spectra were reported for the cyclic compounds [Me$_2$SiO$_{(n-x)/n}$(NH)$_{x/n}$]$_n$, with x = 0–n for n = 3 and 4. The changes in the frequencies corresponding to Si–O and Si–N bonds were particularly studied.[186]

8.7.8 Cyclosilazoxanes with N–N ring units

A bicyclic compound (113) consisting of two fused five-membered rings has been obtained by the reaction of 1,3-dichlorotetramethyldisiloxane O(SiMe$_2$Cl)$_2$ with hydrazine (R = Me):[125]

$$
\begin{array}{c}
\underset{R_2R_2}{\overset{R_2R_2}{O\diagdown_{Si—N—Si}\diagup O}} \\
\underset{Si—N—Si}{} \\
\underset{R_2R_2}{}
\end{array}
$$

(113)

The same reaction carried out with excess of O(SiMe$_2$Cl)$_2$ led to the formation of a ten-membered cyclotetrasila-1,2,6,7-tetraza-4,9-dioxane (31), R = Me, R' = H.[125]

A seven-membered cyclotrisila-1,2-diaza-4,6-dioxane (32), R = R' = Me, could be isolated in the reaction of LiMeN—NMeLi with Me$_2$Si(OSiMe$_2$Cl)$_2$.[126]

8.7.9   Cyclosilazoxanes with Si–Si ring units

A seven-membered cyclotetrasila-1,4-diaza-6-oxane (33), R = R' = Me, and a nine-membered cyclopentasila-1,4-diaza-6,8-dioxane (34), R = R' = Me, were formed by condensation of α,ω-difunctional building units at high dilution, i.e. $O(SiR_2NR'Li)_2$ with ClR_2Si—SiR_2Cl.[139]

8.7.10   Cyclosilazoxanes with N—O ring units

Hydroxylamine hydrochloride reacts with t-butylaminofluorosilanes to give O-fluorosilylsubstituted hydroxylamines, FRR'Si—ONHR″, compounds that contain two functional groups N–H and Si–F. When treated with bases, for example BuLi, six-membered cyclosila-1,4-diaza-3,6-dioxanes (35) were obtained (with R = Me, Bu^t, R' = Me, Bu^t or $N(SiMe_3)_2$ and R″ = H, Me, $SiMe_3$).[187,188]

The crystal structure of $(Bu^tSiONH)_2$ has been determined; the ring is nonplanar, with a twist conformation.[187]

8.7.11   Uses

The polymerization of several cyclic silazoxane compounds was studied in the presence of 5% KOH at 120 °C. The eight- and ten-membered rings were considerably more reactive than the six-membered ones.[189]

The lithium-containing cyclosilazoxanes are useful as polymerization catalysts for silazoxanes. When the cyclic silazoxane contains two lithium atoms it can react with, for example, a silazoxane containing two Si-bonded halogen groups, to give a linear or tridimensional polymer.[190]

Heating a cyclic organo(vinyldisiloxanyl)silazoxane and a methylsiloxane, containing at least two Si-bonded hydrogen atoms per mole, with H_2PtCl_6 in isopropanol gave viscous liquid or gel polymers.[191]

The copolymerization of tetramethyldisiloxane with a cyclotrisila-1,3-diaza-5-oxane gave a viscous liquid polymer, having a molecular weight of c. 25000.[191]

8.8   Silicon–nitrogen–sulphur heterocycles

Sulphur is an interesting heteroatom in silicon–nitrogen chemistry, since it can replace either nitrogen (usually as sulphur(II)) or silicon (usually as sulphur(IV) and sulphur (VI), but also as sulphur(II) in some cases). The first type is illustrated by structures (36) and (37) (p. 224).

The rings in which silicon is replaced by sulphur are illustrated by (114)–(121). Of these, (116) and (117) are not treated here, as they are discussed in the chapter dealing with sulphur–nitrogen rings (Chapter 29).

(114)   (115)   (116)   (117)

(118)   (119)   (120)

8.8.1   Cyclodisilathia(VI)triazanes and related systems

Derivatives of a sulphur(VI) heterocycle $Si_2S^{VI}N_3$ with various substituents (114), R = Me, R' = Me, $SiMe_3$, R" = H, Me, Et, have been prepared by condensation of dichlorodisilazanes $R"N(SiR_2Cl)_2$ with N,N'-substituted sulphamides $SO_2(NHR')_2$.[191]

A spirocyclic ring system, related to the compounds cited in that it contains sulphur(VI), has been obtained as follows (R = Me):[192]

$$2\ SO_2[N(SiR_3)_2]_2 \xrightarrow[-R_3SiCl]{SiCl_4} \quad \xleftarrow[-LiCl]{SiCl_4} 2\ SO_2[NLi(SiR_3)_2]_2 \quad (21)$$

8.8.2   Cyclosilathia(II)azanes

Six-membered rings Si_2N_3S and SiS_2N_3 with sulphur in the oxidation state + 2 have been obtained. Treatment of methylamine with $S(NMeSiMe_2Cl)_2$ gave a cyclodisilathia(II)triazane (115), R = R' = Me, which can be regarded as a ring resulted by replacing silicon in a cyclotrisilazane (2) with sulphur(II).[193]

The reaction of hexamethylcyclotrisilazane $(Me_2SiNH)_3$ (2), R = R' = Me, R" = H, with H_2S passed in large excess over 12 h under reflux gave substitution products by elimination of nitrogen as NH_4SH and formation of heterocycles (36) and (37), R = Me, R' = H.[118]

A specific direct method for the synthesis of organocyclotrisilathia(II)-azanes (36) and (37), R = Me, R' = H, is based on simultaneous exchange reactions of hexamethyldisilathiane $(Me_3Si)_2S$ and hexamethyldisalazane $(Me_3Si)_2NH$ with dichlorodiorganosilanes:[194]

$$n\ (Me_3Si)_2S + (3 - n)\ (Me_3Si)_2NH + MePhSiCl_2 \underset{-Me_3SiCl}{\overset{NEt_3}{\rightleftharpoons}}$$

$$(MePhSiNH)_2(MePhSiS) + (MePhSiNH)(MePhSiS)_2 \qquad (22)$$

$$\underbrace{\hspace{3cm}}_{(37a)} \qquad \underbrace{\hspace{3cm}}_{(36a)}$$

8.8.3   Cyclosilthiazanes with Si–Si ring units

Five-membered rings, cyclodisilathiadiazanes (118) and (119), containing sulphur in oxidation states +4 and +6 respectively have been reported.

Cyclo-2,3-disila-5-thia(IV)-1,4-diazanes (118), R = Me, R' = Me,[133, 135,195] Et,[135,195] Pr,[135,195] Bu,[195] SiMe$_3$,[195] were obtained from 1,2-bis (alkylamino)disilanes $R'NHSiR_2SiR_2NHR'$ and $SOCl_2$. Another derivative, with R' = SiMe$_3$, has been obtained after dilithiation of $(Me_2Si=NHSiMe_3)_2$ and reaction with $SOCl_2$.

Cyclo-2,3-disila-5-thia(VI)-1,4-diazanes (119), R = Me, R' = Et or SiMe$_3$,[135,195] were prepared by a [2 + 3] cyclocondensation from $ClSiR_2SiR_2Cl$ and $SO_2(NHR')_2$ in the presence of triethylamine.[135,195]

A seven-membered cyclo-4,5-disila-2,7-dithia-1,3,6-triazane (120), R = R' = Me, has been obtained by the reaction of $S(NMeSiMe_3)_2$ with $ClMe_2SiSiMe_2Cl$, along with another cyclic compound $(Me_2Si—SiMe_2NMe)_2$.[192]

8.9   Heterocyclosilazoxanes

This family of heterocycles arises by partially replacing the silicon in cyclosilazoxanes with a heteroatom; the rings thus formed contain four different elements. The number of possible combinations of Si, N, O and a heteroelement is very large indeed, but only a few examples are known so far; these are illustrated by (122)–(135) and will be briefly discussed here:

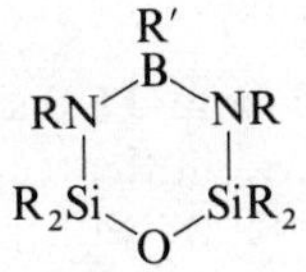

(122)

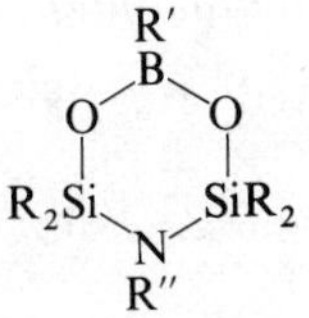

$$R'$$
$$B$$
$$RN \quad NR$$
$$R_2Si \quad SiR_2$$
$$O$$

(122)

(123)

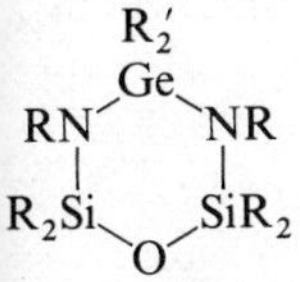

$$R'$$
$$B$$
$$O \quad O$$
$$R_2Si \quad SiR_2$$
$$N$$
$$R''$$

(123)

$$R'_2$$
$$Ge$$
$$RN \quad NR$$
$$R_2Si \quad SiR_2$$
$$O$$

(124)

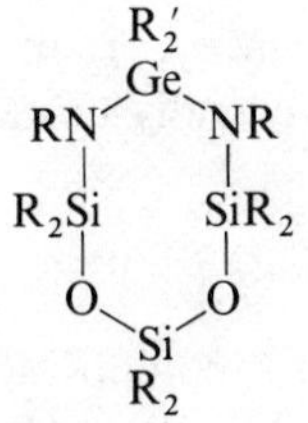

$$R'_2$$
$$Ge$$
$$RN \quad NR$$
$$R_2Si \quad SiR_2$$
$$O \quad O$$
$$Si$$
$$R_2$$

(125)

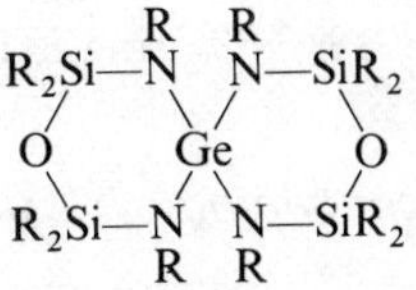

$$R \quad R$$
$$R_2Si—N \quad N—SiR_2$$
$$O \quad Ge \quad O$$
$$R_2Si—N \quad N—SiR_2$$
$$R \quad R$$

(126)

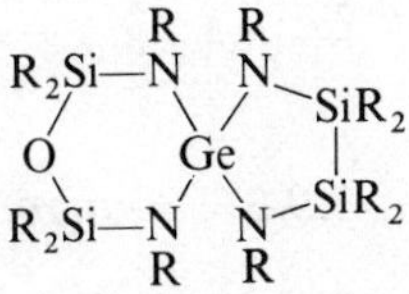

$$R \quad R$$
$$R_2Si—N \quad N—SiR_2$$
$$O \quad Ge \quad$$
$$R_2Si—N \quad N—SiR_2$$
$$R \quad R$$

(127)

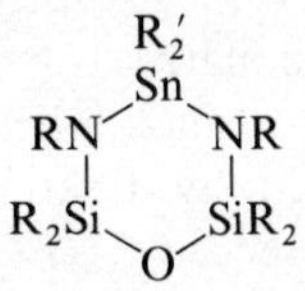

$$R'_2$$
$$Sn$$
$$RN \quad NR$$
$$R_2Si \quad SiR_2$$
$$O$$

(128)

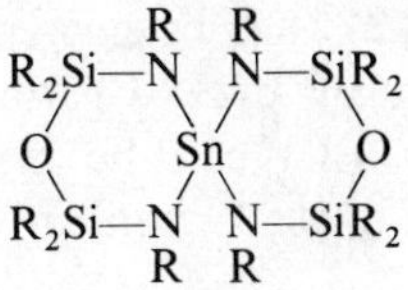

$$R \quad R$$
$$R_2Si—N \quad N—SiR_2$$
$$O \quad Sn \quad O$$
$$R_2Si—N \quad N—SiR_2$$
$$R \quad R$$

(129)

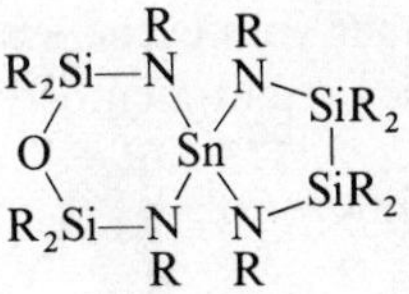

$$R \quad R$$
$$R_2Si—N \quad N—SiR_2$$
$$O \quad Sn \quad$$
$$R_2Si—N \quad N—SiR_2$$
$$R \quad R$$

(130)

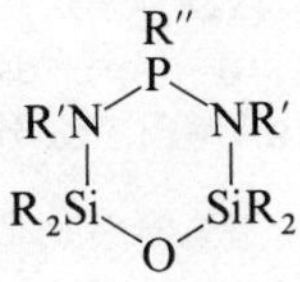

$$R''$$
$$P$$
$$R'N \quad NR'$$
$$R_2Si \quad SiR_2$$
$$O$$

(131)

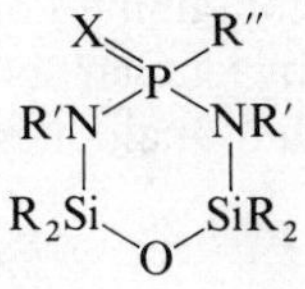

$$X \quad R''$$
$$P$$
$$R'N \quad NR'$$
$$R_2Si \quad SiR_2$$
$$O$$

(132)

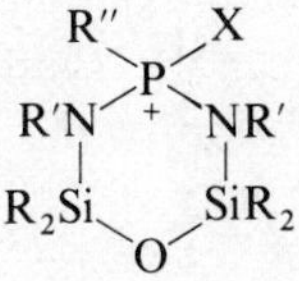

$$R'' \quad X$$
$$P$$
$$R'N \quad^+ \quad NR'$$
$$R_2Si \quad SiR_2$$
$$O$$

(133)

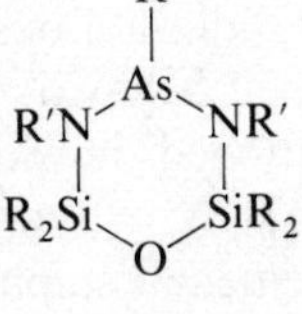

$$R''$$
$$As$$
$$R'N \quad NR'$$
$$R_2Si \quad SiR_2$$
$$O$$

(134)

$$O \quad O$$
$$S$$
$$R'N \quad NR'$$
$$R_2Si \quad SiR_2$$
$$O$$

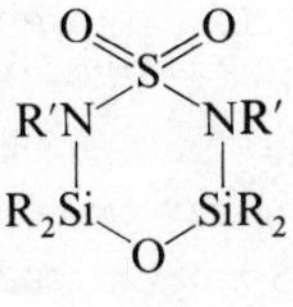

(135)

8.9.1 Boron-containing cyclosilazoxanes

A six-membered ring system, cyclo-1-bora-3,5-disila-2,6-diaza-4-oxane (122), R = Me, R' = Ph, has been obtained in the reaction of a dilithiated 1,5-bis(methylamino)trisiloxane $O(SiR_2NRLi)_2$ with phenylboron dichloride $PhBCl_2$.[179]

A second six-membered ring system, cyclo-1-bora-3,5-disila-4-aza-2,6-dioxane (123), R = Me, R' = Ph, has been isolated in the reaction of $PhB(OH)_2$ with a dichlorodisilazane $MeN(SiMe_2Cl)_2$ in the presence of triethylamine.[199] The BSi_2NO_2 ring is hydrolytically unstable.

8.9.2 Germanium-containing cyclosilazoxanes

A symmetrical six-membered ring of composition $GeSi_2N_2O$, cyclo-1-germa-3,5-disila-2,6-diaza-4-oxane (124), R = Me, R' = Cl, has been obtained by reacting $O(SiMe_2NMeLi)_2$ with $GeCl_4$.[165] If a molar ratio of 1 : 2 is used a spirocyclic compound (126), R = Me, is formed.[144] Another spirocyclic system (127) has been prepared similarly.

The eight-membered ring, cyclo-1-germa-3,5,7-trisila-2,8-diaza-4,6-dioxane (125), R = Me, R' = Cl can be synthesized by treatment of $R_2Si(OSiR_2NHR)_2$ with $GeCl_4$ (1 : 1 ratio) in the presence of NEt_3.[179]

8.9.3 Tin-containing cyclosilazoxanes

Mono- and spirocyclic tin-containing compounds (128)–(130) are known. They can be synthesized by similar procedures to those used for the preparation of the germanium analogues[146,165] and will not be discussed further here.

8.9.4 Phosphorus- and arsenic-containing cyclosilazoxanes

Treatment of $O(SiMe_2NHR)_2$, after lithiation or in the presence of NEt_3, with organophosphorus dichlorides $R''PCl_2$ led to the formation of six-membered cyclo-1-phospha-3,5-disila-2,6-diaza-4-oxanes (131), R = Me, R' = Me,[196] Et,[156] R'' = Me, Et, Ph. Their behaviour is very similar to that of heterocyclic phosphasilazanes mentioned in Section 8.6.1, and the $Si_2P^{III}N_2O$ ring can be converted to (132), X = S and $NSiMe_3$, and (133), X = R or $C(S)S^-$, by reactions with elemental sulphur, trimethylsilylazide Me_3SiN_3, methyl iodide and carbon disulphide respectively.[156,196] The phosphorus(III) compound (131) can act as ligand and forms cobalt(II) complexes.

Derivatives of (132) with R = Me, Et, R' = NH_2, R'' = OPh and X = S or O can be obtained in reactions of phenoxythiophosphoryl dihydrazide

PhOP(S)(NHNH$_2$)$_2$ and phenoxyphosphoryl dihydrazide PhOP(O) (NHNH$_2$)$_2$ with 1,3-dichlorodisiloxanes O(SiR$_2$Cl)$_2$ (R = Me, Et).[197,198] The arsenic-containing heterocycle (134), R = Me, R' = Me and Et, R'' = Cl, has been obtained in the reaction of AsCl$_3$ with O(SiR$_2$NHR')$_2$ in the presence of triethylamine.[165]

8.9.5 Sulphur-containing cyclosilazoxanes

Cyclo-1-thia-3,5-disila-2,6-diaza-4-oxanes (135), R = Me, R' = Me, SiMe$_3$, containing a six-membered ring SSi$_2$N$_2$O with sulphur in oxidation state +6, have been prepared from 1,3-dichlorodisiloxanes O(SiR$_2$Cl)$_2$ with sulphamides SO$_2$(NHR')$_2$ in the presence of triethylamine.[191] These rings are not formed in reactions of SO$_2$Cl$_2$ with O(SiMe$_2$NHR)$_2$.

Bibliography

B 1 I. Haiduc, *The Chemistry of Inorganic Ring Systems*, Part 1, (Wiley-Interscience, London, 1970), p. 364.

B 2 U. Wannagat, *Chem. Ztg.* 97 (1973) 105 (Si–N heterocycles).

B 3 U. Wannagat, M. Schlingmann and H. Autzen, *Z. Naturforsch.* 31b (1976) 621 (Si–N five-membered rings).

B 4 K. A. Andrianov and V. N. Emelyanov, *Russ. Chem. Rev. (Engl. Transl.)* 46 (1977) 1092 (organocyclosilazoxanes).

B 5 Yu. M. Varezhkin, D. Ya. Zhinkin and M. M. Morgunova, *Russ. Chem. Rev. (Engl. Transl.)* 50 (1981) 2212 (organocyclodisilazanes).

B 6 D. Ya. Zhinkin, Yu. V. Varezhkin and M. M. Morgunova, *Russ. Chem. Rev. (Engl. Trans.)* 49 (1980) 1149 (organocyclodisilazanes).

References

1   S. D. Brewer and C. P. Haber, *J. Am. Chem. Soc.* 70 (1948) 3888.

2   L. S. Breed and R. L. Elliott, *Inorg. Chem.* 3 (1964) 1622.

3   U. Wannagat and S. Klemke, *Monatsh. Chem.* 110 (1979) 1089.

4   K. E. Petermann and J. M. Shreeve, *Inorg. Chem.* 15 (1976) 743.

5   T. Kitazume and J. M. Shreeve, *Inorg. Chem.* 16 (1977) 2040.

5a   I. Haiduc and H. Gilman, *Syn. Inorg. Metalorg. Chem.* 1, (1979) 75.

6   R. Appel and M. Montenarh, *Z. Naturforsch.* 31b (1976) 993.

7   (a) W. Wolfsberger, *J. Organomet. Chem.* 88 (1975) 133, 173 (1979) 277; (b) W. S. Sheldrick and W. Wolfsberger, *Z. Naturforsch.* 32b (1977) 22.

8   K. A. Andrianov and N. V. Delazari, *Izv. Akad. Nauk SSSR, Ser. Khim.* 12 (1972) 2748.

9   K. A. Andrianov, V. N. Talanov and L. M. Khananashvili, *Dokl. Akad. Nauk SSSR* 198 (1971) 87.

10   K. A. Andrianov, G. V. Kotrelev, E. F. Tonkikh and N. M. Ivanova, *Dokl. Akad. Nauk SSSR* 201 (1971) 349.

11   K. A. Andrianov, G. V. Kotrelev, V. V. Kazakova and J. E. Rogov, *Izv. Akad. Nauk SSSR, Ser. Khim.* 11 (1975) 2604.

12   K. A. Andrianov, G. V. Kotrelev, V. V. Kazakova and I. E. Rogov, *Izv. Akad. Nauk SSSR, Ser. Khim.* **11** (1975) 2600.

13   K. A. Andrianov, M. J. Shkolnik, Zh. S. Syrtsova, K. J. Petrov, V. M. Kopylov, M. G. Zaitseva and E. A. Koroleva, *Dokl. Akad. Nauk SSSR* **223** (1975) 347.

14   K. A. Andrianov, M. J. Shkolnik, V. M. Kopylov and P. L. Prikhodko, *Dokl. Akad. Nauk SSSR* **227** (1976) 352.

15   C. M. Golino, R. D. Bush and L. H. Sommer, *J. Am. Chem. Soc.* **96** (1974) 614.

16   N. Wiberg and G. Preiner, *Angew. Chem.* **90** (1978) 393.

17   U. Klingebiel and A. Meller, *Chem. Ber.* **109** (1976) 2430

18   U. Klingebiel and A. Meller, *Angew. Chem.* **88** (1976) 307.

19   U. Klingebiel, *Z. Naturforsch,* **33b** (1978) 521.

20   U. Klingebiel and A. Meller, *Z. Naturforsch,* **32b** (1977) 537.

21   U. Klingebiel, H. Hluchy and A. Meller, *Chem. Ber.,* **111** (1978) 906.

22   U. Klingebiel, *J. Organomet. Chem,* **152** (1978) 33.

23   U. Klingebiel D. Bentmann and A. Meller, *J. Organomet. Chem.* **144** (1978) 381.

24   U. Klingebiel and A. Meller, *Z. Anorg. Allg. Chem.* **428** (1977) 27.

25   W. Clegg, U. Klingebiel, G. M. Sheldrick and N. Vater, *Z. Anorg. Allg. Chem.* **482** (1981) 88.

26   W. Clegg, U. Klingebiel, C. Krampe and G. M. Sheldrick, *Z. Naturforsch,* **35b** (1980) 275.

27   U. Klingebiel, *Chem. Ber.* **111** (1978) 2735.

28   U. Klingebiel, J. Neemann and A. Meller, *Z. Anorg. Allg. Chem.* **429** (1977) 63.

29   U. Klingebiel, D. Bentmann and A. Meller, *Monatsh. Chem.* **109** (1978) 1067.

30   U. Klingebiel and A. Meller, *Angew. Chem.* **88** (1976) 647.

31   W. Clegg, U. Klingebiel and G. M. Sheldrick, *Z. Naturforsch.* **37b** (1982) 423.

32   H. Bürger, R. Mellies and K. Wiegel, *J. Organomet. Chem.* **142** (1977) 55.

33   E. Lebedev and V. O. Reikhsfel'd, *Zh. Obsh. Khim.* **40** (1970) 1082.

34   K. A. Andrianov, G. V. Kotrelev and L. J. Nosova, *Zh. Obsh. Khim.* **42** (1972) 854.

35   Zh. V. Gorislavskaya, G. N. Mal'nova, D. Ya. Zhinkin, V. F. Andronov and A. A. Ainshtein, *Zh. Obsh. Khim.* **42** (1972) 2204.

36   N. N. Vlasova, J. J. Tsykhanskaya and Yu. J. Matus, *Izv. Akad. Nauk SSSR, Ser. Khim,* **7** (1973) 1664.

37   A. A. Zhdanov, K. A. Andrianov, T. V. Astapova and B. D. Lavrukhin, *Izv. Akad. Nauk SSSR, Ser. Khim.* **9** (1978) 2104.

38   W. Jansen, H. Kessel and O. Schmitz-Du Mont, *Z. Anorg. Allg. Chem.* **384** (1971) 124.

39   U. Wannagat, O. Smrekar and R. Braun, *Monatsch. Chem.* **100** (1969) 1916.

40   U. Wannagat and D. Labuhn, *Monatsh. Chem.* **105** (1974) 209.

41   U. Wannagat and V. Paul, *Monatsh. Chem.* **105** (1974) 1233.

42   U. Wannagat and D. Labuhn, *Monatsh. Chem.* **104** (1973) 1453.

43   U. Wannagat and L. Gerschler, (a) *Liebigs Ann. Chem.* **744** (1971) 111; (b) *Z. Anorg. Allg. Chem.* **383** (1971) 249.

44   U. Wannagat and L. Gerschler, *Inorg. Nucl. Chem. Lett.* **7** (1971) 285.

45   U. Wannagat, L. Gerschler and H.-J. Wismar, *Monatsch. Chem.* **102** (1971) 1834.

46   U. Wannagat, R. Braun and L. Gerschler, *Z. Anorg. Allg. Chem.* **381** (1971) 168.

47   U. Wannagat, F. Rabet and H.-J. Wismar, *Monatsh. Chem.* **102** (1971) 1429.

48   U. Wannagat, R. Braun, L. Gerschler and H.-J. Wismar, *J. Organomet. Chem.* **26** (1971) 321.

49   L. W. Breed, R. L. Elliott and J. C. Wiley, *J. Organomet. Chem.* **24** (1970) 315.

50   K. A. Andrianov, A. A. Zhdanov, B. A. Astapov, B. D. Lavrukhin, T. V. Streekova and T. V. Astapova, *Izv. Akad. Nauk SSSR, Ser. Khim.* **11** (1979) 2556.

51 U. Klingebiel, J. Neemann and A. Meller, *Monatsh. Chem.* **108** (1977) 1099.
52 U. Wannagat, H. Moretto, P. Schmidt and M. Schulze, *Z. Anorg. Allg. Chem.* **381** (1971) 288.
53 E. P. Lebedev, R. G. Valimukhametova, E. N. Korol and V. O. Reikhsfel'd, *Zh. Obshch. Khim.* **44** (1974) 1941.
54 D. R. Parker and L. H. Sommer, *J. Organomet. Chem.* **110** (1976) C1.
55 W. D. Beiersdorf, D. J. Brauer and H. Bürger, *Z. Anorg. Allg. Chem.* **475** (1981) 56.
56 (a) G. Eisele, H. Autzen and U. Wannagat, *Monatsh. Chem.* **109** (1978) 1267; (b) U. Wannagat and G. Eisele, *Z. Naturforsch.* **33b** (1978) 471.
57 M. Schlingmann and U. Wannagat, (a) *Chem. Ztg* **98** (1974) 458; (b) *Z. Anorg. Allg. Chem.* **419** (1976) 108.
58 M. Schlingmann and U. Wannagat, *Z. Anorg. Allg. Chem.* **429** (1977) 74.
59 K. A. Andrianov, A. B. Zachernyuk and E. A. Zhdanova, *Dokl. Akad. Nauk SSSR* **214** (1974) 325.
60 H. Rosenberg, Tsu-Tzu Tsai, W. W. Adams, M. T. Gehatia, A. V. Fratini and D. R. Wiff, *J. Am. Chem. Soc.* **98** (1976) 8083.
61 K. A. Andrianov, L. M. Khananashvili, V. N. Talanov, M. M. Il'in, J. J. Zhurakovskaya and T. V. Didenko, *Vysokomol. Soedin. B* **21** (1979) 346.
62 K. A. Andrianov, M. M. Il'in, V. N. Talanov, M. V. Konstantinova and V. V. Kazakova, *Zh. Obshch. Khim.* **47** (1977) 1758.
63 K. A. Andrianov, G. V. Kotrelev, V. V. Kazakova and N. A. Tebeneva, *Dokl. Akad. Nauk SSSR* **233** (1977) 353.
64 Ch. A. Pearce and C. Norman, *US Pat.* 3 580 941.
65 U. Wannagat, J. Herzig and H. Bürger, *J. Organomet. Chem.* **23** (1970) 373.
66 L. Párkányi, Gy. Argay, P. Hencsei and J. Nagy, *J. Organomet. Chem.* **116** (1976) 299.
67 V. E. Shklover, P. Adiaasuren, Yu. T. Struckkov, E. A. Zhdanova, V. S. Svistunov, G. V. Kotrelev and K. A. Andrianov, *Dokl. Akad. Nauk SSSR* **241** (1978) 377.
68 V. E. Shklover, P. Adiaasuren, G. V. Kotrelev, E. A. Zhdanova, V. S. Svistunov and Yu. T. Struckov, *Zh. Strukt. Khim.* **21** (1980) 94.
69 B. Rozsondai, J. Hargittai, A. V. Golubinskii, L. V. Vilkov and V. S. Mastryukov, *J. Mol. Struct.* **28** (1975) 339.
70 D. Mootz, J. Fayos and A. Zinnius, *Angew. Chem.* **84** (1972) 27.
71 W. Clegg, M. Noltemeyer, G. M. Sheldrick and N. Vater, *Acta Crystallogr.* **B36** (1980) 2461.
72 G. W. Adamson and J. J. Daly, *J. Chem. Soc. A* **1970**, 2724.
73 (a) V. E. Shklover, Yu. T. Struchkov, G. V. Solomatin, A. B. Zachernyuk and A. A. Zhdanov, *Zh. Strukt. Khim.* **21** (1980) 94; (b) A. A. Zhdanov, A. B. Zachernyuk, G. V. Struchkov and T. Yu, *Dokl. Adad. Nauk SSSR* **251** (1980) 121.
74 V. E. Shklover, Yu. T. Struchkov, B. A. Astapov and K. A. Andrianov, *Zh. Strukt. Khim.* **20** (1979) 102.
75 B. P. E. Edwards, W. Harrison, J. W. Nowell, M. L. Post, H. M. M. Shearer and J. Trotter, *Acta Crystallogr.* **B32** (1976) 648.
76 L. Skoda, U. Klingebiel and A. Meller, *Z. Anorg. Allg. Chem.* **467** (1980) 131.
77 L. Skoda, Dissertation Göttingen (1980).
78 D. M. Adams and W. S. Fernando, *J. Chem. Soc., Dalton Trans.* **1973**, 410.
79 K. A. Andrianov, V. M. Kopylov, M. G. Zaitseva, V. S. Khan and M. J. Shkol'nik, *Zh. Obshch. Khim.* **49** (1979) 2683.
80 E. A. Kirichenko, A. J. Ermakov, N. J. Pimkin, K. A. Andrianov, V. M. Kopylov and M. J. Shkol'nik, *Zh. Obshch. Khim.* **49** (1979) 1333.
81 E. A. Williams and J. D. Cargioli, in *Annual Reports on NMR Spectroscopy*, Vol. 9 (Academic Press, London, 1979), p. 221.

82 H. Marsmann, *NMR 17 Basic Principles and Progress* Springer, Berlin, 1981.
83 B. D. Lavrukhin, K. A. Andrianov and E. J. Fedin, *Org. Magn. Reson.* **7** (1975) 298.
84 B. D. Lavrukhin, T. V. Strelkova, B. A. Astapov and A. A. Zhdanov, *Magnetic Resonance and Related Phenomena: Proc. Congr. AMPERE*, (1979), p. 480.
85 H. Jancke, G. Engelhardt, M. Mägi and E. Lippmaa, *Z. Chem.* **13** (1975) 435.
86 L. Maijs, G. Rumba and J. Yu. Tsereteli, *Latv. PSR. Zinat. Akad. Vestis, Khim. Ser.* **1971**, 92.
87 D. Schmid and U. Wannagat, *Chem. Ztg.* **98** (1974) 575.
88 Yu. V. Kolodyazhnyi, M. G. Gruntfest, V. A. Bren', L. S. Baturina, M. J. Morgunova, D. Ya. Zhinkin, G. S. Gol'din and O. A. Osipov, *Zh. Obshch. Khim.* **45** (1975) 1083.
89 U. Wannagat, M. Moretto and P. Schmidt, *Z. Anorg. Allg. Chem.* **385** (1971) 146.
90 U. Wannagat and M. Schulze, *Inorg. Nucl. Lett.* **5** (1969) 789.
91 U. Wannagat and V. Paul, *Monatsh. Chem.* **105** (1974) 1240.
92 U. Klingebiel, D. Enterling, L. Skoda and A. Meller, *J. Organomet. Chem.* **135** (1977) 167.
93 U. Klingebiel, D. Enterling and A. Meller, *Chem. Ber.* **110** (1977) 1277.
94 U. Klingebiel, J. Deiselmann and A. Meller, *Z. Anorg. Allg. Chem.* **441** (1978) 107.
95 D. Enterling, U. Klingebiel and A. Meller, *Z. Naturforsch.* **33b** (1978) 527.
96 W. Clegg, M. Hesse, U. Klingebiel, G. M. Sheldrick and L. Skoda, *Z. Naturforsch.* **35b** (1980) 1359.
97 L. Skoda, U. Klingebiel and A. Meller, *Chem. Ber.* **113** (1980) 2342.
98 M. Hesse and U. Klingebiel, *J. Organomet. Chem.* **221** (1981) C1.
99 U. Klingebiel, L. Skoda and A. Meller, *Z. Anorg. Allg. Chem.* **441** (1978)£ 113.
100 W. Clegg, U. Klingebiel, G. M. Sheldrick, L. Skoda and N. Vater, *Z. Naturforsch.* **35b** (1980) 1503.
101 M. Hesse, U. Klingebiel and L. Skoda, *Chem. Ber.* **114** (1981) 2287.
102 W. Clegg, M. Noltemeyer, G. M. Sheldrick and N. Vater, *Acta Crystallogr.* **B37** (1981) 986.
103 Tsu-Tzu Tsai and C. J. Marshall, *J. Org. Chem.* **37** (1972) 596.
104 A. D. M. Hailey and G. Nickless, *J. Chromatogr.* **49** (1970) 180.
105 J. Hughes and G. R. Willey, *J. Am. Chem. Soc.* **95** (1973) 8750.
106 J. Dreimanis and G. Rumba, *Latv. PSR Zinat. Akad. Vestis, Khim. Ser.* **1973**, 197.
107 U. Wannagat, E. Bogusch and P. Geymayer, *Monatsh. Chem.* **102** (1971) 1825.
108 L. W. Breed and J. C. Wiley, *Inorg. Chem.* **11** (1972) 1634.
109 W. Fink, *Angew. Chem.* **81** (1969) 499.
110 W. Fink, *Helv. Chim. Acta* **52** (1969) 2261.
111 L. Skoda, U. Klingebiel and A. Meller, *Chem. Ber.* **113** (1980) 1444.
112 W. Clegg, *Acta Crystallogr.* **B36** (1980) 2830.
113 R. P. Bush and C. A. Pearce, *J. Chem. Soc.* A **1969**, 808.
114 W. Clegg, M. Haase, J. Hluchy, U. Klingebiel and G. M. Sheldrick, *Chem. Ber.* **116** (1983) 290.
115 U. Wannagat and E. Bogusch, *Monatsh. Chem.* **102** (1971) 1806.
116 E. P. Lebedev, A. D. Fedorov, V. O. Reikhsfel'd and L. Z. Zakirova, *Zh. Obshch. Khim* **45** (1975) 2645.
117 K. A. Andrianov, Zh. S. Syrtsova and V. M. Kopylov, *Zh. Obshch. Khim.* **40** (1970) 1665.
118 E. P. Lebedev, V. O. Reikhsfel'd and D. V. Fridland, *Zh. Obshch. Khim.* **43** (1973) 683.
119 E. P. Lebedev, D. V. Fridland, V. O. Reikhsfel'd and E. N. Kovl', *Zh. Obshch. Khim.* **46** (1976) 315.
120 H. Nöth, W. Tinhof and T. Taeger, *Chem. Ber.* **107** (1974) 3113.
121 K. Barlos and H. Nöth, *Chem. Ber.* **110** (1977) 2790.

122   Ch. A. Pearce and N. C. Norman, *US Pat.* 3 518 289; 3 518 290.
123   H. Hluchy and U. Klingebiel, *Angew. Chem.* **94** (1982) 292.
124   U. Wannagat, M. Schlingmann and H. Autzen, *Z. Naturforsch.* **31b** (1976) 621.
125   U. Wannagat, E. Bogusch and F. Rabet, *Z. Anorg. Allg. Chem.* **385** (1971) 261.
126   U. Wannagat and M. Schlingmann, *Z. Anorg. Allg. Chem.* **406** (1974) 7.
127   U. Klingebiel, G. Wendenburg and A. Meller, *Monatsh. Chem.* **110** (1979) 289.
128   U. Klingebiel, G. Wendenburg and A. Meller, *Z. Naturforsch.* **32b** (1977) 1482.
129   W. Clegg, H. Hluchy, U. Klingebiel and G. M. Sheldrick, *Z. Naturforsch.* **34b** (1979) 1260.
130   N. Wiberg and G. Ziegleder, *Chem. Ber.* **111** (1978) 2123.
131   F. Höfler and D. Wolfer, *Z. Anorg. Allg. Chem.* **406** (1974) 19.
132   H. Bock, W. Kaim, H. Nöth and A. Semkow, *J. Am. Chem. Soc.* **102** (1980) 4421.
133   U. Wannagat and M. Schlingmann, *Abh. Braunschw. Wiss. Ges.* **24** (1974) 79.
134   U. Wannagat and M. Schlingmann, *Z. Anorg. Allg. Chem.* **419** (1976) 48.
135   U. Wannagat, M. Schlingmann and H. Autzen, *Chem. Ztg* **98** (1974) 372.
136   M. Schlingmann and U. Wannagat, *Z. Anorg. Allg. Chem.* **419** (1976) 115.
137   H. Bürger, M. Schlingmann and G. Pawelke, *Z. Anorg. Allg. Chem.* **419** (1976) 121.
138   U. Wannagat, S. Klemke, D. Mootz and H. D. Reski, *J. Organomet. Chem.* **178** (1979) 83.
139   U. Wannagat and S. Meier, *Z. Anorg. Allg. Chem.* **392** (1972) 179.
140   U. Wannagat, E. Bogusch and R. Braun, *J. Organomet. Chem.* **19** (1969) 367.
141   U. Wannagat, R. Seifert and M. Schlingmann, *Z. Anorg. Allg. Chem.* **439** (1978) 83.
142   R. Seifert and U. Wannagat, *Z. Anorg. Allg. Chem.* **439** (1978) 90.
143   U. Wannagat and R. Seifert, *Monatsh. Chem.* **109** (1978) 209.
144   U. Wannagat, R. Seifert and M. Schlingmann, *Z. Naturforsch.* **32b** (1974) 869.
145   H. W. Roesky, M. Diehl, H. Fuess and J. W. Bats, *Angew. Chem. Int. Ed. Engl.* **90** (1978) 73.
146   U. Wannagat, M. Schlingmann, R. Seifert, *Z. Anorg. Allg. Chem.* **440** (1978) 105.
147   M. Veith, *Angew. Chem.* **87** (1975) 287.
148   O. J. Scherer and M. Puettmann, *Angew. Chem.* **91** (1979) 741.
149   U. Klingebiel, P. Werner and A. Meller, *Chem. Ber.* **110** (1977) 2905.
150   J. Neemann and U. Klingebiel, *Chem. Ber.* **114** (1981) 527.
151   E. Niecke and W. Bitter, *Chem. Ber.* **109** (1976) 415.
152   E. Niecke and O. J. Scherer, *Nach. Chem. Techn.* **23** (1975) 395.
153   U. Klingebiel, P. Werner and A. Meller, *Monatsh. Chem.* **107** (1976) 939.
154   J. O. Scherer, N. T. Kulbach and W. Glaessel, *Z. Naturforsch.* **33b** (1978) 652.
155   M. Kajiwara, M. Makihava and H. Saito, *J. Inorg. Nucl. Chem.* **37** (1975) 2562.
156   U. Wannagat, K. Giesen and F. Rabet, *Z. Ano.*. *Allg. Chem.* **382** (1971) 195.
157   U. Wannagat and H. Autzen, *Z. Anorg. ALLg. Chem.* **420** (1976) 119.
158   U. Wannagat, M. Schlingmann and H. Autzen, *Chem. Ztg* **98** (1974) 111.
159   U. Wannagat and H. Autzen, *Z. Anorg. Allg. Chem.* **420** (1976) 132.
160   O. J. Scherer, W. Glaessel and R. Thalacker, *J. Organomet. Chem.* **70** (1974) 61.
161   H. Autzen and U. Wannagat, *Z. Anorg. Allg. Chem.* **420** (1976) 139.
162   U. Wannagat and M. Schlingmann, *Z. Anorg. Allg. Chem.* **424** (1976) 87.
163   J. Dreimanis and G. Rumba, *Latv. PSR Zinat. Akad. Vestis, Khim. Ser.* **3** (1972) 336.
164   B. Dejak and Z. Lasocki, *J. Organomet. Chem.* **44** (1972) C39.
165   U. Wannagat and F. Rabet, *Inorg. Nucl. Chem. Lett.* **6** (1970) 155.
166   Z. Lasocki and M. Witekowa, *Synth. React. Inorg. Metalorg. Chem.* **4** (1974) 231.
167   E. P. Lebedev, V. G. Zavada and A. D. Fedorov, *Zh. Obshch. Khim.* **49** (1979) 151.
168   E. P. Lebedev, A. D. Fedorov and V. O. Reikhsfel'd, *Zh. Obshch. Khim.* **42** (1972) 2117.
169   R. P. Bush and B. Thomas, *US Pat.* 3 677 977 (*CA* **78** (1973) 4762h)

170 U. Wannagat, E. Bogusch, P. Geymayer and F. Rabet, *Monatsch. Chem.* **102** (1971) 1844.
171 I. Haiduc and H. Gilman, *J. Organomet. Chem.* **18** (1969) P5; *Syn. Inorg. Metalorg. Chem.* **1** (1971) 69.
172 E. P. Lebedev and R. G. Valimukhametova, *Zh. Obshch. Khim.* **47** (1977) 233.
173 Yu. M. Varezhkin, D. Ya. Zhinkin, M. M. Morgunova and V. N. Bochkarev, *Zh. Obshch. Khim.* **45** (1975) 2455.
174 V. N. Bochkarev, A. A. Vernadskii, Yu. M. Varezhkin, M. M. Morgunova and D. Ya. Zhinkin, *Khim. Elementoorg. Soedin.* **31** (1976).
175 A. J. Gusev, M. G. Los, Yu. M. Varezhkin, M. M. Morgunova and D. Ya. Zhinkin, *Zh. Strukt. Khim.* **17** (1976) 378.
176 Yu. M. Varezhkin, M. M. Morgunova and D. Ya. Zhinkin, *Zh. Obshch. Khim.* **47** (1977) 1536.
177 K. A. Andrianov, V. N. Emel'yanov and E. V. Rudman, *Dokl. Akad. Nauk SSSR* **204** (1972) 855.
178 H. W. Roesky and B. Kuhtz, *Chem. Ber.* **109** (1976) 3958.
179 F. Rabet and U. Wannagat, *Z. Anorg. Allg. Chem.* **384** (1971) 115.
180 V. E. Shklover, N. G. Bokii, Yu. T. Struchkov, K. A. Andrianov, A. B. Zachernyuk and E. A. Zhdanova, *Zh. Strukt. Khim.* **15** (1974) 850.
181 Y. M. Varezhkin, D. Y. Zhinkin, M. M. Morgunova and V. P. Anosov, *Zh. Obshch. Khim.* **47** (1977) 2089.
182 M. G. Voronkov, Y. M. Varezhkin, D. Y. Zhinkin, M. M. Morgunova, S. N. Gurkova, A. J. Gusev and N. N. Alekseev, *Dokl. Akad. Nauk SSSR* **227** (1977) 102.
183 D. R. Parker and L. H. Sommer, *J. Organomet. Chem.* **110** (1976) C1.
184 R. P. Bush, N. C. Lloyd and C. A. Pearce, *J. Chem. Soc.* A, **1970**, 1587.
185 R. P. Bush, *UK Pat.* 1 265 001 (*CA* **76** (1972) 141013n).
186 P. Legrand, M. Imbenotte, G. Palavit, J. P. Huvenne and G. Vergoten, *Proc. 6th Int. Conf. Raman Spectroscopy,* Vol. 2 (1978) p. 38.
187 D. Bentmann, W. Clegg, U. Klingebiel and G. M. Sheldrick, *Inorg. Chim. Acta* **45** (1980) 229.
188 D. Bentmann and U. Klingebiel, *Z. Anorg. Allg. Chem.* **477** (1981) 90.
189 K. A. Andrianov, V. N. Talanov, L. M. Khananashvili and T. G. Borodochenkova, *Izv. Akad. Nauk SSSR, Ser. Khim.* **9** (1969) 2004.
190 R. P. Bush, *UK Pat.* 1 256 704 (*CA* 76 (1972) 100331w)
191 U. Wannagat and D. Labuhn, *Z. Anorg. Allg. Chem.* **402** (1973) 147.
192 R. Appel and M. Montenarh, *Z. Naturforsch.* **32b** (1977) 108.
193 O. J. Scherer and G. Wolmershäuser, *Z. Naturforsch.* **29b** (1974) 444.
194 E. P. Lebedev, A. D. Fedorov, M. M. Frenkel', *Zh. Obshch. Khim.* **47** (1977) 961.
195 U. Wannagat, M. Schlingmann and R. Schächter, *Z. Anorg. Allg. Chem.* **445** (1978) 102.
196 U. Wannagat, K.-P. Giesen and H.-H. Falius, *Monatsh. Chem.* **104** (1973) 1444.
197 U. Engelhardt and T. Buenger, *Inorg. Nucl. Chem. Lett.* **14** (1978) 21.
198 U. Engelhardt and T. Buenger, *Z. Naturforsch.* **34b** (1979) 1107.
199 U. Wannagat and G. Eisele, *Z. Naturforsch.* **33b** (1978) 475.
200 J. Dreimanis, I. Strauss, G. Rumba and L. Maijs, *Latv. PSR Zinat. Akad. Vestis, Khim. Ser.* **1972**, 40 (*CA* **76** (1972) 132919k).
201 V. E. Ditsent, I. I. Skorokhodov and M. N. Zolotareva, *Zh. Fiz. Khim.* **46** (1972) 1888.
202 G. P. Brabin, M. Kh. Karapetyants and R. M. Golosova, *Tr. Mosk. Khim. Teknol. Inst.* **81** (1974) 17.
203 W. Fink, *US Pat.* 3 431 222 (1969) (*CA* **70** (1969) 88453d).
204 M. Takamizawa, *Japan Pat.* 79–93 100 (1979) (*CA* **92** (1980) 7173s).

205 G. Rumba, D. Zarina, J. Rozkalns and I. Zviedre, *Kremniiorg. Soedin., Tr. Sovesch.* **3** (1966) 7 (*CA* **73** (1970) 15349b).
206 G. Rumba, J. Dreimanis and A. Vaivads, *USSR Pat.* 280 846 (1970) (*CA* **74** (1971) 64881y).
207 L. W. Breed and R. L. Elliott, *J. Polym. Sci.* A-1, **9** (1971) 183.
208 W. Fink, *US Pat.* 3 553 242 (1971) (*CA* **75** (1971) 20604b).
209 K. A. Andrianov, G. V. Kotrelev, E. F. Tonkina and N. M. Ivanova, *Dokl Akad. Nauk SSSR* **201** (1971) 349 (*CA* **76** (1972) 59698f).
210 F. N. Vishnevskii, D. Ya. Zhinkin, K. K. Popkov, M. M. Morgunov and I. I. Skorokhodov, *Plast. Massy.* **1972**, 18.
211 F. N. Vishnevskii, D. Ya. Zhinkin, B. N. Klimentov, L. M. Konstantinenko and I. I. Skorokhodov, *Plast. Massy* **1972**, 22 (*CA* **77** (1972) 62361s).
212 R. P. Bush and B. Thomas, *US Pat.* 3 677 977 (1972) (*CA* **78** (1972) 4762h).
213 A. M. Wrobel and M. Kryszewski, *Bull. Acad. Pol. Sci., Ser. Sci. Chim.* **22** (1974) 471.
214 W. Fink, *Ger. Offen.* 1 955 906 (1971) (*CA* **75** (1971) 64633q).
215 W. Fink, *US Pat.* 3 575 922 (1971) (*CA* **75** (1971) 64646w).
216 K. A. Andrianov, N. G. Lekishvili, A. I. Noigadeli, G. V. Kotrelev and R. Sh. Tkeshalashvili, *Vysokomol. Soedin.* **B16** (1974) 497.
217 B. Dejak, J. Kulpinski, Z. Lasocki and S. Piechucki, *Polikondens. Protsessy, Tr. Mezhdunar. Simp., 5th, 1975* (1976), p. 213 (*CA* **85** (1976) 143870z).
218 K. A. Andrianov and B. A. Astapov, *Dokl Akad. Nauk SSSR* **227** (1976) 1357.
219 K. A. Andrianov, N. A. Telseneva, I. M. Petrova and G. V. Kotrelev, *Dokl Akad. Nauk SSSR* **222**)1975) 1013.
220 A. A. Khakharov, S. V. Pavutnitskaya, G. V. Ivanova and V. I. Gracgev, *Zh. Prikl. Khim.* **49** (1976) 611 (*CA* **84** (1976) 152046s).
221 A. A. Kharkharov, S. V. Pavutnitskaya, D. Ya. Zhinkin and G. V. Ivanova, *Izv. Vyssh. Uchebn. Zaved., Teknol. Tekst. Prom-sti* No. 5 (1974) 77 (*CA* **82** (1975) 113.059n).
222 A. A. Chviko, O. A. Makarov, D. Ya. Zhinkin, V. A. Sobolev and A. S. Shapatin, *Dopov. Akad. Nauk Ukr. SSR* **B36** (1974) 921.
223 Shin Yuji, *Japan Pat.* 71–26 338 (1971) (*CA* **76** (1972) 15217n).
224 T. Kamaishi and S. Matsuhisa, *Japan Pat.* 69–27 734 (1969) (*CA* **73** (1970) 4358c).
225 K. Hidehiko, O. Chiciro, K. Yasuto and Y. Kenji, *Japan Pat.* 71–27 733 (1971) (*CA* **76** (1972) 4565e).
226 G. N. Antipina, L. N. Smirnov, A. T. Miroshkina, D. Ya. Zhinkin, G. N. Mal'nova and P. V. Krymov, *USSR Pat.* 308 112 (1971) (*CA* **76** (1972) 47248).
227 N. G. Lkyuchnikov, F. I. Karabadzhak and V. B. Losev, *Zh. Prikl. Khim.* **43** (1970) 2763 (*CA* **75** (1971) 9271v).
228 D. Ya. Zhinkin, V. A. Sidnev, L. V. Kozlova and V. E. Miklin, *USSR Pat.* 436 066 (1974) (*CA* **82** (1975) 99742w).
229 K. Uhle and A. Kniting, *Z. Chem.* **14** (1974) 305.
230 K. Uhle and A. Kniting, *DDR Pat.* 112 135 (1975) (*CA* **84** (1976) 90295m).
231 R. L. Ostrozynsky, *US Pat.* 3 843 703 (1974) (*CA* **82** (1975) 86398p).
232 W. E. Dennis, *J. Org. Chem.* **35** (1970) 3253.
233 W. E. Dennis, *Ger. Offen.* 2 110 439 (1971) (*CA* **76** (1972) 59258n).
234 G. Bakassian, *Ger. Offen.* 2 205 360 (1972) (*CA* **77** (1972) 151710x).
235 G. Bakassian and M. Lefort, *US Pat.* 3 884 957 (1975) (*CA* **83** (1975) 58487u).
236 J. B. Fourtillan, J. P. Pometan and R. J. R. Dang, *Fr. Demande* 2 251 311 (1975) (*CA* **84** (1976) 122029w).
237 J. B. Fourtillan, J. P. Pometan and R. J. R. Dang, *Fr. Demande* 2 251 313 (1975) (*CA* **84** (1976) 135812w).

9

Silicon–Phosphorus Heterocycles

G. FRITZ and J. HÄRER

University of Karlsruhe, Federal Republic of Germany

9.1 Introduction

There are two main types of silicon–phosphorus heterocycles: (*a*) cyclophosphapolysilanes, in which one or more silicon atoms of a cyclopolysilane are substituted by phosphorus atoms; and (*b*) cyclosilaphosphanes, in which phosphorus and silicon atoms alternate; occasionally this latter class may contain P–P bonds, while cyclophosphapolysilanes contain in all cases Si–Si bonds.

Silicon–phosphorus heterocycles are prepared by reactions leading to formation of Si–P bonds. These require molecules with functional groups at silicon and phosphorus respectively. Among the phosphorus compounds, phosphides, halogenophosphanes, hydrogenophosphanes and silylphosphanes are the most useful. In these P-functional compounds the P–M (M = alkali metal), P–X (X = halogen), P–H and P–Si bonds are easily cleaved under mild conditions. Since P–C bonds are resistant, alkyl and aryl groups are often used to reduce the number of functional centres and to additionally stabilize the P–Si framework by inductive and steric effects (screening against electrophilic attack). The silicon starting compounds include halogenosilanes, hydrosilanes and metallated silanes; in these compounds alkyl and aryl groups fulfil the same purpose as they do in organophosphorus compounds. Appropriate combinations of these versatile P and Si compounds as starting materials yields new P–Si bonds, and the reactions leading to their formation are briefly mentioned here.

THE CHEMISTRY OF INORGANIC HOMO-
AND HETEROCYCLES VOL 1 ISBN 0-12-655775-6

(a) Salt precipitation

(i) reactions of phosphides with halogenosilanes (very convenient methods) (M = Li, Na, K, Mg):[1]

$$>\!P\!-\!M + X\!-\!Si\!< \quad \longrightarrow \quad -\!Si\!-\!P\!< + MX \qquad (1)$$

(ii) reactions of metallated silanes with halogenophosphanes:[2]

$$>\!P\!-\!X + M\!-\!Si\!< \quad \longrightarrow \quad -\!Si\!-\!P\!< + MX \qquad (2)$$

(b) Elimination of halogenosilanes in reactions of trimethylsilylphosphanes with halogenosilanes:[3]

$$>\!P\!-\!SiMe_3 + X\!-\!Si\!< \quad \longrightarrow \quad -\!Si\!-\!P\!< + Me_3SiX \qquad (3)$$

(c) Condensation with elimination of tris(trimethylsilyl)phosphane:[4]

$$>\!P\!-\!SiMe_3 + (Me_3Si)_2P\!-\!Si\!< \quad \longrightarrow \quad -\!Si\!-\!P\!< + P(SiMe_3)_3 \qquad (4)$$

(d) Dehalogenation of mixtures of halogenosilanes and halogenophosphanes with metals:

$$>\!P\!-\!X + 2\,M + X\!-\!Si\!< \quad \longrightarrow \quad -\!Si\!-\!P\!< + 2\,MX \qquad (5)$$

(e) Elimination of alkyl- and arylphosphanes:[5]

$$2\!-\!Si\!-\!P(H)R \quad \longrightarrow \quad -\!Si\!-\!P(R)\!-\!Si\!< + RPH_2 \qquad (6)$$

(f) Pyrochemical[6] and electric discharge[7] reactions, for example the thermal decomposition of mixtures containing silanes and phosphanes with Si–H and P–H groups:

$$>\!P\!-\!H + H\!-\!Si\!< \quad \longrightarrow \quad -\!Si\!-\!P\!< + H_2 \qquad (7)$$

9.2 Cyclophosphapolysilanes

Cyclophosphapolysilanes are compounds containing Si–Si bonds in the ring; these are produced by the reaction of linear α,ω-dichloropolysilanes with various dilithium phosphides,[8-12] and were discussed in Section 7.2.3 of this book. Therefore these heterocycles will not be treated here.

9.3 Cyclosilaphosphanes

These are rings formed by regular alternation of silicon and phosphorus atoms in the rings. Four- and six-membered rings are known; in addition, polycyclic compounds based upon Si–P alternation have been prepared, and in some cases polycyclic structures containing P–P bonds were also obtained. Formulae (1)–(7) illustrate the known types:

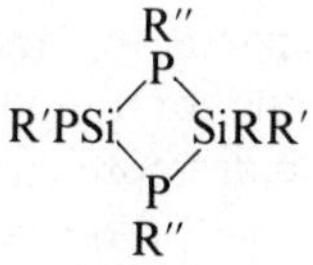

R'PSi⟨P,R''⟩SiRR' (with P–R'' ring)

(1)

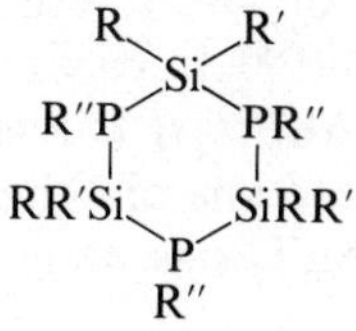

R''P–Si(R,R')–PR''; RR'Si⟨P,R''⟩SiRR'

(2)

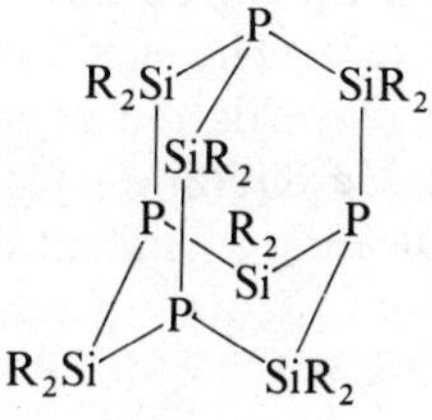

R₂Si⟨P⟩SiR₂ ... SiR₂ ... P–R₂Si–P ... R₂Si⟨P⟩SiR₂

(3)

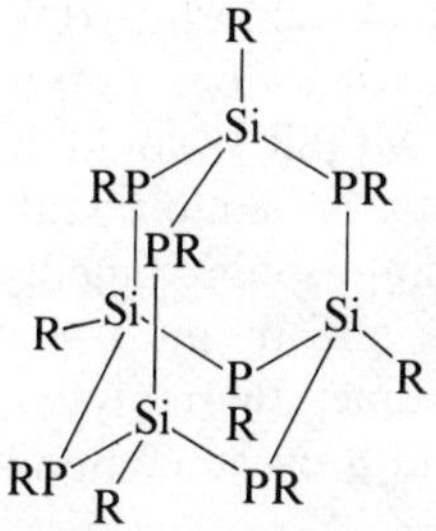

Si(R) cage structure with RP, PR, R–Si, Si–R, RP, PR

(4)

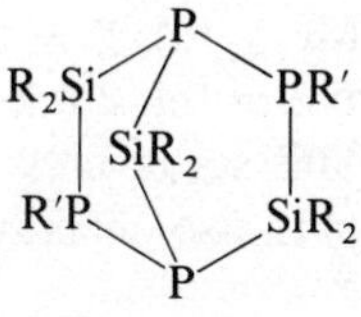

R₂Si⟨P⟩PR' ; SiR₂ ; R'P ; SiR₂ ; P

(5)

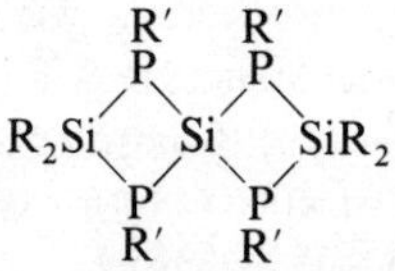

R₂Si–Si(R')–SiR₂ with P, R' substituents

(6)

R₂Si⟨P⟩SiR₂ ; SiR₂ ; P–P ; P

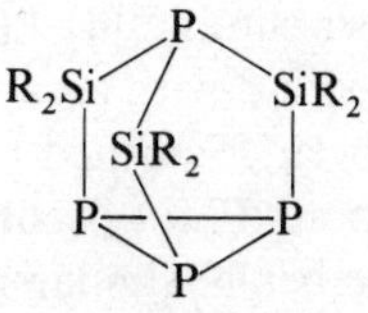

(7)

The synthesis of cyclosilaphosphanes uses reactions of organodichloro-silanes with monophosphides $LiPR_2$ (R = H, Ph, $SiMe_3$) or LiPHR (R = Me, Ph, Bu^t), and with Li_2PR (R = H, Ph, etc.). The halogen atoms of the silanes are substituted stepwise by the phosphido groups, and in many cases the resulting compounds are unstable. The primary reaction step is followed by a series of condensations or rearrangements, leading to four- or six-membered rings. The ring size is determined by the substituents at the Si or P sites. Bulky groups, such as Bu^t favour four-membered rings, and methyl groups six-membered rings. With phenyl substituents both ring sizes are possible.

The reaction of Me_2SiCl_2 with $LiPH_2.DME$ (DME = dimethoxyethane) affords the monosubstituted product $Me_2SiCl(PH_2)$ as well as the disubstituted product $Me_2Si(PH_2)_2$; both these intermediates react further, and a complex mixture of diverse cyclic silaphosphanes is finally formed.[13] The compound $Me_2Si(PH_2)_2$ can also be prepared in larger amounts from Me_2SiCl_2 and $LiAl(PH_2)_4$.[14] Even though the formation of cyclic silaphosphanes by the reaction of lithium phosphides (prepared from PH_3 and $LiBu^n$) with Ph_2SiCl_2 has been claimed,[15] the products should be further investigated.

The reaction of Me_2SiCl_2 with $LiP(SiMe_3)_2.OR_2$ (OR_2 = 2THF or DME) in 1:1 molar ratio in benzene or pentane yields the liquid bis(trimethylsilyl)(dimethylchlorosilyl)phosphane $(Me_3Si)_2P\!-\!SiMe_2Cl$,[16] which can be isolated and does not undergo further reactions (for example, elimination of Me_3SiCl would be expected). The reaction of the same starting materials in a molar ratio of 1:2 yields the crystalline tetrakis (trimethylsilyl)diphosphinodimethylsilane $Me_2Si[P(SiMe_3)_2]_2$, which eliminates $P(SiMe_3)_3$ already at 20–60 °C, indicating a gradual decomposition, whereas short-time thermolysis at about 220 °C produces complete rearrangement to a cyclic compound (1), R = R' = Me, R'' = $SiMe_3$:[16]

$$2\ Me_2Si[P(SiMe_3)_2]_2 \longrightarrow [Me_2Si\!-\!P(SiMe_3)]_2 + 2\ P(SiMe_3)_3 \qquad (8)$$

On prolonged thermolysis (38 h, 200 °C, sealed tube) an additional reaction, further eliminating $P(SiMe_3)_3$, proceeds to yield $P_4(SiMe_2)_6$ (3), R = Me, a compound with an adamantane structure.[4] The structure of this compound has been confirmed by an X-ray diffraction study.[17] This stable crystalline compound (not decomposed at 400 °C in a sealed tube) is also generated in a number of similar reactions (R = Me):[4]

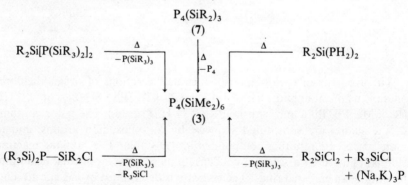

Another silaphosphane with the inverse adamantane structure (4), of composition $(PhSi)_4(PPh)_6$, was formed in the reaction of $PhSiCl_3$ with K_2PPh.[18] However, this compound was characterized only by elemental analysis.

If Me_2SiCl_2 or $MeSiHCl_2$ are reacted with LiHPMe in a molar ratio of
1 : 2, $MePH_2$ and LiCl are eliminated, with formation of compounds
$(Me_2Si—PMe)_2$ (1), $R = R' = R'' = Me$ (in 60% yield) and (MeH-
$Si—PMe)_3$ (2), $R = H$, $R' = R'' = Me$, respectively.[19] The compound (2),
$R = R' = R'' = Me$, is also formed in the reaction of $MePLi_2$ with
Me_2SiCl_2 (24% yield), together with unknown silaphosphanes.[19] The compounds $Me_2Si(PHMe)Cl$ and $Me_2Si(PHMe)_2$ anticipated as intermediates
in the reactions cited could not be isolated.[19]

In the reaction of Me_2SiCl_2 with $LiPHBu^t$ the bulky t-butyl groups afford
the isolation of $Me_2Si(PHBu^t)_2$, in contrast with the unstable
$Me_2Si(PHMe)_2$. The thermolysis of $Me_2Si(PHBu^t)_2$ generates Bu^tPH_2, the
four-membered ring (1), $R = R' = Me$, $R'' = Bu^t$, and a bicyclic compound
with a norbornane structure (5), $R = Me$, $R' = Bu^t$.[19] The molecular
structure of (5) was confirmed by an X-ray diffraction analysis.[20] The
compound $Me_2Si(PHBu^t)_2$ can be lithiated with $LiBu^n$ to form a crystalline
derivative $Me_2Si(PLiBu^t)_2$, which can be used in a variety of reactions,
leading to cyclic and non-cyclic silaphosphanes, for example a spiro compound (6), $R = Me$, $R' = Bu^t$.[19]

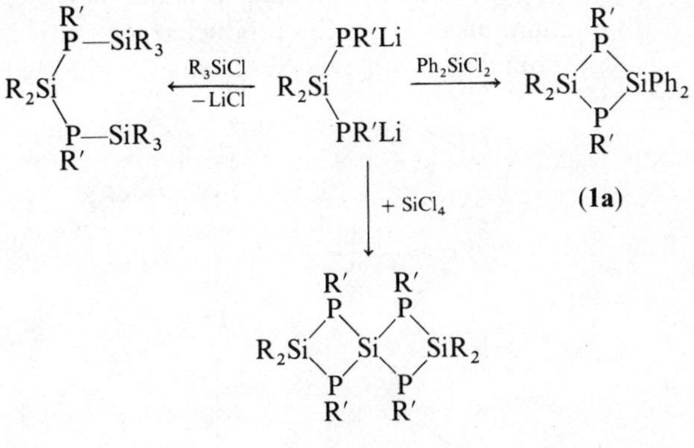

(1a)

(6)

The reaction of $Bu_2^tSiCl_2$ with LiPHMe illustrates again the role of bulky
substituents at silicon sites, which favour the formation of the four-
membered ring (1), $R = R' = Bu^t$, $R'' = Me$.[19] In contrast, in the reaction
of Me_2SiCl_2 with the same phosphide only the six-membered ring compound (2), $R = R' = R'' = Me$, was isolated.

Reactions of Ph_2SiCl_2 with K_2PPh and KHPPh lead to four- and
six-membered ring compounds $(Ph_2Si—PPh)_n$ with $n = 2$ and 3, (1) and (2),
$R = R' = R'' = Ph$.[21]

The reaction of $Bu_2^tSiF_2$ with $LiPH_2$ produces di-t-butylfluorosilylphosphane $Bu_2^tSiF(PH_2)$, which reacts with $LiBu^n$ in hexane to eliminate LiF and butane, with formation of the dimeric compound (1), $R = R' = Bu^t$, $R'' = H^{22}$ in 82% yield.

In the many reactions with monolithiated phosphines described above, the mono- and disubstituted intermediate stages could be identified or isolated; in contrast, the reactions of the dilithiated phosphides with alkyl- and aryldichlorosilanes lead directly to the cyclic compounds. Thus the reaction of Me_2SiCl_2 with Li_2PPh in 1 : 1 molar ratio at $-40\,°C$ yields the four-membered ring compound (1), $R = R' = Me$, $R'' = Ph$. The same reaction at $+40\,°C$ takes another course; in this case the six-membered ring compound (2), $R = R' = Me$, $R'' = Ph$, is formed. The four-membered ring (1) is stable in the solid state at $20\,°C$. In solution an equilibrium between the dimer (1) and trimer (2), $R = R' = Me$, $R'' = Ph$, takes place. Elevated temperatures favour the six-membered-ring derivative, which is observed at $150\,°C$.[8]

The four-membered ring compound (1), $R = R' = Bu^t$, $R'' = H$, mentioned above is also accessible by reaction of $Bu_2^tSiCl_2$ with Li_2PH (formed from PH_3 and 2 LiBu).[23] It can be lithiated with $LiPH_2$.DME in dimethoxyethane, with elimination of PH_3; the product reacts with Me_2SiCl_2 to form a bicyclic compound (8) with two Si_2P_2 rings connected through a $-SiMe_2-$ bridge:[23]

$$(9)$$

(8)

9.4   Silaphosphanes with cage structures

Dodecamethylhexasilatetraphosphaadamantane $P_4(SiMe_2)_6$ (3) has already been mentioned above as an example of polycyclic silaphosphane.[4] Two derivatives of (7) with R = Me[24] and Ph[25] are generated along with other products from phosphorus, sodium–potassium alloy and R_2SiCl_2 (R = Me, Ph) as colourless crystals. The structure of the two compounds was deduced from ^{31}P NMR spectral data, and an X-ray diffraction analysis of (7), R = Me, confirmed the rather unusual cage structure.[26]

The thermal decomposition of (7), R = Me, at 240 °C yields the adamantane (3), R = Me, with elimination of P_4 and other, unidentified, products.[4]

9.5 Chemical behaviour of cyclosilaphosphanes

The chemical behaviour of cyclic Si–P compounds has been little explored. While the chemistry of simple silaphosphanes, dealing with formation and cleavage of Si–P bonds and their applications in synthesis, has been developing continuously,[27] there is only relatively modest knowledge on the reactivity of cyclic silaphosphanes.

Silaphosphanes are extremely sensitive to air and moisture, but can be stored under nitrogen or noble gases. Silaphosphanes with higher vapour pressure ignite spontaneously in air.[19]

The cleavage of Si–P bonds by polar reagents is well known. Therefore it is not surprising that LiMe cleaves Si–P bonds in $(Me_2Si—PMe)_3$ (2), R = R′ = R″ = Me, to yield a lithiated compound $MeP(SiMe_2—PMeLi)_2$, which can be isolated as an etherate containing two moles of DME.[19] This dilithiated compound could be an interesting precursor for many new cyclic compounds, but its reactions with nonmetal halides have not been investigated.

Since cyclosilaphosphanes contain phosphorus(III), these compounds can act as ligands in various transition-metal complexes. The first transition-metal complexes of silaphosphane ligands incorporated noncyclic derivatives, such as $P(SiMe_3)_3$ and $R_nP(SiMe_3)_{3-n}$,[28–30] but a compound containing a cyclic ligand, namely $(Ph_2Si—PPh)_3Mo(CO)_3$ has also been described.[8]

A P-coordinated six-membered ring $PhP(SiMe_2SiMe_2)_2PPh$ has been built up stepwise directly at the metal centre, by reactions of coordinated ligands (R = Ph, methyl groups at silicon are not shown)[31] (next page).

This synthesis was followed by the preparation of other complexes with cyclosilaphosphane ligands. Thus reactions of $Ni(CO)_4$ and $Cr(CO)_5THF$ with four-membered-ring cyclosilaphosphanes (1), R = R′ = Me, R″ = Bu[t], and (1a), R = Me, R′ = Bu[t], leads to the complexes (9) and (10).[32] The binuclear complex (11) is obtained by the reaction of two moles of $Ni(CO)_4$ with (1), R = R′ = Me, R″ = Bu[t].[32]

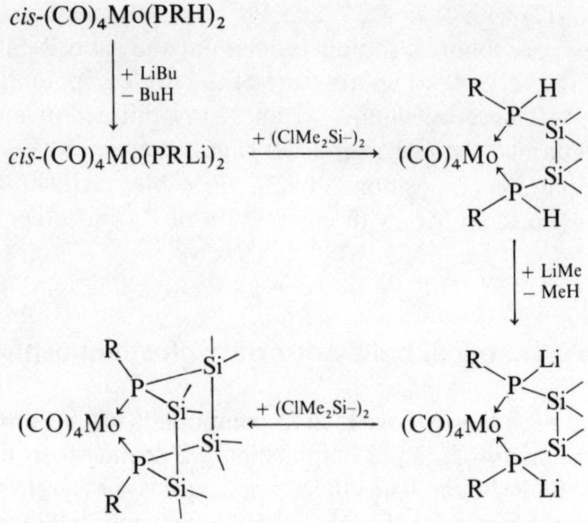

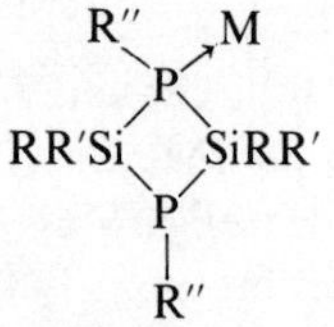

cis-$(CO)_4Mo(PRH)_2$

$\downarrow$ + LiBu / − BuH

cis-$(CO)_4Mo(PRLi)_2$ $\xrightarrow{+ (ClMe_2Si-)_2}$

$(CO)_4Mo$ — structure with R, H, P, Si, Si, P, R, H

$\downarrow$ + LiMe / − MeH

$(CO)_4Mo$ — structure with R, Li, P, Si, Si, P, R, Li $\xleftarrow{+ (ClMe_2Si-)_2}$ $(CO)_4Mo$ — cage structure with R, Si, Si, Si, P, P, R, Si

Structures:

(9) — R″, M, P, RR′Si, SiRR′, P, R″

(10) — R′, M, P, R₂Si, SiPh₂, P, R′

(11) — R″, M, P, RR′Si, SiRR′, P, M, R″

(9) M = Ni(CO)₃
M = Cr(CO)₅

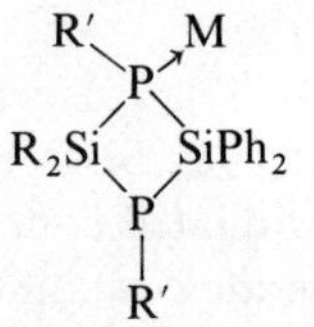

(10) M = Ni(CO)₃

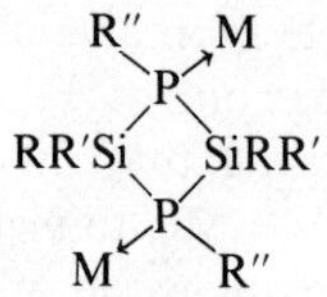

(11) M = Ni(CO)₃

The cage compound (7) can in principle coordinate at several P atoms of the cage. However, the electronic environment of the four P atoms seems to be nonequivalent. There is an obvious tendency to coordinate first at one of the basal phosphorus atoms, incorporated in the three-membered ring, and then to continue the process at the apical P atom. Coordination to all four P atoms has not been observed so far. When two P atoms of the basal P_3 ring coordinate, the cage (7) may act as a bridging ligand. Thus (7), R = Me, reacts with $Mo(CO)_6$, $Cr(CO)_5THF$ or $(\eta^5\text{-}C_5H_5)Mn(CO)_2THF$ to form compounds of type (12), and a dimetallic derivative (13) has also been isolated; the trimetallic compound (14) can be obtained with an excess of $Cr(CO)_5THF$ (methyl groups at silicon are not shown):[33]

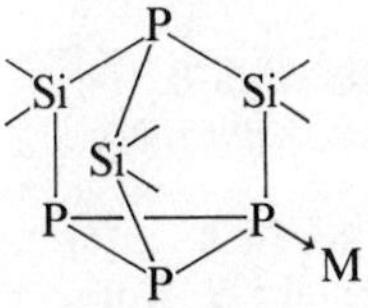

(12) (13) (14)

M = Cr(CO)$_5$ M = CpMn(CO)$_2$ M = Cr(CO)$_5$

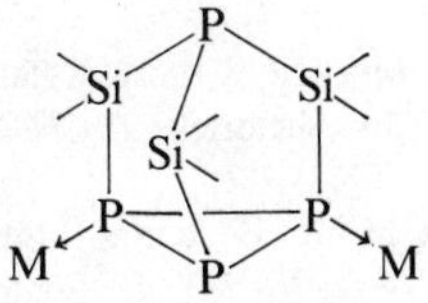

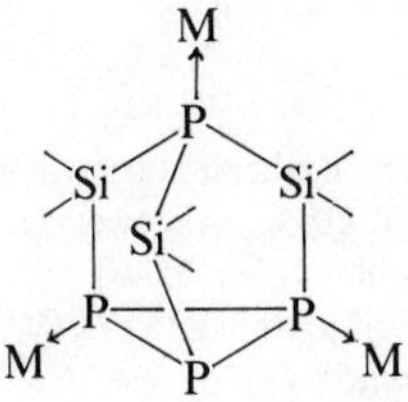

The reaction of the cage (7), R = Me, with tetracarbonylnorborna-dienechromium yields (15),[34] while (C$_6$H$_6$)Cr(CO)$_3$ or (CHT)Cr(CO)$_3$ (CHT = cycloheptatriene) are sources of Cr(CO)$_3$ fragments and yield a more complex compound [P$_4$(SiMe$_3$)$_3$]$_3$Cr$_2$(CO)$_6$ (16) (methyl groups at silicon are not shown):[35]

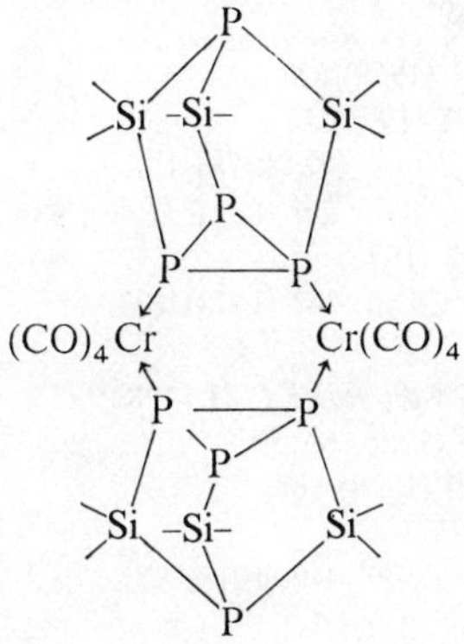

(15) 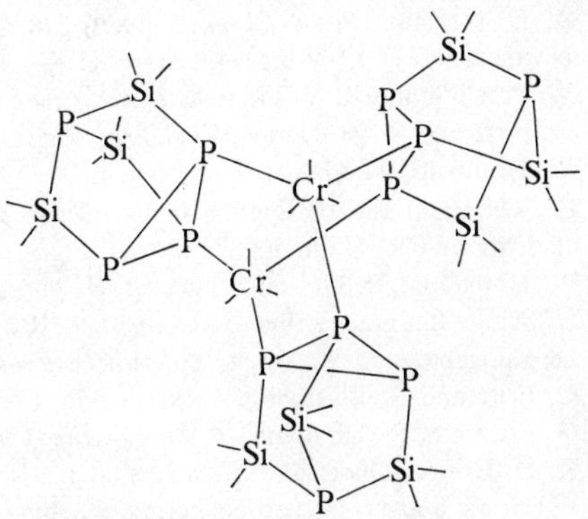 (16)

Bibliography

B 1   G. Fritz, *Angew. Chem.* **78** (1966) 80; Int. Ed. Engl. **5** (1966) 53.

B 2   E. A. Chernyshev, E. F. Bugerenko, *Organomet. Chem. Rev.* **A3** (1968) 469.

B 3   S. N. Borisov, M. G. Voronkov and E. Ya. Lukevits, *Organosilicon Derivatives of Phosphorus and Sulfur* (Plenum Press, New York, 1971).

B 4   I. Haiduc, *The Chemistry of Inorganic Ring Systems*, Part 1, (Wiley-Interscience, London, 1970), p. 592.

References

1 G. Fritz and G. Poppenberg, *Angew. Chem.* **72** (1960) 208; G. Fritz, *Angew. Chem.* **78** (1966) 80; C. Glidewell and G. M. Sheldrick, *J. Chem. Soc.* A **1969**, 350.

2 L. Maier, *Helv. Chim. Acta* **46** (1963) 2667.

3 H. Schumann and W. W. du Mont, *Z. Anorg. Allg. Chem.* **418** (1975) 259; *Angew. Chem.* **87** (1975) 354, *Int. Ed.* **14** (1975) 368; H. Schumann and B. Wöbke, *Chem. Ber.* **109** (1976) 1017.

4 G. Fritz, R. Uhlmann and W. Hölderich, *Z. Anorg. Allg. Chem.* **442** (1978) 86.

5 H. Schäfer, G. Fritz and W. Hölderich, *Z. Anorg. Allg. Chem.* **428** (1977) 222; J. E. Drake and C. Riddle, *Q. Rev.* **24** (1970) 263.

6 G. Fritz, *Z. Anorg. Allg. Chem.* **280** (1955) 332; G. Fritz, *Z. Naturforsch.* **8b** (1953) 776.

7 J. E. Drake and W. L. Jolly, *Chem. Ind. (London)* **1962**, 1470; S. D. Gokhale and W. L. Jolly, *Inorg. Chem.* **3** (1964) 1141; **4** (1965) 596.

8 R. T. Oakley, D. A. Stanislawski and R. West, *J. Organomet. Chem.* **157** (1978) 389.

9 T. H. Newman, R. West and R. T. Oakley, *J. Organomet. Chem.* **197** (1980) 159.

10 A. W. Cordes, P. F. Schubert and R. T. Oakley, *Can. J. Chem.* **57** (1979) 174.

11 T. H. Newman, J. C. Calabrese, R. T. Oakley, D. A. Stanislawski, and R. West, *J. Organomet. Chem.* **225** (1982) 211.

12 E. Hengge, R. Petzold and U. Brychcy, *Z. Naturforsch.* **20b** (1965) 397.

13 R. Uhlmann, Dissertation, Universität Karlsruhe (1979).

14 A. D. Norman, *Inorg. Chem.* **9** (1970) 870; *J. Am. Chem. Soc.* **90** (1969) 6556; A. D. Norman and D. C. Wingeleth, *Inorg. Chem.* **98** (1969).

15 W. Parshall and R. V. Lindsey, *J. Am. Chem. Soc.* **81** (1959) 6273.

16 G. Fritz and W. Hölderich, *Z. Anorg. Allg. Chem.* **431** (1977) 76.

17 W. Hönle and H. G. von Schnering, *Z. Anorg. Allg. Chem.* **442** (1978) 91.

18 H. Schumann and H. Benda, *Angew. Chem.* **24** (1969) 1049; *Int. Ed.* **8** (1969) 989.

19 G. Fritz and R. Uhlmann, *Z. Anorg. Allg. Chem.* **442** (1978) 95.

20 W. Hönle and H. G. von Schnering, *Z. Anorg. Allg. Chem.* **442** (1978) 107.

21 H. Schumann and H. Benda, *Chem. Ber.* **104** (1971) 333.

22 U. Klingebiel and N. Vater, *Angew. Chem.* **94** (1982) 870; *Int. Ed.* **21** (1982) 875.

23 G. Fritz and R. Biastoch, *Z. Anorg. Allg. Chem.* **535** (1986) 63, 95.

24 G. Fritz and R. Uhlmann, *Z. Anorg. Allg. Chem.* **440** (1978) 168.

25 K. D. Hoppe, Dissertation, Universität Karlsruhe (1982).

26 W. Hönle and H. G. von Schnering, *Z. Anorg. Allg. Chem.* **440** (1978) 171.

27 G. Fritz, *Comments Inorg. Chem.* (1982).

28 H. Schumann and O. Stelzer, *Angew. Chem.* **79** (1967) 692; *Int. Ed.* **6** (1967) 701.

29 H. Schumann, O. Stelzer, J. Kuhlmey and U. Niederreuther, *Chem. Ber.* **104** (1971) 993.

30 (a) H. Schumann, L. Rösch, H. Neumann and H. J. Kroth, *Chem. Ber.* **108** (1975) 1630. (b) H. Schumann and H. Benda, *Angew. Chem.* **82** (1970) 46; *Angew. Chem. Int. Ed. Engl.* **9** (1970) 76; *Chem. Ber.* **104** (1971) 333.

31 G. Johannsen and O. Stelzer, *Chem. Ber.* **110** (1977) 3438.

32 G. Fritz and R. Uhlmann, *Z. Anorg. Allg. Chem.* **463** (1980) 149.

33 G. Fritz and U. Kaever-Theobald, unpublished results.

34 G. Fritz and R. Uhlmann, *Z. Anorg. Allg. Chem.* **465** (1980) 59; W. Hönle and H. G. von Schnering, *Z. Anorg. Allg. Chem.* **465** (1980) 72.

35 G. Fritz, R. Uhlmann, K. D. Hoppe, W. Hönle and H. G. von Schnering, *Z. Anorg. Allg. Chem.* **491** (1982) 83.

10

Silicon–Oxygen Heterocycles (Cyclosiloxanes)

V. CHVALOVSKÝ

Institute of Chemical Process Fundamentals, Prague, Czechoslovakia

10.1 Introduction

This chapter deals with inorganic heterocycles made up by alternation of silicon and oxygen; silicon–oxygen heterocycles containing Si–Si bonds have already been discussed in Chapter 7 (Section 7.2.4). The basic types of cyclosiloxanes, their methods of synthesis and their most important reactions have been reported previously and have been fully reviewed.[B 1 – B 7]

Silicon–oxygen heterocycles of various ring sizes are known mainly as organic derivatives $(RR'SiO)_n$; the lower limit of n is 2, in cyclodisiloxanes (**1**), and there seems to be no upper limit. Trimers (**2**), tetramers (**3**), pentamers (**4**) and hexamers (**5**) have been well characterized, and cyclosiloxanes with $n \geqslant 7$ have also been identified.

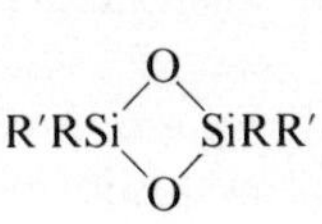

(1)

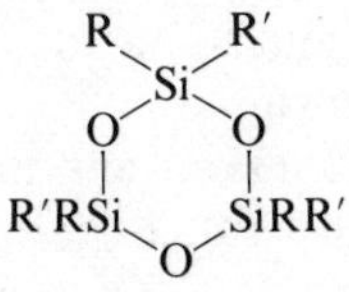

(2)

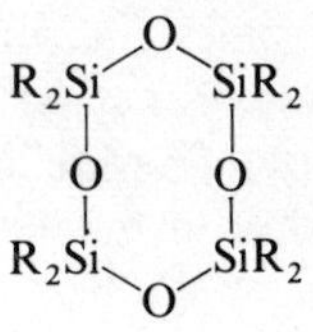

(3)

THE CHEMISTRY OF INORGANIC HOMO-
AND HETEROCYCLES VOL 1 ISBN 0-12-655775-6

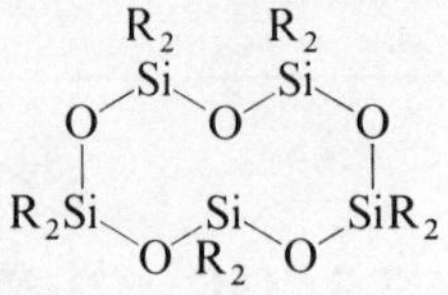

R$_2$ R$_2$
Si   Si
O   O   O
R$_2$Si   Si   SiR$_2$
O   R$_2$ O

(4)

R$_2$
Si
O   O
Si   Si
O   R$_2$ R$_2$ O
R$_2$Si   Si   SiR$_2$
O   R$_2$ O

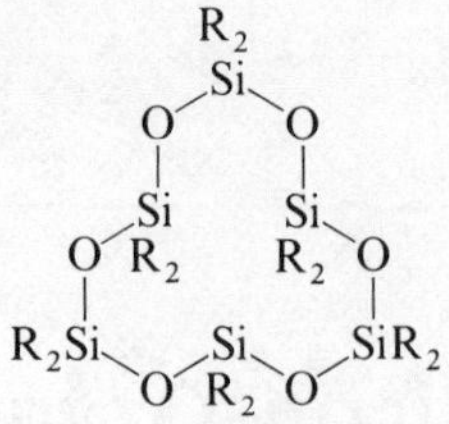

(5)

The cyclodisiloxane ring (1) was reported as a derivative with R = Me, R' = OSiMe$_3$.[104,105] This dimeric structure was not demonstrated and the assignment is doubtful.[936] An interesting development was the synthesis of a nine-membered ring system (6) containing three peroxo groups, alternating with silicon atoms (R = Me):[719]

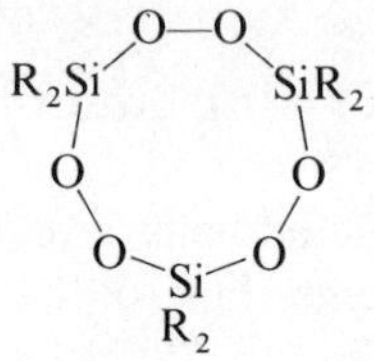

O—O
R$_2$Si   SiR$_2$
O   O
O   O
Si
R$_2$

(6)

The interest in the chemistry of cyclosiloxanes, their structure, physical properties and biological activity has been stimulated by the development of their applications. A large number of publications in this field (over 1200 papers between 1969 and the middle of 1982) documents the effort made in order to get a deeper knowledge of this class of compounds. Not only are the organic substituents on silicon varied, but the skeleton structure is also. Thus, in addition to monocyclic siloxanes, polycyclic derivatives and hetero-cyclosiloxanes (in which silicon atoms are partially substituted for other metalloids or metals) are known.

Because of severe space limitations and the large volume of literature to be covered, in this chapter the physical properties of cyclosiloxanes, their methods of synthesis and reactions are summarized in Tables 10.1–10.5. Therefore only some highlights and typical examples of the preparative works will be briefly commented upon.

From the viewpoint of industrial application, organocyclosiloxanes are important, since in the presence of acidic or basic catalysts they undergo polymerization to produce silicone fluids and elastomers, the linear chains of which are terminated by OSiR$_3$ or hydroxo groups. Much research effort has been spent on development of various techniques of polymerization of cyclosiloxanes, as well as on the study of their physical properties and applications.

The chemistry of organocyclosiloxanes has some biological implications and interest. Thus 2,6-*cis*-diphenylhexamethylcyclotetrasiloxane has been reported to have oestrogenic antigonadotropic activity in some mammalian species, but has no effect on the reproduction organs and fertility of birds.[3,4] Since $(Me_2SiO)_4$ and $(Me_2SiO)_5$ are components of a lotion for skin care,[805] the toxic and allergenic properties of some cyclosiloxanes, for example $(MePhSiO)_3$ and $(MePhSiO)_4$ have been studied in detail.[902]

Other applications use the fact that cyclic organopolysiloxanes behave as ionophores for alkali-metal ions; their cation selectivity depends on their ring size.[645,646]

Carbon-functional organocyclosiloxanes have a broad field of application; the fluoroalkyl derivatives impart good water and oil resistance to textiles,[843] 4-aminobutylheptamethylcyclotetrasiloxane can be used as the extracting phase of ion extraction from aqueous solutions,[263] and 2,4,6,8-tetrakis(mercaptopropyl)-2,4,6,8-tetramethylcyclotetrasiloxane can be used as a component for photocurable printing-ink compositions.[827] Cyclosiloxanes containing an azido group in an aromatic side-chain are used for the preparation of photosensitive plates for lithography.[848,849] Other uses can also be mentioned: diethylaminooxy-substituted cyclotetrasiloxanes are employed for the preparation of room-temperature curing compositions,[305] and small amounts of alkylalkoxy-substituted cyclosiloxanes improve the melt flow and impact properties of polycarbonates.[560]

10.2   Preparation

10.2.1   Hydrolysis

The most frequently used method of synthesis of organocyclosiloxanes is the hydrolysis of diorganochlorosilanes. The effect of reaction conditions on the yield of individual cyclic derivatives, for example the effect of solvents, completely or partially soluble in water, has been studied recently.[13]

The separation of dimethylcyclosiloxanes was developed with the use of preparative gel chromatography.[291]

Much attention has been given to the cohydrolysis of different diorganodichlorosilanes, and the yields of cyclotri- and cyclotetrasiloxanes have been analysed.[665,666,933] Some less common methods of hydrolysis are mentioned, including the reaction of Me_2SiCl_2 with sodium carbonate at 300 °C,[559] the cohydrolysis of diorganodichlorosilanes with organosilazanes,[506] and the vapour-phase hydrolysis of $Cl(SiMe_2O)_nSiMe_2Cl$ with small amounts of water on activated carbon.[320] The same catalyst was applied successfully to the condensation of 1,5-dihydroxo-organocyclosiloxanes, leading to linear organosiloxanes.[28]

Carbon-functional cyclosiloxanes can be prepared in a similar manner. Functional groups at silicon undergo hydrolysis, leading to formation of cyclosiloxanes, while those in the side-groups (e.g. esters) are not affected.[928]

$$(1)$$

10.2.2   Heterofunctional condensation

Several workers developed the synthesis of cyclosiloxanes with a desired sequence of organosiloxane units by means of heterofunctional condensation reactions. In the presence of pyridine, organochlorosilanes react with organosilanols to give cyclosiloxanes; this reaction has been much used.[458,501] It has also been applied to the synthesis of polycyclic siloxanes,[17,90,583,602,603,620,772] including spirocyclic compounds.[930]

10.2.3   Depolymerization

The degradation of linear polydimethylsiloxanes to cyclosiloxanes proceeds at high temperature in an inert atmosphere.[630] This reaction can be used in connection with pyrolytic gas chromatography and mass spectrometry for identification of polydimethylsiloxanes.[440] Thermodynamic studies of this reaction have been reported.[432]

Cyclotrisiloxanes together with cyclotetrasiloxanes are formed as the main products during thermal degradation of poly(methylphenyl)-,[367] dimethyldiphenyl-[336] and dimethylphenylmethyl-[365] siloxane copolymers.

The alkali-catalysed depolymerization of linear polydimethylsiloxanes, proceeding under vacuum at temperatures below 200 °C in the presence of KOH,[456] or alkali-metal silanolates[69] has been reported. Cyclic products are formed under similar conditions from methyl(fluoroalkyl)siloxanes,[441,442] leading to cyclotetrasiloxanes like $(CF_3CF_2CH_2CH_2MeSiO)_4$.

The degradation of cross-linked polydimethylsiloxanes can be performed at about 140 °C both with basic and acidic catalysts (e.g. 65% H_2SO_4).[226]

10.3 Chemical properties and reactions

10.3.1 Polymerization

The most studied reaction of cyclosiloxanes is their anionic and cationic polymerization.[158] Copolymerization of two different cyclosiloxanes, for example $(Me_2SiO)_4$ with $Bu^iMe_7Si_4O_4$ or $Bu^i_2Me_6Si_4O_4$ is also possible.[927] Analogously, diphenyldimethylsiloxane copolymers have been prepared and the sequence distribution of the siloxane units investigated using ^1H NMR as a function of the reaction condition in the reaction of $(Me_2SiO)_3$ with $(Ph_2SiO)_3$.[511]

The cyclotetrasiloxane $(Me_2SiO)_4$ has been copolymerized with a spiro-[5.7]hexasiloxane, and aminofunctional derivatives have been subjected to similar reactions.[307,763]

Acid-catalysed polymerization. The kinetics and mechanism of acid-catalysed polymerization of cyclosiloxanes have been reviewed.[244,584] The most frequently used catalyst is sulphuric acid, for example in the polymerization of dimethylcyclosiloxanes[557] or in copolymerizations of chloromethyl-substituted cyclosiloxanes.[433] Less frequently used are phosphoric acid,[212] trifluoroacetic acid (for examination of the kinetics and mechanism of the ring-opening reactions),[364,889] methanesulphonic acid[363] and $C_4F_9SO_3H$.[277,442] Perfluorinated polymers containing sulphonic-acid groups can be used as equilibration catalysts for the rearrangement of Si–O bonds.[866] In the copolymerization of $(Me_2SiO)_4$ with $(MeViSiO)_4$[613] or with $(MeO)_3Si(CH_2)_3OCH_2(C_2H_3O)$,[652] dodecylbenzenesulphonic acid and octylbenzenesulphonic acid[653] have been used, to give copolymers as stable emulsions.

The siloxane bond can be split on activated acid solid surfaces. In the presence of active carbon, $(Me_2SiO)_4$ disproportionates in the vapour phase in the presence of HCl, to form a mixture of cyclic $(Me_2SiO)_n$, with $n = 3, 5, 6, 7$.[318] On the surface of acid forms of laminated silicates like caolinite and montmorillonite[216–218,284] the polymerization of $(Me_2SiO)_4$ proceeds at 90 °C. The equilibration of methylhydrogencyclosiloxanes has been performed with ion exchangers (wofatite).[211]

Base-catalysed polymerization. Potassium hydroxide was the earliest and is still the most commonly used catalyst for anionic polymerization; preparative and kinetic studies on the KOH-catalysed copolymerization of $(Me_2SiO)_4$ with $(CF_3CH_2CH_2MeSiO)_3$[337] and with $(CF_3CF_2CH_2CH_2Me-SiO)_4$[441] have been reported. The influence of organic substituents on the kinetics of the polymerization of arylorganocyclosiloxanes[635] has been investigated, as has the influence of temperature on the equilibrium concentration of $(Me_2SiO)_4$ and $(Me_2SiO)_5$ during polymerization.[822] Potassium hydroxide was found useful in the copolymerization of $(Me_2SiO)_4$ with alkyltrialkoxysilanes like $MeSi(OMe)_3$[787] or $H_2NCH_2CH_2NH(CH_2)_3$-$Si(OMe)_3$.[235,740] A similar copolymerization of cyclotrisiloxanes has been studied.[607]

In the copolymerization of $(Me_2SiO)_4$ with other cyclotetrasiloxanes, for example $(MeViSiO)_4$, KOH was used in the presence of a heterocyclic compound such as 3'-nonylbenzo-18-crown-6, which increases the polymerization rate.[723]

In comparison with KOH, other anionic polymerization catalysts such as $NaOH$,[191] NH_4OH,[713] NMe_4OH,[556] NR_4F,[83] Me_3SiONa[139] and Me_3SiOK[681] are much less commonly used. The potassium silanolate can be supplied with advantage (at 120 °C) as a cation-complex catalyst containing a molar equivalent of a cyclic (crown) polyether.[312]

During recent years the anionic polymerization of cyclosiloxanes has been frequently induced by organometallic catalysts. Thus BuLi initiates the copolymerization of $(Me_2SiO)_3$ with functional trimethoxysilanes,[564] and lithiostilbene the copolymerization of $(CF_3CH_2CH_2MeSiO)_3$ with $(Ph_2SiO)_3$.[702] Block copolymers from $(Me_2SiO)_3$ and $(Me_2SiO)_4$ with styrene were prepared by anionic polymerization in the presence of lithium biphenyl (or sodium biphenyl),[624] from $(Me_2SiO)_3$ by initiation with a living polymer obtained from $Me_3SiCH{=}CH_2$ and Bu^sLi[624] or sodio-polystyrene,[252,864] and also by copolymerization of $(Me_2SiO)_3$ with $(Ph_2SiO)_3$ using BuLi and $Ph_2Si(OLi)_2$ as initiators, in DMSO or THF as promoters.[402] The litshium-based catalysts minimize the siloxane rearrangement reactions.

Heterogeneous anionic polymerization of $(Me_2SiO)_4$ and $(MePhSiO)_3$ in the presence of intercalation compounds of graphite with K and Li gives high-molecular-weight products.[638] Potassium graphite derivatives, like C_8K, $C_{24}K$ and $C_{34}K$, are stronger catalysts than C_6Li or $C_{18}Li$, probably because of the lower negative charge on the carbon network of the graphite in C_xLi.[414] The use of tetrahydrofuran as promoter proved to be useful in combination with intercalation compounds $C_{6-36}M$ (M = Li, Na, K) as catalysts for polymerization of methyl-, vinyl- or trifluoromethyl-substituted cyclotrisiloxanes.[110]

A completely different type of polymerization is low-temperature plasma polymerization, which can be used, for instance, with $(Me_2SiO)_3$ and $(Me_2SiO)_4$ to prepare silicone-coated membranes with better blood compatibility than silicone-rubber membranes.[236,237] It proceeds through both ring enlargements and dimerization, leading to macrocyclic organosiloxanes.[463]

10.3.2   Reactions on substituents of organocyclosiloxanes

The most studied reaction taking place on substituents of cyclosiloxanes without redistribution of the siloxane bonds, i.e. with preservation of the cyclosiloxane ring, is hydrosilylation of Si–H derivatives with vinyl- and allyl-substituted compounds, in the presence of H_2PtCl_6.[753] Thus the reaction of $(MeViSiO)_4$[611] or $(MeViSiO)_3$[71] with methylhydrogensiloxanes in the presence of this catalyst[428] or $(PPh_3)_3RhCl$[800] leads to cyclolinear products. By hydrosilylation of divinyl-substituted cyclosiloxanes with bis-(dimethylhydrosilyl)carborane, novel carborane–siloxane polymers have been prepared.[925]

Other reactions of the substituents have been reported, for example the thermal oxidation of $(Ph_2SiO)_4$ at 300–350 °C, in which some of the phenyl groups became phenylene groups.[353]

Organocyclosiloxanes containing phosphine-substituted side-chains can act as ligands. Thus siloxane-bound metal complex catalysts can be prepared by exchange reactions of (Ph_2PCH_2) $Me_7Si_4O_4$ with $RhCl(CO)$-$(PPh_3)_2$ or $Rh_2Cl_2(CO)_4$, and the products are catalytically active in hydroformylation reactions.[323]

10.4   Silsesquioxanes

Silsesquioxanes $(RSiO_{1.5})_n$ are polycyclic cage-like siloxanes; these compounds have attracted the attention of both academic and industrial laboratories, owing to their interesting properties and possible applications. For example, the addition of oligomeric methylsilsesquioxanes increases the arc resistance of phenyl–formaldehyde resins,[372] and poly(methylsilsesquioxane)-based composite materials have a high tensile strength and a low water sorption; poly(phenylsilsesquioxanes) are used for the preparation of molecular orientation-controlling layers of liquid-crystal display devices.[391]

The methods of synthesis and the structure of silsesquioxanes have been thoroughly investigated.[399,668] Ethylbicyclosiloxanes and ethyltricyclosiloxanes, containing silsesquioxane units, were prepared by cohydrolysis of $EtSiCl_3$ and $EtHSiCl_2$.[494,495] The cohydrolysis of $MeSiCl_3$ and $PhSiCl_3$ gives mixed methyl–phenyl silsesquioxanes.[796] Alkoxysilanes can also be used as starting materials; thus poly(methylsilsesquioxanes) have been prepared by hydrolysis of $MeSi(OMe)_3$ in the presence of $Ca(OH)_2$.[451]

Much attention has been paid to the physical properties of organosilsesquioxanes,[727] especially their surface properties,[796] dynamic mechanical properties[905] and thermal behaviour. The relatively most stable poly-(phenylsilsesquioxanes) undergo thermal changes due to cross-linking and other defects in the ladder structure.[727] The nature of the organic substituents has a predominant influence on the mass-spectral fragmentation of organosilsesquioxanes;[494] with methyl and ethyl derivatives only the cleavage of the alkyl groups from silicon is involved, with vinyl derivatives Si–O bonds are broken.[872]

The organosilsesquioxanes are often used as starting compounds for the synthesis of high polymers; phenylsilsesquioxanes can be copolymerized with diorganocyclosiloxanes,[920] and cyclolinear block copolymers have also been prepared by polycondensation of dihydroxosilsesquioxanes with diaminodiorganosiloxanes.[29]

10.5   Heterocyclosiloxanes

Organosubstituted heterocyclosiloxanes, with some silicon atoms of the siloxane units replaced by other elements, most frequently boron, phosphorus, aluminium and other metals, have received some attention. Copper-containing methylcyclosiloxanes were claimed[429] to result from methylorganodiacetoxysilanes and $CuCl_2$; their structure, however, is not fully documented. Cyclolinear polymeric structures have been prepared from $EtSiCl_3$ or $PhSiCl_3$ and Mg, Cu, Zn and Cd acetates.[264]

Cycloborasiloxanes and cyclophosphasiloxanes are prepared from boric and phosphoric acids respectively, with diorganosilanediols and diorganocyclosilazanes,[695] or from H_3BO_3 and Ph_2SiCl_2 (heterofunctional condensation)[900] or $Ph_2Si(OH)_2$ (homofunctional condensation)[445]. Aluminium chloride is able to split the siloxane bond, and it reacts with $(Me_2SiO)_4$ at 120 °C to form an aluminocyclosiloxane $ClAl(Me_2Si)_2O_3$,[106] which is catalytically active in the rearrangement of hydrocarbons and can also be used as catalyst in alkylation of benzene by propene.[448]

Table 10.1 Cyclotrisiloxanes

$$\begin{array}{c} R_2 \\ Si \\ O \quad O \\ R_2Si \underset{O}{\quad} SiR_2 \end{array}$$

Compound	Methods of Synthesis[a]	Reactions[a]	B.p.(°C/Torr)	M.p.(°C)	Other physicochemical investigations[b]
$(H_2SiO)_3$		6,24[322]			Q[435]
$[(Me)HSiO]_3$	6,11[293] 6,24[304]	6[293] 6,24[715] 2,15[187] 2,6,15[581]			A[175,176] Q[435]
$[Me_2SiO]_3$	6[422,835,434,660,642] 6,11[912,111] 6,11[482,485,22,297,300,913,294] 6,11[627,267,260,791,543] 6,22[265,360] 6,22[189,424,859,267,339] 6,16[720,729] 6,24[533,457] 6,10,24[357] 6,22,24[823,785] 6,7[396] 7,24[269] 22[166,10] 6,12[837] 6,10[359,358,8] 6,10,22[858,11] 6,14[759] 6,13[81] 6,11,24[528] 6,11,22[881] 6,10,14,22[165,164] 6,11,18,22[259]	6[145,909,379,271,389,516,578] 6[246,173,8] 6,22[792,228,672] 1,6,16[412] 6,16,24[325,439,586] 1,6,16,24[334,512] 22[370,626,643] 22[644,802,267,794] 6,11[133] 6,8[142] 6,8[671] 7,24[269,489,490,409,499] 7,24[55,447,27] 5,7,24[809,810,523] 6,24[919,912,826,825,111,141,580] 6,24[183,204,203,75,508,507,816] 6,24[112,333,410,387,563,385,270] 6,24[562,348,597,577,251,935,272] 6,24[332,163,241,650,651,649,526] 6,24[491,782,659,394,708,250] 6,24[242,24,587,908,248,573,574] 6,24[631,572,148,571,249,568,558] 6,24[737,321,311,566,510,247,492] 6,22,24[430,503,760] 22,24[239,240] 6,22,24[576] 3,6[358]	133–137[133] 134[913,422,561.] 898.457 135[529,909]	60–65[133] 63[111] 62[645] 64.5[913,529] 64[909,578,912.] 858.859.43	O[145,898,346,202,899] O[913,810,538,659,539] O[561,627] M[832,656,860] M[192] N[726,309,411,349] N[461,310,297,890,258] N[522,680] K[489,490,324] K[935,75,267,425,876] K[394,246,56,250,24,27] B[324,357,267,11,164] B[250,249,760] BK[251,586] T[919,623,535,208,209] T[618,425,277,466,465] KO[587] KL[260] OT[529] LS[669] Z[410,387,526,373] I[77,452] IR[1,327] X[640,200] X[733] D[551,550] LO[436] LOZ[542] LOZ[231] LMOZ[672] Q[4,35] IKVX[259]

Table 10.1 *continued*

Compound	Methods of Synthesis[a]	Reactions[a]	B.p.(°C/Torr)	M.p.(°C)	Other physicochemical investigations[b]
		$3,6,24^{273,738}$ $3,6,24^{359}$ $24^{894,575,373,878}$ $6,17^{570}$ $1,7,6^{411}$ $4,6,24^{379,522}$ $4,6,24^{177,178,663}$ $6,16^{324,842}$ $6,23,24^{56}$ $1,6,24^{838,347}$ $1,6,16,18,24^{274}$ $5,24^{525,509}$ $1,6^{453}$ $6,8,24^{670,43}$ $6,7,24^{565,664}$ $6,7,16,24^{524}$			
$Me_5ViSi_3O_3$	6^{140} $6,11^{485,294}$ $6,11,16^{473}$	$2,21^{35,227}$ $2,6,15^{882}$ $2,18,21^{37}$ $2,15,21^{40}$ $18,21,24^{15}$	$57/19^{485,294}$ $145/760^{473}$ $85/80^{140}$	— —	IMN^{485} K^{15} LVZ^{816} N^{680} J^{294}
$(NCCH_2CH_2)Me_5Si_3O_3$ $(MeViSiO)_3$	$6^{545,829}$ $6,24^{546}$ $6,11^{591,253}$ $6,22^{547}$ 23^{647}	24^{678} $6^{271,548}$ $6,24^{296,270,562,577,311}$ $6,22,24^{576}$ $4,24^{178}$ $6,17^{570}$ $19,20^{648}$ $2,15,21,24^{534,521,924}$ $5,7,24^{809,810}$	$79.5/20^{898}$ $58/5^{923}$ $80/20^{561}$	—	O^{538} M^{688} N^{680} K^{534} $O^{561,810,899}$ LO^{898} Z^{296}
$1,3\text{-}(C_4H_3S)_2Me_4Si_3O_3$	$6,11^{851}$				L^{851}
$Me_5PhSi_3O_3$	$6,11^{482,481,485,913,294}$ 6^{835} $6,11,16^{915}$	6^{882} $6,24^{914,227}$ 24^{678}	$94/6^{482}$ $235/760^{915,914}$ $70/15^{913}$	6.5^{485}	$N^{890,680,461}$ $M^{692,196}$ K^{914} K^{678} O^{913} $T^{280,281}$ B^{915} $Y^{179,386}$ ANO^{482} $O^{294,481,813}$
$Me_5EtSi_3O_3$		24^{678}			M^{689}
$\left(\underset{}{\text{SiO}}\right)_3$					

$(C_4H_3S)Me_3Si_3O_3$	6,11[851]				L[851]
$(Cl_2C_6H_3)Me_5Si_3O_3$	22,24[785]				
$trans\text{-}[(CH_3)(CF_3CH_2CH_2)SiO]_3$		22[335]			
$[Me(CF_3CH_2CH_2)SiO]_3$	6,11[685,711]	6[909,685] 6,24[910,919,413,918] 6,24[825,826,748,233,498,215,508] 6,24[736,296,270,911,935,272,828] 6,24[916,781,714,711,715,592,311] 6,24[709,710,724] 1,16,24[331,234] 1,7,24[330] 1,6[453] 6,22,24[675,678]	104/4[909] 104/4[783]	28–34[909]	K[910,935,876,675,916] O[676] N[680] Y[168] T[919] BK[781] YZ[183] Z[423]
$(Et_2SiO)_3$	6[605] 6,11[297]	6[909] 6,24[321,826,616] 4,6,24[178] 6,22[205] 7,24[604]	126/15[909]	—	A[175,176,207] AD[876] Z[605] T[282]
$1,3\text{-}Vi_2\text{-}5\text{-}Ph\text{-}1,3,5\text{-}Me_3Si_3O_3$	6,11,16[47]		80–83/1[47]	—	CN[47]
$(PhCH_2CH_2)Me_5Si_3O_3$	6,11,16[634]		115–117[634]	—	I[634]
$[(C_4H_3S)MeSiO]_3$					LOZ[541]
(structure) $Me_5Si_3O_3$	6,11,16[632] 6,24[633]		110–112[632]		
$1,1\text{-}Ph_2Me_4Si_3O_3$	6,11[111,481] 6,24[129] 22[794] 6,11,16[915]	6[882,786] 6,24[111,914,227] 6,22,23,24[844]	135/2[915] 145/4[129]	65[111] 64.9[15,914] 64.3[129]	B[915] IMN[794] K[914] LO[762] M[692] N[461,680] O[129] T[281,536] X[243]
$1,3\text{-}Ph_2Me_4Si_3O_3$	6[786,835,477] 6,11[482,481,485,377,294]	6,24[914]	153–157/6[482] 123–125/0.5[482] 162/0.03[377] 295/760[914] 150–152/1[47]		AO[482] CO[913] IMNY[377] K[914] M[196,692] Y[176,386] O[294,481]
$1\text{-}Vi\text{-}3,5\text{-}Ph_2\text{-}Me_3Si_3O_3$	6,11,16[47]				N[680] CN[47]
$cis\text{-}[(NaO)PhSiO]_3$				—	octahydrate X[298]
$[\{(NO_2)_2CFCH_2CH_2\}_2SiO]_3$	6,11[169]			207–209[169]	IN[169]
$cis\text{-}(Ph_2C{=}CPh)Me_5Si_3O_3$	6,11[636] 6,11,16[634]		159–161/2[634,636]	—	I[636,634]
$(PhCH_2CHPh)Me_5Si_3O_3$	6,11,16[634]		150–152/1[634]	—	I[634]

Table 10.1 *continued*

Compound	Methods of Synthesis[a]	Reactions[a]	B.p.(°C/Torr)	M.p.(°C)	Other physicochemical investigations[b]
$(Ph_2CHCH_2)Me_5Si_3O_3$	**6,11,16**[634]	**6**[379]	143/1[634]	—	I[634]
$cis\text{-}(MePhSiO)_3$	**6,11**[482,913] **6,22**[793] **22,23**[335] **23**[341]	**6,24**[85,75,54,36,102,63,24] **6,24**[213,430] **23**[7,341,803] **7,24**[54] **6,23,24**[56] **6,10,22,24**[879]	353.4² 178/0.5[482]	97–99.5[85] 99.5[913]	AN[482] CO[913] CY[179,386] C[803] K[85,75,24,56,36] M[196,692] N[461,890,606] O[2,514] B[36] T[601,210] T[56,444,7,341]
$trans\text{-}(MePhSiO)_3$	**6,11**[482,790,913,124] **23**[7,341] **6,22**[793]	**6**[909,379] **6,24**[75,54,625,213] **24**[918] **23**[803,341] **22,23**[335]	372.2² 187–188/0.5[482]	44–46[909] 42[913] 42.5[124]	AN[482] C[803] CY[179,386] CO[913] K[75,24] M[196,692] N[461,890,606] O[2,514,780] T[444,601,7,341] X[768]
$(MePhSiO)_3$	**6**[174,477,835,642,479] **6,24**[633] **6,11**[922,481,476,627] **14,6,24**[857] **6,7,11**[475]	**6**[145,480,68,132,476,271] **6**[64] **6,24**[146,141,62,914,508,296,270] **6,24**[562,577,272,332,30,683,80] **6,11**[253] **6,8,24**[142] **6,7,24**[23] **6,7**[478] **6,22,24**[576] **7,24**[499,27] **1,6**[453] **6,13**[570]	250/1–2[476] 168–200/1[476]	100[477] 70[914]	AT[283] B[30] CN[922] COT[174] K[914,480,876,30,27] LS[552] LO[540] LOZ[541] O[627] O[145,407] L[868] X[769,770] N[680,132,461]
$1\text{-}(PhCH_2CH_2)\text{-}3,5\text{-}Ph_2\text{-}1,3,5\text{-}Me_3Si_3O_3$	**6,11,16**[634]	—	204–205/3[634]		
$1,1\text{-}Cl_2Ph_4Si_3O_3$	**6,11**[115] **6,11,16**[114] **6,11,17**[121]	**6,11**[122,125,824,897] **6,11,16**[123] **6,11,24**[119,126] **5,8,11**[114,113] **6,8,11**[127]	—	83[121]	I[115]
$1,3\text{-}\left(\vcenter{\hbox{⬠}}\right)_2Me_4Si_3O_3$	**6,24**[633] **6,11,16**[632]	—	188–190/1[632]		

Compound			bp (°C/mm)	mp (°C)	References
$1\text{-}(\text{[5-ring]})\text{-}3,5\text{-}Ph_2\text{-}1,3,5\text{-}Me_3Si_3O_3$	**6,24**[633] **6,11,16**[632]	—	154–156/1[632]	—	
$1,1\text{-}Me_2Ph_4Si_3O_3$	**6**[786] **6,11,16**[758]	**6,24**[914]		87–89[758] 88–89[914]	**N**[461] **LO**[762] **K**[914] **T**[536] **X**[243] **M**[692]
$1\text{-}Vi\text{-}1\text{-}Me\text{-}Ph_4Si_3O_3$	**6,11,16**[14]	**6,24**[893] **18**[128]		46–47[14]	
$1\text{-}(Ph_2CHCH_2)\text{-}3,5\text{-}Ph_2\text{-}1,3,5\text{-}Me_3Si_3O_3$	**6,11,16**[634]		246–248/3[634]		
$1\text{-}(PhCH{=}PhC)\text{-}3,5\text{-}Ph_2\text{-}1,3,5\text{-}Me_3Si_3O_3$	**6,11,16**[634]		253–255/3[634]		
$1\text{-}(PhCH_2PhCH)\text{-}3,5\text{-}Ph_2\text{-}1,3,5\text{-}Me_3Si_3O_3$	**6,11,16**[634]		246–247/3[634]		
$1,3\text{-}(PhCH{=}PhC)_2Me_4Si_3O_3$	**9,24**[185]		202–204/2[634]		**I**[634]
$(Bu^tPhSiO)_3$	**24**[185]				
$(Me_2SiO)[(PhCH{=}CPh)MeSiO]_2$	**6,11,16**[634]				
$[(\text{[5-ring]})MeSiO]_3$	**6,11,16**[632] **6,24**[633]		240–242/1–2[632]		
$(Ph_2SiO)_3$	**6,11**[67,68,450] **6,10,18,24**[725] **6,7**[396] **6,13,22**[671] **6,16**[415] **6,24**[129]	**6,1**[45,379,68,271] **6,24**[67,748,204] **6,24**[203,725,562,577,709,710,510] **6,24**[80] **6,17**[570] **24**[12] **6,22,24**[576] **1,6,16,18,24**[274]	300/1[852]	188[67,199] 189[415]	**B**[725,450] **O**[129,896,514] **B**[145,852] **K**[12,450] **N**[680] **N**[310] **X**[199,768] **LO**[762] **T**[535,725]
$(PhMeHSi\text{—}CH_2CH_2)Me_5Si_3O_3$	**2,15,21**[40]		145–148/10[40]		
$(Ph_2HSiCH_2CH_2)Me_5Si_3O_3$	**2,15,21**[40]		165–167/1[40]		
$[(Me_3SiO)_2SiO]_3$	**6**[895] **6,11**[344,345] **6,13**[345]	**6,11**[352]			**MO**[895] **LOZ**[344] **O**[345]

$$R_2Si\!-\!O\!-\!SiR_2$$
$$\begin{array}{ccc} | & & | \\ O & & O \\ | & & | \end{array}$$
$$R_2Si\!-\!O\!-\!SiR_2$$

Table 10.2 Cyclotetrasiloxanes

Compound	Methods of Synthesis[a]	Reactions[a]	B.p.(°C/Torr)	M.p.(°C)	Other physicochemical investigations[b]
$(H_2SiO)_4$	**6,11**[351] **6,14,15,22**[693]	**6,24**[321]	137[87]		X[351] B[693] Q[435]
$[MeHSiO]_4$	**6,11**[297,293,361] **6,24**[304]	**2,15**[721,596,333,743,747] **6**[293,354] **7,11**[667] **7,15**[808] **6,15**[87,86,96,295] **6,24**[880,806,508,816,700,845,488] **6,24**[854,363] **5,15**[612] **2,15,24**[847] **2,15,24**[846,595,149,718,621] **2,6,15**[581] **2,6,15,24**[597,89] **2,15,21**[594,6,187] **1,6,16**[765] **2,15,21,24**[923]	134[923,718]		A[175,176] I[462] K[488] K[86] KT[361] LOZ[437] Z[904,295]
$HMe_7Si_4O_4$	**6,11**[522,728]	**2,15**[109,752,417] **6,24**[880,597,364] **6,15**[608] **2,15,21**[595,50,855] **5,7,9,15,24**[728]	64/20[522]		N[522] Y[386] INO[609] ILZ[460] K[608]
$H_2Me_6Si_4O_4$ $Me_8Si_4O_4$	**6,11**[388] **6**[660,434,8,642] **6,11**[831,886,260,543] **6,11**[791,881,891] **6,14**[759] **6,24**[836] **22**[802,643] **6,13**[81] **6,10,15**[25] **6,11,24**[528,720] **6,24**[85] **6,24,26**[760] **6,11,18,22**[259] **6,11,22**[863] **3,6,10,14,22,24**[164]	**6,24**[364] **24**[555] **6,24**[906,931,699,686,225] **6,24**[355,313,761,855,219,222,754] **6,24**[129,503,836,153,735,131,170] **6,24**[706,617,614,569,449,531,655] **6,24**[697,654,102,63,484,107,661]	176/760[561] 175[529,129,153] 175–177[43,220]	17.5[129]	O[538,299,539,561,676] O[610,861] K[876,675] K[266,587,674,319,907] K[222,153] KT[906] KL[260] I[831] IKVX[259] ILZ[459] T[278,277,384,184,538] T[470,162,316,315,515] A[276] AT[841] V[673,553] B'[64,266,760,153]

Table 10.2 *continued*

Compound	Methods of Synthesis[a]	Reactions[a]	B.p.(°C/Torr)	M.p.(°C)	Other physicochemical investigations[b]
$Me_8Si_4O_4$ *continued*		6,24[587,100,674,554,223,907,631] 6,24[148,854,622,752,417,615,712] 6,24[571,159,568,151,819,403,567] 6,24[321,811,311,438] 6[173,354,446] 24[405,894,732,698,746,438,575] 10[319] 7,24[820,447,682,27] 4,6,24[663,662] 6,10,16,24[220] 4,6[807] 6,7,24[356,565,664,55] 6,22[881] 7,10[834] 6,8,24[722] 23,24[677] 6,16,24[39,731,586] 5,25,10,22[487] 4,6[766] 5,6,24[224] 5,6,24[221,397] 6,17[570] 16[418] 6,22,24[675] 4,6,16,24[741] 20,21,24[266]			BK[220,677,586] LO[395] LOZ[542,807,231] LMOZ[679] OT[529,587] LS[669] N[799,381,680] M[192] Z[696,446,232,406] Z[147,418,150,735] X[733]
$ViMe_7Si_4O_4$	6,11[65,801] 4,6[77]	6[65] 5,10[328] 2,21[798,227,797,35] 21[179,78] 2,15[77] 6,24[161,712,818] 2,15,21[49,40,48] 4,22,24[881] 4,6,24[177] 4,6,21[882] 2,15,21,24[48] 4,18,21[871]	84/20[328] 188.8[77]		Y[386] K[798] IM[801] Z[171] O[561] LO[813]
$1,3\text{-}Vi_2Me_6Si_4O_4$	6,11,16[926]	2,21[38,39,35] 21[41]	77/7[561] 87–88/10[926]		O[561] CN[926] LVZ[815]
$(Me_2SiO)_2(MeViSiO)_2$ $(CF_3CH_2CH_2)Me_2Si_4O_4$	6,11[183,791] 6,16[376]	24[818] 6,24[935]	87/16[376,183]		YZ[183] T[286,285] K[935] N[680] AT[497] O[791] Z[745,582]
$H_3[H_2C\!-\!CH\!-\!CH_2O(CH_2)_3]\text{-}$ $\quad\;\;\backslash O /$ $Me_4Si_4O_4$		6,24[582] 24[744,745]			

Compound				
$1,3\text{-}(Me_2NO)_2Me_6Si_4O_4$		6,24[308]		L[851]
$1,3\text{-}(C_4H_3S)_2Me_6Si_4O_4$				
[structure: Me–Si–O–SiMe$_2$ / Me–Si–O–SiMe$_2$ fused to benzene ring]	6,11[851] 3,10,6[197] 3,6,10,21[835] 3,6,10,21,22[193]			M[692] IMN[193]
$(MeViSiO)_4$	6[545,829] 6,11[591,590,718,548] 6,22[547,704] 6,24[546]	6,24[374,486,400,562,577,160,161] 6,24[757,378,338,404,614,615,712] 6,24[159] 688[7] 24[549,420] 19,20[648] 5,10[328] 6,17[570] 6,26[421] 19[734] 5,7,24[810] 6,7,18,24[814] 2,15,24[847] 2,21,6,24[89] 15,21,24[846] 2,15,21,24[521,924] 2,15,21,24[923,718] 2,21,24[172] 6,22,24[576] 2,21[658] 23[642]	224[898] 224/758[328] 61-63/2[89] 83/5[923,718] 101/7[561]	O[899,810,561] IKX[734] LO[898] LN[382] LVZ[815] K[876] Z[178]
$1,3\text{-}(CF_3CH_2CH_2)_2Me_6Si_4O_4$	6,11[485,294] 4,6,16,19,21,24[691]	6,24[935]	108.3/10[783]	T[285,783] N[680] M[691]
$(CF_3CH_2CH_2)_2Me_6Si_4O_4$	6,11[791]			T[286] K[935] AT[497] O[791]
$Me_7(C_6H_3Cl_2)Si_4O_4$	6,22,24[785]	24[818]		
$(H_2C{=}CHCH_2)_2Me_6Si_4O_4$				
$PhMe_7Si_4O_4$	6[835,786] 6,11[201,182,65,482,481,913] 22[643,644] 5,6,24[884]	6[145,65] 6,24[201,935] 7,10[834] 4,6,24[178]	112–114/7[201,913] 114–115/6[482]	ANO[482] CY[520] K[935] M[196,692] N[890,461,680] O[145,913] T[284] Y[513] O[481] Y[179,386,182]
$(C_4H_3S)_2Me_6Si_4O_4$	6,11[851]			L[851]
$(CF_3CH_2CH_2)_3Me_5Si_4O_4$		6,24[935]		T[285,286] K[935]

Table 10.2 *continued*

Compound	Methods of Synthesis[a]	Reactions[a]	B.p.(°C/Torr)	M.p.(°C)	Other physicochemical investigations[b]
O(CH₂)₃Me₇Si₄O₄ ring with S, O₂ (structure)	2,15,21[752,417,855]	6,24[752,417]			
(Et₂NO)₂Me₆Si₄O₄		9,69[306,307]			M[689]
[SiO]₄ (structure)					
(Vi₂SiO)₄		6,7,24[527]			T[287,288] X[243] M[192] SX[371]
[SiO]₄ (structure)					
1,3,5-Vi₃-7-PhMe₄Si₄O₄	6,11,16[47]	6[909,705]	109–111/1[47]		CN[47]
[(CF₃CH₂CH₂)MeSiO]₄	6,11[485,294]	6,24[910,311,724,919,706] 24[707]	148.5/4[909,783]		K[910,876] T[919] Z[423] N[680]
(Et₂SiO)₄	6,11[297]	6,24[826,321] 6,22[205] 4,6,24[178]			A[175,176] K[876] T[282]
1,3,5-(Et₂NO)₃Me₅Si₄O₄		9,69[307]			
[HS(CH₂)₃MeSiO]₄		6,24[403,637] 16,18[431]			
trans-1,5-Ph₂Me₆Si₄O₄	6,11,16[401,890] 22,23[335] 23[803]	6,24[401] 22,23[335]		55[401,890]	C[803,336] CY[179,386] Y[180] CSX[229] N[401,804,890] T[284] N[680] M[692]
1,3-Ph₂Me₆Si₄O₄	6[786]	—			

cis-1,5-Ph$_2$Me$_6$Si$_4$O$_4$	6,11[181] 6,11,16[401,890] 22,23[335]	3,6,10[197] 22,23[335] 23[803]	101–102/2.10[890]		AILMNOV[804] C[803,336] CSX[229] CY[179,386] KY[687] HY[756] M[375] N[890,401] Y[185,628.180,629,518] Y[517,9,181,833]
1,1-Ph$_2$Me$_6$Si$_4$O$_4$	6[786] 6,11[201.751,111,481,749] 6,24[129] 6,16[750] 6,22[532]	6[145] 6,24[201,111] 4,6,24[178] 6,22,23,24[844]	170–172/7[201] 170/8[852,129]	31[111] 305.09 ± 0.04[599]	A[537] M[692] N[680,461] O[145,852,129] T[536,598,599] Y[386,179]
1,5-Ph$_2$Me$_6$Si$_4$O$_4$	6[786]				SX[230] X[186] N[680] M[692] LMO[865]
Ph$_2$Me$_6$Si$_4$O$_4$	6[477] 6,11[482,481,913]	6[835] 6,24[935] 3,6,10,21,22[198,193]	137–138/0.5[482] 168–170/7[913]		A[482] N[461] Y[179] CY[386] CO[913] M[196] O[481] K[935]
1,5-Ph$_2$-3,7-Vi$_2$Me$_4$Si$_4$O$_4$	6,11,16[921,47]	2,21[47] 21[41]	150–153/1–2[921]		IN[921] LVZ[815] CN[47]
1,3-Ph$_2$-5,7-Vi$_2$Me$_4$Si$_4$O$_4$	6,11,16[38,47]		164–165/2[38,47]		I[38] CN[47] LOZ[541]
[Me(C$_4$H$_3$S)SiO]$_4$		9,69[763] 6,24[742,763]			Z[742]
1,3-(Et$_2$NO)$_2$-5,7-Bu$_2$Me$_4$Si$_4$O$_4$					
(Et$_2$NO)$_3$BuMe$_4$Si$_4$O$_4$	6,11[20] 6,11,16[136]	5,11[136]	206/0.002[136]	142[20]	
1,1,5,5-Cl$_4$-3,3,7,7-Ph$_4$Si$_4$O$_4$	6[114,477] 6,11[481,913,482]	6[145,835] 6,24[935] 7,10[834] 3,10,21,22[198]	187–188/0.5[482] 209–210/4[913]		AO[482] CO[913] K[935] O[145,514] M[262,196] Y[167,179] T[284]
1,3,5-Ph$_3$Me$_5$Si$_4$O$_4$					
[(KO)PhSiO]$_4$	7[95]	6[95] 24[92]			I[51] O[739] AX[98]
cis-[(HO)PhSiO]$_4$	6,11[88]	6[96] 6,16[88,52,51,26] 6,16,24[61] 6,24[60,98,739] 6,22,24[89]		165–166[88] 160–162[51] 176–176.5[89]	
[(HO)PhSiO]$_4$	6,11[90]	6[58,57] 6,24[21,53,97,42,90,108] 6,16[619,771,32] 6,11[33] 5,6[87] 6,15[86] 16,24[31]		164–165[87,53] 165–166[32,90]	K[86] T[850]
1,1,5,5-(HO)$_4$Ph$_4$Si$_4$O$_4$	5,11[134,136]	6,16[135]		157[136]	
1,3,5-Ph$_3$-7-Vi-Me$_4$Si$_4$O$_4$	6,11,16[47]	—	180–183/1[47]		CN[47]

Table 10.2 continued

Compound	Methods of Synthesis[a]	Reactions[a]	B.p.(°C/Torr)	M.p.(°C)	Other physicochemical investigations[b]
$1,5\text{-}(NH_2)_2\text{-}1,5\text{-}Me_2\text{-}Ph_4Si_4O_4$	8,11[505]	6,13,24[504]	—	107–109[505]	I[505]
$1,5\text{-}(HO)_2\text{-}1,5\text{-}Me_2\text{-}Ph_4Si_4O_4$	5,11[19,505]	6,16,24[504]	—	140[19], 137–139[505]	I[505]
$1,5\text{-}H_2\text{-}1,5\text{-}Me_2\text{-}Ph_4Si_4O_4$	6,11,16[921] 6,11,13,16[500]	2,15,24[847,846] 6,15[608]	245–250/3–4[921] 215–217/2[500]	87–88[921] 85–86[500]	IN[921] INO[609] ILZ[459]; I[500] K[608]
$1,5\text{-}Cl_2\text{-}1,5\text{-}Me_2\text{-}Ph_4Si_4O_4$	6,11[19]	6,11[505] 6,11,24[59,504] 5,11[19,505] 9,11[505,502] 8,11[505] 7,11[505]	251/1.5[19]	126–127[19], 120[505]	IN[505]
$1,3\text{-}(Et_3NO)_2Bu_4\text{-}1,3\text{-}Me_2Si_4O_4$		6,24[742]			Z[742]
$(MePhSiO)_4$	6[174,379,835,477] 6,11[627,790,922] 6,11[67,482,481,476] 6,24[642] 6,22[793] 6,7,11[475] 3,10,21[23] 3,10,21,22[198]	6[145,68] 6,24[401,161,67] 7,10[834] 5,10[328] 6,25,10,24[474]	208–215/2[67] 500[2] 237/1.5[328] 230–245/1[476]		CN[922] COT[174] O[145,407,2,476] O[627,790] LN[382] LOZ[541] ILZ[459] T[600,639,444,210,284,7,536] L[853] M[196,692] Y[179,386] X[243]
$(PhMeSiO)_4$ cis-cis-trans,trans-	6,11[890] 22,23[335]	22,23[335]			N[890] X[392]
$(PhMeSiO)_4$ cis-trans-cis-trans-	22,23[335]				N[890]
$(PhMeSiO)_4$ cis-cis-cis-trans-					N[890]
syn-$1,1\text{-}3,3\text{-}Ph_4Me_4Si_4O_4$	6,24[129]			73.4[129]	O[129]
anti-$1,1\text{-}3,3\text{-}Ph_4Me_4Si_4O_4$	6,24[129]			131–132[129]	O[129]
$(Ph_2SiO)_2(Me_2SiO)_2$	6[786]	6,24[99] 3,10,23[7]		72[773]	T[7,536,583,444,467] M[692] N[680] K[99] R[99] X[773]
$1,1,5,5\text{-}Ph_4Me_4Si_4O_4$	6[786]	6[145]			O[145] A[537] T[536,464] N[680] M[692]

1,3-Me$_2$-1,3-Vi$_2$Ph$_4$Si$_4$O$_4$	6,11,16[14]			57–60[14]	
1,5-(C$_5$H$_5$N.HO)$_2$-1,5-(HO)$_2$-Ph$_4$Si$_4$O$_4$	6,16[135]	18[128]		119[135]	
1,5-(Et$_2$N)$_2$-1,5-Me$_2$Ph$_4$Si$_4$O$_4$	8,11[505]	6,13,24[504]	268–270/1[505]		I[505]
1,5-H$_2$Ph$_6$Si$_4$O$_4$	6,11,16[926]	6,15[608]	270–280/3[926]	60–61[926]	CN[926] INO[609] ILZ[459] K[608,9,32]
(Ph$_2$SiO)$_3$(Me$_2$SiO)	6[786] 6,11[481] 6,24[129] 6,11,16[758]		300/0.6[852]	114–116[758]	O[852,129] N[461] A[537]
1,3-(Ph$_2$CHCH$_2$)-5,7-Ph$_2$Me$_4$-Si$_4$O$_4$	6,11,16[634]		328–331/4[634]		
(MePhSiO)$_2$[Ph$_2$CHCH$_2$(Me)SiO]$_2$	6,11,16[634]		232–234/1[634]		
(Ph$_2$SiO)$_4$	6[379,703] 6,11[450,831,544] 6,13[856] 5,6[67,68] 6,15[84,82] 6,18[701]	6,24[67,144,262] 6,24[143,160,161,757,712,129,131] 6,7,24[527,23]	330–340/1[129]	200[67] 200–201[129]	K[84,101] A[537] L[144] LO[762] O[145,426,514,340] OT[789] ILZ[459] Z[755,150] T[535] LS[552] I[831] X[867]
(Me$_2$SiO)$_3$(Me$_3$SiMeSiO)	22[644] 5,6,24[884]				IMN[644]
(Me$_2$SiO)$_3$[Cl$_2$MeSiCH$_2$CH$_2$(Me)-SiO]	2,15,21[49,50]				
(Me$_2$SiO)$_3$(ClMe$_2$SiCH$_2$CH$_2$Me-SiO)	2,15,21[49,50]				
(HMeSiO)$_3$[(EtO)$_3$SiCH$_2$CH$_2$Me-SiO]	2,15,21[40]				
(Me$_2$SiO)$_3$(HPhMeSiCH$_2$CH$_2$Me-SiO)	2,15,21[40]	6,24[582]	162–164/10[40]		Z[582]
1,5(H$_2$C—CHCH$_2$O(CH$_2$)$_3$-3,7—H$_2$Me$_4$Si$_4$O$_4$		6,7,24[582]			Z[582]
(Me$_2$SiO)$_3$[(HPh$_2$Si(CH$_2$)$_2$MeSiO]	2,15,21[40]		173–174/1[40]	38–39[40]	
1,1,5,5-(ClMe$_2$SiO)$_4$Ph$_4$Si$_4$O$_4$	6,11,16[135]		220–224/0.004[135]		
1,1,5,5-(ClMeEtSiO)$_4$Ph$_4$Si$_4$O$_4$	6,11,16[135]		237–242/0.002[135]		

Table 10.2 *continued*

Compound	Methods of Synthesis[a]	Reactions[a]	B.p.(°C/Torr)	M.p.(°C)	Other physicochemical investigations[b]
1,1,5,5-(ClEt$_2$SiO)$_4$Ph$_4$Si$_4$O$_4$	6,11,16[135]		255–260/0.002[135]		
1,1,5,5-(ClMePhSiO)$_4$Ph$_4$Si$_4$O$_4$	6,11,16[135]		315–320/0.002[135]		
1,1,5,5-(Cl$_2$PhSiO)$_4$Ph$_4$Si$_4$O$_4$	6,11,16[135]		327–335/0.004[135]		
1,1,5,5-(EtOMe$_2$SiO)$_4$Ph$_4$Si$_4$O$_4$	6,11,16[135]		220–222/0.001[135]		
1,1,5,5-(EtOMePhSiO)$_4$Ph$_4$Si$_4$O$_4$	6,11,16[135]		334–336/0.002[135]		
[Me(Me$_3$SiO)SiO]$_4$	6[862] 6,11[228,105] 4,6[66]				
[(Me$_3$SiO)PhSiO]$_4$	6[95] 6,16[94] 6,11,24[90] 6,13,16[32]		278–288/6.8[94] 290/2[32]	265–266[94] 256–257[90]	DIX[32] I[95] AINX[90]
[(Me$_3$SiO)$_2$SiO]$_4$	6[895,903,303] 6,11[344,870,345] 6,13[345] 6,11,24[874]	6,11[352]		162[870]	MO[895,303] LOZ[344] I[869] O[345] N[383] LO[903]
[(Me$_3$SiOSiMe$_2$O)PhSiO]$_4$	6,15,24[90]				AINX[90]

a,b See footnotes to Table 10.1.

Table 10.3 Higher cyclosiloxanes $(R_2SiO)_n$, $n = 5,6,7,8,9,10$.

Compound	Methods of Synthesis[a]	Reactions[a]	B.p.(°C/Torr)	M.p.(°C)	Other physicochemical investigations[b]
$(H_2SiO)_5$		6,24[321]			
$(HMeSiO)_5$	6,11[297,293,361] 6,24[304]	6[293] 6,24[880,806] 1,6,16[765] 2,15,21[595] 2,6,15,24[597] 2,15,24[831]			A[175,176] I[462] LOZ[437] KT[361]
$(Me_2SiO)_5$	6[786,70,660,642,8,835] 6,11[482,297,300,913,80,260,543,791] 6,24[533,457,129,836] 6,16[729,730] 6,16[837] 6,14[759] 6,13[81] 6,22[360,189] 6,22,24[265,823,785] 22[166] 6,11,24[720] 6,11,24[528] 6,11,18,22[259] 6,11,22[881]	6[173,516] 4,6[716] 6,24[597,917,935] 6,24[587,674,148,836,321,892,795] 6,24[533,580,919] 7,24[490,489] 6,22[792,532] 22[370] 5,7,24[809,810] 6,7,16,24[483] 24[349] 10,24[317] 4,6,24[178] 6,21[830] 6,22,24[675] 23,24[677] 3,6,24[529] 6,16,24[586]	210[898,529] 210[913,129] 210[457] 72/2[483] 101/20[716,561]	−44[129]	A[175,176] AT[279] O[861,346] O[899,892,913,810,538,539,111] O[561,676,129,587] BK[586] BK[917] K[490,489,483,716] K[935,876,675,677,587] K[674] LS[669] LO[813,395,898] LOZ[542,231] M[860] IKVX[259] KL[260] N[309,310,297,890] N[680] L[343,342] IR[327] T[919] T[208,209,536,277] X[640] I[462] IM[801] V[673] OT[529] Z[406] K[935] AT[497]
$(Me_2SiO)_2[(CF_3CH_2CH_2)MeSiO]_3$	6,11[791]	6,24[935]			
$(Me_2SiO)[(CF_3CH_2CH_2)MeSiO]_4$					AT[497]
$[Me(CF_3CH_2CH_2)SiO]_5$		6,24[919]			T[919] Z[423] N[680] L[342]
$[(C_4H_3)MeSiO](Me_2SiO)_4$	6,11[851]				L[851]
$[(C_4H_3)MeSiO]_2(Me_2SiO)_3$	6,11[851]				L[851]
$(Me_2SiO)_2(Ph_2SiO)_3$	6,24[129]				O[129]
$(Me_2SiO)_3(Ph_2SiO)_2$	6[786] 6,24[129]			67–68.5[129]	O[129]

Table 10.3 *continued*

Compound	Methods of Synthesis[a]	Reactions[a]	B.p.(°C/Torr)	M.p.(°C)	Other physicochemical investigations[b]
$(Me_2SiO)_4(Ph_2SiO)$	6^{786} $6,24^{129}$ $6,11^{481,751,749}$ $6,16^{750}$	$4,6,24^{178}$	$125/0.2^{129}$		N^{461} $T^{536,468,469}$ $O^{129,481}$
$(Et_2SiO)_5$		$6,24^{321}$ $4,6,24^{178}$			
$(Cl_2C_6H_3)Me_9Si_5O_5$		$6,22,24^{785}$			
$(H_2SiO)_6$		$6,24^{321}$			
$(HMeSiO)_6$	$6,11^{293,361}$ $6,24^{304}$	6^{293} $6,24^{806}$ $2,6,15,24^{597}$ $1,6,16^{765}$			$A^{175,176}$ I^{462} LOZ^{437} KT^{361}
$(Me_2SiO)_6$	$6^{70,642,8,660}$ $6,11^{913,482,485,294}$ $6,24^{457,533}$ $6,16^{729,730}$ $6,11,24^{720}$ $6,11,24^{528}$ $6,11,22^{881}$ $6,22^{189}$ $4,25,6,22^{471}$ $6,22,24^{785,823}$ $6,14^{759}$	$6,24^{587,148,321,892,935,533}$ 6^{516} $6,22^{532,792}$ $7,24^{188,490,489}$ $5,6^{716}$ $5,7,24^{810}$ $6,7,16,24^{483}$ 22^{370}	$170/100^{898}$ $245^{913,529}$ $97/2^{483}$ $128/20^{716,561}$ $179/100^{457}$		$A^{175,176}$ AT^{279} I^{462} IR^{327} $K^{490,489,483,716,935,587}$ $L^{343,342}$ LM^{684} $LO^{395,898}$ LOZ^{542} LS^{669} M^{860} N^{309} N^{310} $O^{892,561,810,913,539}$ $O^{346,899}$ $T^{489,529,587,209}$ X^{640} Z^{697}
$(Et_2SiO)_6$		$6,24^{321}$			T^{282}
$[(C_4H_3S)MeSiO](Me_2SiO)_5$	$6,11^{851}$				L^{851}
$[(C_4H_3S)MeSiO]_2(Me_2SiO)_4$	$6,11^{851}$				L^{851}
$(Ph_2SiO)(Me_2SiO)_5$	6^{786} $6,24^{129}$				O^{129}
$(Cl_2C_6H_3)Me_{11}Si_6O_5$	$6,22,24^{785}$				
$(HO)_2Si \begin{smallmatrix} (O-SiMePh)_2O \\ (O-SiMePh)_2O \end{smallmatrix} Si(OH)_2$	$6,16^{137}$ $5,6,16^{138}$	$6,16^{138}$			I^{138}

Compound				
$(H_2SiO)_7$				
$(HMeSiO)_7$	6,11[361]	6,24[321]		I[462] LOZ[437] KT[361]
$(Me_2SiO)_7$	6[8] 6,11[482,485,294] 6,22[189] 6,14[759] 6,24[533,457] 6,11,24[528] 6,11,22[881] 6,22,24[823,785]	1,6,16[765] 2,15,21[595] 2,6,15,24[597] 6,24[893,916,587,148,321] 7,24[490,489] 22[370] 5,10[328] 6,22[532] 6,16,24[586]	154/20[898,328] 154/20[457]	A[175,176] I[462] K[490,489] K[587] L[343,342] LO[898,395] O[899,893,539] T[208,209] LM[684] LOZ[542] LS[669] BK[586] OT[587] Z[916]
$[(C_4H_3S)MeSiO](Me_2SiO)_6$	6,11[851]			L[851]
$[(C_4H_3S)MeSiO]_2(Me_2SiO)_5$	6,11[851]			L[851]
$(Et_2SiO)_7$		6,24[321]		T[282]
$(Cl_2C_6H_3)Me_{13}Si_7O_7$	6,22,24[785]			
$(H_2SiO)_8$		6,24[321]		
$(Me_2SiO)_8$	6[8] 6,24[457] 6,22[189] 6,11,24[528] 6,11,22[881] 6,22,24[823]	6,24[321] 6,24[148,587,321] 4,24[916] 6,16,24[586]	290[898,457]	A[175,176] K[587] O[587,899] LO[898,395] LOZ[542] LM[684] LS[669] BK[586]
$[(C_4H_3S)MeSiO](Me_2SiO)_7$	6,11[851]			L[851]
$[(C_4H_3S)MeSiO]_2(Me_2SiO)_6$	6,11[851]			L[851]
$(Et_2SiO)_8$		6,24[321]		T[282]
$(Me_2SiO)_9$	6[303] 6,22[189] 6,24[457] 6,11[268] 6,11,24[528]	24[268,575] 6,24[587,148] 6,16,24[586]	188/20[898,457]	A[175,176] BK[586] LO[898,395] LM[684] LS[669] K[587] O[587,899] MO[303] T[208,209]
$[(C_4H_3S)MeSiO](Me_2SiO)_8$	6,11[851]			L[851]
$[(C_4H_3S)MeSiO]_2(Me_2SiO)_7$	6,11[851]			L[851]
$(Me_2SiO)_{10}$	6,11[485,294] 6,24[457] 6,11,24[528]	6,24[148]	150/1[898,457]	LM[684] LO[395,898] LS[669] O[899] T[208,209]
$[(C_4H_3S)MeSiO](Me_2SiO)_9$	6,11[851]			L[851]

[a,b] See footnotes to Table 10.1.

Table 10.4 Polycyclic siloxanes.

Compound	Methods of Synthesis[a]	Reactions[a]	B.p.(°C/Torr)	M.p.(°C)	Other physicochemical investigations[b]
R_2Si—O—SiR_2 / O—Si—O / R_2Si—O—SiR_2					
$Me_8Si_5O_6$	6,16[73]	6,24[12,120,116,118]		121[73]	I[73] M[657] X[200] K[116,12]
1,1,3,3-$Me_4Ph_4Si_5O_6$	6,11[114,125,897] 6,11,16[123]	6[114] 6,24[117,116,118,12, 16,120]	230/0.5[123,125] 230/0.5[114]	151[117,114,125] 149[123] 155–156[897]	IN[123] J[897] K[116,12] N[125]
					X[780]
trans,trans,d,1- 1,3,6,8-$Me_4Ph_4Si_5O_6$	6,11[72] 6,16[73]	6,24[12,116,118,120] 24[72]			I[73] X[18] AT[74] K[116,12]
1,3,6,8-$Me_4Ph_4Si_5O_6$	6,11,16[123]	6,24[12,120,116,118]	265/0.005[114]		IN[123] K[116,12]
1,3-$Me_2Ph_6Si_5O_6$			265/0.005[123]	107[123]	
(structure: R''R''Si—O—SiR''R'' bridged cage with SiRR')					
R = R' = Me; R'' = Ph	6,11,16[52]	19[875]		220–222[51,52]	IN[51,52] X[778]
(cis,syn,cis)					
R = R' = Me; R'' = Ph	6,13,16[32]	6,24[34]		162.5[776]	X[776]
R = Me; R' = R'' = Ph		6,24[34]			
R = Me; R' = Vi; R'' = Ph	6,11[33]	6,24[34] 2,15,21,24[91]		150–153[33]	IN[33] LVZ[815]

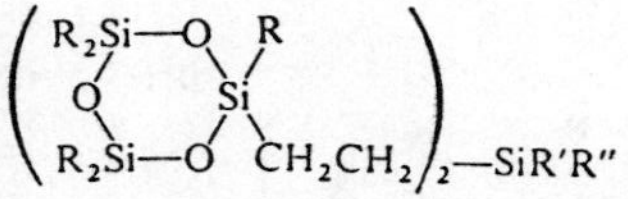

$R = Me; R' = Me; R'' = Ph$ **2,15,21**[40]   162–165/2[40]   44–45[40]

$R = Me; R' = R'' = Ph$ **2,15,21**[40]   190–195/1[40]   38–40[40]

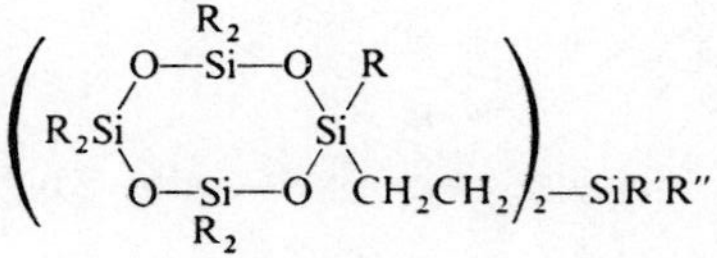

$R = Me; R' = Me; R'' = Ph$ **2,15,21**[40]   172–174/1[40]   33–35[40]

$R = Me; R' = Ph; R'' = Ph$ **2,15,21**[40]   200–203/1[40]

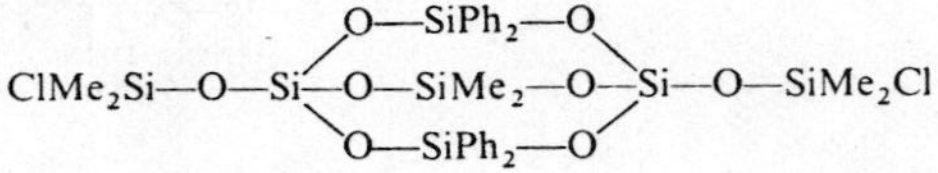

6,11,16[135]   212–214/0.002[135]

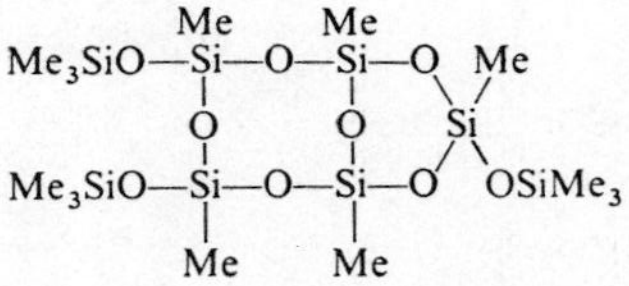

6,11[105]

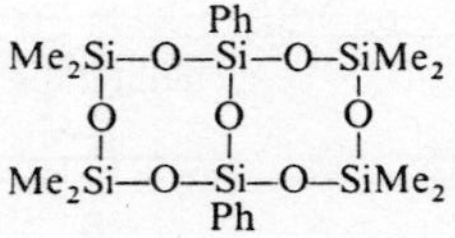

X[779]

Table 10.4 continued

Compound	Methods of Synthesis[a]	Reactions[a]	B.p.(°C/Torr)	M.p.(°C)	Other physicochemical investigations[b]
(structure: Me$_3$SiO–Si–O–Si–O–Si–OSiMe$_3$ / Me$_3$SiO–Si–O–Si–O–Si–OSiMe$_3$ with Me groups)	6,11[104,105] 6,4[66] 6[103]				
(structure: Me$_2$Si–O–Si–O–Si–SiMe$_2$ / Me$_2$Si–O–Si–O–Si–SiMe$_2$ with Ph groups)	6,11[57,58] 6,16[771]		235–240/1 × 10^{-3} [58]		CSX[771]
(structure: (Me$_3$Si)$_2$Si–O–Si–O–Si–Si(OSiMe$_3$)$_2$ / (Me$_3$Si)$_2$Si–O–Si–O–Si–Si(OSiMe$_3$)$_2$ with OSiMe$_3$ groups)	6,11[344,345] 6,13[345]				LOZ[344] O[345]
R = C$_6$H$_4$Cl-m (structure: R–Si–OH / HO–Si–R with R and O bridges)	5,6,11,24[93]	6,16[93]			
R = Ph	5,6[87] 6,11[90]	6,24[90]			

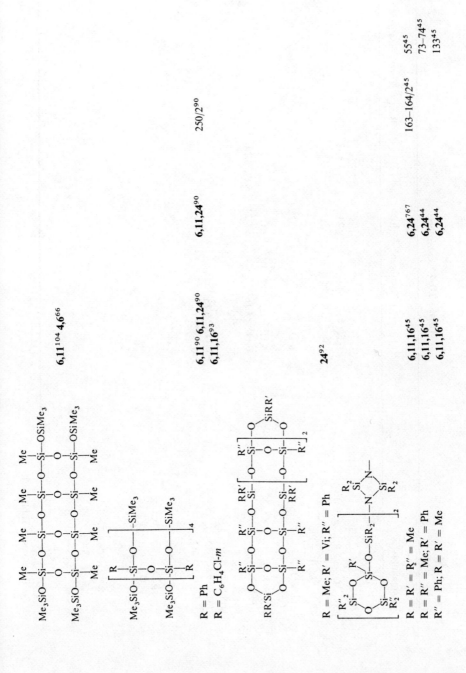

AINX[90]

250/2[90]

6,11,24[90]

6,11[104] 4,6[66]

6,11[90] 6,11,24[90]
6,11,16[93]

24[92]

R = Me; R' = Vi; R'' = Ph

R = Ph
R = C₆H₄Cl-m

R = R' = R'' = Me
R = R'' = Me; R' = Ph
R'' = Ph; R = R' = Me

6,11,16[45]
6,11,16[45]
6,11,16[45]

6,24[767]
6,24[44]
6,24[44]

163–164/2[45]

55[45]
73–74[45]
133[45]

IN[45] X[767]
IN[45]
I[45]

Table 10.4 *continued*

Compound	Methods of Synthesis[a]	Reactions[a]	B.p.(°C/Torr)	M.p.(°C)	Other physicochemical investigations[b]
	6,11[72]	24[72]			
R = Me; R' = Ph					
$(MeSiO_{1.5})_8$	6,11[105]	22,24[416]			IK[416]
$(MeSiO_{1.5})_{10}$					X[155]
$(ViSiO_{1.5})_8$	6,11[873]				SX[156] X[154,873]
$(PhSiO_{1.5})_8$	6,11[408] 6,16[771]	6,24[408] 6,19,20,24[579]			BOT[408] T[840] B[579] X[771,398] CSX[777]
$(PhSiO_{1.5})_{12}$					X[254]
$(Me_3SiO—SiO_{1.5})_6$	6,11[345] 6,13[345]				O[345]
$(Me_3SiO—SiO_{1.5})_8$	6,11[344]				M[393] LOZ[344] X[694] N[530]
$(Me_3SiO—SiO_{1.5})_{10}$	6,11[344]				LOZ[344] N[530]
					N[350]
R = OSiMe_3					

[a,b] See footnotes to Table 10.1.

Table 10.5 Heterocyclosiloxanes.

Compound	Methods of Synthesis[a]	Reactions[a]	M.p.(°C)	Other physicochemical investigations[b]
Me$_2$Si—SiMe$_2$ ring with O, B—Ph	6,7,11[883]	6,7,24[883]		
SiMe$_2$, O, B—Ph ring (Me$_2$Si)	6,7,11[883] 6,7,15[717]		37[883] (b.p. 77/0.8)[883]	MN[883]
R$_2$Si—Si, B—Ph ring (R$_2$Si, R$_2$, O); R = Me	6,7,24[883]			MN[883]
Me—N, SiMe$_2$, B—Ph ring (Me$_2$Si, O)	6,7,11[883]		71[883]	MN[883]

Table 10.5 *continued*

Compound	Methods of Synthesis[a]	Reactions[a]	M.p.(°C)	Other physicochemical investigations[b]
Cl–Al ring with O, SiMe₂, O, Me₂Si, Si	6,7[72]	6,7[448]		
MePhSi–Ti–SiMePh (with O bridges)	6,7[72]	24[72]		
RR'Si–Ti–Si RR' (R = Me; R' = Ph)	6,7[72]	24[72]		
(CO)₄Fe–SiMe₂ / (CO)₄Fe–SiMe₂ (O bridged ring)	6,15[369]		84[369]	
(CO)₄Ru–SiMe₂ / (CO)₄Ru–SiMe₂ (O bridged ring)	6,15[369]		130[369]	IN[369]

Structure				
Me₂Si–O–Me₂Si–RhH(PPH₃)ₙ n = 2	18[261] 14,15,24[261]	18[261]		**IN**[261]
n = 3	18[261]	18[261]		
[(Ph₃P)₂HRh(SiMe₂–O–SiMe₂)]₂	18[261]			
Me₂Si(O)Me₂Si–Pd(PPh₃)₂	9,14,24[368] **14,15**[261]		142[368,261]	**N**[368] **IN**[261]
Me₂Si(O)Me₂Si–Pt(PPh₃)₂	6,15[301,302]		172–174[301,302]	**I**[302] **IN**[261]
Me₂Si–O–SiMe₂ / Me₂Si–O–SiMe₂ · Ir(CO)H(PPh₃)₂	14,15[261]	6,24[261]		**IN**[261]
Me₂Si(O)Me₂Si–Ir(CO)H(PPh₃)₂	6,14,24[368] **14,15**[261]		150–154[261] 150[368]	**INX**[368] **IN**[261]

---

a,b See footnotes to Table 10.1.

Bibliography

B 1 V. Bazant, J. Hetflejs, V. Chvalovský *et al.*, *Handbook of Organosilicon Compounds*. *Advances since 1961*, Vols. 2–4 (Dekker, New York, 1975–1980); V. Chvalovský and J. Rathouský, *Organosilcon Compounds*, Vols. 5, 6, 7, 8, 9, 10 (Czechoslovak Academy of Sciences Publishing House, Prague, 1981–1983).

B 2 M. G. Voronkov, V. P. Mileshkevich and Yu. A. Yuzhelevskii, *The Siloxane Bond. Physical Properties and Chemical Transformations* (Plenum Press, New York, 1978).

B 3 I. Haiduc, *The Chemistry of Inorganic Ring Systems*, Part 1 (Wiley-Interscience, London, 1970), p. 443.

B 4 S. Pawlenko, *Methoden der Organischen Chemie (Houben-Weyl)*, 4th Edn, Band XII/5. *Silicium-Organische Verbindungen*, (G. Thieme Verlag, Stuttgart, 1980).

B 5 O. K. Johannson and C. Lee, in *Cyclic Monomers*, edited by K. C. Frisch (Wiley, New York, 1972).

B 6 M. G. Voronkov and V. I. Lavrentyev, *Top. Curr. Chem.*, **102** (1982) 201.

B 7 V. E. Shklover and Yu. T. Struchkov, *Usp. Khim.* 49 (1980) 518; *Russ. Chem. Rev.* **49** (1980) 272.

References

1 D. M. Adams and W. S. Fernando, *J. Chem. Soc.* A 4 (1973) 410.

2 A. A. Ainohtein, G. P. Pavlikova, S. V. Syavtsillo and E. P. Tseitlin, *Khim. Prom. (Moscow)* 48 (1972) 395. (*CA* 77 (1972) 63051c).

3 T. A. Aire and F. I. Ikegwuonu, *IRCS Med. Sci. Libr. Compend* 7 (1979) 186.

4 T. A. Aire, F. I. Ikegwuonu and E. H. Heath, *Arch. Androl.* 2 (1979) 371.

5 I. Aishima, H. Sakurai, J. Takashi, H. Morita and Y. Hirotsu, *US Pat.* 3 661 878 (*CA* 77 (1972) 49196x).

6 M. Aisman, B. Kysela and M. Schaetz, *Czech. Pat.* 150 735 (*CA* 81 (1974) 41352g).

7 R. M. Aizatullova and V. G. Genchel, *6. Vses. Konf. Kalorim. 1973, Tbilisi, Metsniereba* (1973), p. 133.

8 I. S. Akhrem, N. M. Christovalova, E. I. Mysov and M. E. Vol'pin, *Izv. Akad. Nauk SSSR, Ser. Khim.* 7 (1978) 1598.

9 L. Albanus, N. E. Bjorklund, B. Gustafsson and M. Jonsson. *Acta Pharmacol. Toxicol. Suppl.* **36** (1975) 93.

10 W. Ando and A. Sekiguchi. *J. Organomet. Chem.* **133** (1977) 219.

11 W. Ando, A. Sekiguchi and T. Migita. *J. Am. Chem. Soc.* **97** (1975) 7159.

12 K. A. Andrianov, *Vysokomol. Soedin.* **A20** (1978) 243 (*CA* **88** (1978) 121953p).

13 K. A. Andrianov, M. S. Aslanov, B. I. D'yachenko, B. D. Nedorosol, T. V. Vasil'eva, L. M. Khananashvili, B. A. Golovnya and B. V. Molchanov, *USSR Pat.* 663 700.

14 K. A. Andrianov, O. G. Blokhina and L. M. Volkova, *Zh. Obshch. Khim.* **46** (1976) 1537.

15 K. A. Andrianov, O. G. Blokhina, B. G. Zavin and N. V. Petrsova, *Vysokomol. Soedin.* **A19** (1977) 434 (*CA* **86** (1977) 156049j).

16 K. A. Andrianov, E. A. Burlova and A. B. Zachernyuk, *Vysokomol. Soedin.* **B19** (1977) 211 (*CA* **87** (1977) 6703z).

17 K. A. Andrianov, A. I. Chernyavskii and N. N. Makarova, *Izv. Akad. Nauk SSSR, Ser. Khim.* **8** (1979) 1835.

18 K. A. Andrianov, V. I. Elinek, M. A. Sipyagina and E. P. Agapova, *Izv. Akad. Nauk SSSR, Ser. Khim.* **10** (1971) 2145.

19 K. A. Andrianov, V. N. Emel'yanov and A. M. Muzafarov, *Dokl. Akad. Nauk SSSR, Ser. Khim.* **226** (1976) 827.

20 K. A. Andrianov, V. N. Emel'yanov, A. M. Muzafarov and A. Tot, *Dokl. Akad. Nauk SSSR, Ser. Khim.* **222** (1975) 603.

21 K. A. Andrianov, V. N. Emel'yanov and E. V. Rudman, *Dokl. Akad. Nauk SSSR, Ser. Khim.* **212** (1973) 872.

22 K. A. Andrianov, M. A. Ezerets, I. I. Khazanov, V. G. Serov and V. T. Kaufman, *Izv. Akad. Nauk SSSR, Ser. Khim.* **8** (1973) 1894.

23 K. A. Andrianov, L. A. Gavrikova and E. F. Rodionova, *Vysokomol. Soedin.* **A13** (1971) 937 (*CA* **75** (1971) 49817j).

24 K. A. Andrianov, Yu. K. Godovskii, V. S. Svistunov, V. S. Papkov, A. A. Zhdanov and G. L. Slonimskii, *Dokl. Akad. Nauk SSSR, Ser. Khim.* **234** (1977) 1326.

25 K. A. Andrianov, S. A. Igonina and V. I. Sidorov, *J. Organomet Chem.* **128** (1977) 43.

26 K. A. Andrianov, N. N. Kakarova and K. N. Raspopova, *USSR Pat.* 550 406 (*CA* **86** (1977) 190649g).

27 K. A. Andrianov, L. M. Khananashvili, V. M. Kopylov, P. L. Prikhod'ko and V. A. Temnikovskii, *1 Vsesoyuz. Simp., Irkutsk* (1977), p. 264.

28 K. A. Andrianov, L. M. Khananashvili, N. N. Makarova, O. V. Mukbaniani, S. M. Meladze and N. A. Koyava, *USSR Pat.* 794 029.

29 K. A. Andrianov, L. M. Khananashvili, N. N. Makarova, O. V. Mukbaniani, S. M. Meladze and N. A. Koyava, *USSR Pat.* 757 555.

30 K. A. Andrianov, E. I. Khubulava, V. M. Kopylov, V. A. Temnikovskii, A. I. Nogaideli and L. M. Khananashvili, *Dokl. Akad. Nauk SSSR, Ser. Khim.* **229** (1976) 614.

31 K. A. Andrianov, I. Yu. Klement'ev, N. A. Babkina and V. S. Tikhonov, *Tr. Mosk. Inst. Tonkoi Khim. Tekhnol.* **3** (1973) 67 (*CA* **82** (1975) 4617x).

32 K. A. Andrianov, I. Yu. Klement'ev, G. N. Kartsev and V. S. Tikhonov, *Zh. Obshch. Khim.* **42** (1972) 1342.

33 K. A. Andrianov, I. Yu. Klement'ev, B. D. Lavrukhin and V. S. Tikhonov, *Zh. Obshch. Khim.* **45** (1975) 2658.

34 K. A. Andrianov, I. Yu. Klement'ev and V. S. Tikhonov, *Dokl. Nauk SSSR, Ser. Khim.* **241** (1978) 834.

35 K. A. Andrianov, V. M. Kopylov, M. I. Shkol'nik and L. M. Khananashvili, *1. Vsesoyuz. Simp., Irkutsk* (1977), p. 291.

36 K. A. Andrianov, V. M. Kopylov, V. A. Temnikovskii and L. M. Khananashvili *Vysokomol. Soedin.* **A18** (1976) 1714 (*CA* **85** (1976) 143632y).

37 K. A. Andrianov, V. M. Kotov and T. A. Pryakhina, *Izv. Akad. Nauk SSSR, Ser. Khim.* **9** (1975) 2055.

38 K. A. Andrianov, V. M. Kotov and T. A. Pryakhina, *Izv. Akad. Nauk SSSR, Ser. Khim.* **1** (1975) 129.

39 K. A. Andrianov, V. M. Kotov and T. A. Pryakhina, *Vysokomol. Soedin.* **B18** (1976) 254 (*CA* **85** (1976) 21951f).

40 K. A. Andrianov, V. M. Kotov and T. A. Pryakhina, *Izv. Akad. Nauk SSSR, Ser. Khim.* **12** (1978) 2789.

41 K. A. Andrianov, V. M. Kotov, T. A. Pryakhina and N. N. Lyapunova, *Izv. Akad. Nauk SSSR, Ser. Khim.* **2** (1977) 410.

42 K. A. Andrianov, G. V. Kotrelev, A. I. Nogaideli, I. V. Zhuravleva, N. G. Lekishvili, Yu. I. Tolchinskii and V. I. Pushich, *Vysokomol. Soedin.* **A19** (1977) 451 (*CA* **87** (1977) 53893w).

43 K. A. Andrianov, G. V. Kotrelev, N. A. Tebenova, N. V. Pertsova, I. I. Tverdokhlebova, P. A. Kurginyan and T. A. Larina, *Vysokomol. Soedin.* **A20** (1978) 692 (*CA* **88** (1978) 153104d).

44 K. A. Andrianov, G. V. Kotrelev, E. A. Zhdanova, T. V. Strelkova, E. S. Obolonkova, V. A. Martirosov and N. I. Pankov, *Vysokomol. Soedin.* A20 (1978) 2355 (*CA* 90 (1979) 6931j).

45 K. A. Andrianov, G. V. Kotrelev, E. A. Zhdanova, T. V. Strelkova and N. I. Pankov, *Dokl. Akad. Nauk SSSR, Ser. Khim.* 233 (1977) 349.

46 K. A. Andrianov, B. D. Lavrukhin, T. V. Birynukova, B. G. Zavin and V. S. Svistunov, *Dokl. Akad. Nauk SSSR, Ser. Khim.* 207 (1972) 1113.

47 K. A. Andrianov, B. D. Lavrukhin, V. M. Kotov, T. A. Pryakhina and T. V. Strelkova, *Izv. Akad. Nauk SSSR, Ser. Khim.* 4 (1978) 843.

48 K. A. Andrianov, I. A. Lavygin, V. E. Ditsent, V. M. Kotov, T. A. Pryakhina, M. N. Zolotareva and N. A. Terenk'eva, *Vysokomol. Soedin.* A19 (1977) 76 (*CA* 86, 107123v).

49 K. A. Andrianov, G. I. Magomedov, O. V. Shkol'nik, B. A. Izmailov, L. V. Morozova, V. N. Kalinin and V. G. Syrkin, *USSR Pat.* 632 699 (*CA* 90 (1979) 87654w).

50 K. A. Andrianov, G. J. Magomedov, O. V. Shkol'nik, B. A. Izmailov, L. V. Morozova, V. N. Kalinin and V. G. Syrkin, *USSR Pat.* 632 700 (*CA* 90 (1979) 87655x).

51 K. A. Andrianov and N. N. Makarova, *Izv. Akad. Nauk SSSR, Ser. Khim.* 9 (1972) 2046.

52 K. A. Andrianov, N. N. Makarova and N. V. Chizhova, *Izv. Akad. Nauk SSSR, Ser. Khim.* 11 (1975) 2554.

53 K. A. Andrianov, K. I. Makarova, V. A. Dmitriev and N. N. Makarova: *USSR* 331 072 (*CA* 77 (1972) 49146f).

54 K. A. Andrianov, A. I. Nogaideli, E. I. Khubulava, V. M. Kopylov, L. M. Khananashvili, N. N. Bravina and V. A. Temnikovskii, *Dokl. Akad. Nauk SSSR, Ser. Khim.* 222 (1975) 1339.

55 K. A. Andrianov, A. I. Nogaideli, E. I. Khubulava, V. M. Kopylov, L. M. Khananashvili, A. G. Kolchina and N. G. Sycheva, *Vysokomol. Soedin.* A18 (1976) 2619 (*CA* 86 (1977) 73344a).

56 K. A. Andrianov, A. I. Nogaideli, E. I. Khubulava, V. M. Kopylov, A. G. Kolchina, L. M. Khananashvili and N. G. Sycheva, *Soobshch. Akad. Nauk Gruz. SSR* 82 (1976) 361 (*CA* 85 (1976) 144358a).

57 K. A. Andrianov, A. I. Nogaideli, N. N. Makarova and O. V. Mukbaniani, *Soobshch. Akad. Nauk Gruz. SSR* 81 (1976) 349 (*CA* 85 (1976) 5753n).

58 K. A. Andrianov, A. I. Nogaideli, N. N. Makarova and O. V. Mukbaniani, *Dokl. Akad. Nauk SSSR, Ser. Khim.* 230 (1976) 593.

59 K. A. Andrianov, A. I. Nogaideli, N. N. Makarova and O. V. Mukbaniani, *Izv. Akad. Nauk SSSR, Ser. Khim.* 6 (1977) 1393.

60 K. A. Andrianov, A. I. Nogaideli, N. N. Makarova, O. V. Mukbaniani and K. N. Raspopova, *Dokl. Akad. Nauk SSSR, Ser. Khim.* 224 (1975) 825.

61 K. A. Andrianov, A. I. Nogaideli, D. Ja. Tsvankin, N. N. Makarova and O. V. Mukbaniani, *Dokl. Akad. Nauk SSSR, Ser. Khim.* 229 (1976) 1353.

62 K. A. Andrianov, S. A. Pavlova, I. I. Tverdokhlebova, N. V. Pertsova and V. A. Temnikovskii, *Vysokomol. Soedin.* A14 (1972) 1816 (*CA* 77 (1972) 165185s).

63 K. A. Andrianov, S. A. Pavlova, I. P. Tverdokhlebova, N. V. Pertsova, V. A. Temnikovskii and L. N. Pronina, *Vysokomol. Soedin.* A19 (1977) 466 (*CA* 86 (1977) 156125f).

64 K. A. Andrianov, S. A. Pavlova, I. I. Tverdokhlebova, N. V. Pertsova, B. G. Zavin and I. I. Mamaeva, *Vysokomol. Soedin.* B16 (1974) 665 (*CA* 82 (1975) 43838g).

65 K. A. Andrianov, I. M. Petrova and S. E. Jakushkina *Vysokomol. Soedin.* A12 (1970) 1683 (*CA* 73 (1970) 120967z).

66 K. A. Andrianov, N. N. Petrovnina, T. V. Vasil'eva, V. E. Shklover and B. I. D'yachenko, *Zh. Obshch. Khim.* 48 (1978) 2692.

67 K. A. Andrianov and E. F. Rodionova, *Vysokomol. Soedin.* B13 (1971) 829 (*CA* 76 (1972) 86223d).

68 K. A. Andrianov, E. F. Rodionova and G. M. Luk'yanova, *Vysokomol. Soedin.* **B14** (1972) 331 (*CA* **77** (1972) 88908a).

69 K. A. Andrianov, A. S. Shapatin, G. Ya. Zhigalin, I. M. Golyaeva, M. A. Kleinovskaya, O. A. Safronova, A. G. Trufanov, N. G. Ufimtsev and V. M. Kopylov *USSR Pat.* 683 206.

70 K. A. Andrianov, M. I. Shkol'nik, V. M. Kopylov and N. N. Bravina *Vysokomol. Soedin.* **B16** (1974) 893 (*CA* **82** (1975) 98516p).

71 K. A. Andrianov, M. I. Shkol'nik, V. M. Kopylov, L. M. Khananashvili, V. T. Minakov and M. G. Zaitseva, *Dokl. Akad. Nauk SSSR* **249** (1979) 363.

72 K. A. Andrianov, M. A. Sipyagina and N. A. Dyudina, *Temat. Sb. Mosk. Int. Tonk. Khim. Tekhnol.* **7** (1977) 67 (*CA* **89** (1978) 130222x).

73 K. A. Andrianov, M. A. Sipyagina and N. P. Gashnikova, *Izv. Akad. Nauk SSSR, Ser. Khim.* **11** (1970) 2493 (*CA* **76** (1972) 46243d).

74 K. A. Andrianov, M. A. Sipyagina, V. Ya. Kovalenko, A. G. Kolchina and T. S. Bebchuk, *Izv. Akad. Nauk SSSR, Ser. Khim.* **6** (1972) 1318.

75 K. A. Andrianov, G. L. Slonimskii, Yu. K. Godovskii, A. A. Zhdanov, B. G. Zavin and V. S. Svistunov, *Vysokomol. Soedin.* **B15** (1973) 837 (*CA* **80** (1974) 71212h).

76 K. A. Andrianov, G. L. Slonimskii, V. Yu. Levin, Yu. K. Godovskii, I. K. Kuznetsova, D. Ya. Tsvankin, V. A. Moskalenko and L. I. Kuteinikova, *Vysokomol. Soedin.* **A12** (1970) 1268 (*CA* **73** (1970) 56470c).

77 K. A. Andrianov, J. Soucek, J. Hetflejs and L. M. Khananashvili, *Zh. Obshch. Khim.* **45** (1975) 2215.

78 K. A. Andrianov, J. Soucek, J. Hetflejs and L. M. Khananashvili, *Izv. Akad. Nauk SSSR, Ser. Khim.* **4** (1975) 965.

79 K. A. Andrianov, J. Soucek, L. M. Khananashvili and L. Ambrus, *Izv. Akad. Nauk SSSR, Ser. Khim.* **3** (1975) 606.

80 K. A. Andrianov, V. S. Svistunov, Yu. K. Godovskii, B. G. Zavin and V. S. Papkov, *5th Int. Symp. on Organosilicon Compounds, Karlsruhe 1978*, p. 140.

81 K. A. Andrianov, V. N. Talanov, M. M. Il'in, A. I. Chernyshev, V. V. Kazakova and E. E. Stepanova, *Zh. Obshch. Khim.* **47** (1977) 2071.

82 K. A. Andrianov and L. M. Tartakovskaya *Izv. Akad. Nauk SSSR, Ser. Khim.* **11** (1972) 2631.

83 K. A. Andrianov, L. M. Tartakovskaya and V. M. Kopylov, *USSR Pat.* 465 931.

84 K. A. Andrianov, L. M. Tartakovskaya and B. I. Shapiro, *Zh. Obshch. Khim.* **42** (1972) 176.

85 K. A. Andrianov, V. A. Temnikovskii, L. M. Khananashvili and N. P. Lyapina, *Vysokomol. Soedin.* **A14** (1972) 2235 (*CA* **78** (1973) 72673e).

86 K. A. Andrianov and V. S. Tikhonov, *Zh. Obshch. Khim.* **43** (1973) 2449.

87 K. A. Andrianov, V. S. Tikhonov and B. A. Astapov, *Vysokomol. Soedin.* **B12** (1970) 577 (*CA* **73** (1970) 110197j).

88 K. A. Andrianov, V. S. Tikhonov and I. Yu. Klement'ev, *Uch. Zap. Mosk. Inst. Tonkoi Khim. Tekhnol.* **1** (1970) 70.

89 K. A. Andrianov, V. S. Tikhonov and I. Yu. Klement'ev *Vysokomol. Soedin.* **B16** (1974) 67 (*CA* **81** (1974) 13830t).

90 K. A. Andrianov, V. S. Tikhonov, I. Yu. Klement'ev and Phan Minh Ngoc, *Dep. VINITI* 4139 (1977) (*CA* **90** (1979) 187027n).

91 K. A. Andrianov, V. S. Tikhonov, I. Yu. Klement'ev and M. N. Rozhnova, *Vysokomol. Soedin.* **A18** (1976) 2288 (*CA* **86** (1977) 30191f).

92 K. A. Andrianov, V. S. Tikhonov and Phan Minh Ngoc, *Temat. Sb. Mosk. Int. Tonk. Khim. Tekhnol.* **7** (1977) 72 (*CA* **89** (1978) 130223y).

93 K. A. Andrianov, V. S. Tikhonov, Phan Minh Ngoc and E. I. Voronina, *Dep. VINITI* 4138 (1977).

94 K. A. Andrianov, V. S. Tikhonov, S. A. Tverskaya, V. N. Alanichev and M. N. Chernobrovkina, *Zh. Obshch. Khim.* **40** (1970) 339.

95 K. A. Andrianov, V. S. Tikhonov, Le Lien Viet and Phan Minh Ngoc, *Zh. Obshch. Khim.* **46** (1976) 324.

96 K. A. Andrianov, V. S. Tikhonov, G. P. Makhneva and G. S. Chernov, *Izv. Akad. Nauk SSSR, Ser. Khim.* **4** (1973) 956.

97 K. A. Andrianov, V. N. Tsvetkov, D. Ya. Tsvankin, A. I. Nogaideli, N. N. Makarova, O. V. Mukbaniani, M. G. Vitovskaya, Ya. V. Genin and G. F. Kolbina, *Vysokomol. Soedin.* **A18** (1976) 890 (*CA* **85** (1976) 124729u).

98 K. A. Andrianov, I. I. Tverdokhlebova, N. N. Makarova, I. I. Mamaeva, A. K. Chekalov, V. M. Men'shov and S. A. Pavlova, *Vysokomol. Soedin.* **A20** (1978) 377 (*CA* **88** (1978) 137273c).

99 K. A. Andrianov, I. I. Tverdokhlebova, S. A. Pavlova, T. A. Larina and I. M. Petrova, *Vysokomol. Soedin.* **A18** (1976) 1117 (*CA* **85** (1976) 47168f).

100 K. A. Andrianov, I. I. Tverdokhlebova, S. A. Pavlova, T. A. Larina, P. A. Kurginyan, G. V. Kotrelev, T. A. Tebeneva, N. V. Pertsova and E. G. Khoroshilova, *Vysokomol. Soedin.* **A19** (1977) 2300 (*CA* **88** (1978) 74626x).

101 K. A. Andrianov, I. I. Tverdokhlebova, S. A. Pavlova, T. A. Larina, B. G. Zavin, G. F. Sablina, I. M. Petrova and T. N. Balykova, *4th Int. Symp. on Organosilicon Chemistry, Moscow* (1975), Vol. 2, p. 16.

102 K. A. Andrianov, I. I. Tverdokhlebova, S. A. Pavlova, N. V. Pertsova and N. L. Orlovskaya *Vysokomol. Soedin.* **A19** (1977) 19 (*CA* **86** (1977) 73354d).

103 K. A. Andrianov, T. V. Vasil'eva, V. D. Deeva, N. M. Katashuk and B. I. D'yachenko, *Vysokomol. Soedin.* **A19** (1977) 167 (*CA* **86** (1977) 90292z).

104 K. A. Andrianov, T. V. Vasil'eva, N. M. Katashuk, T. V. Snigireva and B. I. D'yachenko *Vysokomol. Soedin.* **A18** (1976) 1270 (*CA* **86** (1977) 5512d).

105 K. A. Andrianov, T. V. Vasil'eva and N. M. Petrovnina, *1. Vsesoyuz. Simp., Irkutsk 1977, Tezisy Dokl.* (1977), p. 287.

106 N. N. Belov, I. M. Kolesnikov and T. V. Zaitseva, *Zh. Fiz. Khim.* **56** (1982) 2103.

107 K. A. Andrianov, G. V. Vinogradov, Yu. G. Yanovskii, A. B. Zachernyuk, E. A. Burlova and S. I. Sergeenkov, *Dokl. Akad. Nauk SSSR, Ser. Khim.* **232** (1977) 810.

108 K. A. Andrianov, M. G. Vitovskaya, S. V. Bushin, V. N. Emel'yanov, A. M. Muzafarov, D. Ya. Tsvankin and V. N. Tsvetkov, *Vysokomol. Soedin* **A20** (1978) 1277 (*CA* **89** (1978) 60182h).

109 K. A. Andrianov, S. A. Volkov, V. I. Sidorov and L. M. Tartakovskaya, *Zh. Obshch. Khim.* **30** (1970) 2049.

110 K. A. Andrianov, M. E. Vol'pin, A. I. Nogaideli, Ts. V. Kakuliya, E. I. Khubulava, S. L. Zakoyan, N. D. Lapkina, V. A. Postnikov, V. M. Kopylov and Yu. N. Novikov, *USSR Pat.* 681 070.

111 K. A. Andrianov, S. E. Yakushkina, I. I. Koretko, B. D. Lavrukhin and I. I. Petrova, *Vysokomol. Soedin.* **A13** (1971) 2754.

112 K. A. Andrianov, S. E. Yakushkina, I. M. Petrova and L. M. Chebysheva, *USSR Pat.* 303 332 (*CA* **75** (1971) 152750f).

113 K. A. Andrianov and A. B. Zachernyuk, *Khim. Geterotsikl. Soedin.* **9** (1972) 1183 (*CA* **78** (1973) 11038e).

114 K. A. Andrianov and A. B. Zachernyuk, *Vysokomol. Soedin.* **A16** (1974) 1435 (*CA* **81** (1974) 169927r).

115 K. A. Andrianov and A. B. Zachernyuk, *Zh. Vses. Khim. Obshchest.* **19** (1974) 107 (*CA* **80** (1974) 12105f).

116   K. A. Andrianov and A. B. Zachernyuk, *Am. Chem. Soc. Polym. Prepr.* **16** (1975) 442.
117   K. A. Andrianov and A. B. Zachernyuk, *Vysokomol. Soedin.* **B16** (1974) 307 (*CA* **81** 92010b).
118   K. A. Andrianov and A. B. Zachernyuk, *1st IUPAC Int. Symp. on Polymeric Hetero-cycles (Ring-Opening), Warsaw–Jablonna* (1975), S.1.
119   K. A. Andrianov, A. B. Zachernyuk and E. A. Burlova, *USSR Pat.* 602 510 (*CA* **89** (1978) 7011a).
120   K. A. Andrianov, A. B. Zachernyuk and E. A. Burlova, *Dokl. Akad. Nauk SSSR, Ser. Khim.* **236** (1977) 873.
121   K. A. Andrianov, A. B. Zachernyuk and S. A. Danielov, *Khim. Geterotsikl. Soedin.* **7** (1972) 893 (*CA* **77** (1972) 140213m).
122   K. A. Andrianov, A. B. Zachernyuk and B. D. Lavrukhin, *Dokl. Akad. Nauk SSSR, Ser. Khim.* **207** (1972) 593.
123   K. A. Andrianov, A. B. Zachernyuk, B. D. Lavrukhin and N. G. Vasilenko, *Dokl. Akad. Nauk SSSR, Ser. Khim.* **204** (1972) 1112.
124   K. A. Andrianov, A. A. Zhdanov, E. F. Rodionova and N. G. Vasilenko, *Vysokomol. Soedin.* **B19** (1977) 806 (*CA* **88** (1978) 74584g).
125   K. A. Andrianov, A. B. Zachernyuk and G. V. Solomatin, *Dokl. Akad. Nauk SSSR, Ser. Khim.* **236** (1977) 601.
126   K. A. Andrianov, A. B. Zachernyuk and G. V. Solomatin, *USSR Pat* 603 645 (*CA* **89** (1978) 43760c).
127   K. A. Andrianov, A. B. Zachernyuk and E. A. Zhdanova, *Dokl. Akad. Nauk SSSR. Ser. Khim.* **214** (1974) 325.
128   K. A. Andrianov, B. G. Zavin, O. G. Blokhina and V. S. Svistunov, *1. Vsesoyuz. Simp., Irkutsk 1977, Tezisy Dokl.* (1977), p. 112.
129   K. A. Andrianov, B. G. Zavin and G. F. Sablina, *Vysokomol. Soedin.* **A20** (1978) 1099 (*CA* **89** (1978) 44286g).
130   K. A. Andrianov, B. G. Zavin and G. F. Sablina, *Vysokomol. Soedin.* **A14** (1972) 1156 (*CA* **77** (1972) 48912r).
131   K. A. Andrianov, B. G. Zavin and G. F. Sablina, *5th Int. Symp. on Organosilicon Compounds, Karlsruhe* (1978), p. 142.
132   K. A. Andrianov, B. G. Zavin, A. A. Zhdanov, A. M. Evdokimov, T. V. Biryukova and B. D. Lavrukhin, *Vysokomol. Soedin.* **B14** (1972) 327 (*CA* **77** (1972) 89194b).
133   K. A. Andrianov, A. A. Zhdanov, I. V. Karpova and V. A. Odinets, *Izv. Akad. Nauk SSSR, Ser. Khim.* **3** (1971) 582.
134   K. A. Andrianov, A. A. Zhdanov, N. A. Kurasheva and E. S. Khynku, *5th. Int. Symp. on Organosilicon Compounds, Karlsruhe* (1978), p. 72.
135   K. A. Andrianov, A. A. Zhdanov, N. A. Kurasheva and E. S. Khynku, *Dokl. Akad. Nauk SSSR, Ser. Khim.* **240** (1978) 591.
136   K. A. Andrianov, A. A. Zhdanov, N. A. Kurasheva and E. S. Khynku, *Dokl. Akad. Nauk SSSR, Ser. Khim.* **236** (1977) 354.
137   K. A. Andrianov, A. A. Zhdanov, N. A. Kurasheva and E. S. Khynku, *1. Vsezoyuz. Simp., Irkutsk 1977, Tezisy Dokl.* (1977), p. 108.
138   K. A. Andrianov, A. A. Zhdanov, N. A. Kurasheva, E. S. Khynku and T. V. Popova, *Izv. Akad. Nauk SSSR, Ser. Khim.* **3** (1975) 703.
139   K. A. Andrianov, A. A. Zhdanov, N. A. Kurasheva and L. I. Kuteinikova, *Vysokomol. Soedin.* **A21** (1979) 1640.
140   K. A. Andrianov, A. A. Zhdanov, E. F. Rodionova and E. G. Vasilenko, *Zh. Org. Khim.* **45** (1975) 2444.
141   K. A. Andrianov, A. A. Zhdanov, B. G. Zavin and A. M. Evdokimov, *Dokl. Akad. Nauk SSSR* **199** (1971) 597.

142 K. A. Andrianov, A. A. Zhdanov, B. G. Zavin and A. M. Evdokimov, *Vysokomol. Soedin.* **B13** (1971) 397 (*CA* **75** (1971) 88968d).

143 K. A. Andrianov, A. A. Zhdanov, B. G. Zavin and G. F. Sablina, *4th Int. Symp. on Organosilicon Chemistry, Moscow* (1975), Vol. 2, p. 8.

144 K. A. Andrianov, A. A. Zhdanov, B. G. Zavin and G. F. Sablina, *Vysokomol. Soedin.* **A14** (1972) 1855 (*CA* **77** (1972) 165109v).

145 K. A. Andrianov, A. A. Zhdanov, B. G. Zavin and T. I. Sunekants, *Vysokomol. Soedin.* **A12** (1970) 20.

146 K. A. Andrianov, A. A. Zhdanov, B. G. Zavin, V. S. Svistunov and S. E. Yakushkina, *Vysokomol. Soedin.* **B13** (1971) 598. (*CA* **76** (1972) 4656k).

146a M. Arai, M. Seino and Y. Koike, *Japan Kokai* 79 74 900.

147 V. G. Aristova, N. A. Ivanova, I. M. Zimmer, I. I. Skorokhodov, A. I. Gorbunov and E. P. Belozerova, *Dep. VINITI* 2459 (1977) (*CA* **90** (1979) 72614j).

148 B. A. Ashby, *US Pat.* 3 903 047.

149 B. A. Ashby, *Ger. Offen.* 1 495 921.

150 K. Ashida, M. Ohtani, T. Yokoyama and S. Ohkubo, *J. Cell. Plast.* **14** (1978) 255.

151 W. H. Atwell and G. N. Bokerman, *Fr. Demande* 2 163 579 (*CA* **80** (1974) 3625v).

152 W. A. Aue and P. P. Wickramanayake, *J. Chromatogr.* **200** (1980) 3.

153 N. N. Baglei and M. T. Bryk, *Vysokomol. Soedin.* **A20** (1978) 2473 (*CA* **90** (1979) 39317s).

154 I. A. Baidina, N. V. Podberezskaya, V. I. Alekseev, T. N. Martynova, S. V. Borisov and A. N. Kanev, *Zh. Strukt. Khim.* **20** (1979) 648.

155 I. A. Baidina, N. N. Podberezskaya, S. V. Borisov, V. I. Alekseev, T. N. Martynova and A. N. Kanev, *Zh. Strukt. Khim.* **21** (1980) 125.

156 I. A. Baidina, N. V. Podberezskaya, T. N. Martynova, S. V. Borisov, A. N. Kanev and Yu. V. Basikhin, *1. Vsesoyuz. Simp., Irkutsk 1977, Tezisy Dokl.* (1977), p. 239.

157 P. Bajaj, S. K. Varshney and A. Misra, *J. Polym. Sci., Polym. Chem. Ed.* **18** (1980) 295.

158 T. N. Baratova, V. P. Mileshkevich, V. S. Fikhtengol'ts, V. F. Gridina and O. N. Dolgov, *Khim. i Praktich. Primenenie Kremnii—i Fosfororganich. Soedin.* **1** (1979) 31.

159 M. Bargain and C. Millet, *Ger. Offen.* 2 724 194 (*CA* **88** (1978) 90806m).

160 J. H. Baronnier and G. L. Pagni, *Ger. Offen.* 1 745 319.

161 J. H. Baronnier and G. L. Pagni, *Ger. Offen.* 1 745 320.

162 J. A. Barrie, R. B. Dawson and R. N. Sheppard, *J. Chem. Soc. Faraday Trans.* I **74** (1978) 490.

163 J. T. Barry: *US Pat.* 3 780 025.

164 T. J. Barton and J. A. Kilgour, *J. Am. Chem. Soc.* **98** (1976) 7231.

165 T. J. Barton and J. A. Kilgour, *J. Am. Chem. Soc.* **96** (1974) 2278.

166 T. J. Barton, G. Marquardt and J. A. Kilgour, *J. Organomet. Chem.* **85** (1975) 317.

167 Yu. M. Batulin, A. L. Klyashchitskaya and N. K. Kulagina, *4th Int. Symp. on Organosilicon Chemistry, Moscow* (1975) Vol. 1, p. 14.

168 Yu. M. Batulin, A. L. Klyashchitskaya, M. G. Domilak and E. M. Chirkova, *Sb. 2—II. Vsesoyuz. Simpoz. Biol. Akt. Soedin. Elem. IV. B Gruppy* (1977) p. 68 (*CA* **89** (1978) 85446k).

169 K. Baum, D. A. Lerdal and J. S. Horn, *J. Org. Chem.* **43** (1978) 203.

170 Bausch and Lomb Inc, *Neth. Appl.* 78 07 833.

171 A. N. Bashkirov, R. A. Fridman, S. M. Nosakova, L. G. Liberov and S. I. Beilin, *Izv. Akad. Nauk* **9** (1975) 2132.

172 A. G. Bayer, *Fr. Demande* 2 203 821.

173 A. G. Bayer, *Fr. Demande* 2 335 513.

174 M. S. Beevers and J. A. Semlyen, *Polymer* **12** (1971) 373.

175　M. A. A. Beg and I. A. Khan, *Rev. Roum. Chim.* **17** (1972) 1579.
176　M. A. A. Beg and I. A. Khan, *Rev. Roum. Chim.* **17** (1972) 1599.
177　E. W. Bennett, *Ger. Offen.* 1 964 553; *Fr. Demande* 2 027 189; *US Pat.* 3 642 851, 3 686 253; *UK Pat.* 1 306 810 (*CA* **73** (1970) 67100h).
178　E. W. Bennett, *US Pat.* 3 646 090 (*CA* **77** (1972) 6912k).
179　D. R. Bennett, S. J. Gorzinski and J. E. Lebeau, *Toxicol. Appl. Pharmacol.* **21** (1972) 55.
180　D. R. Bennett and R. R. Levier, *US Pat* 3 821 373 (*CA* **82** (1975) 26147w).
181　D. R. Bennett and J. A. Mchard, *US Pat.* 3 975 521 (*CA* **85** (1976) 130538x).
182　D. R. Bennett and J. A. Mchard, *US Pat.* 3 919 417 (*CA* **84** (1976) 39201a).
183　D. R. Bennett, J. A. Mchard, J. F. Hampton and D. R. Chapman, *Ger. Offen.* 1 933 110; *UK Pat.* 1 233 851, 1 312 255; *Fr. Demande* 2 012 560, 2 012 147, 2 012 148 (*CA* **72** (1970) 103722c).
184　M. S. Benson and J. Winnick, *J. Chem. Engng Data* **21** (1976) 432.
185　D. Bentmann and U. Klingebiel, *J. Fluorine Chem.* **15** (1980) 519.
186　K. Bergstrom and J. Gurtler, *Acta Chem. Scand.* **25** (1971) 175.
187　F. A. Bergstrom and M. T. Maxson, *Ger. Offen.* 2 529 781.
188　A. E. Bey, D. R. Weyenberg and L. Seibles, *Am. Chem. Soc. Polym. Prepr.* **11** (1970) 995.
189　M. Blazsov, G. Garzo and T. Szekely, *Chromatografia* **5** (1972) 485 (*CA* **77** (1972) 165231d).
190　B. A. Bluestein: *US Pat.* 4 111 973 (*CA* **90** (1979) 138010t).
191　B. A. Bluestein and E. R. Evans, *Ger. Offen.* 2 945 786.
192　V. N. Bochkarev, A. A. Bernadskii, M. B. Lotarev, M. V. Sobolevskii and D. V. Nazarova, *Zh. Obshch. Khim.* **47** (1977) 1524.
193　V. N. Bochkarev, A. V. Kisin, V. G. Osipov, N. P. Telegina, N. G. Komalenkova, A. N. Polivanov and E. A. Chernyshev, *Khim. Geterotsikl. Soedin.* **1** (1978) 24 (*CA* **88** (1978) 121292x).
194　V. N. Bochkarev, A. N. Polivanov, A. A. Bernadskii and N. E. Rodzevich, *Zh. Obshch. Khim.* **50** (1980) 1074.
195　V. N. Bochkarev, A. N. Polivanov, A. A. Bernadskii, T. F. Slyusarenko, N. N. Silkina and B. N. Klimentov, *Fiz.-khim. Metody Issled. Elementoorgan. Soedin.*, M (1980) 63.
196　V. N. Bochkarev, A. N. Polivanov, N. P. Telegina. A. G. Kuznetsova and E. A. Chernyshev, *Zh. Obshch. Khim.* **44** (1974) 1061.
197　V. N. Bochkarev, A. N. Polivanov, N. P. Telegina, N. A. Palamarchuk, V. S. Lozovskaya, E. A. Chernyshev and T. I. Agapova, *Zh. Obshch. Khim.* **44** (1974) 1775.
198　V. N. Bochkarev, N. P. Telegina, A. N. Polivanov, A. V. Kisin, N. G. Komalenkova and E. A. Chernyshev, *4th Int. Symp. on Organosilicon Chemistry, Moscow* (1975), Vol. 1 p. 112.
199　N. G. Bokii, G. N. Zakharova and Yu. T. Struchkov, *Zh. Strukt. Khim.* **13** (1972) 291.
200　M. Bordeau, J. Dedier, E. Frainnet, J. P. Fayet and P. Mauret, *J. Organomet. Chem.* **59** (1973) 125.
201　S. N. Borisov, Z. V. Kurlova, Yu. A. Yuzhelevskii and E. A. Chernyshev, *Vysokomol. Soedin.* **B12** (1970) 332 (*CA* **73** (1970) 45899y).
202　S. N. Barisov, N. P. Timofeeva, Yu. A. Yuzhelevskii, E. G. Kogan and N. V. Kozlova, *Zh. Obshch. Khim.* **42** (1972) 873.
203　E. E. Bostick and W. A. Fessler, *Ger. Offen.* 2 049 547 (*CA* **75** (1971) 21478a).
204　E. E. Bostick and J. J. Zdaniewski, *Ger. Offen.* 2 048 914 (*CA* **75** (1971) 21496e).
205　D. C. Bradley and C. Prevedorou-Dumas, *Chem. Ind. (London)* **52** (1970) 1659.
206　D. Braga and G. Zanotti, *Acta Crystallogr.* **B36** (1980) 950.
207　G. P. Bragin and M. Kh. Karapet'yants, *Tr. Mosk. Khim.-Tekhnol. Inst.* **75** (1973) 8 (*CA* **81** (1974) 159129t).

208 G. P. Bragin and M. Kh. Karapet'yants, *Dep. VINITI* **73** (1973) 7654.
209 G. P. Bragin and M. Kh. Karapet'yants, *Zh. Fiz. Khim.* **48** (1974) 1049.
210 G. P. Bragin and M. Kh. Karapet'yants *Tr. Khim. Khim. Tekhnol.* **4** (1975) 76 (*CA* **85** (1976) 166765h).
211 I. Braun, H. Hamann and G. Tschernko, *Plaste Kautsch.* **26** (1979) 603.
212 A. F. Bresak and E. Tolgyesi, *US Pat.* 4 208 471.
213 L. S. Bresler, V. P. Mileshkevich, Yu. A. Yuzhelevskii and N. P. Timofeeva, *Zh. Strukt. Khim.* **19** (1978) 453.
214 E. D. Brown, *Fr. Demande* 1 579 860 (*CA* **72** (1970) 113510w).
215 E. D. Brown and L. A. Haluska, *Ger. Offen.* 2 047 490; *UK Pat.* 1 275 003; *Fr. Demande* 2 062 738; *US Pat.* 3 607 899 (*CA* **76** (1972) 15166b).
216 M. T. Bryk and N. N. Baglei, *Kinet. Katal.* **21** (1980) 963.
217 M. T. Bryk, N. N. Baglei, I. E. Skobets, N. G. Vasil'ev and F. D. Ovcharenko, *Dokl. Akad. Nauk SSSR* **254** (1980) 1144.
218 M. T. Bryk, N. N. Baglei, N. G. Vasil'ev and F. D. Ovcharenko, *Dokl. Akad. Nauk SSSR* **253** (1980) 376.
219 M. T. Bryk, N. N. Baglei, N. G. Vasil'ev and F. D. Ovcharenko, *Dokl. Akad. Nauk SSSR, Ser. Khim.* **239** (1978) 113.
220 M. T. Bryk, I. E. Skobets' and N. G. Vasil'ev, *Kinet. Katal.* **19** (1978) 904.
221 M. T. Bryk, I. E. Skobets', N. G. Vasil'ev and O. D. Kurilenko, *Dopov. Akad. Nauk Ukr. SSR* **B2** (1978) 124.
222 M. T. Bryk and I. A. Varavko, *Plast. Massy* **8** (1978) 18 (*CA* **89** (1978) 147307b).
223 M. T. Bryk, I. A. Varavko and O. D. Kurilenko, *Dopov. Akad. Nauk Ukr. SSR* **B6** (1977) 513 (*CA* **87** (1977) 118171f).
224 M. T. Bryk, I. A. Varavko and O. D. Kurilenko, *Vysokomol. Soedin.* **A20** (1978) 1015 (*CA* **89** (1978) 24910k).
225 W. Buechner, B. Degen, L. Fries, H. Judat, R. Mundil and K. H. Rudolph, *Ger. Offen.* 2 705 563 (*CA* **89** (1978) 216055g).
226 J. Burkhardt, E. Louis and W. Rouchbetger, *Ger. Offen.* 2 839 652.
227 J. Burkhardt, S. Nitzsche and K. H. Wegehaupt, *Ger. Offen.* 1 794 219.
228 V. T. Bykov, T. P. Avilova and N. P. Shapkin, *Vysokomol. Soedin.* **A12** (1970) 724 (*CA* **73** (1970) 26141d).
229 D. Carlstrom, *Nobel Symp. 1977* **40** (1978) 523 (*CA* **90** (1979) 71257h).
230 D. Carlstrom and G. Falkenberg, *Acta Chem. Scand.* **27** (1973) 1203.
231 R. M. Cassidy, M. T. Hurtean, J. P. Mislan and R. W. Ashley, *J. Chromatogr. Sci.* **14** (1976) 444.
232 A. D. Caunt, *Def. Publ. U.S. Pat. Office* 965 004 (*CA* **90** (1979) 138432g).
233 D. R. Chapman, *Ger. Offen.* 2 021 304; *UK Pat.* 1 258 843; *Fr. Demande* 2 047 216; *US Pat* 3 661 847 (*CA* **74** (1971) 54440q).
234 D. R. Chapman, *Ger. Offen.* 2 021 438; *UK Pat.* 1 258 844; *Fr. Demande* 2 047 217; *US Pat.* 3 689 455 (*CA* **74** (1971) 64685u).
235 D. R. Chapman, *US Pat.* 4 230 632.
236 A. S. Chawla, *Artif. Organs* **3** (1979) 92.
237 A. S. Chawla, *Trans. Am. Soc. Artif. Intern. Organs* **25** (1979) 287.
238 A. S. Chawla and L. E. St Pierre, *Adv. Chem. Ser.* **91** (1969) 229.
239 A. S. Chawla and L. E. St Pierre, *J. Polym. Sci.*, Part A–1, **10** (1972) 2691.
240 A. S. Chawla and L. E. St Pierre, *J. Appl. Polym. Sci.* **16** (1972) 1887.
241 A. S. Chawla and L. E. St Pierre, *J. Appl. Polym. Sci.* **19** (1975) 353.
242 P. J. Chiesa, *US Pat.* 3 957 657 (*CA* **85** (1976) 65185p).
243 O. I. Cherenkova, N. V. Alekseev and A. I. Gusev, *Zh. Strukt. Khim.* **16** (1975) 504.
244 J. Chojnowski, *Polimery* **24** (1979) 285.

245   J. Chojnowski, *4–I Mezhdunarodyi Mikrosimp. "Uspekhi v Oblasti Ion. Polimerizatsii",* Ufa (1979), p. 19

246   J. Chojnowski and M. Mazurek, *Adv. Ionic Polym., Proc. 1972 Int. Symp.* (1975), p. 71.

247   J. Chojnowski, S. Penczek, M. Mazurek, M. Sciborek and L. Wilczek, *Pol. Pat.* 94 212 (*CA* **90** (1979) 122270s).

248   J. Chojnowski, S. Penczek, M. Mazurek, M. Sciborek and L. Wilczek, *Pol. Pat.* 85 600 (*CA* **87** (1977) 152723x).

249   J. Chojnowski, M. Sciborek and J. Kowalski, *Makromol. Chem.* **178** (1977) 1351.

250   J. Chojnowski and M. Sciborek, *1st IUPAC Int. Symp. on Polymer Heterocycles (Ring-Opening), Warsaw–Jablonna* (1975), S.1., p. 117.

251   J. Chojnowski, M. Scibiorek and L. Wilczek, *4th Int. Symp. on Organosilicon Chemistry, Moscow* (1975), Vol. 2, p. 7.

252   J. Chojnowski and J. Zietera, *Bull. Acad. Pol. Sci., Ser. Sci. Chim.* **27** (1979) 555.

253   G. I. Chogovadze, V. I. Topschiaskvili, L. M. Khananashvili, *Plast. Massy* **1982** No. 10, 12.

254   W. Clegg, G. M. Sheldrick and N. Vater, *Acta Crystallogr.* **B36** (1980) 3162.

255   H. L. Clever, *Solubility Data Ser.* **1** (1979) 109.

256   H. L. Clever, *Solubility Data Ser.* **1** (1979) 248.

257   H. L. Clever, *Solubility Data Ser.* **2** (1979) 107.

258   A. F. Cockerill, R. C. Harden, G. L. O. Davies and D. M. Rackham, *Org. Magn. Reson.* **6** (1974) 452.

259   B. Csakvari, L. Fabry, P. Gomory and K. Ujszaszy, *Acta Chim. (Budapest)* **90** (1976) 233.

260   B. Csakvari, L. Fabry, P. Gomory and K. Ujszaszy, *Acta Chim. (Budapest)* **91** (1976) 1.

261   M. D. Curtis and J. Greene, *J. Am. Chem. Soc.* **100** (1978) 6362.

262   Dai Nippon Printing Co. Ltd, *Fr. Demande* 2 145 954.

263   R. J. Dain, G. D. Manning and A. R. Burkin, *PCT Int. Appl.* 80 00 796.

264   A. D. Damaeva, G. G. Mashutiva, E. A. Kirichenko, A. P. Yakimchuk and G. R. Bagantdinova: *Dep. VINITI* (1979) 3991.

265   I. M. T. Davidson and J. F. Thompson, *J. Chem. Soc., Chem. Commun.* **1971**, 251.

266   I. M. T. Davidson and J. F. Thompson, *J. Chem. Soc., Faraday Trans.* I **1976**, 1088.

267   I. M. T. Davidson and J. F. Thompson, *J. Chem. Soc., Faraday Trans.* I **1975**, 2260.

268   W. G. Davies and H. V. A. Beedle, *Ger. Offen.* 2 542 425 (*CA* **85**, 22248u).

269   W. G. Davies and D. P. Jones, *Am. Chem. Soc. Prepr.* **11** (1970) 447.

270   W. G. Davies, T. C. Kendrick and D. P. Jones, *UK Pat.* 1 308 459.

271   W. G. Davies, T. C. Kendrick and D. P. Jones, *Fr. Demande* 2 046 978 (*CA* **76** (1972) 25796w).

272   V. P. Davydova, Z. S. Lebedeva and A. V. Karlin, *Vysokomol. Soedin* **B17** (1975) 89 (*CA* **83** (1975) 98786x).

273   J. W. Dean, *Ger. Offen.* 2 116 837 (*CA* **76** (1972) 73475n).

274   J. W. Dean *US Pat.* 3 673 272 (*CA* **77** (1972) 89111x).

275   A. De Montigny, H. H. Moretto and K. F. Thom, *Ger. Offen.* 2 802 667.

276   E. Dickinson, *J. Phys. Chem.* **81** (1977) 2108.

277   E. Dickinson, D. C. Hunt and I. A. Mclure, *J. Chem. Thermodyn.* **7** (1975) 731.

278   E. Dickinson and I. A. Mclure, *J. Chem. Thermodyn.* **7** (1975) 725.

279   V. E. Ditsent, I. I. Skorokhodov, N. A. Terent'eva and M. N. Zolotareva, *Zh. Fiz. Khim.* **45** (1971) 1587.

280   V. E. Ditsent, I. I. Skorokhodov, N. A. Terent'eva and M. N. Zolotareva, *Zh. Fiz. Khim.* **48** (1974) 771.

281   V. E. Ditsent, I. I. Skorokhodov, N. A. Terent'eva and M. N. Zolotareva, *Dep. VINITI* (1973) 7263.

282 V. E. Ditsent, I. I. Skorokhodov, N. A. Terent'eva and M. N. Zolotareva, *Zh. Fiz. Khim.* **52** (1978) 1820.

283 V. E. Ditsent, I. I. Skorokhodov, N. A. Terent'eva and M. N. Zolotareva, *Zh. Fiz. Khim.* **46** (1972) 544.

284 V. E. Ditsent, I. I. Skorokhodov, N. A. Terent'eva and M. N. Zolotareva, *Zh. Fiz. Khim.* **48** (1974) 771.

285 V. E. Ditsent, I. I. Skorokhodov, N. A. Terent'eva, M. N. Zolotareva and M. A. Kleinovskaya, *Zh. Fiz. Khim.* **48** (1974) 780.

286 V. E. Ditsent, I. I. Skorokhodov, N. A. Terent'eva, M. N. Zolotareva and M. A. Kleinovskaya, *Dep. VINITI* **73** (1973) 7286.

287 V. E. Ditsent, I. I. Skorokhodov, N. A. Terent'eva, M. N. Zolotareva and M. B. Lotarev, *Zh. Fiz. Khim.* **48** (1974) 2152.

288 V. E. Ditsent, I. I. Skorokhodov, N. A. Terent'eva, M. N. Zolotareva and M. B. Lotarev, *Dep. VINITI* 1284 (1974).

289 S. Dobos, G. Fogarasi and E. Castellucci, *Spectrochim. Acta*, **A28** (1972) 877.

290 K. Dodgson, D. J. Bannister and J. A. Semlyen, *Polymer* **21** (1980) 663.

291 K. Dodgson, D. Sympson and J. A. Semlyen, *Polymer* **19** (1978) 1285.

292 O. N. Dolgov, A. V. Karlin, L. Z. Marmur and T. N. Baratova, *Khim. i Praktich. Primenenie Kremnii-i Fosfororganich. Soedin.* L (1979) 73.

293 Dow Corning Corp., *Japan Pat.* 73 34 382.

294 Dow Corning Corp., *Japan Pat.* 73 30 480.

295 Dow Corning Corp., *Fr. Demande* 2 143 386.

296 Dow Corning Ltd, *Fr. Demande* 2 145 701.

297 V. A. Drozdov, A. P. Kreshkov, N. D. Rumyantseva and V. F. Andronov, *Plast. Massy* **3** (1973) 70 (*CA* **79** (1973) 19985c).

298 I. L. Dubchak, V. E. Shklover, M. M. Levitskii, A. A. Zhdanov and Yu. T. Struchkov, *Zh. Strukt. Khim.* **21** (1980) 103.

299 G. M. Dugacheva and A. G. Anikin, *Zh. Anal. Khim.* **30** (1975) 632.

300 A. P. Dushechkin, V. V. Ponomarev, S. A. Golubtsov, B. S. Kara-Georgiev, B. S. Kirillov and A. G. Kuznetsova, *USSR Pat.* 374 351.

301 C. Eaborn, T. N. Metham and A. Pidcock, *J. Organomet. Chem.* **54** (1973) C3.

302 C. Eaborn, T. N. Metham and A. Pidcock, *J. Organomet. Chem* **63** (1973) 107.

303 G. Eglinton, J. N. M. Firth and B. L. Welters, *Chem. Geol.* **13** (1974) 125.

304 I. Endo, *Japan Kokai* 77 69 500 (*CA* **87** (1977) 118455b).

305 I. Endo, *Japan Kokai* 79 129 052.

306 I. Endo, *US Pat.* 4 233 428.

307 I. Endo and M. Hashimoto, *Japan Kokai* 79 90 350.

308 I. Endo and C. Shimizu, *Japan. Kokai* 78 79 929 (*CA* **89** (1978) 181365d).

309 G. Engelhardt, H. Jancke, M. Magi, T. Pehk and E. Lippmaa, *J. Organomet. Chem.* **28** (1971) 293.

310 G. Engelhardt, M. Magi and E. Lippmaa, *J. Organomet. Chem.* **54** (1973) 115.

311 E. R. Evans, *Fr. Demande* 2 362 885.

312 R. E. Evans, *US Pat.* 4 157 337.

313 E. R. Evans, *US Pat.* 4 122 247 (*CA* **90** (1979) 88570c).

314 E. R. Evans, *Ger. Offen.* 2 737 698 (*CA* **88** (1978) 153289t).

315 M. B. Ewing and K. N. Marsh, *J. Chem. Thermodyn.* **10** (1978) 267.

316 M. B. Ewing and K. N. Marsh, *J. Chem. Thermodyn.* **9** (1977) 371.

317 M. A. Ezerets, I. I. Khazanov, K. A. Andrianov and V. G. Serov, *4th Int. Symp. on Organosilicon Chemistry, Moscow* (1975), Vol. 1, p. 96.

318 M. A. Ezerets, I. I. Khazanov, K. A. Andrianov and N. T. Vetrova, *Izv. Akad. Nauk SSSR, Ser. Khim.* **12** (1979) 2735.

319   M. A. Ezerets, I. I. Khazanov, N. T. Vetrova and K. A. Andrianov, *Izv. Akad. Nauk SSSR, Ser. Khim.* **12** (1977) 2707.

320   M. A. Ezerets, N. T. Vetrova, L. N. Ignatova and I. I. Khazanov, *Zh. Obshch. Khim.* **50** (1980) 604.

321   Z. Faix, F. Jirat, J. Mares and K. Svehla, *Czech. Pat.* 173 005 (*CA* **89** (1978) 199233n).

322   Z. Faix, F. Jirat, J. Mares and K. Svehla, *Czech. Pat.* 173 005 (*CA* **89** (1978) 199233u).

323   M. O. Farrell, Ch. H. Van Dyke, L. J. Boucher and S. J. Metlin, *J. Organomet. Chem.* **169** (1979) 199.

324   W. A. Fessler and P. C. Juliano, *Ind. Engng Chem. Prod. Res. Develop.* **11** (1972) 407.

325   W. A. Fessler and P. C. Juliano, *Am. Chem. Soc. Prepr.* **12** (1971) 150.

326   D. V. Flatt, *Fr. Demande* 1 602 163.

327   G. Fogarasi, H. Hacker, V. Hoffmann and S. Dobos, *Spectrochim. Acta* **A30** (1974) 629.

328   J. Franc and K. Placek, *Collect. Czech. Chem. Commun.* **38** (1973) 513.

329   V. Frey, P. John. G. Kalchgruber, R. Leser, R. Riedle and H. Spoerk, *Ger. Offen.* 2 630 744 (*CA* **89** (1978) 43752b).

330   C. L. Frye, *Fr. Demande* 2 047 218.

331   C. L. Frye, *Ger. Offen.* 2 021 437 (*CA* **74** (1971) 64301j).

332   C. L. Frye and Chi-Long Lee, *Ger. Offen.* 1 720 498.

333   C. L. Frye and W. T. Collins, *J. Org. Chem.* **35** (1970) 2964.

334   C. L. Frye, R. M. Salinger, F. W. G. Fearon, J. M. Klosowski and T. Deyoung, *J. Org. Chem.* **35** (1970) 1308.

335   C. L. Frye and D. E. Spielvogel, *Swiss Pat.* 603 671.

336   C. L. Frye and D. E. Spielvogel, *US Pat.* 3 954 823 (*CA* **85** (1976) 108762u).

337   H. Fujiki, *Japan Kokai* 79 143 499.

338   T. Fujino and T. Takahashi, *Japan Kokai* 73 04 367 (*CA* **80** (1974) 49432h).

339   J. M. Funt, R. D. Parekh, J. H. Magill and Y. T. Shah, *J. Polym. Sci.* **13** (1975) 2181.

340   S. N. Gadzhiev, V. A. Klyuchnikov, G. A. Lobanov, K. K. Kerimov, A. G. Kuznetsov and L. N. Martynovskaya, *Vses. Konf. Kalorim., Rasshir. Tezisy Dokl.* (1977), Vol. 1, p. 114.

341   S. N. Gadzhiev, G. G. Murullaev, A. G. Kuznetsova and K. A. Karasharli, *6 Vses. Konf. Kalorim., 1973, Tbilisi, Metsniereba* (1973), p. 155.

342   G. Garzo and G. Alexander, *Chromatographia* **4** (1971) 554.

343   G. Garzo and G. Alexander, *Proc. Conf. Appl. Phys. Chem., Ind., Vesprem* (1971), Vol. 1, p. 91. (*CA* **76** (1972) 58802y).

344   G. Garzo and D. Hoebbel, *J. Chromatogr.* **119** (1976) 173.

345   G. Garzo, D. Hoebbel, Z. J. Ecsery and K. Ujszaszi, *J. Chromatogr.* **167** (1978) 321.

346   G. Garzo, J. Tamas, T. Szekely and K. Ujszaszi, *Acta Chim. (Budapest)* **69** (1971) 273.

347   General Electric Co., *UK Pat.* 1 353 502.

348   General Electric Co., *Fr. Demande* 2 108 040.

348a J. C. Getson, *US Pat.* 4 166 078.

349   M. G. Gibby, A. Pines and J. S. Wangh, *J. Am. Chem. Soc.* **94** (1972) 6231.

350   L. S. D. Glasser, E. E. Lachowski, R. K. Harris and J. Jones *J. Mol. Struct.* **51** (1979) 239.

351   C. Glidewell, A. G. Robiette and G. M. Sheldrick, *J. Chem. Soc., Chem. Commun.* **1970**, 931.

352   J. Goetz, W. D. Jamieson and C. R. Masson, *Front. Glass Sci. Technol., Proc. Annu. Meet. Int. Comm. Glass 1969* (1970), p. 69 (*CA* **77** (1972) 78827j).

353   E. A. Goldovskii, L. A. Korotkikh and A. A. Dontsov, *Vysokomol. Soedin.* **B22** (1980) 512.

354   Th. Goldschmidt, *Fr. Demande* 2 338 295.

354a Th. Goldschmidt, *Belg. Pat.* 874 404.

355 Th. Goldschmidt, *UK Pat.* 1 514 226 (*CA* **90** (1979) 6885x).

356 Th. Goldschmidt, *Fr. Demande* 2 347 399.

357 C. M. Golino, R. D. Bush and L. H. Sommer, *J. Am. Chem. Soc.* **96** (1974) 614.

358 C. M. Golino, R. D. Bush and L. H. Sommer, *J. Am. Chem. Soc.* **97** (1975) 7371.

359 C. M. Golino, R. D. Bush, P. On and L. H. Sommer, *J. Am. Chem. Soc.* **97** (1975) 1957.

360 S. A. Golubtsov, K. A. Andrianov, A. G. Kuznetsova and V. I. Ivanov, *USSR Pat.* 301 067 (*CA* **76** (1972) 25417y).

362 T. Graczyk and Z. Lasocki, *Bull. Acad. Pol. Sci., Ser. Sci. Chim.* **26** (1978) 919.

363 T. Graczyk and Z. Lasocki, *Bull. Acad. Pol. Sci., Ser. Sci. Chim.* **27** (1979) 185.

364 T. Graczyk and Z. Lasocki, *Bull. Acad. Pol. Sci., Ser. Sci. Chim.* **27** (1979) 181.

365 N. Grassie and K. F. Francey, *Polym. Degradation Stab.* **2** (1980) 53.

366 N. Grassie, K. F. Francey and I. G. Macfarlane, *Polym. Degradation Stab.* **2** (1980) 67.

367 N. Grassie, I. G. Macfarlane and K. F. Francey, *Eur. Polym. J.* **15** (1979) 415.

368 J. Greene and M. D. Curtis, *J. Am. Chem. Soc.* **99** (1977) 5176.

369 J. Greene and M. D. Curtis, *Inorg. Chem.* **17** (1978) 2324.

370 L. E. Gusel'nikov, N. S. Nametkin, T. Kh. Islamov, A. A. Sobtsov and V. M. Vdovin, *Izv. Akad. Nauk SSSR, Ser. Khim.* **1** (1971) 84 (*CA* **74** (1971) 140628p).

371 A. I. Gusev, D. Yu. Nesterov, N. V. Alekseev, V. V. Zverev, M. B. Lotarev, D. V. Nazarova, N. E. Rodzevich and M. V. Sobolevskii, *1 Vsesoyuz. Simp., Irkutsk 1977, Tezisy Dokl.* (1977), p. 56.

372 D. V. Gvozdev, M. S. Akutin, N. T. Andrianov, Yu. M. Budnitskii, B. V. Yakobson, I. Ya. Faidel, A. V. Timofeev, N. M. Slozhenikina, N. I. Bychkov and V. P. Kostikov, *USSR Pat.* 755 820.

373 J. R. Hahn *Ger. Offen.* 2 816 638 (*CA* **90**, 7394u).

374 M. Hamada, H. Koyanagi and Y. Adachi, *Japan Kokai* 79 32 200 (*CA* **79** (1973) 127216m).

375 C. G. Hammar, J. G. Frei, S. Stromberg and J. Vessman, *Acta Pharmacol. Toxicol. Suppl.* **36** (1975) 33.

376 J. F. Hampton, *US Pat.* 3 555 065 (*CA* **74** (1971) 112209j).

377 J. F. Hampton and C. W. Lacefield, *Ger. Offen.* 2 036 687 (*CA* **74** (1971) 142053w).

378 N. Harada and M. Tanaka, *Japan Kokai* 75 132 062 (*CA* **84** (1976) 60611u).

379 D. Harber, A. Holt and A. W. P. Jarvie, *J. Organomet. Chem.* **38** (1972) 255.

380 R. K. Harris, *Proc. 20th Colloq. Spectrosc. Int. Invited Lect.* (1977), Vol. 2, p. 183.

381 R. K. Harris and B. J. Kimber, *Appl. Spectrosc. Rev.* **10** (1975) 117.

382 R. K. Harris, B. J. Kimber, M. C. Wood and A. Holt, *J. Organomet. Chem.* **116** (1976) 291.

383 R. K. Harris and R. H. Newman, *Org. Magn. Reson.* **9** (1977) 426.

384 R. K. Harris, M. I. M. Wazeer, O. Schlak and R. Schmutzler, *J. Chem. Soc., Dalton Trans.* **1976**, 17.

385 J. M. Hartwell, *UK Pat.* 1 289 526 (*CA* **78** (1973) 30496u).

386 J. F. Hayden and S. A. Barlow, *Toxicol. Appl. Pharmacol.* **21** (1972) 68.

387 J. R. Heffel and A. E. Bey, *Ger. Offen.* 2 038 782; *UK Pat.* 1 301 158; *Fr. Demande* 2 056 989; *US Pat.* 3 729 444 (*CA* **74** (1971) 113500j).

388 J. Heimberger, H. Cheerschmidt and K. Ruehlmann, *Plaste Kaut.* **25** (1978) 386 (*CA* **89** (1978) 163976z).

389 J. S. Higgins, K. Dodgson and J. A. Semlyen, *Polymer* **20** (1979) 553.

390 K. Hirakawa and M. Honda, *Ger. Offen.* 2 908 249.

391 Hitachi, Ltd, *Japan Kokai* 80 62 427.

392 P. B. Hitchcock and R. Mason, *Acta Crystallogr.* **B34** (1978) 694.

393 D. Hoebbel and W. Wieker, *Z. Anorg. Allg. Chem.* **384** (1971) 43.

394   H. J. Hoelle and B. R. Lehnen, *Eur. Polym. J.* **11** (1975) 663.
395   R. A. Holroyd, T. E. Gangwer and A. O. Allen, *Chem. Phys. Lett.* **3** (1975) 520.
396   A. Holt, A. W. P. Jarvie and J. J. Mallabar, *J. Organomet. Chem.* **59** (1973) 141.
397   T. Hongu, *Japan Kokai* 78 00 800.
398   M. A. Hossain, M. B. Hursthouse and K. M. A. Malik, *Acta Crystallogr.* **B35** (1979) 2258.
399   D. Ch. Hrncir, *Texas A and M. Univ., College Station, Texas, USA* (1979), 153.
400   W. E. Hutchinson, *US Pat.* 3 647 741.
401   J. F. Hyde and D. E. Spielvogel, *Ger. Offen.* 2 032 826; *UK Pat.* 1 264 948; *US Pat* 3 652 628 (*CA* **75** (1971) 49315u).
402   J. A. Ibemesi and D. J. Meier, *Report* UCRL–15159 (1979).
403   K. Ito and T. Fukuda, *Japan Kokai* 78 12 958 (*CA* **89** (1978) 110794g).
404   K. Ito, N. Kuga and T. Fukuda, *Japan Kokai* 75 129 653 (*CA* **84** (1976) 123136r).
405   K. Ito and T. Sato, *Japan Kokai* 76 26 961 (*CA* **85** (1976) 22613j).
406   M. Ito, E. Takano, K. Matsumoto and Y. Miyoshi, *Japan Kokai* 78 142 542 (*CA* **90** (1979) 142080h).
407   N. T. Ivanova, S. V. Syavtsillo and L. D. Prigozhina, *Zavod. Lab.* **39** (1973) 1455 (*CA* **81** (1974) 26080y).
408   C. Janin and A. Guyot, *J. Chim. Phys. Physicochim. Biol.* **69** (1972) 1125 (*CA* **78** (1973) 16531v).
409   P. C. Juliano, *US Pat.* 3 663 650 (*CA* **77** (1972) 127417m).
410   P. C. Juliano, *Ger. Offen.* 2 164 469; *Fr. Demande* 2 121 029; *UK Pat.* 1 347 922 (*CA* **77** (1972) 115360z).
411   P. C. Juliano, W. A. Fessler and J. D. Cargioli, *23rd Int. Congr. Pure Appl. Chem. Macromol. Prepr.* (1971), Vol. II, p. 1212.
412   P. C. Juliano, W. A. Fessler and J. D. Cargioli, *Am. Chem. Soc. Prepr.* **12** (1971) 158.
413   E. G. Kagan, Yu. A. Yuzhelevskii, N. N. Fedoseeva and N. M. Geller, *Vysokomol. Soedin.* **B13** (1971) 798 (*CA* **76** (1972) 86222c).
414   Ts. V. Kakuliya, L. M. Khananashvili, L. S. Dneprovskaya, Yu. N. Novikov and M. E. Vol'pin, *Vysokomol. Soedin.* **A22** (1980) 1690.
415   G. S. Kalinina, T. A. Basalgina, N. S. Vyazankin, G. A. Razuvaev, V. A. Jablokov and N. V. Jablokova, *J. Organomet. Chem.* **117** (1976) 231.
416   A. N. Kanev, S. A. Prokhorova, G. A. Kokovin, V. E. Fedorov, A. S. Chernobrov and T. N. Martynova, *1. Vsesoyuz. Simp., Irkutsk 1977, Tezisy Dokl.* (1977), p. 243.
417   B. Kanner and B. Prokai, *Ger. Offen.* 2 629 138 (*CA* **86** (1977) 140986b).
418   J. Karger and A. Szafner, *Makromol. Chem.* **179** (1978) 519.
419   B. D. Karstedt, *US Pat.* 3 775 452.
420   B. D. Karstedt, *Ger. Offen.* 2 307 085; *Fr. Demande* 2 172 340 (*CA* **80** (1974) 16134j).
421   B. D. Karstedt, *Ger. Offen.* 1 941 411; *Fr. Demande* 2 016 946; *US Pat.* 3 715 334 (*CA* **74** (1971) 100519x).
422   B. L. Kaufman and A. V. Karlin, *Zh. Obshch. Khim.* **40** (1970) 1568.
423   B. L. Kaufman, V. M. Krasikova, A. I. Nikitina and M. P. Grinblat, *USSR Pat.* 473 948 (*CA* **83** (1975) 188046k).
424   T. C. Kendrick, D. R. Thomas and I. Francis, *UK Pat.* 1 293 944 (*CA* **78** (1973) 44193e).
425   R. E. Kesting, K. F. Jackson and J. M. Newman, *J. Appl. Polym. Sci.* **15** (1971) 1527.
426   P. H. Keyes and W. B. Daniels, *J. Chem. Phys.* **62** (1975) 2000.
427   L. M. Khananashvili, V. M. Kopylov, E. I. Khubulava and T. G. Stakhrovskaya, *Soobshch. Akad. Nauk Gruz. SSR* **98** (1980) 333.
428   L. M. Khananashvili, V. M. Kopylov, M. I. Shkol'nik, M. G. Zaitseva, J. Schraml, J. Soucek and V. Chvalovsky, *Zh. Obshch. Khim.* **50** (1980) 1565.

429 D. A. Khankhodzhaeva, V. O. Reikhsfel'd, E. P. Lebedev and V. I. Krikunenko, *USSR Pat.* 765 273.

430 E. I. Khubulava, V. M. Kopylov, A. I. Nogaideli, L. M. Khananashvili and K. A. Andrianov, *1. Vsesoyuz. Simp., Irkutsk* (1977), p. 274.

431 Y. K. Kim, *US Pat.* 3 794 672; *Ger. Offen.* 2 409 383; *Fr. Demande* 2 235 945 (*CA* 81 (1974) 38880x).

432 Sh. G. Kaplan, V. F. Aleksandrova, M. V. Sobolevskij, *Plast. Massy* 1982 No. 8, 53.

433 V. V. Kireev, S. O. Shumakova and A. A. Izyneev, *Dep. VINITI* 1729 (1979).

434 E. A. Kirichenko, A. I. Ermakov and L. A. Kulyamina, *Zh. Obshch. Khim.* 46 (1976) 712.

435 E. A. Kirichenko, A. I. Ermakov and I. N. Samsonova, *Zh. Fiz. Khim.* 51 (1977) 2506.

436 E. A. Kirichenko, B. A. Markov and V. A. Kochetov, *Izv. Vyssh. Uchebn. Zaved., Khim. Khim. Teknol.* 18 (1975) 1504 (*CA* 84 (1976) 69075e).

437 E. A. Kirichenko, B. A. Markov, V. A. Kochetov and A. P. Chuguev, *Zh. Anal. Khim.* 31 (1976) 2021.

438 T. Kirino, *Japan Kokai* 78 119 955 (*CA* 90 (1979) 40172x).

439 K. Kishimoto, S. Sasaki, Y. Koda and M. Suzuki, *Ger. Offen.* 2 554 498 (*CA* 85 (1976) 79089x).

440 J. C. Kleinert and C. J. Weschler, *Anal. Chem.* 52 (1980) 1245.

441 R. Kleinstneck, M. Leupold and G. Marquardt, *Ger. Offen.* 2 834 172.

442 R. Kleinstneck, M. Leupold and G. Marquardt, *Ger. Offen.* 2 834 171.

443 L. A. Klimov, N. E. Norden, G. A. Alekseichuk and Z. A. Baskova, *USSR Pat.* 464 601.

444 V. A. Klyuchnikov, L. N. Martynovskaya and G. A. Lobanov, *1st Czech. Conf. Calorimetry, Lect. Short. Commun.* (1977), D 15/1 (*CA* 87 (1977) 173705h).

445 H. Kobayashi, *Japan Kokai* 79 83 100.

446 V. A. Kochetov, J. Soucek, B. A. Markov, E. A. Kirichenko, K. A. Andrianov and L. M. Khananashvili *Zh. Anal. Khim.* 31 (1976) 2252.

447 I. Koda and M. Suzuki, *Japan Kokai* 78 00 980.

448 I. M. Kolesnikov, N. N. Belov and V. V. Lopatin, *Zh. Fiz. Khim.* 53 (1979) 2982.

449 H. J. Kollmeier and G. Rossmy, *Ger. Offen.* 2 431 394 (*CA* 84 (1976) 106508w).

450 I. G. Kolokol'tseva, V. D. Lobkov, A. V. Karlin, Yu. A. Yuzhelevskii and N. A. Silina, *V Sb., Issled. v Obl. Fiz. i Khimii Kauchukov i Rezin.* 95 (1975) (*CA* 84 (1976) 164929v).

451 K. Kondo, H. Terao and S. Tanaka, *Japan Kokai* 79 72 300.

452 R. Konopka and B. Stojczyk, *Acta Phys. Pol.* A40 (1971) 537.

453 V. V. Korol'ko, E. G. Kagan, G. V. Dotsenko, T. I. Saratovkina and G. A. Ivanova, *USSR Pat.* 491 638 (*CA* 84 (1976) 44340z).

454 K. Koseki, T. Jamaoka, T. Tsunoda, Y. Funahashi and M. Hatonaka, *Nippon Insatsu Gakkai Ronbunshu* 16 (1976) 126.

455 K. Koseki, T. Yamaoka, T. Tsunoda, Y. Funahashi and M. Hatanaka, *Nippon Insatsu Gakkai Ronbunshu* 16 (1976) 131.

456 J. Kosina, *Czech. Pat.* 176 702.

457 J. Kowalski, J. Chojnowski, W. Stanczyk and M. Mazurek, *Polimery* 19 (1974) 77.

458 N. A. Koyava, O. V. Mukbaniani, L. M. Khananashvili and V. G. Tsitsishvili, *Zh. Obshch. Khim.* 50 (1980) 1793.

459 V. M. Krasikova, L. A. Klimov and G. A. Nikolaev, *Zh. Anal. Khim.* 32 (1977) 127.

460 V. M. Krasikova, V. P. Mileshkevich and A. N. Kaganova, *Zh. Anal. Khim.* 29 (1974) 1199.

461 A. P. Kreshkov, V. F. Andronov and V. A. Drozdov, *Zh. Anal. Khim.* 27 (1972) 2244.

462 A. P. Kreshkov, E. A. Kirichenko and A. I. Ermakov, *Tr. Mosk. Khim.-Tekhnol. Inst.* 80 (1974) 77.

463 M. Kryszewski, A. M. Wrobel and J. Tyczkowski, *ACS Symp. Ser.* (1978) 108 (Plasma Polym.) p. 219.

464 A. M. Kuliev, O. I. Dzhafarov, K. A. Karasharli and A. G. Kuznetsova, *Zh. Fiz. Khim.* **51** (1977) 1798.

465 A. M. Kuliev, O. I. Dzhafarov, K. A. Karasharli and A. G. Kuznetsova, *Dep. VINITI* 987 (1977) (*CA* **90** (1979) 77363m).

466 A. M. Kuliev, O. I. Dzhafarov, K. A. Karasharli and A. G. Kuznetsova, *Zh. Fiz. Khim.* **51** (1977) 1564.

467 A. M. Kuliev, O. I. Dzhafarov, K. A. Karasharli and A. G. Kuznetsova, *Zh. Fiz. Khim.* **50** (1976) 1903.

468 A. M. Kuliev, O. I. Dzhafarov, K. A. Karasharli and A. G. Kuznetsova, *Zh. Fiz. Khim.* **51** (1977) 1564.

469 A. M. Kuliev, D. I. Dzhafarov, K. A. Karasharli and A. G. Kuznetsova, *Dep. VINITI* 986 (1977) (*CA* **90** (1979) 77362k).

470 A. M. Kuliev, S. A. Mekhtiev, O. I. Dzhafarov and K. A. Karasharli, *Sb. 2. Vses. Konf. Termodin. Org. Soedin., Gorkii, Tezisy Dokl.* (1976), p. 16.

471 M. Kumada and M. Ishikawa, *Japan Kokai* 72 38 930 (*CA* **78** (1973) 58615f).

472 S. Kumagui, Y. Okunuki and M. Nara, *Japan Kokai* 79 46 847.

473 T. V. Kurlova, Yu. A. Yuzhelevskii, E. G. Kagan, L. N. Golubeva, A. V. Karlin and V. P. Mileshkevich, *USSR Pat.* 372 913 (*CA* **80** (1974) 15065a).

474 A. G. Kuznetsova, V. N. Bochkarev, V. I. Ivanov, A. V. Kisin, A. N. Polivanov, V. V. Sokolov and G. N. Turkel'taub, *Zh. Obshch. Khim.* **46** (1976) 2025.

475 A. G. Kuznetsova, S. A. Golubtsov and V. I. Ivanov, *US Pat.* 3 558 681 (*CA* **74** (1971) 88122c).

476 A. G. Kuznetsova, S. A. Golubtsov and V. I. Ivanov, *Fr. Demande* 1 596 033 (*CA* **74** (1971) 142043t).

477 A. G. Kuznetsova, S. A. Golubtsov, V. I. Ivanov, T. I. Gerasimova and G. V. Kaznina, *Zh. Obshch. Khim.* **43** (1973) 1498.

478 A. G. Kuznetsova, V. I. Ivanov and S. A. Golubtsov, *Zh. Obshch. Khim.* **40** (1970) 706.

479 A. G. Kuznetsova, V. I. Ivanov and S. A. Golubtsov, *USSR Pat.* 287 940 (*CA* **74** (1971) 112204d).

480 A. G. Kuznetsova, V. I. Ivanov and S. A. Golubtsov, *Khim. Prom. (Moscow)* **47** (1971) 268 (*CA* **75** (1971) 6006b).

481 A. G. Kuznetsova, V. I. Ivanov, S. A. Golubtsov, N. P. Telegina and T. I. Gerasimova, *USSR Pat.* 285 922 (*CA* **75** (1971) 20591v).

482 A. G. Kuznetsova, N. P. Telegina, S. A. Golubtsov, V. F. Andronov, V. I. Ivanov and T. I. Gerasimova, *Zh. Obshch. Khim.* **42** (1972) 1756.

483 A. G. Kuznetsova, N. P. Telegina, S. A. Golubtsov, T. I. Gerasimova and V. I. Ivanov, *Plast. Massy* **15** (1973) 11 (*CA* **80** (1974) 83741t).

484 V. P. Kuznetsova, K. V. Zapunnaya and P. I. Goroshko, *USSR Pat.* 260 176 (*CA* **73** (1970) 440lm).

485 C. W. Lacefield, *Ger. Offen.* 2 036 616; *UK Pat.* 1 287 555; *Fr. Demande* 2 055 486. (*CA* **74** (1971) 127151h).

486 Z. Laita, E. Nejedly and M. Jelinek, *Czech. Pat.* 140 244 (*CA* **76** (1972) 100372k).

487 R. Lagarde and J. Lahaye, *Bull. Soc. Chim. Fr.*, Pt 1 **1977**, 825 (*CA* **88** (1978) 198179y).

488 Z. Lasocki and T. Graczyk, *1st IUPAC Int. Symp. on Polymeric Heterocycles (Ring-Opening), Warsaw–Jablonna* (1975), S.1., p. 138.

489 Z. Lasocki, J. Kulpinski and W. Gador, *Polimery* **15** (1970) 508 (*CA* **74** (1971) 126317f).

490 Z. Lasocki, J. Kulpinski and W. Gador, *Polimery* **15** (1970) 442 (*CA* **74** (1971) 11259e).

491 Z. Lasocki and M. Witekowa, *1st IUPAC Int. Symp. Polymer Heterocycles (Ring-Opening), Warsaw–Jablonna* (1975), S.1., p. 69.

492 Z. Lasocki and M. Witekowa, *J. Macromol. Sci., Chem.,* **A11** (1977) 457.

493 V. I. Lavrent'ev and V. G. Kostrovskii, *Zh. Obshch. Khim.* **49** (1979) 2013.

494 V. I. Lavrent'ev, V. M. Kovrigin and V. G. Kostrovskii, *Izv. Sib. Otd. Akad. Nauk SSSR, Ser. Khim.* **5** (1979) 126.

495 V. I. Lavrent'ev, M. G. Voronkov and V. M. Kovrigin, *Zh. Obshch. Khim.* **50** (1980) 382.

496 I. A. Lavygin, B. A. Izmailov, A. A. Zhdanov, V. D. Myakushev, I. I. Skorokhodov and V. M. Migalina, *Zh. Prikl. Khim. (Leningrad)* **53** (1980) 1155.

497 I. A. Lavygin, I. I. Skorokhodov, M. A. Kleinovskaya, G. A. Potashova and A. M. Yur'eva, *Zh. Fiz. Khim.* **52** (1978) 1542.

498 D. C. Law, *Ger. Offen.* 2 037 146 (*CA* **74** (1971) 127195a).

499 E. P. Lebedev, V. A. Baburina, D. V. Fridland and V. I. Krikunenko, *USSR Pat.* 454 230 (*CA* **83** (1975) 80311d).

500 E. P. Lebedev and A. D. Fedorov, *Zh. Obshch. Khim.* **48** (1978) 931.

501 E. P. Lebedev, A. D. Fedorov and V. O. Reikhsfel'd, *Zh. Obshch. Khim.* **49** (1979) 147.

502 E. P. Lebedev and M. M. Frenkel', *Zh. Obshch. Khim.* **47** (1977) 2275.

503 B. V. Lebedev, N. N. Mukhina and T. G. Kulagina, *Vysokomol. Soedin.* **A20** (1978) 1297 (*CA* **89** (1978) 60183j).

504 E. P. Lebedev and L. Z. Zakirova, *USSR Pat.* 604 854 (*CA* **89** (1978) 25351r).

505 E. P. Lebedev, L. Z. Zakirova, I. E. Saratov and V. D. Reikhsfel'd, *Zh. Obshch. Khim.* **47** (1977) 1772.

506 E. P. Lebedev, V. G. Zavada and A. D. Fedorov, *Zh. Obshch. Khim.* **49** (1979) 151.

507 C. L. Lee, *Ger. Offen.* 1 953 231.

508 C. L. Lee, *Ger. Offen.* 1 720 496.

509 C. J. Lee and O. K. Johannson, *J. Polym. Sci., Polym. Chem. Ed.* **14** (1976) 729.

511 C. L. Lee and O. W. Marko, *Polym. Prepr., Am. Chem. Soc., Div. Polym. Chem.* **19** (1978) 250.

512 P. M. Lefebvre, R. Jerome and P. Teyssie, *Macromolecules* **4** (1977) 871.

513 R. Le Fevre, F. Coulston and L. Golberg, *Toxicol. Appl. Pharmacol.* **21** (1972) 29.

514 V. E. Lelikova, V. N. Knyazev, E. G. Vlasova and V. A. Drozdov, *Zh. Anal. Khim.* **30** (1975) 793.

515 M. T. Letcher, *J. Chem. Thermodyn.* **12** (1980) 297.

516 C. J. Letteral, *US Pat.* 3 694 405; *Can. Pat.* 970 095. (*CA* **78** (1973) 17047d).

517 R. R. Le Vier, D. R. Bennett and M. J. Hunter, *Acta Pharmacol. Toxicol. Suppl.* **36** (1975) 68.

518 R. R. Le Vier and W. F. Boley, *Acta Pharmacol. Toxicol. Suppl.* **36** (1975) 55.

519 R. R. Le Vier and M. E. Jankowiak, *Biol. Reprod.* **7** (1972) 260 (*CA* **78** (1973) 509h).

520 R. R. Le Vier and M. E. Jankowiak, *Toxicol. Appl. Pharmacol.* **21** (1972) 80.

521 V. Yu. Levin, K. A. Andrianov, G. L. Slonimskii, A. A. Zhdanov, E. A. Lyubavskaya and A. P. Malykhin, *Vysokomol. Soedin.* **A16** (1974) 1951 (*CA* **82** (1975) 4870z).

522 G. C. Levy, J. D. Gargioli, P. C. Juliano and T. D. Mitchell, *J. Am. Chem. Soc.* **95** (1973) 3445.

523 R. N. Lewis, *US Pat.* 4 113 690. (*CA* **90** (1979) 56098p).

524 R. N. Lewis, *US Pat.* 3 979 546. (*CA* **85** (1976) 194255k).

525 R. N. Lewis, *US Pat.* 3 925 285. (*CA* **84** (1976) 91379d).

526 R. N. Lewis, *US Pat.* 3 956 166. (*CA* **85** (1976) 64470r).

527 R. N. Lewis and E. R. Martin, *US Pat.* 4 066 680. (*CA* **88** (1978) 89837r).

528 S. Lin and J. M. Miller, *J. Fluorine Chem.* **9** (1977) 161.

529 J. Lipowitz and M. J. Ziemelis, *J. Flammability* **7** (1976) 504.

530 E. T. Lippmaa, M. A. Alla, T. J. Pehk and G. Engelhardt, *J. Am. Chem. Soc.* **100** (1978) 1929.

531 W. W. Limburg, *US Pat.* 3 957 725.

532   C. J. Litteral, *Fr. Demande* 2 111 629.
533   C. J. Litteral, *Ger. Offen.* 2 152 270 (*CA* **77** (1972) 63098y).
534   V. M. Litvinov, B. D. Lavrukhin and A. A. Zhdanov, *Vysokomol. Soedin.* **A18** (1976) 2515 (*CA* **86** (1977) 122107m).
535   I. L. Lopatkina, L. A. Kucherskaya, A. G. Kuznetsova and Yu. Kh. Shaulov, *Zh. Fiz. Khim.* **47** (1973) 2900.
536   I. L. Lopatkina, L. A. Kucherskaya, A. G. Kuznetsova and Yu. Kh. Shaulov, *Zh. Fiz. Khim.* **49** (1975) 251.
537   I. L. Lopatkina, L. A. Kucherskaya, A. G. Kuznetsova and Yu. Kh. Shaulov, *Zh. Fiz. Khim.* **48** (1974) 768.
538   B. M. Luskina and N. N. Bravina, *Zh. Anal. Khim.* **30** (1975) 399.
539   B. M. Luskina, N. N. Bravina, M. V. Sobolevskii, T. V. Koroleva, M. A. Kleinovskaya and S. M. Galanina, *Zh. Anal. Khim.* **30** (1975) 1419.
540   B. M. Luskina and V. V. Masina, *Metody Anal. Kontroly a Proizvod. Khim. Prom-sti* **12** (1977) 1 (*CA* **89** (1978) 52903d).
541   B. N. Luskina and N. N. Troitskaya, *Zh. Anal. Khim.* **33** (1978) 1435.
542   B. M. Luskina, N. N. Troitskaya and V. V. Mosina, *Zh. Anal. Khim.* **3** (1976) 779.
543   G. Maass, H. J. Luecking, W. Buechner and B. Degen, *Ger. Offen.* 2 557 624. (*CA* **87** (1977) 184669a).
544   G. Maass, H. J. Luecking, J. Maas and K. Seyfried, *Ger. Offen.* 2 536 010 (*CA* **87** (1977) 23480s).
545   R. J. Macher, *US Pat.* 3 607 898 (*CA* **75** (1971) 140973t).
546   R. J. Macher, *Ger. Offen.* 2 131 827 (*CA* **78** (1973) 72356d).
547   R. J. Macher, *UK Pat.* 1 343 686 (*CA* **80** (1974) 121091r).
548   R. J. Macher, *Fr. Demande* 2 144 065 (*CA* **79** (1973) 67250f).
549   M. R. MacLaury *Proc. Int. Conf. on Fire Safety* **4** (1979) 260.
550   L. A. Mai, *4th Int. Symp. on Organosilicon Chemistry, Moscow,* (1975), Vol. 1, p. 49.
551   L. A. Mai, *Latv. PSR Zinot. Akad. Vestis, Kim. Ser.* **1** (1974) 39 (*CA* **80** (1974) 132616s).
552   N. A. Makulov, D. Ya. Zhinkin, N. A. Gradskova and O. P. Trokhachenkova, *Zavod. Lab.* **41** (1975) 180 (*CA* **83** (1975) 71188s).
553   K. C. Malhotra, *Aust. J. Chem.* **29** (1976) 1185.
554   L. Mandik, A. Kotlanova and J. Foltyn, *Chem. Prům* **26** (1976) 87 (*CA* **84** (1976) 180757t).
555   R. A. Mansfield, *Macromol. Synth.* **4** (1972) 65.
556   M. Marcu, S. E. Lazarescu and G. Stiubianu, *Rom. Pat.* 63 878.
557   M. Marcu, M. Spiratos and G. Stiubianu, *Rom. Pat.* 63 851.
558   M. Marcu, M. Spiratos, G. Stiubianu and S. Lazarescu, *Rom. Pat.* 61 910 (*CA* **89** (1978) 90953c).
559   M. Marcu, G. Stiubianu, I. Vosniuc and M. Spiratos, *Rom. Pat.* 65 111.
560   V. Mark and P. S. Wilson, *Ger. Offen.* 2 832 339.
561   B. A. Markov, V. A. Kochetov, J. Soucek, E. A. Kirichenko, K. A. Andrianov and L. M. Khananashvili, *Zh. Anal. Khim.* **30** (1975) 2466.
562   E. R. Martin, *Fr. Demande* 2 192 137.
563   E. R. Martin, *Ger. Offen.* 2 335 118.
564   E. R. Martin, *US Pat.* 4 208 503.
565   E. R. Martin, *Ger. Offen.* 2 556 432 (*CA* **85** (1976) 95182c).
566   E. R. Martin, *US Pat.* 4 046 795.
567   E. R. Martin, *US Pat.* 4 078 104 (*CA* **89** (1978) 112836w).
568   E. R. Martin, *Fr. Demande* 2 330 715.

569   E. R. Martin, *US Pat.* 3 960 575 (*CA* **85** (1976) 79886e).

570   E. R. Martin, *Swiss Pat.* 598 271; *Australian Pat.* 343 906.

571   E. R. Martin, *Ger. Offen.* 2 339 761 (*CA* **81** (1974) 106584f).

572   E. R. Martin, *US Pat.* 3 878 263.

573   E. R. Martin, *Ger. Offen.* 2 649 584 (*CA* **87** (1977) 54645d).

574   E. R. Martin, *Ger. Offen.* 2 615 077 (*CA* **86** (1977) 17594v).

575   E. R. Martin, *US Pat.* 3 890 269.

576   E. R. Martin, *Fr. Demande* 2 195 650.

577   E. R. Martin, *Fr. Demande* 2 229 735; *Ger. Offen.* 2 423 369; *US Pat.* 3 883 623; *Belg. Pat.* 814 909 (*CA* **82** (1975) 126203e).

578   N. I. Martyakova, E. G. Kagan, S. B. Dolgoplosk and L. A. Mai, *Zh. Obshch. Khim.* **46** (1976) 2038 (*CA* **86** (1977) 30127q).

579   T. N. Martynova and Yu. N. Nikonorov, *1. Vsesoyuz. Simp., Irkutsk 1977, Tezisy Dokl.* (1977), p. 241.

580   H. Maruyama and T. Ento, *Japan Kokai* 71 21 602 (*CA* **76** (1972) 155267k).

581   J. E. Matherly, *US Pat.* 3 642 685; *Ger. Offen.* 2 132 960; *UK Pat.* 1 296 302; *Fr. Demande* 2 098 278 (*CA* **76** (1972) 142125n).

582   Y. Matsumoto, B. Murai and I. Endo, *Japan Kokai* 78 33 256 (*CA* **89** (1978) 111954w).

583   O. V. Mukhaniani, S. M. Meladze, I. G. Esartiya, L. M. Khananashvili, O. A. Tskhakaya, *Izv. Akad. Nauk Gruz. SSR, Ser. Khim.* **8** (1982) 75.

584   K. Matyjaszewski, M. Zielinski, P. Kubisa, S. Slomkowski, J. Chojnowski and S. Penczek, *Makromol. Chem.* **181** (1980) 1469.

585   I. Maxim and M. Spiratos, *Rom. Pat.* 63 889.

586   M. Mazurek and J. Chojnowski, *Macromolecules* **11** (1978) 347.

587   M. Mazurek and J. Chojnowski, *Makromol. Chem.* **178** (1977) 1005.

588   M. Mazurek, M. Scibiorek, J. Chojnowski, B. G. Zavin and A. A. Zhdanov, *Eur. Polym. J.* **16** (1980) 57.

589   M. Mazurek, J. Zietera, W. Sadowska and J. Chojnowski, *Makromol. Chem.* **181** (1980) 777.

590   H. R. McEntee and J. S. Razzano, *Fr. Demande* 2 215 425 (*CA* **82** (1975) 73636q).

591   H. R. McEntee and J. S. Razzano, *Ger. Offen.* 2 303 155; *US Pat.* 3 763 212 (*CA* **80** (1974) 4037s).

592   D. E. Mcvannel, *US Reissue* 26 697 (*CA* **72** (1970) 67481u).

593   P. Mehta and M. Zeldin, *Inorg. Chim. Acta* **22** (1977) L33.

594   F. Meiller, *US Pat.* 3 798 251.

595   F. Meiller, *UK Pat.* 1 333 455; *Ger. Offen.* 2 133 397; *Fr. Demande* 2 097 406.

596   F. Meiller, *Ger. Offen.* 2 133 397; *Fr. Demande* 2 097 406; *UK Pat.* 1 333 455 (*CA* **76** (1972) 127704g).

597   F. Meiller, *Fr. Demande* 2 178 724 (*CA* **81** (1974) 65142q).

598   S. A. Mekhtiev, K. A. Karasharli and O. I. Dzhafarov, *Zh. Fiz. Khim.* **49** (1975) 259 (*CA* **82** (1975) 145945q).

599   S. A. Mekhtiev, K. A. Karasharli and O. I. Dzhafarov, *Dep. VINITI* 2588 (1974), p. 74.

600   S. A. Mekhtiev, K. A. Karasharli and O. I. Dzhafarov, *Zh. Fiz. Khim.* **49** (1975) 2578.

601   S. A. Mekhtiev, K. A. Karasharli, O. I. Dzhafarov and A. G. Kuznetsova, *Zh. Fiz. Khim.* **49** (1975) 1914 (*CA* **83** (1975) 210092w).

602   S. M. Meladze, O. V. Mukbaniani, N. N. Makarova and L. M. Khananashvili, *Soobshch. Akad. Nauk Gruz. SSR* **99** (1980) 105.

603   S. M. Meladze, O. V. Mukbaniani, N. N. Makarova and L. M. Khananashvili, *Zh. Obshch. Khim.* **50** (1980) 2493.

604   M. O. Melikyan, D. A. Tergazarova and M. M. Vartanyan, *Arm. Khim. Zh.* **23** (1970) 74.

605 A. M. Mezheritskii, N. N. Vdovenko, V. I. Chirtsov and B. A. Sharov, *USSR Pat.* 540 881 (*CA* **86** (1977) 90913c).

606 V. P. Mileshkevich, L. S. Bresler, Yu. A. Yuzhelevskii, N. P. Timofeeva and Yu. V. Tsyganov, *1. Vsesoyuz. Simp., Irkutsk, 1977, Tezisy Dokladov* (1977), p. 71.

607 V. P. Mileshkevich, A. V. Karlin, T. N. Baratova, S. N. Mikhailov, Yu. A. Yuzhelevskii and T. V. Kurlova, *USSR Pat.* 435 673.

608 V. P. Mileshkevich and N. F. Novikova, *Zh. Obshch. Khim.* **48** (1978) 1125.

609 V. P. Mileshkevich, N. F. Novikova, L. S. Bresler, N. B. Zaitsev, V. V. Korol'ko and G. A. Nikolaev, *Zh. Obshch. Khim.* **47** (1977). 2564.

610 R. Mills and H. G. Hertz, *J. Phys. Chem.* **84** (1980) 220.

611 F. J. Modic, *UK Pat.* 2 016 493.

612 F. J. Modic, *US Pat.* 3 504 006 (*CA* **72** (1970) 112520n).

613 R. E. Moeller, *Braz. Pedido* PI 7703 261.

614 R. E. Moeller, *Ger. Offen.* 2 556 559 (*CA* **85** (1976) 79881z).

615 R. E. Moeller *US Pat.* 4 008 346 (*CA* **86** (1977) 123152j).

616 E. L. Morehouse, *Ger. Offen.* 2 063 522; *Fr. Demande* 2 074 303; *US Pat.* 3 657 305; 3 686 254 (*CA* **75** (1971) 152551s).

617 H. H. Moretto, A. De Montigny, K. F. Thom and R. Schliebs, *Ger. Offen.* 2 453 482 (*CA* **85** (1976) 78927g).

618 A. M. Mosin, *Zh. Fiz. Khim.* **49** (1975) 750.

619 O. V. Mukbaniani, N. N. Makarova, A. I. Nogaideli and K. A. Andrianov, *1. Vsesoyuz. Simp., Irkutsk, 1977, Tezisy Dokl.* (1977), p. 273.

620 O. V. Mukbaniani, S. M. Meladze, N. N. Makarova and L. M. Khananashvili, *Soobshch. Akad. Nauk Gruz. SSR* **99** (1980) 109.

621 B. Murai, *Japan Kokai* 76 139 854 (*CA* **86** (1977) 91506j).

622 B. Murai, S. Yamanouti and M. Onori, *Japan Kokai* 76 30 920.

623 E. Murrill and L. W. Breed, *Inorg. Chem.* **10** (1971) 641.

624 N. S. Nametkin, S. G. Durgaryan, E. G. Novitskii, V. G. Filippova, N. K. Gladkova and V. V. Teplyakov, *Ger. Offen.* 2 905 939.

625 N. S. Nametkin, I. A. Grushevenko, V. N. Perchenko, G. L. Kamneva and M. E. Kuzovkina, *Dokl. Akad. Nauk SSSR, Ser. Khim.* **207** (1972) 1358.

626 N. S. Nametkin, T. Kh. Islamov, L. E. Gusel'nikov, A. A. Sobtsov and V. M. Vdovin, *Izv. Akad. Nauk SSSR, Ser. Khim.* **1** (1971) 90.

627 N. P. Nazarenko, M. B. Fromberg, A. G. Grozdov and N. P. Gashnikova, *Ukr. Khim. Zh.* **41** (1975) 1060 (*CA* **83** (1975) 206607a).

628 L. Nicander, *Acta Pharmacol. Toxicol. Suppl.* *I* **31** (1972) 17.

629 L. Nicander, *Acta Pharmacol. Toxicol. Suppl.* **36** (1975) 40.

630 J. M. Nielsen, *J. Appl. Polym. Sci., Appl. Polym. Symp.* **35** (1979) 223.

631 S. Nitzsche, M. Wick, P. Hittmair and E. Wolfarth, *Ger. Offen.* 1 745 556.

632 A. I. Nogaideli and T. V. Chogovadze, *Soobshch. Akad. Nauk Gruz. SSR* **87** (1977) 365 (*CA* **88** (1978) 37885k).

633 A. I. Nogaideli, L. M. Khananashvili, L. I. Nakaidze, T. V. Chogovadze, V. S. Tskhovrebashvili, R. V. Volkova and G. A. Nogaideli, *Soobshch. Akad. Nauk Gruz. SSR* **92** (1978) 341 (*CA* **91** (1979) 21497m).

634 A. I. Nogaideli, L. I. Nakaidze and V. S. Tskhovrebashvili, *Zh. Obshch. Khim.* **48** (1978) 1344.

635 A. I. Nogaideli, L. I. Nakaidze and V. S. Tskhovrebashvili, *Izv. Akad. Nauk Gruz. SSR, Ser. Khim.* **6** (1980) 50.

636 A. I. Nogaideli, D. Ya. Zhinkin, L. I. Nakaidze, A. S. Shapatin and V. S. Tskhovrebashvili, *Zh. Obshch. Khim.* **46** (1976) 1048.

637 A. Noshiro, M. Takamisawa, Y. Inoue, T. Sugita and H. Fujii, *Japan Kokai* 78 17 405 (*CA* **90** (1979) 79149v).
638 Yu. N. Novikov, Ts. V. Kakuliya, L. M. Khananashvili, V. M. Kopylov and M. E. Vol'pin, *Dokl. Akad. Nauk SSSR* **245** (1979) 848.
639 G. G. Nurullaev and S. N. Gadzhiev, *Azerb. Khim. Zh.* **2** (1975) 28 (*CA* **84** (1976) 120890w).
640 H. Oberhammer, W. Zeil and G. Fogarasi, *J. Mol. Struct.* **18** (1973) 309.
641 H. Okamoto and I. Yanagisawa, *Japan Kokai* 74 92 025 (*CA* **82** (1975) 98810e).
642 H. Okamoto and I. Yanagisawa, *Ger. Offen.* 2 363 539 (*CA* **81** (1974) 136716t).
643 H. Okinoshima and W. P. Weber, *J. Organomet. Chem.* **149** (1978) 279.
644 H. Okinoshima and W. P. Weber, *J. Organomet. Chem.* **155** (1978) 165.
645 C. J. Olliff and P. Ladbrook, *Bioelectrochem. Bioenerg.* **6** (1979) 105.
646 C. J. Olliff, G. R. Pickering and K. J. Rutt, *J. Inorg. Nucl. Chem.* **42** (1980) 1201.
647 C. J. Olliff, G. R. Pickering and K. J. Rutt, *J. Inorg. Nucl. Chem.* **42** (1980) 288.
648 G. M. Omietonski, *Fr. Demande* 2 011 518.
649 G. M. Omietonski and V. T. Chuang, *Fr. Demande* 2 212 370.
650 G. M. Omietonski and V. T. Chuang, *Fr. Demande* 2 212 369.
651 G. M. Omietonski, H. D. Furbee and V. T. Chuang, *Fr. Demande* 2 212 368.
652 I. Ona, M. Ozaki and K. Usui, *Japan Kokai* 79 131 661.
653 I. Ona, M. Ozaki and K. Usui, *UK Pat.* 2 018 271.
654 Y. Ono, *Japan Kokai* 76 119 752 (*CA* **86** (1977) 74363t).
655 Y. Ono, *Japan Kokai* 76 134 745 (*CA* **86** (1977) 91031a).
656 V. Yu. Orlov, N. S. Nametkin, L. E. Gusel'nikov and T. Kh. Islamov, *Org. Mass. Spectrom.* **4** (1970) 195.
657 V. Yu. Orlov, A. B. Zachernyuk and K. A. Andrianov, *Izv. Akad. Nauk SSSR, Ser. Khim.* **2** (1974) 330.
658 A. Ossko, H. Goller, J. Herzig, W. Buechner and A. De Montigny, *Ger. Offen.* 2 251 297 (*CA* **81** (1974) 106571z).
659 V. V. Ostrovskii, V. V. Golubkov, N. P. Kharitonov and T. A. Vyglazova, *Vsb. Issled. v Obl. Fiz. i Khimii Kauchukov i Rezin* **1975**, 180 (*CA* **84** (1976) 180841r).
660 V. V. Ostrovskii and N. P. Kharitonov, *Zh. Obshch. Khim.* **47** (1977) 606.
661 F. D. Ovcharenko, M. T. Bryk, N. N. Baglei and N. G. Vasil'ev, *Dokl. Akad. Nauk SSSR, Ser. Khim.* **235** (1977) 865.
662 T. J. Pacansky, *Am. Chem. Soc. Polym. Prepr.* **17** (1976) 564.
663 S. C. Pace and J. C. Riess, *J. Organomet. Chem.* **121** (1976) 307.
664 S. C. Pace and J. G. Reiss, *J. Polym. Sci., Polym. Chem. Ed.* **16** (1978) 1627.
665 D. Paetzelt and K. Ruehlmann, *Plaste Kautsch.* **26** (1979) 606.
666 D. Paetzelt, R. Schiedewitz and K. Ruehlmann, *Plaste Kautsch.* **26** (1979) 610.
667 K. C. Pande and R. E. Ridenour, *Chem. Ind. (London),* **2** (1970) 56.
668 V. S. Papkov, E. S. Obolonkova, M. N. Il'ina, A. A. Zhdanov and G. L. Slonimskii, *Vysokomol. Soedin.* **A22** (1980) 117.
669 R. D. Parker, *Fresenius Z. Anal. Chem.* **292** (1978) 362.
670 D. R. Parker and L. H. Sommer, *J. Am. Chem. Soc.* **98** (1976) 618.
671 D. R. Parker and L. H. Sommer, *J. Organomet. Chem.* **110** (1976) C1.
672 V. V. Pavlov, G. Ya. Guba, Yu. I. Gorlov and A. S. Shapatin, *1.Vsesoyuz. Simp., Irkutsk (1977),* p. 284.
673 V. V. Pchelintsev, E. G. Kagan and Yu. A. Yuzhelevskii, *5 Sb., Issled. v Obl. Fiz. i Khim. Kauchukov i Rezin* (1975) p. 103 (*CA* **84** (1976) 151820c).
674 V. V. Pchelintsev, E. G. Kagan, Yu. A. Yuzhelevskii and M. A. Volkova, *Vysokomol. Soedin.* **B18** (1976) 237 (*CA* **85** (1976) 6174m).

675 V. V. Pchelintsev and Yu. A. Yuzhelevskii, *Prom. Sint. Kauch. Ref. Sb.* **10** (1975) 8 (*CA* **85** (1976) 33449m).

676 V. V. Pchelintsev, Yu. A. Yuzhelevskii, G. V. Ivanova and N. S. Smirnova, *Zh. Obshch. Khim.* **46** (1976) 1189.

677 V. V. Pchelintsev, Yu. A. Yuzhelevskii and E. G. Kagan, *5 Sb., Issled. v Obl. Fiz. i Khim. Kauchukov i Rezin.* (1975), p. 99.

678 V. V. Pchelintsev, Yu. A. Yuzhelevskii, V. P. Mileshkevich and T. I. Baratova, *Prom. Sint. Kauch.* **10** (1978) 9 (*CA* **90** (1979) 104390h).

679 E. D. Pellizzari, J. E. Bunch, R. E. Berkley and J. McRae, *Anal. Chem.* **48** (1976) 803.

680 V. A. Pestunovich, M. F. Larin, M. G. Voronkov, G. Engelhardt, H. Jancke, V. P. Mileshkevich and Yu. A. Yuzhelevskii, *Zh. Strukt. Khim.* **18** (1977) 578.

681 L. P. Petersen, *Belg. Pat.* 883 860.

682 L. P. Petersen, *US Pat.* 4 125 551 (*CA* **90** (1979) 88257f).

683 W. R. Peterson, *Rev. Silicon, Germanium, Tin, Lead Compds* **1** (1974) 193.

684 G. R. Pickering, C. J. Olliff and K. J. Rutt, *Org. Mass Spectrom.* **10** (1975) 1035.

685 O. R. Pierce and R. C. Hedlund, *Am. Chem. Soc., Div. Org. Coatings Plast. Chem.*, Paper 29 (1969) 250.

686 O. R. Pierce and K. M. Lee, *Report* AFOSR–TR–80–0469 (1980).

687 A. Pilbrant and B. Strindberg, *Acta Pharmacol. Toxicol. Suppl.* **36** (1975) 139.

688 A. N. Polivanov, A. A. Bernadskii, Z. V. Belyakova, M. G. Pomerantseva, D. V. Nazarova and V. N. Bochkarev, *Zh. Obshch. Khim.* **47** (1977) 2561.

689 A. N. Polivanov, A. A. Bernadskii and V. N. Bochkarev, *Zh. Obshch. Khim.* **48** (1978) 1662.

690 A. N. Polivanov, A. A. Bernadskii, V. N. Bochkarev, V. I. Zhun and V. D. Sheludyakov, *Zh. Obshch. Khim.* **50** (1980) 614.

691 A. N. Polivanov, A. A. Bernadskii, V. N. Bochkarev, V. V. Zverev, N. V. Ivanova, N. E. Rodzevich and M. V. Sobolevskii, *1. Vsesoyuz. Simp., Irkutsk, 1977, Tezisy Dokl.* (1977), p. 69.

692 A. N. Polivanov, A. A. Bernadskii, V. S. Fal'ko, N. P. Telegina, V. I. Khvostenko and V. N. Bochkarev, *Zh. Obshch. Khim.* **48** (1978) 399.

693 T. L. Pollock, H. S. Sandhu, A. Jodhon and O. P. Strausz, *J. Am. Chem. Soc.* **95** (1973) 1017.

694 H. R. Pomez, *Rev. Cenic., Cienc. Fis.* **7** (1976) 259 (*CA* **89** (1978) 68934a).

695 A. I. Ponomarev, I. V. Sevast'yanova, V. N. Gruber, A. L. Klebanskii, N. N. Stepanova, N. N. Bausheva, L. A. Aver'yanova and S. V. Aver'yanov, *USSR Pat.* 367 713.

696 B. Prokai, *US Pat.* 3 952 038 (*CA* **85** (1976) 33957a).

697 B. Prokai, *US Pat.* 4 003 847 (*CA* **86** 1977) 91115f).

698 B. Prokai and B. Kanner, *US Pat.* 3 979 419.

699 B. Prokai and B. Kanner, *Ger. Offen.* 2 365 802 (*CA* **85** (1976) 125000m).

700 W. J. Raleigh, *UK Pat.* 1 354 311; *Ger. Offen.* 2 125 841; *US Pat.* 3 654 195; *Fr. Demande* 2 090 330.

701 J. S. Razzano, *Ger. Offen.* 2 448 033 (*CA* **83** (1975) 59782y).

702 J. S. Razzano, *Can. Pat.* 1 082 396.

703 J. S. Razzano, *US Pat.* 3 842 110 (*CA* **82** 57933d).

704 J. S. Razzano, *US Pat.* 3 846 464; *Fr. Demande* 2 252 347; *Ger. Offen.* 2 455 502; *Japan Kokai* 75 1000 75 (*CA* **82** (1975) 86400h).

705 J. S. Razzano, *Ger. Offen.* 2 517 475 (*CA* **84** (1976) 60868h).

706 J. S. Razzano, *US Pat.* 2 619 187 (*CA* **86** (1977) 74156c).

707 J. S. Razzano, *US Pat.* 3 937 684.

708 J. S. Razzano, *US Pat.* 3 853 932; *Ger. Offen.* 2 460 085 (*CA* **82** (1975) 74218s).

709 J. S. Razzano, *US Pat.* 4075 169.
710 J. S. Razzano, *US Pat.* 4028 338.
711 J. S. Razzano, *Ger. Offen.* 2621 751 (*CA* **86** (1977) 73670d).
712 J. S. Razzano, *Ger. Offen.* 2350 369.
713 J. S. Razzano, G. P. De Zuba and N. J. Reo, *US Pat.* 4222 917.
714 J. S. Razzano and V. G. Simpson, *Ger. Offen.* 2618 852 (*CA* **86** (1977) 44577y).
715 J. S. Razzano and V. G. Simpson, *US Pat.* 3974 120 (*CA* **85** (1976) 160844m).
716 L. P. Razumovskii, Yu. V. Moiseev, A. G. Kuznetsova and G. E. Zaikov, *Izv. Akad. Nauk SSSR, Ser. Khim.* **7** (1975) 1502.
717 G. A. Razuvaev, V. A. Dodonov, O. S. D'yachkovskaya, V. I. Shcherbakov and E. U. Chistova, *4th Int. Symp. on Organosilicon Chemistry, Moscow* (1975), Vol. 1, p. 97.
718 G. A. Razuvaev, L. M. Terman, S. D. Yul'chevskaya, O. D. Sutina, V. A. Chudakova and Yu. A. Dimant, *Vysokomol. Soedin.* **B18** (1976) 812 (*CA* **86** (1977) 55751b).
719 G. A. Razuvaev, V. A. Yablokov, A. V. Ganyushkin, V. E. Shklover, I. Tsinker and Yu. T. Struchkov, *Dokl. Akad. Nauk SSSR, Ser. Khim.* **242** (1978) 132.
720 J. D. Reedy and H. D. Furbee, *US Pat.* 3983 148 (*CA* **86** (1977) 121510a).
721 V. O. Reikhsfel'd, E. P. Lebedev and A. M. Evdokimov, *Zh. Obshch. Khim.* **40** (1970) 1078.
722 V. O. Reikhsfel'd, V. N. Vinogradov, E. P. Lebedev and N. A. Filippov, *Issled. Obl. Fiz. Khim. Kauch. Rezin.* **3** (1973) 21 (*CA* **86** (1977) 121890z).
723 Rhone-Poulenc Industries S. A., *UK Pat.* 1542 187.
724 M. O. Riley, Y. K. Kim and O. R. Pierce, *J. Polym. Sci., Polym. Chem. Ed.* **16** (1978) 1929.
725 D. N. Roark and L. H. Sommer, *Chem. Anal. (Warsaw)* **5** (1973) 167.
726 E. G. Rochow, *All. Prakt. Chem.* **17** (1966) 43.
727 M. D. Romanova and I. A. Metkin, *Zh. Prikl. Khim. (Leningrad)* **54** (1981) 212.
728 P. Rosciszewski, E. Jagielska, T. Strojny and B. Becker, *4th Int. Symp. on Organosilicon Chemistry, Moscow* (1975), Vol. 1, p. 92.
729 G. Rossmy, *UK Pat.* 1306 312; *Ger. Offen.* 2059 546; *Fr. Demande* 2077 374; *US Pat.* 3655 712 (*CA* **75** (1971) 118898a).
730 G. Rossmy, *UK Pat.* 1306 311; *Ger. Offen.* 2059 554; *US Pat.* 3652 624; *Fr. Demande* 2077 623 (*CA* **75** (1971) 130435e).
731 G. Rossmy and R. D. Langenhagen, *DDR Pat.* 129 659.
732 G. Rossmy and R. D. Langenhagen, *Ger. Offen.* 2714 807 (*CA* **87** (1977) 202569z).
733 B. Rozsondai, I. Hargittai and T. Garzo, *Magy. Kem. Foly.* **82** (1976) 515.
734 E. Ruiz-Hitzky and J. J. Fripiat, *Bull. Soc. Chim. Fr.* Pt 1 (1976) 1341.
735 E. Ruiz-Hitzky and A. Van Meerbeek, *Colloid. Polym. Sci.* **256** (1978) 135.
736 J. C. Saam and F. W. G. Fearon, *Fr. Demande* 2105 983 (*CA* **78** (1973) 16772z).
737 J. C. Saam and A. H. Ward, *Belg. Pat.* 858 032 (*CA* **89** (1978) 110779f).
738 J. C. Saam, A. H. Ward and F. W. G. Fearon, *Adv. Chem. Ser.* **129** (1973) 239.
739 S. M. Samoilov and V. A. Aulov, *Vysokomol. Soedin.* **A18** (1976) 984 (*CA* **85** (1976) 6369k).
740 A. J. Sanders, *US Pat.* 4246 029.
741 Yu. A. Sangalov, A. I. Il'yasova and K. S. Minsker, *Vysokomol. Soedin.* **B18** (1976) 554 (*CA* **85** (1976) 124438y).
742 T. Sato, *Japan Kokai* 78 31 762 (*CA* **89** (1978) 147959r).
743 Y. Sato, H. Inomata and T. Shibara, *Japan Kokai* 75 64 393 (*CA* **83** (1975) 148272z).
744 Y. Sato, A. Komiya and Y. Fujimura, *Japan Kokai* 77 146 499.
745 Y. Sato, A. Komiya and Y. Fujimura, *Japan Kokai* 77 146 499 (*CA* **92** (1980) 24487q).
746 Y. Sato and H. Okinojima, *Japan Kokai* 77 98 798 (*CA* **88** (1978) 74734f).

747 Y. Sato, N. Shinohara, S. Mori, H. Inomata and S. Hamada, *Japan Kokai* 75 65 596 (*CA* **83** (1975) 132551d).

748 J. C. Saylor, *Ger. Offen.* 1 924 743 (*CA* **72** (1970) 123694y).

749 R. L. Schank, *Fr. Demande* 2 004 165 (*CA* **72** (1970) 121691w).

750 R. L. Schank, *US Pat.* 3 546 265 (*CA* **74** (1971) 100214u).

751 R. L. Schank, *UK Pat.* 1 262 149 (*CA* **76** (1972) 113365k).

752 C. L. Schilling, *Ger. Offen.* 2 629 137 (*CA* **86** (1977) 140919g).

753 C. L. Schilling, *US Pat.* 4 160 775.

754 S. Schindler and K. Ruehlmann, *Plaste Kaut.* **25** (1978) 384.

755 J. G. Schindler, G. Stork, H. J. Struch and W. Schael, *Fresenius Z. Anal. Chem.* **290** (1978) 45.

756 C. G. Schmiterlow and C. Sjogren, *Acta Pharmacol. Toxicol. Suppl.* **36** (1975) 131.

757 K. Schnurrbusch, R. Erdmender and W. Kniege *US Pat.* 3 803 084.

758 T. G. Selin, *US Pat.* 3 557 177 (*CA* **74** (1971) 76517x).

759 P. K. Sen, T. B. Brennan and H. Gilman, *Indian J. Appl. Chem.* **35** (1972) 121.

760 D. Seyferth, T. F. O. Lim and D. P. Duncan, *J. Am. Chem. Soc.* **100** (1978) 1626.

761 V. D. Sheludyakov, S. S. Mkhitaryan, D. Ya. Zhinkin and V. F. Mironov, *Zh. Obshch. Khim.* **48** (1978) 2506.

762 Z. A. Shevchenko, T. I. Lezhen, A. I. Faworskaya and Yu. A. Yuzhelevskii, *Vestn. Leningrad. Univ., Fiz. Khim.* **2** (1972) 146 (*CA* **77** (1972) 147325h).

763 Ch. Shimizu, *Japan Kokai* 79 90 349.

764 Shin-Etsu Chemical Industry Co. Ltd, *Japan Kokai* 80 92 761.

765 Shin-Etsu Chemical Co. Ltd, *Fr. Demande* 2 223 377; *Ger. Offen.* 2 414 878; *US Pat.* 3 898 256; *Japan Kokai* 74 124 032.

766 N. Shinohara, M. Arai and S. Ichinohe, *Japan Kokai* 77 65 226 (*CA* **87** (1977) 135905r).

767 V. E. Shklover, P. Ad'yaasuren, Yu. T. Struchkov, E. A. Zhdanova, V. S. Svistunov, G. V. Kotrelev and K. A. Andrianov, *Dokl. Akad. Nauk SSSR, Ser. Khim.* **241** (1978) 377.

768 V. E. Shklover, N. G. Bokii and Yu. T. Struchkov, *4th Int. Symp. on Organosilicon Chemistry, Moscow* (1975), Vol. 1, p. 18.

769 V. E. Shklover, N. G. Bokii, Yu. T. Struchkov, K. A. Andrianov, B. G. Zavin and V. S. Svistunov, *Zh. Strukt. Khim.* **15** (1974) 90.

770 V. E. Shklover, N. G. Bokii, Yu. T. Struchkov, K. A. Andrianov, B. G. Zavin and V. S. Svistunov, *Zh. Strukt. Khim.* **15** (1974) 841.

771 V. E. Shklover, A. N. Chekhlov, Yu. T. Struchkov, N. N. Makarova and K. A. Andrianov, *Zh. Strukt. Khim.* **19** (1978) 1091.

772 V. E. Shklover, T. V. Timofeeva, Yu. T. Struchkov and A. A. Zhdanov, *Vysokomol. Soedin.* **A25** (1983) 1406.

773 V. E. Shklover, A. E. Kalinin, A. I. Gusev, N. G. Bokii, Yu. T. Struchkov, K. A. Andrianov and I. M. Petrova, *Zh. Strukt. Khim.* **14** (1973) 692.

774 V. E. Shklover, I. Yu. Klement'ev and Yu. T. Struchkov, *Dokl. Akad. Nauk SSSR* **250** (1980) 877.

775 V. E. Shklover and Yu. T. Struchkov, *Usp. Khim.* **49** (1980) 518.

776 V. E. Shklover, Yu. T. Struchkov, I. Yu. Klement'ev, V. S. Tikhonov and K. A. Andrianov, *Zh. Strukt. Khim.* **20** (1979) 302.

777 V. E. Shklover, Yu. T. Struchkov, N. N. Makarova and K. A. Andrianov, *Zh. Strukt. Khim.* **19** (1978) 1107.

778 V. E. Shklover, Yu. T. Struchkov, N. N. Makarova and K. A. Andrianov, *5th Int. Symp. on Organosilicon Compounds, Karlsruhe* (1978), p. 64.

779 V. E. Shklover, Yu. T. Struchkov, N. N. Makarova and A. A. Zhdanov, *Cryst. Struct. Commun.* **9** (1980) 1.

780 V. E. Shklover, Yu. T. Struchkov, A. B. Zachernyuk and K. A. Andrianov, *Zh. Strukt. Khim.* **19** (1978) 116.
781 P. P. Shpakov, A. V. Zak, E. G. Kagan and Yu. A. Yuzhelevskii, *Kinet. Katal.* **16** (1975) 891.
782 G. R. Siciliano, *Ger. Offen.* 2 609 681 (*CA* **85** (1976) 178199q).
783 E. V. Sivtsova, G. V. Burova and M. P. Sidorova, *Zh. Prikl. Khim. (Leningrad)* **50** (1977) 173.
784 I. E. Skobets, F. D. Ovcharenko, M. T. Bryk and N. G. Vasil'ev, *Kolloid. Zh.* **41** (1979) 501.
785 I. I. Skorokhodov, V. E. Ditsent and G. N. Levitanskaya, *Vysokomol. Soedin.* **B20** (1978) 250 (*CA* **89** (1978) 24956e).
786 I. I. Skorokhodov, V. E. Ditsent and E. I. Vovshin, *Vysokomol. Soedin.* **B16** (1974) 716 (*CA* **82** (1975) 43795y).
787 S. Sliwinski and H. Hamann, *DDR Pat.* 137 720.
788 O. K. Smirnov, V. N. Zhdamarova, *USSR Pat.* 697 616.
789 G. W. Smith, *Phase Transitions* **1** (1979) 107.
790 M. V. Sobolevskii, L. M. Blekh, L. V. Sobolevskaya, G. A. Potashova, M. A. Kleinovskaya and Yu. K. Molokanov, *Plast. Massy* **7** (1976) 30 (*CA* **85** (1976) 143185e).
791 M. V. Sobolevskii, M. A. Kleinovskaya, N. E. Rodzevich, Yu. K. Molokanov, V. V. Zverev, N. P. Grinevich and A. M. Yur'eva, *Khim. Prom. (Moscow)* **11** (1977) 848 (*CA* **89** (1978) 164223p).
792 M. V. Sobolevskii, I. I. Skorokhodov, V. E. Ditsent, L. V. Sobolevskaya and G. M. Moiseeva, *Vysokomol. Soedin.* **A12** (1970) 2714 (*CA* **74** (1971) 42782q).
793 M. V. Sobolevskii, I. I. Skorokhodov, V. E. Ditsent, L. V. Sobolevskaya, E. I. Vovshin and L. M. Blekh, *Vysokomol. Soedin.* **A16** (1974) 729 (*CA* **81** (1974) 121330p).
794 L. H. Sommer and J. Mclick, *J. Organomet. Chem.* **101** (1975) 171.
795 M. E. Sorkin, *Ger. Offen.* 2 057 121; *UK Pat.* 1 289 748; *Fr. Demande* 2 069 772; *US Pat.* 3 624 017 (*CA* **75** (1971) 118900v).
796 J. M. Sosa, *Macromolecules* **13** (1980) 1260.
797 J. Soucek, K. A. Andrianov, L. M. Khananashvili and V. M. Myasina, *Dokl. Akad. Nauk SSSR, Ser. Khim.* **227** (1976) 98.
798 J. Soucek, K. A. Andrianov, L. M. Khananashvili and V. M. Myasina, *Dokl. Akad. Nauk SSSR, Ser. Khim.* **222** (1975) 128.
799 J. Soucek, G. Engelhardt, K. Stransky and J. Schraml, *Collect. Czech. Chem. Commun.* **41** (1976) 234.
800 J. Soucek, A. A. Kuzma, L. M. Khananashvili and V. M. Myasina, *Czech. Pat.* 179 188.
801 J. Soucek, K. Stransky, R. Rericha and J. Hetflejs, *Collect. Czech. Chem. Commun.* **40** (1975) 2611.
802 H. S. D. Soysa, H. Okinoshima and W. P. Weber, *J. Organomet. Chem.* **133** (1977) C17.
803 D. E. Spielvogel and C. L. Frye, *J. Organomet. Chem.* **16** (1978) 165.
804 D. E. Spielvogel and L. F. Hanneman, *Acta Pharmacol. Toxicol. Suppl.* **36** (1975) 25.
805 M. S. Starch, *Belg. Pat.* 882 409.
806 F. O. Stark and E. Georg, *Ger. Offen.* 570 448.
807 I. V. Starodubtseva, M. A. Ezerets and A. S. Bulycheva, *Zh. Anal. Khim.* **31** (1976) 586 (*CA* **85** (1976) 6282v).
808 Stauffer Chemical Co., *UK Pat.* 1 176 205; *Ger. Offen.* 1 745 468; *US Pat.* 3 519 600 (*CA* **72** (1970) 56503v).
809 Stauffer Chemical Co., *Ger. Offen.* 2 263 819; *Fr. Demande* 2 166 184; *US Pat.* 3 799 962 (*CA* **79** (1973) 92376i).
810 Stauffer Chemical Co., *Fr. Demande* 2 166 184; *Ger Offen.* 2 263 819; *US Pat.* 3 799 962 (*CA* **79** (1973) 92376j).

811 H. H. Steinbach, K. Schnurrbusch and M. Rieder, *Ger. Offen.* 2 658 115 (*CA* **89** (1978) 181346y).

812 G. Stiubianu, S. E. Lazarescu and M. Marcu, *Rom. Pat.* 64 439.

813 K. Stransky, J. Kohoutova and J. Soucek, *Collect. Czech. Chem. Commun.* **41** (1976) 2523.

814 I. M. Strukova, I. B. Klimenko, N. G. Shelkunov, L. A. Vol'f and Yu. K. Kirilenko, *Zh. Prikl. Khim.* **45** (1972) 913.

815 M. P. Strukova, G. I. Veslova and S. I. Kobelevskaya, *Zh. Anal. Khim.* **32** (1977) 626.

816 R. S. Stuart, *Ger. Offen.* 2 100 329; *Fr. Demande* 2 075 930; *US Pat.* 3 717 665 (*CA* **75** (1971) 152614q).

817 T. Suminoe, Y. Matsumura and O. Tomomitsu, *Japan Kokai* 79 83 957.

818 V. A. Sutyagin and A. K. Tsapuk, *Khim. Vys. Energ.* **11** (1977) 308.

819 J. Suzuki, T. Fujino and K. Kondo, *Japan Kokai* 73 13 396 (*CA* **79** (1973) 92843j).

820 N. G. Sviridova, E. G. Kagan, M. G. Voronkov, S. B. Dolgoplosk and N. A. Zhukova, *Zh. Obshch. Khim.* **46** (1976) 130.

821 N. G. Sviridova, G. I. Karpov, E. G. Kagan, S. B. Dolgoplosk and V. A. Slizkova, *USSR Pat.* 455 600.

822 V. S. Svistunov, V. S. Papkov and A. A. Zhdanov, *Vysokomol. Soedin.* **A22** (1980) 1316.

823 T. Szekely, M. Blazso, K. A. Andrianov, A. A. Zhdanov and E. A. Kashutina, *Magyar Kem. Foly.* **80** (1974) 517.

824 T. Szekely, M. Lengyel, V. S. Papkov, A. E. Zachernyuk, A. A. Zhdanov and K. A. Andrianov, *Acta Chim. (Budapest)* **89** (1976) 307.

825 J. P. Szendrey, *US Pat.* 3 560 436 (*CA* **74** (1971) 100522t).

826 J. P. Szendrey, *US Pat.* 3 560 437 (*CA* **74** (1971) 112668h).

827 M. Takamizawa, Y. Inoue, A. Noshiro, T. Fujii and M. Watanabe, *Japan Kokai* 79 136 910.

828 M. Takamizawa and K. Koya, *Japan Kokai* 75 132 100 (*CA* **84** (1976) 75061p).

829 M. Takamizawa, Y. Yamamoto and H. Okamoto, *Japan Kokai* 74 124 067 (*CA* **82** (1975) 125477k).

830 T. Takiguchi and E. Asada, *Japan Kokai* 72 29 300 (*CA* **78** (1973) 150533a).

831 T. Takiguchi, E. Fujikawa, Y. Yamamoto and M. Ueda, *Nippon Kagaku Kaishi* **1** (1974) 108 (*CA* **81** 64024r).

832 J. Tamas, K. Ujszaszy and G. Bujtas, *Kem. Kozlem.* **32** (1969) 185 (*CA* **72** (1970) 131686n).

833 C. D. Tarasek, K. J. Zimmermann, T. A. Aire and B. G. Crabo, *Int. J. Androl.* **1** (1978) 378.

834 N. P. Telegina, A. I. Demchenko, M. V. Sobolevskii, K. P. Grinevich and N. N. Teplova, *1. Vsesoyuz. Simp., Irkutsk, 1977, Tezisy Dokl.* (1977), p. 226.

835 N. P. Telegina, A. G. Kuznetsova, V. N. Bochkarev, T. A. Agapova, N. A. Palamarchuk, V. S. Lozovskaya, N. A. Kondratova and A. N. Polivanov, *Zh. Obshch. Khim.* **44** (1974) 2688.

836 J. S. Thayer, *Syn. Inorg. Metalorg. Chem.* **8** (1978) 371.

837 D. R. Thomas and J. Francis, *UK Pat.* 1 231 448; *US Pat.* 3 627 805 (*CA* **75** (1971) 36967h).

838 J. Thompson, *Ger. Offen.* 2 121 786; *Fr. Demande* 2 091 121; *UK Pat.* 1 342 244 (*CA* **76** (1972) 154755f).

839 J. Thompson, *UK Pat.* 1 342 244; *Ger. Offen.* 2 121 786; *Fr. Demande* 2 091 121 (*CA* **70** (1969) 154755f).

840 O. A. Tikhonova, L. Ya. Tsvetkova, B. V. Lebedev and I. B. Rabinovich, *Tr. Khim. Khim. Tekhnol.* **1** (1973) 21 (*CA* **80** (1974) 108964h).

841 R. P. Tomlins and K. N. Marsh, *J. Chem. Thermodyn.* **8** (1976) 1185.

842 Toray Silicone Co. Ltd, *Fr. Demande* 2 213 316.
843 Toray Silicone Co. Ltd, *Japan Kokai* 80 137 275.
844 A. G. Trufanov, V. E. Ditsent, M. A. Kleinovskaya, Yu. K. Molokanov and I. I.
     Skorokhodov, *4th Int. Symp. on Organosilicon Chemistry, Moscow* (1975), Vol. 2, p. 22.
845 J. Tsuji and M. Hara, *Japan Kokai* 73 42 959 (*CA* **81** (1974) 121448h).
846 J. Tsuji, M. Hara and K. Ohno, *Ger. Offen.* 1 942 798; *UK Pat.* 1 278 072; *Fr. Demande*
     2 016 385; *US Pat.* 3 658 866 (*CA* **73** (1970) 15497y).
847 J. Tsuji, M. Hara and K. Ohno, *Ger. Offen.* 1 942 798.
848 T. Tsunoda, T. Yamaoka, K. Koseki, M. Hatanaka and Y. Funabashi, *Japan Kokai*
     79 30 250.
849 T. Tsunoda, T. Yamaoka, K. Koseki, M. Hatanaka and Y. Funabashi, *Japan Kokai*
     79 30 300.
850 L. Ya. Tsvetkova and B. V. Lebedev, *Tr. Khim. Khim. Tekhnol.* **1** (1973) 19 (*CA* **80** (1974)
     36603q).
851 G. N. Turkel'taub, V. N. Bochkarev, M. L. Galashina and E. I. Golysheva, *Zavod. Lab.*
     44 (1978) 278 (*CA* **89** (1978) 122671c).
852 G. N. Turkel'taub and B. M. Luskina, *Zh. Anal. Khim.* **26** (1971) 2243.
853 G. N. Turkel'taub and B. M. Luskina, *Metody Anal. Kontr. Proizvod. Khim. Prom.* **5**
     (1972) 5.
854 S. Ueno, *Japan Kokai* 76 33 840.
855 Union Carbide Corp., *UK Pat.* 1 515 958 (*CA* **90** (1979) 106092e).
856 A. V. Uvarov, V. D. Sheludyakov and E. G. Gorlov, *1. Vsesoyuz. Simp., Irkutsk, 1977,
     Tezisy Dokl.* (1977), p. 278.
857 B. I. Vainshtein, E. A. Vaks and L. P. Bogovtseva *Zh. Obshch. Khim.* **46** (1976) 137.
858 P. B. Valkovich and W. P. Weber, *J. Org. Chem.* **40** (1975) 229.
859 P. B. Valkovich and W. P. Weber, *J. Organomet. Chem.* **99** (1975) 231.
860 W. J. A. Vandenhenvel, J. L. Smith, R. A. Firestone and J. L. Beck, *Anal. Lett.* **5** (1972)
     285.
861 F. J. Van Lenten, J. E. Conaway and L. B. Rogers, *Separ. Sci.* **12** (1977) 1.
862 N. V. Varlamova, A. I. Sidnev, V. V. Severnyi and K. A. Andrianov, *Vysokomol. Soedin.*
     A12 (1970) 2685 (*CA* **74** (1971) 5432w).
863 R. Varma, P. Orlancer and A. K. Ray, *J. Inorg. Nucl. Chem.* **37** (1975) 1797.
864 S. K. Varshney and D. N. Khanna, *J. Appl. Polym. Sci.* **25** (1980) 2501.
865 J. Vessman, C. G. Hammar, B. Lindeke, S. Stromberg, R. Levier, R. Robinson, D.
     Spielvogel and L. Hanneman, *Nobel. Symp. 1977* **40** (1978) 535 (*CA* **90** (1979) 114769w).
866 S. C. Vick *US Pat.* 4 222 952.
867 F. Volino and A. J. Dianoux, *Ann. Phys. (Paris)* **3** (1978) 151.
868 M. A. Volodina, T. A. Gorshkova and K. T. Maky Abubakr, *Anal. Lett.* **10** (1977) 1189.
869 M. G. Voronkov, E. I. Brodskaya, V. V. Keiko, S. G. Shevchenko, T. N. Bazhenova,
     S. F. Pavlov, V. B. Modonov, E. I. Dubinskaya and Yu. L. Frolov, *Dokl. Akad. Nauk
     SSSR, Ser. Khim.* **232** (1977) 1100.
870 M. G. Voronkov, E. I. Cubinskaya and N. A. Chuikova, *Zh. Obshch. Khim.* **47** (1977)
     2395.
871 M. G. Voronkov, S. V. Kirpichenko, V. V. Keiko, V. A. Pestunovich, E. O. Tsetlina,
     V. Chvalovsky and J. Vcelak, *Izv. Akad. Nauk SSSR, Ser. Khim.* **9** (1975) 2052.
872 M. G. Voronkov, V. I. Lavrent'ev, A. N. Kanev, V. G. Kostrovskii and S. A.
     Prokhorova, *Dokl. Akad. Nauk SSSR* **249** (1979) 106.
873 M. G. Voronkov, T. N. Martynova, R. G. Mirskov and V. I. Belyi, *Zh. Obshch. Khim.* **49**
     (1979) 1522.
874 M. G. Voronkov, S. F. Pavlov and E. I. Dubinskaya, *USSR Pat.* 575 350 (*CA* **88** (1978)
     51013f).

875 M. G. Voronkov, L. Steiling, S. V. Kirpichenko and E. E. Kuznetsova, *5th. Int. Symp. on Organosilicon Compounds, Karlsruhe* (1978), p. 62.
876 M. G. Voronkov and N. G. Sviridova, *Zh. Obshch. Khim.* **46** (1976) 126.
877 I. Vosniuc, *Rom. Pat.* 62 940.
878 I. Vosniuc, *Rom. Pat.* 62 940 (*CA* **92** (1980) 77811x).
879 E. I. Vovshin, N. P. Telegina, I. I. Luchko, A. A. Bystrov and V. E. Ditsent, *Khim. Prom. (Moscow)* **2** (1976) 105 (*CA* **85** (1976) 96198z).
880 N. V. Vvedenskii, V. O. Reikhsfel'd and A. G. Ivanova, *Kremniiorgan. Mater.* (1971) p. 255 (*CA* **77** (1972) 154285y).
881 Wacker-Chemie GMBH, *Fr. Demande* 2 357 570.
882 Wacker-Chemie GMBH, *Fr. Demande* 2 018 841 (*CA* **74** (1971) 143107d).
883 U. Wannagat and G. Eisele, *Z. Naturforsch.* **33b** (1978) 475.
884 W. P. Weber and H. Okinoshima, *5th. Int. Symp. on Organosilicon Compounds, Karlsruhe* (1978), p. 200.
885 K. Weinberg and G. C. Johnson, *Ger. Offen.* 2 802 485 (*CA* **89** (1978) 197709e).
886 R. Wenske, S. Berwald and H. Teichmann, *DDR Pat.* 68 772 (*CA* **72** (1970) 90614s).
887 M. L. Wheeler, *US Pat.* 3 674 891.
888 W. Wieder, J. Witte and H. H. Moretto, *Ger. Offen.* 2 856 836.
889 L. Wilczek and J. Chojnowski, *Macromolecules* **14** (1981) 9.
890 D. E. Williams, G. M. Ronk and D. E. Spielvogel, *J. Organomet. Chem.* **69** (1974) 69.
891 R. Witlib and M. Ksiazek, *Pol. Pat.* 92 673 (*CA* **90** (1979) 104883q).
892 P. V. Wright, *J. Polym. Sci., Polym. Phys. Ed.* **11** (1973) 51.
893 T. C. Wu, *US Pat.* 3 586 699 (*CA* **75** (1971) 110907g).
894 T. C. Wu, *US Pat.* 3 487 098 (*CA* **72** (1970) 55640a).
895 F. F. H. Wu, J. Goetz, W. D. Jamieson and C. R. Masson, *J. Chromatogr.* **48** (1970) 515.
896 S. K. Wu, Y. C. Jiang and J. F. Rabek, *Polym. Bull. (Berlin)* **3** (1980) 319.
897 T. C. Wu and P. J. Launer, *J. Chem. Engng Data* **18** (1973) 350.
898 M. Wurst and J. Churacek, *Collect. Czech. Chem. Commun.* **36** (1971) 3497.
899 M. Wurst and J. Churacek, *J. Chromatogr. Sci.* **70** (1972) 1.
900 S. Yajima, M. Hamano, Y. Hasegawa, K. Kamino and T. Yamane, *UK Pat.* 2 020 676.
901 Yu. G. Yanovskii and G. V. Vinogradov, *Vysokomol. Soedin.* **A22** (1980) 2567.
902 E. B. Yanushevshaya, A. P. Loburenko, A. G. Zherko and A. O. Marchenko, *Khim. Prom-st., Ser.: Toksikol. Sanit. Khim. Plastmass* **4** (1979) 40.
903 H. Yokoi and S. Enomoto, *Chem. Pharm. Bull.* **26** (1978) 1846.
904 E. Yashii, H. Ikeshima and K. Ozaki, *Chem. Pharm. Bull.* **20** (1972) 1827.
905 Yu. Tung-Yin and Bao Chi-Nai, *Kao Fen Tzu Tung Hsun* **2** (1979) 112.
906 C. G. Yun and D. H. Chang, *Hwahak Kwa Hwahak Kongop* **18** (1975) 88 (*CA* **84** (1976) 5456x).
907 C. G. Yun and D. H. Chang, *Hwahak Kwa Hwahak Kongop* **19** (1976) 261 (*CA* **87** (1977) 39888s).
908 Yu. A. Yuzhelevskii, N. N. Fedoseva, K. N. Makarov and L. L. Gervits, *Vysokomol. Soedin.* **B19** (1977) 874 (*CA* **88** (1978) 90066v).
909 Yu. A. Yuzhelevskii, E. G. Kagan and N. N. Fedoseeva, *Vysokomol. Soedin.* **A12** (1970) 1585.
910 Yu. A. Yuzhelevskii, E. G. Kagan and N. N. Fedoseeva, *Dokl. Akad. Nauk SSR, Ser. Khim.* **190** (1970) 647.
911 Yu. A. Yuzhelevskii, E. G. Kagan, O. N. Larionova and A. L. Klebanskii, *USSR Pat.* 238 780 (*CA* **81** (1974) 50274k).
912 Yu. A. Yuzhelevskii, E. G. Kagan, N. P. Timofeeva, T. D. Doletskaya and A. L. Klebanskii, *Vysokomol. Soedin.* **A13** (1971) 183.

348 V. CHVALOVSKÝ

913 Yu. A. Yuzhelevskii, T. V. Kurlova and V. N. Churmaeva, *Gazov. Khromatogr.* **15** (1971) 56 (*CA* **80** (1974) 43763v).

914 Yu. A. Yuzhelevskii, T. V. Kurlova, E. A. Chernyshev and A. G. Kuznetsova, *Vysokomol. Soedin.* **B15** (1973) 827 (*CA* **82** (1975) 86672y).

915 Yu. A. Yuzhelevskii, T. V. Kurlova, E. G. Kagan and M. V. Suvorova, *Zh. Obshch. Khim.* **42** (1972) 2006.

916 Yu. A. Yuzhelevskii, V. V. Pchelintsev and N. N. Fedoseeva, *Vysokomol. Soedin.* **B18** (1976) 873 (*CA* **86** (1977) 73181v).

917 Yu. A. Yuzhelevskii, V. V. Pchelintsev and E. G. Kagan, *4th Int. Symp. on Organosilicon Chemistry, Moscow* (1975), Vol. 2, p. 10.

918 Yu. A. Yuzhelevskii, A. S. Shapatin, B. L. Kaufman, N. N. Fedoseeva, P. G. Ten and D. Ya. Zhinkin, *Vysokomol. Soedin.* **B13** (1971) 904 (*CA* **76** (1972) 113638b).

919 Yu. A. Yuzhelevskii, V. V. Sokolov, L. V. Tagieva and E. G. Kagan, *Vysokomol. Soedin.* **B13** (1971) 95.

920 B. G. Zavin, A. Yu. Rabkina and A. A. Zhdanov, *4-I Mezdunar. Mikrosimpoz. "Uspekhi v Oblasti Ion. Polimerizatsii", Ufa* (1979), p. 30.

921 A. A. Zhdanov, K. A. Andrianov, T. V. Astapova and G. F. Vinogradova, *Izv. Akad. Nauk SSSR, Ser. Khim.* **11** (1973) 2628.

922 A. A. Zhdanov, K. A. Andrianov, M. M. Levitskii and B. A. Lavrukhin, *Zh. Obshch. Khim.* **40** (1970) 2577.

923 A. A. Zhdanov, K. A. Andrianov and A. P. Malykhin, *Vysokomol. Soedin.* **A16** (1974) 2345 (*CA* **82** (1975) 44202q).

924 A. A. Zhdanov, K. A. Andrianov and A. P. Malykhin, *Vysokomol. Soedin.* **A16** (1974) 1765 (*CA* **82** (1975) 43782s).

925 A. A. Zhdanov, K. A. Andrianov, A. P. Malykhin, I. L. Dubchak and T. V. Astapova, *USSR* 657 039.

926 A. A. Zhdanov, K. A. Andrianov, A. P. Malykhin, B. D. Lavrukhin, I. L. Dubchak and T. V. Strelkova, *Izv. Akad. Nauk SSSR, Ser. Khim.* **5** (1976) 1113.

927 A. A. Zhdanov, G. M. Ignat'eva and V. A. Odinets, *Vysokomol. Soedin.* **B22** (1980) 621.

928 A. A. Zhdanov, E. A. Kashutina, B. D. Lavrukhin and O. I. Shchegolikhina, *Zh. Obshch. Khim.* **49** (1979) 1302.

929 A. A. Zhdanov, E. S. Khynku, N. A. Kurasheva and V. E. Shklover, *Dokl. Akad. Nauk SSSR* **250** (1980) 1392.

930 A. A. Zhdanov, N. A. Kurasheva and E. S. Khynku, *Vysokomol. Soedin.* **B22** (1980) 624.

931 A. A. Zhdanov, E. F. Rodionova and L. A. Gavrikova, *Vysokomol. Soedin.* **B18** (1976) 494 (*CA* **85** (1976) 124437x).

932 A. A. Zhdanov, A. B. Zachernyuk, G. V. Solomatin and Yu. T. Struchkov, *Dokl. Akad. Nauk SSSR* **251** (1980) 121.

933 G. Ya. Zhigalin, A. S. Shapatin, A. L. Gilyazetdinov, L. V. Pshenichnaya and L. M. Antipin, *Vysokomol. Soedin.* **B21** (1979) 844.

934 M. J. Ziemelis and J. Maris, *Ger. Offen.* 2 925 305.

935 V. V. Zverev, N. P. Telegina, A. G. Kuznetsova, I. P. Kharlamova, V. V. Stegalkina and M. V. Sobolevskii, *4th Int. Symp. on Organosilicon Chemistry, Moscow* (1975) Vol. 2, p. 12.

936 *Added in proof:* Authentic cyclodisiloxanes have now been reported. See: R. West, M. J. Fink, M. J. Michalczyk, D. J. De Young and J. Michl, *7th Intl. Symp. Organosilicon Chem.*, Kyoto, Sep. 9–14, 1984, Abstr. p. 1; *J. Am. Chem. Soc.*, **106**, (1984) 822.

11

Silicon–Sulphur Heterocycles

IONEL HAIDUC

Babes-Bolyai University, Cluj-Napoca, Romania

11.1 Introduction

Progress in silicon–sulphur heterocyclic chemistry has been rather slow; however, since 1970 some new rings have been reported and new derivatives prepared. Presently, the heterocycles **(1)**–**(11)** are known. Some silicon–sulphur heterocycles containing Si–Si bonds are treated in Section 7.2.4 of this book because of their closer relationship with cyclopolysilanes.

(1)   (2)   (3)   (4)

(5)   (6)

THE CHEMISTRY OF INORGANIC HOMO-
AND HETEROCYCLES VOL 1 ISBN 0-12-655775-6

R₂Si structures (7), (8), (9), (10), (11)

(7)   (8)   (9)   (10)   (11)

Replacement of either silicon or sulphur in cyclosilathianes results in mixed heterocycles, illustrated by formulae (5)–(11). Some nitrogen-containing heterocyclosilathianes based on the rings Si_3S_2N and Si_3SN_2 are described in Section 8.8.

For reviews dealing with silicon–sulphur heterocycles see the Bibliography.[B 1 – B 5]

11.2 Cyclosilathianes

These are rings formed by regular alternation of silicon and sulphur atoms (1) and (2).

11.2.1 Preparation

The parent compound of cyclotrisilathianes (2), R = R′ = H, was prepared in 55% yield by reacting diiodosilane with mercury(II) sulphide in refluxing benzene, followed by depolymerization of the oligomeric mixture thus formed in vacuum at 210 °C:[1]

$$H_2SiI_2 + HgS \xrightarrow{80\,°C} (H_2SiS)_x \xrightarrow{210\,°C} (H_2SiS)_3 \qquad (1)$$

Tetrafluorocyclodisilathiane (1), R = R′ = F, was formed in the reaction of difluorodiiodosilane with mercury(II) sulphide,[2] and by cocondensation of a mixture of :SiF_2 and SiF_4 with S_2Cl_2 at −196 °C.[3]

Tetrabromocyclodisilathiane (1), R = R′ = Br, was formed by heating Si_2Br_6 with SF_4 at 120 °C or with SF_6 at 450 °C.[4] The cyclodisilathianes (1), R = R′ = Cl or Br, and the adamantane-like compound $Br_4Si_4S_6$ (4), R = Br, were prepared by the reaction of silicon disulphide with silicon tetrahalides.[54]

Organic derivatives of cyclosilathianes have been much more thoroughly investigated than purely inorganic compounds. The main preparative procedure continues to be the reaction of organodichlorosilanes $RR'SiCl_2$ with hydrogen sulphide, preferably in the presence of an acid scavenger.[5–11]

Only new dimers (**1**), $R = Et$, $R' = CH_2CH_2Ph$, $CH_2CHMePh$, have been obtained by this route,[10] and several trimeric derivatives (**2**), $R = H$, $R' = Me$, Et;[5] $R = Me$, $R' = Me$,[6] Et, Vi,[7] Ph,[8–10] CH_2CH_2Ph, $CH_2CHMePh$, $CH_2CH_2CH_2CN$,[10] Et, Pr^n, Bu^n, C_5H_{11}, C_6H_{13};[11] $R = Et$, $R' = CH_2CH_2CH_2CN$,[10] have been also obtained. Generally, the trimers are formed when the reaction is carried out under mild conditions, at room temperature or below; if the temperature is raised or the trimers (**2**) are heated or distilled, they are slowly converted to dimers (**1**).[7–9, 12]

When alkylalkoxydichlorosilanes $R(R'O)SiCl_2$ are reacted with hydrogen sulphide, the formation of adamantane-like tetramers (**4**) as by-products, in addition to cyclosilathianes (**1**) or (**2**), was observed; with $MeSi(OMe)Cl_2$ the tetramer (**4**), $R = Me$, was the exclusive product.[11]

Cyclosilathianes containing bulky substituents, for example Pr^i and Bu^t, cannot be obtained by thiohydrolysis of organodichlorosilanes; in this case silver sulphide must be used instead of hydrogen sulphide.[19]

The use of solid NH_4SH instead of gaseous hydrogen sulphide in the reaction with $RR'SiCl_2$ may be advantageous and higher yields of cyclosilathianes ($R = Me$, $R' = Me$ or Ph) are obtained.[13]

Treatment of a mixture of cyclosilazanes $(Me_2SiNH)_n$, with $n = 3$ and 4, and diorganodichlorosilanes R_2SiCl_2 with hydrogen sulphide also yields cyclosilathianes (predominantly the trimer). In a similar manner, mixtures of diethylaminosilanes and dichlorosilanes yield predominantly trimeric cyclosilathianes,[14] as shown in Table 11.1. This procedure affords unsymmetrically substituted compounds, which would be difficult to obtain otherwise.[14]

Table 11.1 Formation of cyclosilathianes from Si–N compounds.

Si–N compound	Chlorosilane	Cyclosilathianes formed (% yield)	
$(Me_2SiNH)_{3,4}$	Me_2SiCl_2	$(Me_2SiS)_2$	(6.6)
		$(Me_2SiS)_3$	(86.2)
$Me_2Si(NEt_2)_2$	Me_2SiCl_2	$(Me_2SiS)_3$	(52.5)
	Et_2SiCl_2	$(Me_3SiS)_3$	(14.4)
		$(Et_2SiS)_2$	(21.7)
		gem-$Et_2Me_4Si_3S_3$	(44.7)
	$MePhSiCl_2$	$(Me_2SiS)_3$	(9.2)
		$Me_5PhSi_3S_3$	(64.5)
		$(MePhSiS)_3$	(22.5)
	$Me(EtO)SiCl_2$	$(EtO)Me_5Si_3S_3$	(23.6)
		$(Me_2SiS)_3$	(21.2)
	Ph_2SiCl_2	$(Me_2SiS)_3$	(34.6)
		$(Ph_2SiS)_3$	(46.6)
$MePhSi(NEt_2)_2$	$MePhSiCl_2$	$(MePhSiS)_3$	(72.2)

Cyclosilazanes can be converted directly to cyclosilathianes, on treatment with hydrogen sulphide, under heating without solvent.[15,16] Thus $(Me_2SiNH)_3$ gave 70% $(Me_2SiS)_2$ after 25 h of heating to reflux in the presence of hydrogen sulphide; shorter reaction times yield partially substituted rings, containing cyclosilathiazane rings.[16]

The reaction of organodichlorosilanes $RR'SiCl_2$ with hexamethyldisilathiane $(Me_3Si)_2S$ in the presence of pyridine as catalyst, under non-equilibrium conditions (continuous distillation of Me_3SiCl from the system) gives cyclosilathianes (1), R = R' = Et or Ph,[17,17a] and (2), R = Me, R' = Vi, Ph, $CH_2CH_2CF_3$, $CH_2CHMeCOOMe$,[17] in good yields.

Some thermal reactions were reported to produce cyclosilathianes. Thus heating diphenylsilane Ph_2SiH_2 with elemental sulphur in refluxing decalin gave 29% $(Ph_2SiS)_3$,[18] while copyrolysis of hexaorganodisilanes with sulphur (equation 2) or with SF_6, and still better the thermal reaction of R_2HSi—$SiHR_2$ (R = Me, Pr^i, cyclo-C_6H_{11}, Bu^t) with sulphur (equation 3) gave cyclodisilathianes (1):[4,19]

$$R_3Si\text{—}SiR_3 + S \xrightarrow{320\,°C} (R_2SiS)_2 + R_3Si\text{—}S\text{—}SiR_3 + SiR_4 \qquad (2)$$

$$R_2HSi\text{—}SiHR_2 + S \xrightarrow{320-400\,°C} (R_2SiS)_2 + R_2SiH_2 \qquad (3)$$

The formation of some $R_3HSi_2S_2$, partially substituted (1), was observed as by-product; it is believed that the reaction occurs via silylenes $:SiR_2$ and $:SiHR$, which react with sulphur to form the cyclodisilathiane.[19]

Cyclodisilathianes were also formed on heating sulphur with $SiBu_4^t$ (with considerable isomerization of Bu^t to Bu^i groups) or with di-t-butylsilacyclobutane at 260 °C.[19] Generation of monomeric silathione $R_2Si{=}S$ in various reactions, for example in the pyrolysis of $(R_2SiCH_2)_2$ with $Ph_2C{=}S$,[20] also results in the formation of cyclosilathianes.

Alkoxy- and phenoxycyclodisilathianes (1), R = R' = OBu^t,[21] and 2,3-$Me_2C_6H_3O$,[22] are formed on heating silicon disulphide (a polymeric compound containing Si_2S_2 rings) with alcohols or phenols.[21,22] The main product of this reaction is, however, $(RO)_3Si$—SH.

Monoorgano-substituted silicon derivatives in several reactions similar to those described above lead to tetrameric compounds $(RSiS_{1.5})_4$ with the adamantane structure (4). Thus such compounds were formed in the reactions of $MeSi(OR)Cl_2$ with hydrogen sulphide,[11] $PhSiCl_3$ with $(Me_3Si)_2S$[17] and $PhSiH_3$ with elemental sulphur in refluxing decalin.[18]

11.2.2 Structure

Various methods have been used to characterize or to confirm the structures of cyclosilathianes mentioned above. Spectroscopic techniques, like ^1H

NMR and infrared spectroscopy, have been used routinely. In addition, ^{29}Si NMR spectroscopy was used for $(H_2SiS)_3$[1] and mass spectra for $(H_2SiS)_3$,[1] $(Ph_2SiS)_3$[18] and several cyclodisilathianes.[19] Proton NMR spectroscopy revealed that the trimers $(RR'SiS)_3$ (12), R = Me, R' = Et, Vi, Ph, occur as mixtures of conformational isomers (chair and boat), while the dimers $(RR'SiS)_2$ are a mixture of geometrical isomers (*cis* and *trans*).[7-9] Detailed ^{1}H, ^{13}C and ^{29}Si NMR spectroscopic data have been reported for several cyclodi- and cyclotrisilathianes (1) and (2), and for adamantane-like compounds (4).[55]

The molecular structures of several cyclosilathianes have been determined by single-crystal X-ray diffraction. Thus 1,3,5-trimethyl-1,3,5-triphenyl-cyclotrisilathiane (2), R = Me, R' = Ph, was found to be a *trans*-isomer, with a stable twist boat conformation of the ring.[23] The adamantane-like tetrameric compound $(MeSiS_{1.5})_4$ has also been confirmed by X-ray diffraction.[24,25]

The crystal structures of the inorganic derivatives $(Cl_2SiS)_2$, $(Br_2SiS)_2$ and $Br_4Si_4S_6$ have also been determined.[54] In the crystal structure of the dimeric $(Me_2SiS)_2$ at $-120\,°C$, the four-membered ring Si_2S_2 was found to be planar,[56] with Si–S 2.152 Å, SiSSi 82.6° and SSiS 97.5°. The nonbonded transannular distances Si $\cdots$ Si 2.837 Å and S $\cdots$ S 3.327 Å are shorter than the sum of the van der Waals radii, suggesting some interaction.

A discussion of the vibrational spectra of cyclosilathianes $(Ph_2SiS)_{2,3}$ (1) and (2), R = Ph, was published.[26]

11.2.3 Chemical reactions

Cyclosilathianes are reactive compounds, sensitive to reagents containing mobile hydrogen. As a result, few reactions occuring with *ring preservation* are known. Thus adduct formation between $(Me_2SiS)_3$ and ethylenediamine or aminoethoxytrimethylsilane was reported to occur at room temperature, preceding the ring-cleavage reactions, which start at 80 °C.[27]

The *ring-cleavage* reactions of cyclosilathianes with water (hydrolysis),[10] alcohols, aminoalcohols and glycols,[28-32] phenols,[33-36] carboxylic acids[33-35,37] and amines,[35,38] leading to siloxanes and difunctional organo-silanes, R_2SiX_2 (X = OR', OCOR', NHR'), were investigated in detail and kinetic data were reported, as well as reactivity orders for these reagents. The hydrolytic stability is influenced by the nature of the organic substituents:[10]

$$(EtHSiS)_3 < [Et(NCCH_2CH_2CH_2)SiS]_3 < [Et(PhCH_2CH_2)SiS]_2$$

Cyclosilathianes also react with aldehydes to form trithioaldehydes $(RCHS)_3$ and siloxanes. The reaction requires heating to 140 °C and the presence of $ZnCl_2$ as catalyst.[53]

The silathiane rings are also cleaved by organometallic reagents. Thus the trimer $(Me_2SiS)_3$ (2), R = Me, reacts with trimethylaluminium in refluxing benzene to give Me_2Al—S—$SiMe_3$.[39] Perhaps this type of reaction could be extended to other organometallic compounds.

Organocyclosilathianes readily undergo *redistribution reactions* with a variety of reagents. Thus heating $(Me_2SiS)_3$ with $MePhSiCl_2$ in the presence of pyridine as catalyst yields 82% $(MePhSiS)_3$ and Me_2SiCl_2.[17a] The redistribution between $(Me_2SiS)_3$ and Me_2SiBr_2, resulting in ring-opening with formation of bromine-terminated poly(dimethylsilathianes), was investigated in detail by ^{1}H NMR spectroscopy; the equilibrium is shifted in favour of the ring compound.[40] Cyclic and short-chain linear compounds were observed in the redistribution reactions in the systems $(Me_2SiS)_3$—Me_2GeCl_2 and $(Me_2SiS)_3$—$(Me_2GeS)_3$—Me_2SiCl_2.[41]

In the absence of another reagent, four- and six-membered cyclosilathianes can be readily interconverted by heating; the equilibrium between (1) and (2), R = Me, was investigated by ^{1}H NMR spectroscopy at various temperatures and dilutions. It was found that higher temperatures and dilutions favour the formation of the smaller ring.[42] This reaction explains the presence of cyclodisilathianes in many reactions in which a cyclotrisilathiane is heated.

A redistribution reaction also occurs on heating mixtures of cyclosilathianes and cyclosilazanes (R = Me):[43]

These redistributions involve monomeric silathiones $R_2Si{=}S$, which were trapped in the pyrolysis products of cyclosilathianes with hexamethylcyclotrisiloxane to yield mixed Si—S—Si—O rings (see Section 11.5).[44,45] Other redistribution reactions of cyclosilathianes, leading to various inorganic heterocycles, are discussed in Sections 11.3 and 11.4.

11.2.4 Uses

Cyclosilathianes have few practical uses. They seem to inhibit the UV and γ-ray initiated polymerization of methylmethacrylate[47] and to increase the stability of poly(methylmethacrylate) to γ-radiation.[48] Cyclosilathianes can be used as reagents for facile reduction of sulphoxides RSOR′ to organic sulphides RSR′,[49] and for introducing –SiR_2– units into a polysiloxyphosphazene, prepared by heating $(R_2SiS)_3$ (R = Me, Ph) with a hydroxocyclotriphosphazene $(HO)_3(PhO)_3P_3N_3$.[50]

11.3 Cyclo-1,2,4-trisila-3,5-dithianes

A five-membered Si_3S_2 ring with structure (3), R = Me, was prepared in 75% yield by redistribution reactions between cyclodi- and cyclotrisilathianes (1) and (2) and cyclotetrasiladithiane $R_8Si_4S_2$, catalysed by pyridine at 70 °C:[46]

R₂Si–S–SiR₂ / R₂Si–S–SiR₂ (six-membered ring) + $(R_2SiS)_{2\ or\ 3}$, 70 °C

R₂Si–S–SiR₂ / R₂Si–S (ring) — 215 °C, 5 days — SiR_2, $-H_2C=CH_2$ →

R₂Si–S–SiR₂ / R₂Si–S (ring) + $(R_2SiS)_2$, 215 °C, 5h — SiR_2

$R_2Si—SiR_2$ / S, S / Si / R_2

(3)

The same compound is formed in the pyrolysis of 1,1,2,2-tetramethyl-1,2-disila-3,6-dithiacyclohexane alone or in the presence of tetramethylcyclodisilathiane (1),[45] or by irradiation of $(Me_2Si)_6$ (which generates the silylene :SiMe₂) in the presence of $(Me_2SiS)_2$.[45] These reactions occur via dimethylsilathione $Me_2Si=S$ as intermediate.[45]

11.4 Cycloborasilathianes

The five-membered ring (5), R = Me, R′ = Ph, was prepared by reacting a cyclotetrasiladithiane (12) with phenylboron dichloride in CH_2Cl_2 at low temperature:[45]

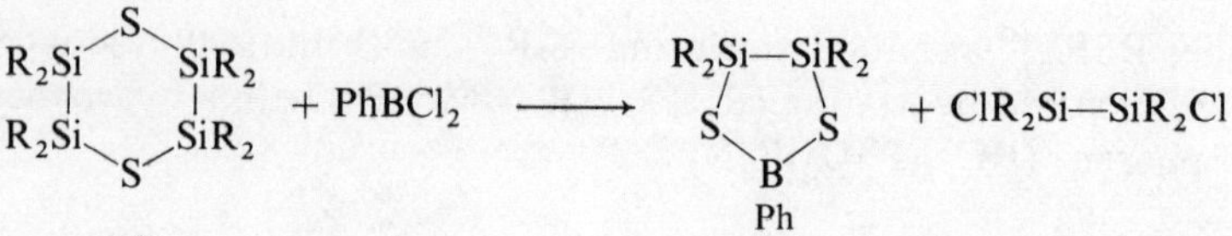

$$R_2Si\underset{S}{\overset{S}{\diamond}}SiR_2 \;\; + \; PhBCl_2 \;\; \longrightarrow \;\; \underset{S}{\overset{R_2Si-SiR_2}{\diamond}}\underset{\underset{Ph}{\overset{|}{B}}}{S} \;\; + \; ClR_2Si-SiR_2Cl \quad (5)$$

(12)

(5)

On heating (slow sublimation), the ring (5) undergoes a rearrangement, with the formation of (12) and (PhBS)$_3$.[46]

Another five-membered ring (6) can be obtained by three different ring-interconversion reactions, based on structure-unit redistribution.[46] The ring is labile and undergoes further redistributions (R = Me):[46]

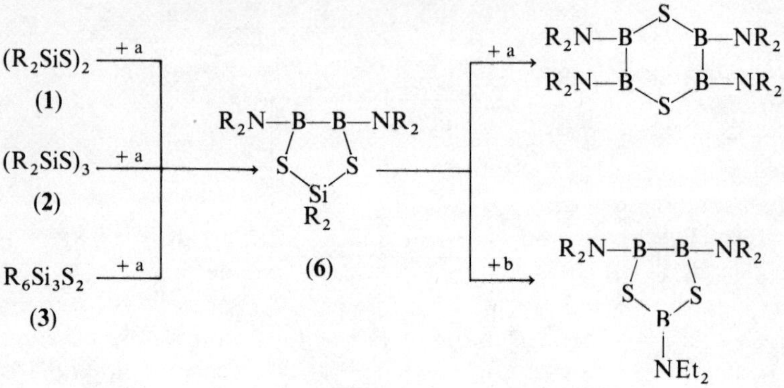

a = Cl(Me$_2$N)BB(NMe$_2$)Cl   b = Br$_2$B—NEt$_2$

The structure of (6), R = Me, was established by single-crystal X-ray diffraction; the five-membered ring is planar.[46]

11.5 Cyclosilathioxanes

Cyclosilathioxanes (7)–(9) were prepared by reacting dichlorodisiloxanes O(SiR$_2$Cl)$_2$, or their mixtures with dichlorosilanes R$_2$SiCl$_2$, with stoichiometric amounts of (Me$_3$Si)$_2$S, in the presence of a tertiary-amine catalyst (pyridine, triethylamine) without solvent.[51,52] Thus the ring (7) was prepared from a 1:1 mixture of disiloxane and dichlorosilane (R = Me, R' = Me or Ph, R" = Ph):[51,52]

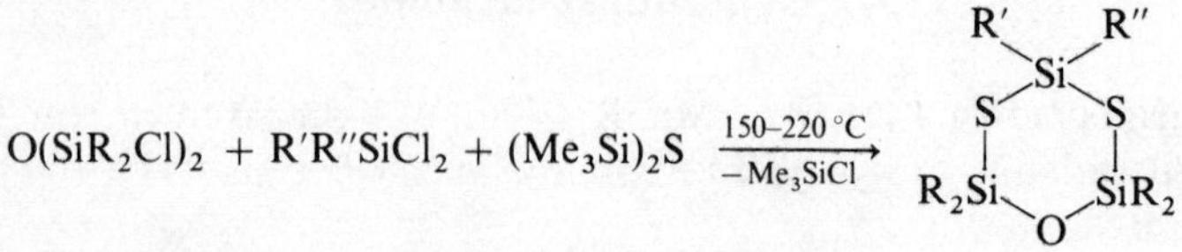

$$O(SiR_2Cl)_2 \; + \; R'R''SiCl_2 \; + \; (Me_3Si)_2S \; \xrightarrow[-\,Me_3SiCl]{150-220\,^\circ C} \; \underset{R_2Si\diagdown_O\diagup SiR_2}{\overset{R'\diagdown_{Si}\diagup R''}{\underset{S\quad\quad S}{}}} \quad (6)$$

If a longer chain is present in the starting siloxane then the expected eight-membered ring (10) is not formed, and the reaction proceeds with separate cyclization of each starting organosilicon unit (R = Me):[52]

$$Cl(R_2SiO)_2SiR_2Cl + RPhSiCl_2 + (R_3Si)_2S \xrightarrow[-Me_3SiCl]{Py}$$

$$\longrightarrow \quad \begin{array}{c} R_2 \\ Si \\ O \qquad O \\ | \qquad | \\ R_2Si \qquad SiR_2 \\ S \end{array} \quad + (RPhSiS)_3 \qquad (7)$$

(8)

Derivatives of (8) were prepared in nearly quantitative yield from 1,5-dichlorotrisiloxanes $R_2'Si(OSiR_2Cl)_2$ and $(Me_3Si)_2S$ in the presence of pyridine at 150–220 °C.[52,52a]

In a similar manner, 1,3-dichlorotetramethyldisiloxane $O(SiMe_2Cl)_2$ reacted with $(Me_3Si)_2S$ to form the eight-membered ring (9), R = Me.[52,52a]

Small amounts of cyclotrisilathiadioxane derivatives (8), R = Me, Et, were formed as by-products in the reaction of cyclotrisilathianes with benzaldehyde in the presence of $ZnCl_2$ catalyst at 140 °C.[53]

The eight-membered ring (10) is formed by insertion of dimethylsilathione $Me_2Si{=}S$ (generated in pyrolysis reactions) into the cyclotrisiloxane ring (which acts as a trapping reagent).[44,45] The reaction takes place by copyrolysis of two cyclic compounds (R = Me):[44]

$$(R_2SiS)_3 + (R_2SiO)_3 \xrightarrow[7\,h]{200\,°C} (10) + (R_2SiS)_2 \qquad (8)$$
$$(2) \qquad\qquad (1)$$

The formation of the monomeric silathione $R_2Si{=}S$ also explains the reaction between a cyclotrisilathiane (2) with 1,1,3,3-tetramethyl-2-oxa-1,3-disilacyclopentane to yield a novel five-membered ring (11) (R = Me):[44]

$$(R_2SiS)_3 + \begin{array}{c} H_2C{-}CH_2 \\ / \qquad \backslash \\ R_2Si \qquad SiR_2 \\ O \end{array} \xrightarrow[-H_2C{=}CH_2]{\Delta} \begin{array}{c} R_2Si{-}SiR_2 \\ \backslash \qquad / \\ O \qquad S \\ Si \\ R_2 \end{array} + (R_2SiS)_2 \qquad (9)$$
$$(2) \qquad\qquad\qquad\qquad (1)$$
$$(11)$$

The presence of (1), R = Me, in the product suggests that the four-membered ring Si_2S_2 is considerably more thermally stable than the six-membered Si_3S_3 ring.

Bibliography

B 1 I. Haiduc, *The Chemistry of Inorganic Ring Systems*, Part 1. (Wiley-Interscience, London, 1970), p. 579.

B 2 D. Brandes, *J. Organomet. Chem. Libr.* **7** (1979) 257 (organosilicon derivatives of sulphur, selenium, tellurium).

B 3 V. Chvalovsky *et al.*, *Organosilicon Compounds* (Czechoslovak Academy of Sciences), Vols. 3 (1973), 4 (1973), 5 (1977), 6 (1978/9), 7 (1980) (register of all known organosilicon compounds, including those with sulphur and selenium).

B 4 S. N. Borisov, M. G. Voronkov and E. Ya. Lukevits, *Organosilicon Derivatives of Phosphorus and Sulphur* (Plenum Press, New York, 1971).

B 5 B. W. Glawincewski and J. E. Drake, *Rev. Silicon, Germanium, Tin, Lead Compounds* **3** (1978) 279 (organosilicon compounds containing bonds to Group VI elements, including cyclosilathianes).

B 6 B. Krebs, *Angew. Chem.* **95** (1983) 113; *Angew. Chem. Int. Ed. Engl.* **22** (1983) 113 (thio- and seleno-compounds of main group elements).

B 7 A. Haas and R. Hitze, in *Sulphur in Organic and Inorganic Chemistry*, Vol. 4, edited by H. Senning (Dekker, New York, 1982), p. 1.

References

1   A. Haas and M. Vongehr, *Chem. Ztg* **99** (1975) 432.
2   B. J. Aylett, I. A. Ellis and J. R. Richmond, *J. Chem. Soc., Dalton Trans.* **1973**, 981.
3   C. Lau and J. C. Thompson, *Inorg. Nucl. Chem. Lett.* **13** (1977) 433.
4   M. Weidenbruch and G. Roettig, *Inorg. Nucl. Chem. Lett.* **13** (1977) 85.
5   V. O. Reikhsfeld and E. P. Lebedev, *Zh. Obshch. Khim.* **37** (1967) 1412.
6   D. A. Armitage, M. J. Clark and A. W. Sinden, *Inorg. Synth.* **15** (1974) 207.
7   M. M. Millard, L. J. Pazdernik, W. F. Haddon and R. E. Lundin, *J. Organomet. Chem.* **52** (1973) 283.
8   M. M. Millard, K. Steel and L. R. Pazdernik, *J. Organomet. Chem.* **13** (1968) P 7.
9   M. M. Millard and L. J. Pazdernik, *J. Organomet. Chem.* **51** (1973) 135.
10 V. O. Reikhsfeld and E. P. Lebedev, *Zh. Obshch. Khim.* **40** (1970) 615.
11 E. P. Lebedev, D. V. Fridland and V. O. Reikhsfeld, *Zh. Obshch. Khim.* **45** (1975) 2641.
12 L. R. J. Pazdernik, Ph. D. thesis, University of Iowa (1970) (*Diss. Abstr.* **31** (1970) 3244B).
13 E. P. Lebedev, D. V. Fridland and V. O. Reikhsfeld, *Zh. Obshch. Khim.* **44** (1974) 2784.
14 E. P. Lebedev, D. V. Frilland, V. O. Reikhsfeld and E. N. Korol, *Zh. Obshch. Khim.* **46** (1976) 315.
15 E. P. Lebedev and D. V. Fridland, *USSR Pat.* 395 368 (1973) (*CA* **80** (1974) 27371k).
16 E. P. Lebedev, V. O. Reikhsfeld and D. V. Fridland, *Zh. Obshch. Khim.* **43** (1973) 683.
17 E. P. Lebedev, M. M. Frenkel and E. N. Korol, *Zh. Obshch. Khim.* **47** (1977) 1510.
17a E. P. Lebedev, M. M. Frenkel, V. O. Reikhsfeld and D. V. Fridland, *Zh. Obshch. Khim.* **47** (1977) 1424.
18 F. Fehér and R. Lüpschen, *Z. Naturforsch.* **26b** (1971) 1191.
19 M. Weidenbruch, A. Schäfer and R. Rankers, *J. Organomet. Chem.* **195** (1980) 171.
20 L. H. Sommer and J. McLick, *J. Organomet. Chem.* **101** (1975) 171.
21 W. Wojnowski and M. Wojnowska, *Z. Anorg. Allg. Chem.* **389** 1972) 302.
22 W. Wojnowski and M. Wojnowska, *Z. Anorg. Allg. Chem.* **397** (1973) 69.
23 L. R. J. Pazdernik, F. Brisse and R. Rivest, *Acta Crystallogr.* **33B** (1977) 1780.

24 J. C. J. Bart and J. J. Daly, *Chem. Commun.* **1968**, 1207.
25 J. C. J. Bart and J. J. Daly, *J. Chem. Soc., Dalton Trans.* **1975** 2063.
26 F. Fehér and H. Goller, *Z. Naturforsch.* **25b** (1970) 250.
27 E. P. Lebedev and V. A. Baburina, *Zh. Obshch. Khim.* **45** (1975) 1648.
28 V. O. Reikhsfeld and E. P. Lebedev, *Zh. Obshch. Khim.* **39** (1969) 221.
29 V. O. Reikhsfeld and E. P. Lebedev, *Zh. Obshch. Khim.* **40** (1970) 2052.
30 D. V. Fridland, E. P. Lebedev and V. O. Reikhsfeld, *Zh. Obshch. Khim.* **46** (1976) 326.
31 D. V. Fridland, E. P. Lebedev and V. O. Reikhsfeld, *Zh. Obshch. Khim.* **47** (1977) 1504.
32 E. P. Lebedev, V. O. Reikhsfeld and I. E. Saratov, *Khimiya i Prakt. Primenenie Kremnii i Fosfororganich. Soedinenii,* **1977** 96 (*CA* **89** (1978) 215485k).
33 E. P. Lebedev and V. O. Reikhsfeld, *Zh. Obshch. Khim.* **40** (1970) 2056.
34 V. A. Baburina and E. P. Lebedev, *Zh. Obshch. Khim.* **46** (1976) 1782.
35 V. A. Baburina and E. P. Lebedev, *Zh. Obshch. Khim.* **48** (1978) 125.
36 E. P. Lebedev and V. A. Baburina, *USSR Pat.* 503 880 (1976) (*CA* **84** (1976) 135299j).
37 W. Wojnowski and M. Wojnowska, *Z. Anorg. Allg. Chem.* **398** (1973) 167.
38 E. P. Lebedev and V. A. Baburina, *Zh. Obshch. Khim.* **45** (1975) 2431.
39 T. Sakakibara, T. Hirabayashi and Y. Ishii, *J. Organomet. Chem.* **46** (1972) 231.
40 K. Moedritzer, *Inorg. Chim. Acta* **29** (1978) 249.
41 K. Moedritzer, J. R. Van Wazer and D. W. Matula, *Inorg. Chim. Acta* **3** (1969) 559.
42 K. Moedritzer, *J. Organomet. Chem.* **21** (1970) 315.
43 A. D. M. Hailey and G. Nickless, *J. Inorg. Nucl. Chem.* **31** (1971) 657.
44 H. S. D. Soysa and W. P. Weber, *J. Organomet. Chem.* **165** (1979) C1.
45 H. S. D. Soysa, I. N. Jung and W. P. Weber, *J. Organomet. Chem.* **171** (1979) 177.
46 H. Nöth, H. Fußstetter, H. Pommerening and T. Taeger, *Chem. Ber.* **113** (1980) 342.
47 G. V. Leplanin, V. N. Salimgareeva and S. R. Rafikov, *Dokl. Akad. Nauk SSSR* **240** (1978) 633.
48 S. R. Rafikov, S. S. Ibragimov, G. V. Leplyanin, V. N. Salimgareeva, S. P. Pivovarov, A. I. Polyakov and N. L. Filippov, *Vysokomol. Soedin.* **A23** (1981) 540.
49 H. S. D. Soysa and W. P. Weber, *Tetrahedron Lett.* **1978** 235.
50 M. D. Mizhiritskii and E. P. Lebedev, *USSR Pat.* 785 326 (1980) (*CA* **94** (1981) 104351b).
51 E. P. Lebedev and M. M. Frenkel, *USSR Pat.* 536 186 (1976) (*CA* **86** (1977) 171531y).
52 E. P. Lebedev and M. M. Frenkel, *Zh. Obshch. Khim.* **47** (1977) 2275.
52a E. P. Lebedev and M. M. Frenkel, *USSR Pat.* 514 839 (1976) (*CA* **85** (1976) 63164g).
53 E. P. Lebedev, M. D. Mizhiritskii, V. A. Baburina and S. I. Zaripov, *Zh. Obshch. Khim.* **49** (1979) 1084.
54 J. Peters, J. Mandt, M. Meyring and B. Krebs, *Z. Kristallogr.* **156** (1981) 90.
55 H. G. Horn and M. Hemeke, *Chem. Ztg* **106** (1982) 263.
56 W. E. Shklover, Yu. T. Struchkov, L. E. Guselnikov, W. W. Wolkowa and W. G. Awakyan, *Z. Anorg. Allg. Chem.* **501** (1983) 153.

12

Germanium Homocycles (Cyclopolygermanes) and Related Heterocycles

IONEL HAIDUC

Babes-Bolyai University, Cluj-Napoca, Romania

MARTIN DRÄGER

Gutenberg University, Mainz, Federal Republic of Germany

Cyclopolygermanes of varying ring sizes, containing from three to seven germanium atoms, (1)–(5), are presently known. The progress made in the field since 1970 involves mainly determination of molecular structures by diffraction methods, in addition to the synthesis of some novel rings, especially heterocycles. Although the chemistry of cyclogermanes is very similar to that of cyclopolysilanes, the germanium chemistry has been much less thoroughly explored. The known cyclopolygermanes are illustrated by (1)–(5). A variety of heterocyclogermanes can be obtained by introducing heteroatoms and preserving some Ge–Ge bonds, as illustrated by formulae (6)–(11).

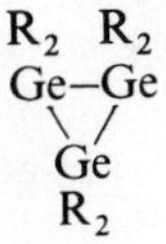

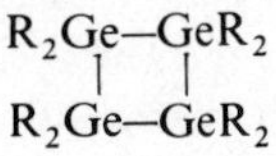

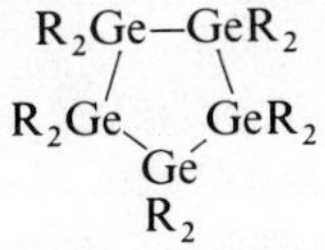

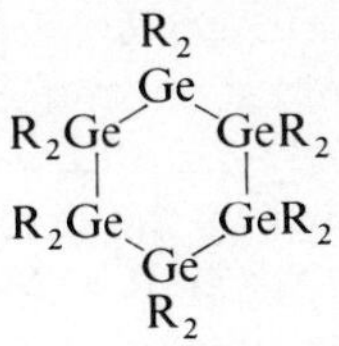

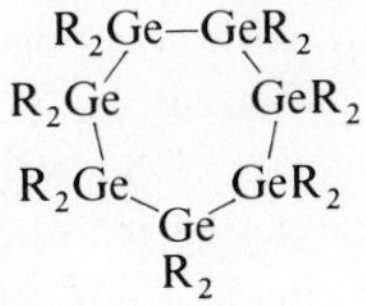

$$\begin{array}{ccccc}
(1) & (2) & (3) & (4) & (5)
\end{array}$$

THE CHEMISTRY OF INORGANIC HOMO-
AND HETEROCYCLES VOL 1 ISBN 0-12-655775-6

$$
\begin{array}{cccc}
R_2Ge\!-\!GeR_2 & R_2Ge\!-\!GeR_2 & R_2Ge\!-\!GeR_2 & R_2Ge\!-\!GeR_2 \\
R_2Ge\diagdown\!\diagup GeR_2 & R_2Ge\diagdown\!\diagup GeR_2 & R_2Ge\diagdown\!\diagup GeR_2 & R_2Ge\diagdown\!\diagup GeR_2 \\
O & S & Se & Te \\
(6) & (7) & (8) & (9)
\end{array}
$$

$$
\begin{array}{cc}
R_2Ge\!-\!GeR_2 & R_2Ge \overset{O}{\diagup\diagdown} GeR_2 \\
R_2Ge\diagdown\!\diagup GeR_2 & R_2Ge\diagdown_{\;O}\!\diagup GeR_2 \\
\underset{R}{As} & (11) \\
(10)
\end{array}
$$

The absence of five-membered rings with X = N or P and of six-membered rings with X = S, Se, N, P, As as heteroatoms is surprising. These rings should exist.

The chemistry of cyclopolygermanes and related heterocycles has been reviewed.[B 1 – B 4]

12.1   Cyclopolygermanes

Cyclotrigermanes Ge_3R_6 (**1**) are difficult to obtain and can be prepared only with bulky groups (e.g. R = 3,5-dimethylphenyl) at germanium by reducing the corresponding chloride.[1] See also ref. 20.

Higher cyclogermanes are frequently formed as mixtures of derivatives with various ring sizes and will be discussed together. Permethylated derivatives $(GeMe_2)_n$ are known for $n = 4$–7. The tetramer Ge_4Me_8 (**2**), R = Me, is formed in the reaction of $HGeX_3$ with methyllithium,[2] while higher members (**3**)–(**5**), R = Me, were prepared by reduction of Me_2GeCl_2 with lithium metal in tetrahydrofuran.[2–4] The reduction of dihalides was also used for the preparation of octaphenylcyclotetragermane Ge_4Ph_8 (**2**), R = Ph,[5] and of the *p*-tolyl analogue (**2**), R = *p*-tolyl, using sodium metal.[6] The pentamer (**3**), R = *p*-tolyl, and hexamer (**4**), R = *p*-tolyl, were obtained by reducing R_2GeCl_2 with sodium naphthalenide.[6]

The isopropyl tetramer $Ge_4Pr^i_8$ (**2**), R = Pr^i, was formed along with other compounds in the reaction of $GeCl_4$ with Pr^iMgCl, magnesium metal and $LiAlH_4$, but identified only through its mass spectrum.[7]

Cyclopolygermanes are stable toward atmospheric oxygen and water. Only the permethylated pentamer Ge_5Me_{10} is reported to undergo oxidation in air, while the methylated hexamer and heptamer are stable. The cyclogermanes can be cleaved by halogens; thus controlled cleavage of Ge_4Ph_8 by iodine gives 1,4-diiodooctaphenyltetragermane $I(GePh_4)_2I$, which is useful for the synthesis of heterocyclic polygermanes.[8]

Cyclopolygermanes are colourless crystalline solids, and exhibit a tendency to form several polymorphic modifications. Thus the pentamer Ge_5Ph_{10} is known in monoclinic, triclinic and tetragonal forms.[9,10] Both the pentamer Ge_5Ph_{10} and the hexamer Ge_6Ph_{12} exhibit plastic crystalline states before melting.[9–11]

Several molecular structures of cyclopolygermanes have been determined by X-ray diffraction (the illustrations can be found in ref. B 3). Octaphenyl-cyclotetragermane Ge_4Ph_8 contains a planar Ge_4 ring, and the phenyl groups are situated in perpendicular planes. The Ge–Ge bond distance is 2.458–2.472 Å and the GeGeGe bond angles in the ring are in the range 89.4–90.2°.[5] In decaphenylcyclopentagermane Ge_5Ph_{10}[9] and in its benzene adduct $Ge_5Ph_{10}.C_6H_6$[12] different molecular conformations were found. The ring conformation can be either envelope (C_s) or half-chair (C_2) and the Ge–Ge interatomic distances are in the range 2.440–2.473 Å in Ge_5Ph_{10}[9] and 2.438–2.473 Å in $Ge_5Ph_{10}.C_6H_6$.[12] The bond angles GeGeGe are in the ranges 103.5–107.5° and 102.5–106.1° respectively.

Dodecamethylcyclohexagermane Ge_6Me_{12} contains a Ge_6 ring in a chair conformation.[13] The Ge–Ge bond lengths appear to be somewhat shorter than in other cyclogermanes (2.365–2.377 Å), but this may be due to the difficulties encountered during the solving of the structure. The GeGeGe bond angles are in the range 111.5–113.2°.[13] The perphenylated hexamer crystallizes with solvent molecules, and diffraction investigations of $Ge_6Ph_{12}.7C_6H_6$[11] and $Ge_6Ph_{12}.2C_6H_5CH_3$[14] have been reported. In both compounds the Ge_6 ring has the chair conformation. The Ge–Ge interatomic distances are c. 2.46 Å, and the angles are slightly larger than tetrahedral (111.4–114.7°).

The crystal packing of cyclogermane molecules has been analysed and discussed in terms of the theory of closest packing.[10,14,15]

Spectroscopic techniques have been routinely used for characterization of cyclopolygermanes. The cyclic compounds $(GePh_2)_n$ ($n = 4, 5, 6$) exhibit a ring-size dependence of vibrational bands[15] and the ^{13}C NMR chemical shifts for the phenyl substituents,[5] which is useful in identifying these compounds. The ^{1}H NMR spectra are complex and difficult to interpret.

12.2 Heterocyclogermanes

Most germanium heterocycles with Ge–Ge bonds are five-membered rings. *Octaphenylcyclooxatetragermane* (6) is formed in the hydrolysis of 1,4-diiodooctaphenylcyclotetragermane $I(GePh_2)_4I$ in the presence of a base.[8] It is a crystalline solid (m.p. 220 °C).

Octaphenylcyclothiotetragermane (7) (m.p. 179 °C) and *octaphenylcyclo-selenatetragermane* (8) (m.p. 189 °C, dec.) are formed in similar reactions of the same diiodide $I(GePh_2)_4I$ with sodium sulphide and hydrogen selenide respectively.[8] The structures of the two compounds have been determined by X-ray diffraction. In Ph_8Ge_4S the five-membered ring exhibits an envelope conformation with Ge–Ge bonds 2.443–2.454 Å, GeGeGe bond angle 100.1–102.5°, GeGeS 105.4–109.5°.[8] In Ph_8Ge_4Se the ring is not far from planarity and exhibits a twist conformation, with a twofold axis, passing through the selenium atom and the opposite Ge–Ge bond.[16] The Ge–Ge distances are in the range 2.437–2.448 Å and the angles are GeGeGe 105.8° and GeGeSe 110.4°.[16]

Octaphenylcyclotelluratetragermane (9) (dec. 50 °C), a very unstable and air-sensitive compound, was formed in the reaction of $I(GePh_2)_4I$ with NaHTe in alcoholic solution.[8]

Nonaphenylcycloarsatetragermane (10) (m.p. 100 °C, dec.) is similarly obtained in the reaction of $I(GePh_2)_4I$ with $PhAsLi_2$ in toluene–THF. It hydrolyses slowly in air to form Ph_8Ge_4O (6).[10]

Two derivatives of the six-membered heterocycle *cyclo-1,4-dioxatetra-germane* (11), with $R = Bu^n$ and $R = Ph$, were obtained by hydrolysis of 1,2-dihalodigermanes XR_2Ge—GeR_2X ($R = Bu^n$, $X = Cl$;[17] $R = Ph$, $X = Br^{18}$) or by hydrolysis of the analogous bis(trichloroacetate) Cl_3CCOO—$GePh_2$—$GePh_2OCOCCl_3$.[19] The n-butyl derivative (11), $R = Bu^n$, is a liquid (b.p. 132 °C) and the phenyl derivative (11), $R = Ph$, a crystalline solid (m.p. 216–217 °C).

Bibliography

B 1   I. Haiduc, *The Chemistry of Inorganic Ring Systems*, Part 1 (Wiley-Interscience, London, 1970) p. 66.

B 2   A. L. Rheingold, *Homoatomic Rings, Chains and Macromolecules of Main Group Elements* (Elsevier, Amsterdam, 1977).

B 3   M. Dräger, L. Ross and D. Simon, *Rev. Silicon, Germanium, Tin, Lead Compounds* **7** (1983) 299.

B 4   K. M. MacKay and R. Watt, *Organomet. Chem. Rev.* **A4** (1969) 137.

References

1 S. Masamune, Y. Hanzawa and D. J. Williams, *J. Am. Chem. Soc.* **104** (1982) 6136.
2 E. Carberry and B. D. Dombek, *Organosilicon Symp. Pittsburg, Pa., 27 March 1971.*
3 E. Carberry and B. D. Dombek, *J. Organomet. Chem.* **22** (1970) C 43.
4 E. Carberry and B. D. Dombek, *J. Organomet. Chem.* **36** (1972) 61.
5 L. Ross and M. Dräger, *J. Organomet. Chem.* **199** (1980) 195.
6 M. Richter and W. P. Neumann, *J. Organomet. Chem.* **20** (1969) 81.
7 A. Carrick and F. Glockling, *J. Chem. Soc.* A **1966**, 623.
8 L. Ross and M. Dräger, *J. Organomet. Chem.* **194** (1980) 23.
9 M. Dräger and L. Ross, *Z. Kristallogr.* **154** (1980) 265; *Z. Naturforsch.*, **38b** (1983) 665.
10 L. Ross, Thesis, Universität Mainz (1980).
11 M. Dräger, L. Ross and D. Simon, *Z. Anorg. Allg. Chem.* **466** (1980) 145.
12 L. Ross and M. Dräger, *Z. Anorg. Allg. Chem.* **519** (1984) 225.
13 W. Jensen, R. Jacobson and J. Benson, *Cryst. Struct. Commun.* **4** (1975) 299.
14 M. Dräger and L. Ross, *Z. Anorg. Allg. Chem.* **476** (1981) 95.
15 M. Dräger and L. Ross, *Z. Anorg. Allg. Chem.* **515** (1984) 141.
16 L. Ross and M. Dräger, *Z. Anorg. Allg. Chem.* **472** (1981) 109.
17 E. J. Bulten and J. G. Noltes, *Tetrahedron Lett.* **1966**, 3471.
18 K. Kühlein, Ph. D. thesis, Universität Giessen (1966).
19 F. Glockling and R. E. Houston, *J. Chem. Soc. Dalton Trans.*, **1973**, 1357.
20 J. T. Snow, S. Murakani, S. Masamune and D. J. Williams, *Tetrahedron Lett.* **25** (1984) 4191.

13

Germanium-Containing Heterocycles

IONEL HAIDUC

Babes-Bolyai University, Cluj-Napoca, Romania

13.1 Introduction

This chapter deals with heterocycles in which germanium atoms alternate with nitrogen, phosphorus, oxygen, sulphur, selenium and tellurium. Those containing Ge–Ge bonds are regarded as derivatives of germanium homocycles and are discussed in Chapter 12.

An overview of germanium-containing heterocycles is given by the following formulae, which illustrate the known rings of various sizes (1)–(16) and polycyclic systems (17)–(19) to be discussed in this chapter:

THE CHEMISTRY OF INORGANIC HOMO-
AND HETEROCYCLES VOL 1 ISBN 0-12-655775-6

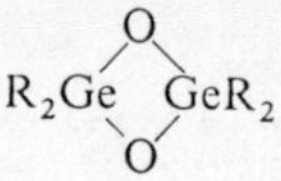

$$R_2Ge \underset{O}{\overset{O}{<>}} GeR_2$$

(6)

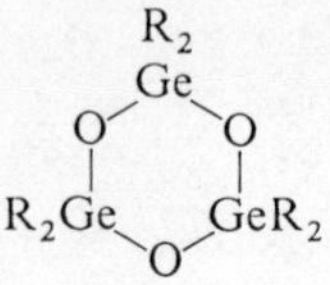

$$\begin{array}{c} R_2 \\ Ge \\ O \quad O \\ R_2Ge \underset{O}{\qquad} GeR_2 \end{array}$$

(7)

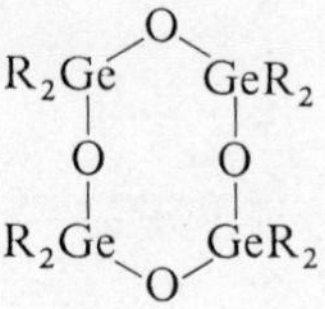

$$\begin{array}{c} R_2Ge \overset{O}{\qquad} GeR_2 \\ O \qquad O \\ R_2Ge \underset{O}{\qquad} GeR_2 \end{array}$$

(8)

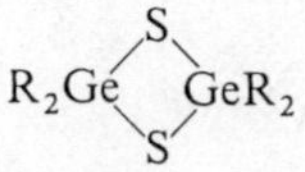

$$R_2Ge \underset{S}{\overset{S}{<>}} GeR_2$$

(9)

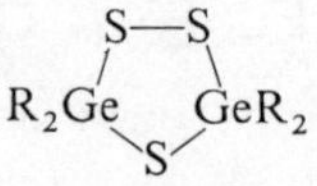

$$R_2Ge \overset{S-S}{<\qquad>} GeR_2 \\ S$$

(10)

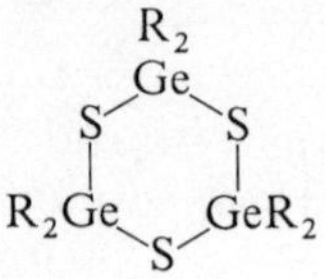

$$\begin{array}{c} R_2 \\ Ge \\ S \quad S \\ R_2Ge \underset{S}{\qquad} GeR_2 \end{array}$$

(11)

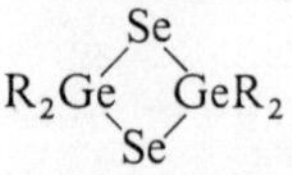

$$R_2Ge \underset{Se}{\overset{Se}{<>}} GeR_2$$

(12)

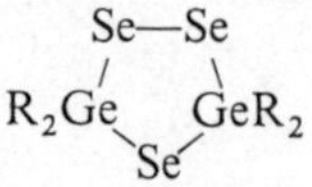

$$R_2Ge \overset{Se-Se}{<\qquad>} GeR_2 \\ Se$$

(13)

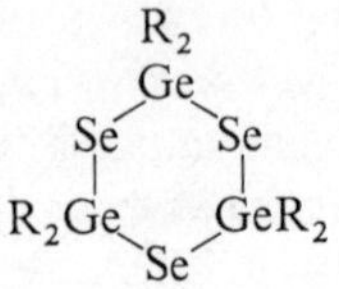

$$\begin{array}{c} R_2 \\ Ge \\ Se \quad Se \\ R_2Ge \underset{Se}{\qquad} GeR_2 \end{array}$$

(14)

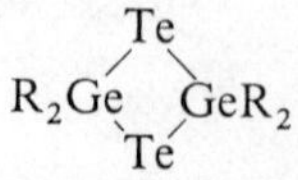

$$R_2Ge \underset{Te}{\overset{Te}{<>}} GeR_2$$

(15)

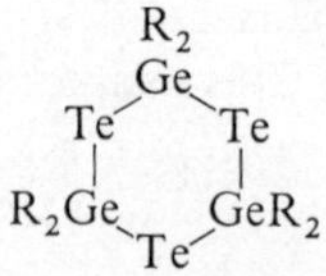

$$\begin{array}{c} R_2 \\ Ge \\ Te \quad Te \\ R_2Ge \underset{Te}{\qquad} GeR_2 \end{array}$$

(16)

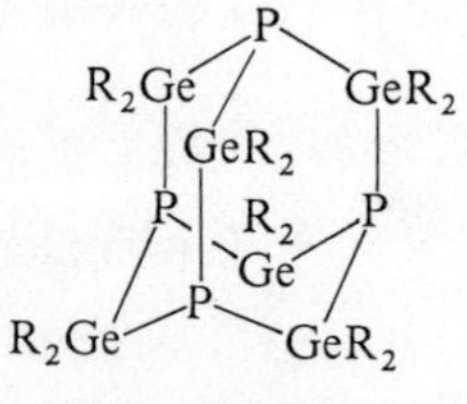

$$\begin{array}{c} P \\ R_2Ge \qquad GeR_2 \\ GeR_2 \\ P \quad \underset{Ge}{R_2} \quad P \\ R_2Ge \underset{P}{\qquad} GeR_2 \end{array}$$

(17)

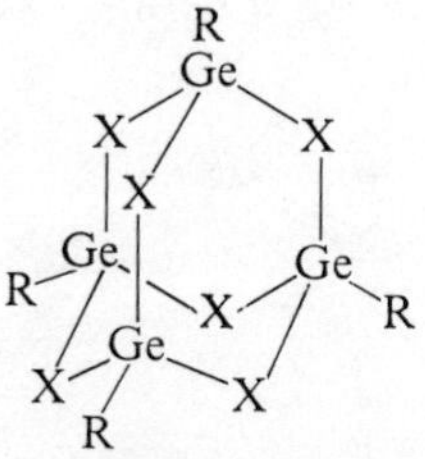

$$\begin{array}{c} R \\ Ge \\ X \qquad X \\ X \\ R Ge \qquad Ge R \\ Ge \quad X \\ X \quad Ge \quad X \\ R \end{array}$$

(18a) X = O
(18b) X = S
(18c) X = Se

$$\begin{array}{c} R' \quad R' \\ N \qquad N \\ R_2Ge \quad Ge \quad GeR_2 \\ N \qquad N \\ R' \quad R' \end{array}$$

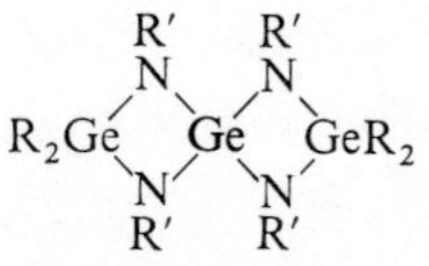

(19)

The early chemistry of germanium-containing rings has been reviewed,[B1–B7] but progress in the field has been significant in recent years.

13.2 Germanium–nitrogen heterocycles

Cyclodigermazanes have been prepared by treating dichlorodiphenylgermane, Ph_2GeCl_2 with N-dilithiopentafluoroaniline $C_6F_5NLi_2$, when (1),

$R = Ph$, $R' = C_6F_5$, was isolated in good yield.[1] Treatment of $Me_2Ge(NLi-SiEt_3)_2$ with Me_2GeCl_2 gave (1), $R = Me$, $R' = SiEt_3$,[2] whereas the reaction with $GeCl_4$ gave a spirobicyclic compound (19), $R = Me$, $R' = SiEt_3$.[2]

The monomer $Ph_2Ge{=}NMe$, formed as a transient intermediate in the pyrolysis of the appropriate azagermacyclobutane (with elimination of ethylene), dimerizes with formation of (1), $R = Ph$, $R' = Me$; the same compound was prepared by a direct route from Ph_2GeCl_2 and methylamine.[3] Cyclic $(F_2GeNPh)_n$ derivatives (dimer and trimer) are formed by cyclization of the transient monomer $F_2Ge{=}NPh$, generated in the reaction of GeF_2 with phenyl azide.[4]

The halogermylphosphinimines, formed as shown in equation (1), were found to dimerize (equation 2). Two forms of the cyclic dimers (1a), $R = Me$, Bu or Cl, $R' = Me$ or Et, $X = Cl$ or Br, and (1b), $R = Me$, $R' = Me$, Et, $X = Cl$, Br or I, can be isolated. A monomer–dimer equilibrium occurs in solution, but the formation of the ionic compounds (1b) is irreversible.[5,6]

$$2\,R_2GeX_2 + Me_3Si{-}N{=}PR_3' \xrightarrow[-Me_3SiCl]{} XR_2Ge{-}N{=}PR_3' \qquad (1)$$

(1a)   (1b)

$$\qquad (2)$$

An X-ray diffraction study of (1a), $R = X = Cl$, $R' = Me$, confirmed the presence of trigonal bipyramidal, five-coordinate germanium in a four-membered planar Ge_2N_2 ring.[7]

A six-membered cyclotrigermazane (2), $R = Cl$, $R' = Me$, was obtained in the reaction of $GeCl_4$ with methylamine,[8-10] and similar derivatives (2), $R = R' = Me$,[11,12] $R = Et$, $R' = Me$,[12] $R = Bu$, $R' = H$,[13] and $R = Me$, $R' = Ph$,[14] were prepared in reactions of R_2GeCl_2 with amines or ammonia. The reaction of Me_2GeCl_2 with a lithiated 1,1-dimethylhydrazine $LiNHNMe_2$ also gave a cyclotrigermazane (2), $R = Me$, $R' = NMe_2$.[15]

The ammonolysis of $GePh_4$ in liquid ammonia in the presence of KNH_2 gave a product of composition $K_3[Ge_4N_8H_5]$, believed to contain an eight-membered Ge_4N_4 ring; intermediate formation of the tetramer $(Ph_2GeNH)_4$ was also postulated.[16] However, there is no other proof of the existence of an eight-membered Ge_4N_4 ring, and this interpretation must be viewed with caution.

The molecular structure of $(Cl_2GeNMe)_3$ (2), R = Cl, R' = Me, was determined by X-ray diffraction; the compound contains a nearly planar six-membered Ge_3N_3 ring with Ge–N 1.780–1.864 Å, GeNGe 126.3°, NGeN 114.9° and ClGeCl 103.0°.[17]

The basic properties of two cyclotrigermazane derivatives (2), R = Me, Et, R' = Me, were estimated by measuring the infrared $\Delta\nu_{C-D}$ frequency shift of deuterochloroform upon hydrogen-bond formation. The conclusion was that (p–d)π bonding in Ge–N is less than in Si–N bonds.[12]

No uses are known for cyclogermazanes, but the compound $(Cl_2GeNMe)_3$ has been patented as a potential pesticide and herbicide.[10]

Several ring systems containing germanium, nitrogen and a third heteroatom have been reported. Thus the reaction of $Me_2Ge(NSiMe_3Li)_2$ with organoelement dihalides and with Group IV tetrahalides gave four-membered (20) and spirocyclic (21) compounds respectively:[2,18]

$$R_2Ge \underset{\underset{R'}{N}}{\overset{\overset{R'}{N}}{\diagup\diagdown}} ER''_n \qquad R_2Ge \underset{\underset{R'}{N}}{\overset{\overset{R'}{N}}{\diagup\diagdown}} E \underset{\underset{R'}{N}}{\overset{\overset{R'}{N}}{\diagdown\diagup}} GeR_2$$

(20a) E = B
(20b) E = Si
(20c) E = Sn

(21a) E = Si
(21b) E = Sn

Other monocyclic[19-21] and spirocyclic[22-24] compounds containing Ge–N–Si units have already been mentioned in Section 8.5.1 and 8.9.2. of this book.

13.3 Germanium–phosphorus heterocycles

A four-membered cyclodigermaphosphane (3), R = R' = Ph, has been prepared from Ph_2GeCl_2 and KHPPh or K_2PPh; the latter reagent favours the formation of the trimer (4), R = R' = Ph.[25] A trimer (4), R = Me, R' = Ph, was also obtained from Me_2GeCl_2 and phenylphosphine in the presence of NEt_3.[25] An isomeric four-membered ring system (5), R = Ph, R' = Bu^t, was obtained from Ph_2GeCl_2 and $K(Bu^tP—PBu^t)K$.[26] An X-ray diffraction study confirmed the structure (Ge–Ge 2.421 Å, Ge–P 2.332–2.347 Å, GeGeP 82.8–83.1°).[27]

A mixture of 75% dimeric $(Me_2GePPh)_2$ (3), R = Me, R' = Ph, and 25% trimeric $(Me_2GePPh)_3$ (4), R = Me, R' = Ph, was found to result from the reaction of Me_2GeCl_2 with 2-phenyl-1,3-tetramethyl-1,3-disila-2-phospha-cyclopentane, via the intermediate formation of the transient monomer $Me_2Ge\!=\!PPh$.[28]

Two polycyclic (cage) hexagermatetraphosphanes $(R_2Ge)_6P_4$ with adamantane structures (17), R = Me and Et, have been formed in high yield in
the mercury-catalysed thermal decomposition of $R_2Ge(PH_2)_2$.[29,30] The
molecular structure of $(Me_2Ge)_6P_4$ has been resolved by an X-ray diffraction study (Ge–P 2.322 Å, PGeP 120.7°).[30]

A germanium–phosphorus compound $(PhGeP)_7$, probably having a polycyclic (cage) structure, has also been reported.[31]

13.4 Germanium–oxygen heterocycles

Four-membered ring cyclodigermoxanes (6), first reported in 1961, have
since not been mentioned in the literature. Only an X-ray investigation of
sodium metagermanate Na_2GeO_3 showed the presence of the $[Ge_2O_6]^{4-}$
anion, containing a Ge_2O_2 ring (6), R = O^-.[32]

Six-membered ring cyclotrigermoxanes (7), R = Bu^n,[33] Ph,[34] C_6F_5[35]) are
formed in the hydrolysis of diorganogermanium dihalides, a reaction that
also gives tetrameric germoxanes (8), R = Ph,[36] Pr^i.[36] The reported
tetramer $[(C_6F_5)_2GeO]_4$[37] was later identified as a trimer (7), R = C_6F_5.[35]

The mass spectrum of $(Ph_2GeO)_3$ (7), R = Ph, was investigated and was
one of the early proofs of the six-membered ring structure.[34] A molecular-
structure determination of $(Ph_2GeO)_3$ performed later[38] showed it to
contain a nonplanar ring, intermediate between twisted-boat and chair
conformation. The tetramer (8), R = Ph, was also investigated by X-ray
diffraction, either as pure $(Ph_2GeO)_4$[39,40] or as a benzene adduct
$(Ph_2GeO)_4 \cdot 1/2C_6H_6$.[39] In both molecules the eight-membered Ge_4O_4 ring
is nonplanar, with S_4 symmetry, Ge–O c. 1.75 Å and OGeO 107–110°.

Some thermal properties of certain organocyclogermoxanes have been
investigated. Thus the enthalpy of formation of octamethylcyclotetrager-
moxane $(Me_2GeO)_4$ has been determined.[41,42] The heat capacity of
$(R_2GeO)_4$ (8), R = Me, Et, molecular crystals was measured at liquid-
nitrogen temperature. No phase transition was observed for $(Me_2GeO)_4$.[43]

Some cyclic germanium–oxygen anions were discovered by X-ray diffraction studies of germanates. Thus the barium germanate $BaGe_4O_9$ was found
to contain $[Ge_3O_9]^{6-}$ ring anions, consisting of three GeO_4 tetrahedra
sharing corners, in addition to GeO_6 octahedra.[44] A similar structure was
found in α-$PbGeO_4$[45] and in $K_2Ge_4O_9$.[46] The $[Ge_3O_9]^{6-}$ anion is also
present in $Bi_2[Ge_3O_9]$.[47] Polycyclic germanates are also possible, and one
such structure has been found in $Li_2[Ge_7O_{15}]$.[48] A bicyclic germoxane
structure (22) was found in the anion $[Ge_5O_{14}]^{8-}$, discovered in the
thallium salt $Tl_8Ge_5O_{14}$.[49] These few structure determinations promise a
large variety of germanate structures, including many possibly cyclic ones;

the tendency of germanium to achieve sixfold coordination will, however, complicate the matter.

Few chemical reactions of cyclogermoxanes have been investigated. The tetramer (8), R = Me, reacts with arsenic(III) fluoride to form Me_2GeF_2.[50] Several organofunctional "sesquigermoxanes" $(RGeO_{1.5})_n$ claimed to have antitumour and other therapeutical properties,[51-54] have been reported. The t-butyl derivative was found to have a hexameric cage structure.[76]

Few heterocyclogermoxanes were reported so far, but the oxidation of $Me_2Ge(PH_2)_2$ and the reaction of the same phosphide with H_3PO_4 gave an eight-membered cyclogermaphosphoxane (23), R = Me:[55]

(22)   (23)

13.5 Germanium–sulphur heterocycles

Cyclic dimers $(R_2GeS)_2$ are formed in the reaction of dibutyldiethoxygermanium $Bu_2Ge(OEt)_2$ with hydrogen sulphide[33] or by heating bis(pentafluorophenyl)germanium hydride $(C_6F_5)_2GeH_2$ with elemental sulphur.[56] The dimer (9), R = Bu^t, was obtained by several procedures, including the reaction of $Bu_2^tGeCl_2$ with KSH, condensation of $Bu_2^tGe(SH)_2$ and the reaction of $(Bu_2^tGeH)_2$ with elemental sulphur;[57] the five-membered ring compound $Bu_4^tGe_2S_3$ (10) R = Bu^t was formed in the reaction of $Bu_2^tGeCl_2$ with H_2S in the presence of imidazole, and as a by-product in the reaction with KSH.[57] The molecular structure of (9), R = Bu^t, was determined by X-ray diffraction.[57]

Thiogermanates, containing the $[Ge_2S_6]^{4-}$ anion (9), R = S$^-$, were prepared by dissolving GeS_2 in aqueous sodium sulphide, at pH 7.5–8.[58,59] These compounds were reported earlier in the literature as trithiogermanates $M_2^IGeS_3$.

A cyclotrigermathiane (11), R = Et, was formed in the reaction of $Et_2Ge.NEt_3$ with thiirane by trimerization of the monomeric intermediate $Et_2Ge{=}S$.[60,61]

$$Et_2Ge.NEt_3 + H_2C{-\!-}CH_2 \longrightarrow \left[\begin{matrix} Et_2Ge{-\!-}S \\ | \quad\quad | \\ H_2C{-\!-}CH_2 \end{matrix} \right] \xrightarrow{-C_2H_4} [Et_2Ge{=}S] \longrightarrow$$

$$\longrightarrow \tfrac{1}{3}\,(Et_2GeS)_3 \tag{3}$$

The same cyclic compound (11), R = Et, is obtained from the monomeric $Et_2Ge{=}S$ generated by treatment of 1,3-dithia-2-germacyclopentane with tributylphosphine.[62] The reverse reaction, i.e. the conversion of the cyclic trimer (11), R = Et, into the corresponding monomer $Et_2Ge{=}S$ occurs on heating at 160 °C[63] (see also refs. 64–66).

Monoorganogermanium moieties combine with sulphur to form poly-cyclic compounds with the adamantane structure (18b). Thus treatment of $PhGeCl_3$ with H_2S in the presence of triethylamine gave $(PhGe)_4S_6$ (18b), R = Ph,[67] and heating $C_6F_5GeH_3$ with elemental sulphur produced $(C_6F_5Ge)_4S_6$ (18b), R = C_6F_5.[56] This type of structure has been confirmed by an X-ray diffraction investigation of the methyl derivative $(MeGe)_4S_6$ (18b), R = Me; the molecular parameters are Ge–S 2.202–2.224 Å, SGeS 110.8–112.5°.[68]

Germanium thiohalides with an adamantane structure (18b, R = Br, I) are formed in the reactions of $GeBr_4$ or GeI_4 with H_2S in carbon disul-phide.[69,70] An anion $Ge_4S_{10}{}^{4-}$ with a similar structure (18b), R = S^-, has been identified in the caesium salt $Cs_4[Ge_4S_{10}].4H_2O$ (prepared from GeS_2 and Cs_2S in aqueous solution). Other alkali metal salts of this anion were prepared similarly.[71,72] The vibrational spectra (IR–Raman) of this anion have been thoroughly investigated.[73]

13.6 Germanium–selenium heterocycles

The four-membered ring compound $(Bu_2^tGeSe)_2$ (12), R = Bu^t, was obtained in the reaction of $(Bu_2^tGeH)_2$ with elemental selenium; in this reaction the five-membered ring (13), R = Bu^t, was formed as a by-product.[57]

A six-membered ring derivative (14), R = Ph, was prepared by the reaction of Ph_2GeCl_2 with H_2Se in the presence of triethylamine, while similar treatment of $PhGeCl_3$ gave the adamantane-like compound (18c), R = Ph.[74] Both compounds were analysed by X-ray diffraction. The trimer

(14) has a twisted-boat conformation and Ge–Se 2.342–2.363 Å, SeGeSe 112.1–114.2°. In the adamantane structure (18c) Ge–Se is 2.339–2.360 Å and SeGeSe 111.3–117.3°.[74]

An inorganic anion $Ge_4Se_{10}{}^{4-}$ (18c), R = Se⁻, was discovered in the thallium compound $Tl_4Ge_4Se_{10}$ (prepared by melting Tl_2Se with $GeSe_2$), and a full X-ray diffraction analysis was performed.[75]

13.7 Germanium–tellurium heterocycles

The four-membered ring compound (15), R = Bu^t, was formed in the reaction of $(Bu_2^tGeH)_2$ with elemental tellurium.[57]

A six-membered tellurium-containing ring is also known, namely the unstable trimer (16), R = Ph, prepared from Ph_2GeCl_2 and NaHTe, under rigorously anhydrous and oxygen-free conditions.[39] The compound decomposes on standing at room temperature, with liberation of free tellurium.[39]

Bibliography

B 1 I. Haiduc, *The Chemistry of Inorganic Ring Systems*, Part 1 (Wiley-Interscience, London, 1970), p. 594.

B 2 M. Dräger, L. Ross and D. Simon, *Rev. Silicon, Germanium, Tin, Lead Compounds* 7 (1983) 299 (germanium-containing inorganic rings).

B 3 M. Lesbré, P. Mazerolles, and J. Satgé, *The Organic Chemistry of Germanium* (Wiley-Interscience, New York, 1971).

B 4 O. J. Scherer, *Angew. Chem. Int. Ed. Engl.* 8 (1969) 861 (germanium–nitrogen rings).

B 5 H. Schumann, *Angew. Chem. Int. Ed. Engl.* 8 (1969) 937 (germanium–phosphorus rings).

B 6 J. Satgé, *Adv. Organomet. Chem.* 21 (1982) 241 (multiply bonded species with Ge=O, Ge=S, Ge=N, Ge=P bonds and their relationship with cyclic species).

B 7 B. Krebs, *Angew. Chem. Int. Ed. Engl.* 22 (1983) 113 (thio- and selenocompounds).

References

1 I. Haiduc and H. Gilman, *Syn. Inorg. Metalorg. Chem.* 1 (1971) 75.

2 O. J. Scherer and D. Biller, Unpublished results cited in ref. B 4.

3 M. Rivière-Baudet, P. Rivière and J. Satgé, *J. Organomet. Chem.* 154 (1978) C 23.

4 P. Rivière, A. Caze, A. Castel, M. Rivière-Baudet and J. Satgé, *J. Organomet. Chem.* 155 (1978) C 58.

5 W. Wolfsberger and H. H. Pickel, *J. Organomet. Chem.* 54 (1973) C8.

6 W. Wolfsberger and H. H. Pickel, *J. Organomet. Chem.* 145 (1978) 29.

7 W. S. Sheldrick, D. Schomburg and W. Wolfsberger, *Z. Naturforsch* 33b (1978) 493.

8 W. Eisenhut and J. R. Van Wazer, *Inorg. Nucl. Chem. Lett.* 3 (1967) 359.

9 W. Eisenhut and J. R. Van Wazer, *Inorg. Chem.* 7 (1968) 1642.

10   W. Eisenhut, K. Moedritzer and J. R. Van Wazer, *US Pat.* 3 487 096 (1969) (*CA* **72** (1970) 66367t).
11   I. Schumann-Ruidisch and B. Jutzi-Mebert, *J. Organomet. Chem.* **11** (1968) 77.
12   Z. Pacl, M. Jakoubková, V. Bažant, and V. Chvalovský, *Coll. Czech. Chem. Commun.* **36** (1971) 1682.
13   F. Rijkens, M. J. Janssen and G. J. M. Van der Kerk, *Rec. Trav. Chim. Pays-Bas* **84** (1965) 1597.
14   A. H. Bates, Ph.D. thesis, Harvard University (1970) (*Diss. Abstr.* **31** (1971) 6473B).
15   L. K. Peterson and K. I. The, *Can. J. Chem.* **50** (1972) 553.
16   O. Schmitz-Dumont and W. Jansen, *Z. Anorg. Allg. Chem.* **363** (1968) 140.
17   M. Ziegler and J. Weiss, *Z. Naturforsch.* **26b** (1971) 735.
18   O. J. Scherer and D. Biller, *Z. Naturforsch.* **22b** (1967) 1079; *Angew. Chem.* **79** (1967) 410.
19   U. Wannagat, E. Bogusch and R. Braun, *J. Organomet. Chem.* **19** (1969) 367.
20   U. Wannagat and L. Gerschler, *Z. Anorg. Allg. Chem.* **383** (1971) 249.
21   U. Wannagat and F. Rabet, *Inorg. Nucl. Chem. Lett.* **6** (1970) 155.
22   M. Schlingmann and U. Wannagat, *Chem. Ztg* **98** (1974) 458.
23   U. Wannagat, R. Seifert and M. Schlingmann, *Z. Naturforsch.* **32b** (1974) 869.
24   U. Wannagat and R. Seifert, *Monatsh. Chem.* **109** (1978) 209.
25   H. Schumann and H. Benda, *Chem. Ber.* **104** (1971) 333.
26   M. Baudler and H. Benda, Unpublished results cited in ref. B 2.
27   R. Fröhlich and K. F. Tebbe, *Z. Anorg. Allg. Chem.* **506** (1983) 27.
28   C. Couret, J. Satgé, J. D. Andriamizaka and J. Escudié, *J. Organomet. Chem.* **157** (1978) C 35.
29   A. Dahl and A. D. Norman, *J. Am. Chem. Soc.* **92** (1970) 5525.
30   A. Dahl, A. D. Norman, H. Shenav and R. Schaeffer, *J. Am. Chem. Soc.* **97** (1975) 6367.
31   H. Schumann and H. Benda, *J. Organomet. Chem.* **21** (1970) P12.
32   D. W. Cruickshank, A. Kalman and J. S. Stephens, *Acta Crystallogr.* **B34** (1978) 1333.
33   S. Mathur, G. Chandra, A. K. Rai and R. C. Mehrotra, *J. Organomet. Chem.* **4** (1965) 294.
34   F. Glockling and R. E. Houston, *J. Chem. Soc., Dalton Trans.* **1973**, 1357.
35   M. Weidenbruch and N. Wessal, *Chem. Ber.* **105** (1972) 173.
36   A. Carrick and F. Glockling, *J. Chem. Soc.* A **1966**, 623.
37   D. E. Fenton, A. G. Massey and D. S. Urch, *J. Organomet. Chem.* **6** (1966) 352.
38   L. Ross and M. Dräger, *Chem. Ber.* **115** (1982) 615.
39   L. Ross, Ph.D. thesis, Universität Mainz (1980).
40   L. Ross and M. Dräger, *Z. Naturforsch.* **39b** (1984) 868.
41   E. A. Volchkova, D. D. Smolyaninova, V. G. Genchel, I. L. Lopatkina and Yu. Kh. Shaulov, *Zh. Fiz. Khim.* **46** (1972) 1837 (*CA* **77** (1972) 106145f).
42   V. G. Genchel, A. I. Toropkova, Yu. Kh. Shaulov and D. D. Smolyaninova, *Zh. Fiz. Khim.* **48** (1974) 1837 (*CA* **81** (1974) 177820r).
43   D. D. Smolyaninova, and Yu. Kh. Shaulov, *Zh. Fiz. Khim.* **49** (1975) 1401 (*CA* **83** (1975) 153583f).
44   Yu. I. Smolin, *Dokl. Akad. Nauk SSSR* **181** (1968) 595.
45   A. Yu. Shashkov, N. V. Rannev and Yu. N. Venevtsev, *Koord. Khim.* **10** (1984) 1420.
46   H. Völlenkle and A. Wittmann, *Monatsh. Chem.* **102** (1971) 1245.
47   B. C. Grobmaier, S. Haussühl and P. Klüfers, *Z. Kristallogr.* **149** (1979) 261.
48   H. Völlenkle, A. Wittmann and H. Nowotny, *Monatsh. Chem.* **101** (1970) 46.
49   M. Touboul and Y. Feutelais, *Acta Crystallogr.* **B35** (1979) 810.
50   S. C. Pace, *Synth. React. Inorg. Metalorg. Chem.* **5** (1975) 373.
51   T. Sato, *Japan Kokai* 81 118 015 (1981) (*CA* **95** (1981) 225683n).

52   Yoshitomi Pharmaceutical Inds. Ltd, *Japan Kokai* 81 108 708 (1981) (*CA* **96** (1982) 859u).
53   Teijin Ltd, *Japan Kokai* 83 21 686 (1983) (*CA* **98** (1983) 179646p).
54   Asai Germanium Research Institute, *Japan Kokai* 84 31 785 (1984) (*CA* **101** (1984) 91237a); 84 79 197 (1984) (*CA* **101** (1984) 171517g); 84 95 293 (1984) (*CA* **101** (1984) 152092y).
55   A. R. Dahl and A. D. Norman, *Inorg. Chem.* **14** (1975) 1093.
56   M. N. Bochkarev, L. P. Maiorova, N. S. Vyazankin and G. A. Razuvaev, *J. Organomet. Chem.* **82** (1974) 65.
57   M. Wojnowska, M. Noltemeyer, H. J. Füllgrabe and A. Meller, *J. Organomet. Chem.* **228** (1982) 229.
58   B. Krebs, S. Pohl and W. Schiwy, *Angew. Chem.* **82** (1970) 884; *Angew. Chem. Int. Ed. Engl.* **9** (1970) 897.
59   B. Krebs, S. Pohl and W. Schiwy, *Z. Anorg. Allg. Chem.* **393** (1972) 241.
60   J. Barrau, M. Bouchaut, A. Castel, A. Cazes, G. Dousse, H. Lavayssière, P. Rivière and J. Satgé, *Synth. React. Inorg. Metalorg. Chem.* **9** (1979) 273.
61   J. Barrau, M. Bouchaut, H. Lavayssière, G. Dousse and J. Satgé, *Helv. Chim. Acta* **62** (1979) 152.
62   J. Barrau, H. Lavayssière, G. Dousse, C. Couret and J. Satgé, *J. Organomet. Chem.* **221** (1981) 271; P. Rivière, A. Castel and J. Satgé, *J. Organomet. Chem.* **212** (1981) 351.
63   H. Lavayssière, G. Dousse, J. Barrau, J. Satgé and M. Bouchaut, *J. Organomet. Chem.* **161** (1978) C 59.
64   C. Guimon, G. Pfister-Guillozo, H. Lavayssière, G. Dousse, J. Barrau and J. Satgé, *J. Organomet. Chem.* **249** (1983) C 17.
65   H. Lavayssière, J. Satgé, J. Barrau and M. Traore, *J. Organomet. Chem.* **240** (1982) 335.
66   J. Barrau, G. Rima, H. Lavayssière, G. Dousse and J. Satgé, *J. Organomet. Chem.* **246** (1983) 227.
67   A. Müller, P. Christophliemk and H. P. Ritter, *Z. Naturforsch.* **28b** (1973) 519.
68   R. H. Benno and C. F. Fritchie, *J. Chem. Soc., Dalton Trans.* **1973**, 543.
69   S. Pohl, *Angew. Chem.* **88** (1976) 162.
70   S. Pohl, U. Seyer and B. Krebs, *Z. Naturforsch.* **36b** (1981) 1432.
71   E. Philippot, M. Ribes and O. Lindquist, *Rev. Chim. Minér.* **8** (1971) 477.
72   B. Krebs and S. Pohl, *Z. Naturforsch.* **26b** (1971) 853.
73   A. Müller, B. N. Cyvin, S. J. Cyvin, S. Pohl and B. Krebs, *Spectrochim. Acta* **32A** (1976) 67.
74   H. J. Jacobsen, Thesis, Universität Bielefeld (1977) (cited in ref. B 2).
75   G. Eulenberger, *Z. Naturforsch.* **36b** (1981) 521.
76   H. Puff, S. Franken and W. Schuh, *J. Organomet. Chem.,* **256** (1983) 23.

Cyclostannanes

P. G. HARRISON

University of Nottingham, England

Although tin has more pronounced metallic character than either silicon or germanium, it exhibits somewhat similar behaviour in that it forms homocyclic systems, made up of tin atoms. Their size may vary from four to seven R_2Sn units:

$$
\begin{array}{cccc}
R_2Sn{-}SnR_2 & R_2Sn{-}SnR_2 & \overset{\displaystyle R_2}{\underset{}{Sn}} & R_2Sn{-}SnR_2 \\
| \quad | & & & \\
R_2Sn{-}SnR_2 & R_2Sn \quad SnR_2 & R_2Sn \quad SnR_2 & R_2Sn \quad SnR_2 \\
& Sn & Sn & Sn \\
& R_2 & R_2 & R_2 \\
(1) & (2) & (3) & (4)
\end{array}
$$

Tin–tin bonds are also present in some tin-rich heterocycles, which are discussed in Chapters 15 and 16.

The chemistry of cyclostannanes has been much less investigated in recent years than that of analogous silicon or germanium rings.

14.1 Syntheses

The reaction of diorganostannanes R_2SnH_2 (R = Et, Bu, Bu^i, Hex^c, Ph) with Bu_2^iHg in hexane at $-30\,°C$ followed by warming to room temperature gives the corresponding cyclostannanes (5), usually as mixtures of products:

$$nR_2SnH_2 + n\,Bu_2^iHg \xrightarrow[-Bu^iH]{-30\,°C} (R_2Sn{-}Hg)_n \xrightarrow{-nHg} (R_2Sn)_n \qquad (1)$$

$$(6) \qquad\qquad (5)$$

THE CHEMISTRY OF INORGANIC HOMO-
AND HETEROCYCLES VOL 1 ISBN 0-12-655775-6

Only in the case of R = Bu^t is the dialkylmercurial intermediate (6) isolable. The ring size of the cyclostannane formed is dependent upon the nature of the organic group R. When R = Hex^c, the product is exclusively the cyclopentastannane, but for R = Et, Bu or Bu^i only small amounts of the five-membered ring are obtained, and the major product is the cyclohexastannane. In the case of R = Et, the product mixture also contains the cycloheptastannane (Et$_2$Sn)$_7$. Only the hexamer occurs with the phenyl homologue.[1,2] Smaller rings cannot be prepared by this method. However, two examples of cyclotetrastannanes, in which the Sn$_4$ ring is stabilized by very bulky organic groups, have been sythesized by different routes. Tetrakis{bis(trimethylsilylmethyl)tin} (7) is formed in high yield by the reaction of the corresponding dihydride and diethylamino derivatives:[3]

$$2(Me_3SiCH_2)_2SnH_2 + 2(Me_3SiCH_2)_2Sn(NEt_2)_2$$

$$\longrightarrow [(Me_3SiCH_2)_2Sn]_4 + 2Et_2NH \quad (2)$$
$$(7)$$

The very similar compound tetrakis[meso-o-bis(trimethylsilylmethyl)phenyltin] (8) is a minor product accompanying the meso,meso,spiro-metallobicycle Sn[CHSiMe$_3$C$_6$H$_4$CHSiMe$_3$-o]$_2$ (9) in the substitution of tin(IV) chloride with [{Li(tmeda)}$_2${o-C$_6$H$_4$(CHSiMe$_3$)$_2$}] at −78 °C in diethyl ether:

$$SnCl_4 + 2[\{Li(tmeda)\}_2\{o\text{-}C_6H_4(CHSiMe_3)_2\}]$$

(3)

(8)   (9)
15%   52%

Higher yields (> 40%) of (8) may be achieved from the reaction of the tin(II) aryloxide Sn(OC$_6$H$_2$Me-4-Bu$^t_2$-2,6) with the lithium reagent, but, surprisingly, no substitution takes place with tin(II) chloride.[4]

14.2 Structural data

Both tetrakis [bis(trimethylsilylmethyl)tin][3] (Fig. 14.1) and tetrakis[meso-o-bis(trimethylsilylmethyl)phenyltin][4] (Fig. 14.2) are characterized by a central square Sn$_4$ four-membered ring, in which the Sn–Sn distances (2.83(3) Å

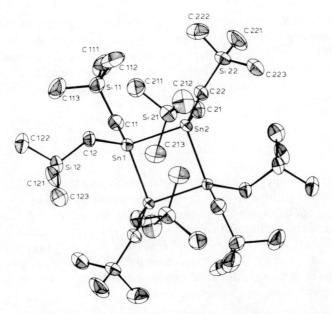

Figure 14.1 Molecular structure of tetrakis[bis(trimethylsilylmethyl)tin]. (Reproduced with permission from *J. Organomet. Chem* **215** (1981) 41.)

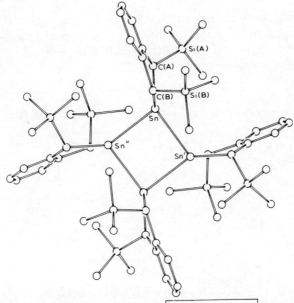

Figure 14.2 Molecular structure of [*meso*-Sn(CHSiMe₃C₆H₄CHSiMe₃-*o*]₄ (Reproduced with permission from *J. Organomet. Chem.* **233** (1982) C28.)

and 2.852(3) Å respectively) are somewhat longer than in either the six-membered ring compound $(Ph_2Sn)_6$ (2.78 Å)[5] or the dimeric $[Sn\{CH(SiMe_3)_2\}_2]_2$ (2.764(2) Å).[6] Because of the rather rigid Sn_4 framework, the coordination polyhedron about the tin atoms is considerably distorted from ideal tetrahedral, with valence-bond angles falling in the range 90–124°.

14.3 Spectra

Mass spectroscopy has been employed widely in the characterization of both homo- and heterocyclostannanes, but is complicated by the complex nature of polyisotopic fragments. Indeed, fractional-evaporation and mass-chromatographic techniques within the mass spectrometer have been used to identify the compounds of crude dialkyltin $(R_2Sn)_n$ mixtures. In general, both loss of alkyl groups from tin and reduction in ring size are observed. For example, the mass spectrum of $(Et_2Sn)_6$ comprises fragments with ring sizes ranging from six to three.

14.4 Chemical reactions

Photolysis (high-pressure mercury or "daylight" lamps) of cyclostannanes results in degradation to the corresponding dialkylstannanediyl (dialkyl-stannylene), which may undergo further reaction if a substrate (for example an alkyl halide) is present, or recyclize:[7-9]

$$(R_2Sn)_n \underset{hv}{\rightleftharpoons} R_2Sn—(R_2Sn)_{n-2}—SnR_2 \underset{-R_2Sn}{\rightleftharpoons} R_2Sn—(R_2Sn)_{n-3}—SnR_2$$

R = Me, Bu, Ph

$$nR_2Sn: \qquad (4)$$

$$\downarrow R'X$$

$$R'R_2SnX$$

$(Ph_2Sn)_6$ is also cleaved by lead(IV) acetate in benzene:[10]

$$(Ph_2Sn)_6 + 3Pb(OAc)_4 \longrightarrow 3Ph_2(AcO)SnSn(OAc)Ph_2 + 3Pb(OAc)_2 \qquad (5)$$

References

1   U. Blaukat and W. P. Neumann, *J. Organomet. Chem.* **63** (1973) 27.
2   J. P. Ritter and W. P. Neumann, *J. Organomet. Chem.* **56** (1973) 199.
3   V. K. Belsky, N. N. Zemlyansky, N. D. Kolosova and I. V. Borisova, *J. Organomet. Chem.* **215** (1981) 41.
4   M. F. Lappert, W. P. Leung, C. L. Raston, A. J. Thorne, B. W. Skelton and A. H. White, *J. Organomet. Chem.* **233** (1982) C28.
5   D. H. Olson and R. E. Rundle, *Inorg. Chem.* **2** (1963) 1310.
6   P. J. Davidson, D. H. Harris and M. F. Lappert, *J. Chem. Soc., Dalton Trans.* **1976**, 2268.
7   W. P. Neumann and A. Schwarz, *Angew. Chem. Int. Ed. Engl.* **14** (1975) 812.
8   S. Kozima, K. Kobayashi and M. Kawanisi, *Bull. Chem. Soc. Jpn* **49** (1976) 2837.
9   K. Kobayashi, K. Kuno, M. Kawanisi and S. Kozima, *Bull. Chem. Soc. Jpn* **50** (1977) 1353.
10 U. Christen and W. P. Neumann, *J. Organomet. Chem.* **39** (1972) C58.

Tin–Nitrogen and Tin–Phosphorus Heterocycles

M VEITH

Technische Universität, Braunschweig, Federal Republic of Germany

In this chapter, rings and cages are described that contain the elements tin and nitrogen, or tin and phosphorus. Ring systems containing elements other than these three are only considered if their chemistry and structure are dominated by stannazane or stannaphosphane units. Polycyclic compounds and cages that are exclusively based upon tin and nitrogen or tin and phosphorus have been known for only a short time and are also dealt with.

15.1 Cyclostannazanes

Alternation of tin and nitrogen atoms leads to formation of four-membered (1) and six-membered (2) ring systems, known as cyclodistannadiazanes and cyclotristannatriazanes respectively:

(1)　　　　　　　　(2)

15.1.1 Preparation and structure

Several methods have been developed for the synthesis of tin–nitrogen ring systems based upon the formation of tin–nitrogen bonds. This may be achieved by reactions that take place with elimination of a salt (equation

THE CHEMISTRY OF INORGANIC HOMO-
AND HETEROCYCLES VOL 1 ISBN 0-12-655775-6

(1), $n = 2$, $R = Me$, $R' = Bu^t$;[1] $n = 3$, $R_2 = MeCl$, $R' = Et$;[2] and equation (2), $n = 2$, $R_2 = MeBu^t$, $R' = Bu^t$;[3,4] $R = Me$, $R' = CMe(CF_3)_2$[5]), elimination of a volatile amine, i.e. transamination (equation (3), $n = 2$, $R = Bu^t$, $R' = Me$, CH_2Ph;[6] $n = 3$, $R = Me$, $R' = Me$, Et[7]) or the elimination of tetramethylstannane (equation (4), $n = 3$, $R' = SO_2Me$, SO_2CF_3[8,9]):

$$R_2SnCl_2 + Li_2NR' \longrightarrow (R_2SnNR')_n + 2LiCl \tag{1}$$

$$R_2Sn(NR'Li)_2 + Cl_2SnR_2 \longrightarrow (R_2SnNR')_n + 2LiCl \tag{2}$$

$$R_2Sn(NR''_2)_2 + H_2NR' \longrightarrow (R_2SnNR')_n + 2HNR''_2 \tag{3}$$

$$R'N(SnMe_3)_2 \xrightarrow{\Delta} (Me_2Sn\text{---}NR')_n + SnMe_4 \tag{4}$$

The transamination reaction (3) seems to proceed in two steps, as has been shown for the reaction of bis(dimethylamino)bis(t-butyl)tin with primary aliphatic amines.[6] In the first step the amino groups on the tin atom are interchanged (equation 5), while in the second the primary amine is excluded, probably by hydrogen transfer from one nitrogen atom to the other (equation 6):[6]

$$Bu_2^tSn(NMe_2)_2 + 2\ R'NH_2 \longrightarrow Bu_2^tSn(NHR')_2 + 2\ Me_2NH \tag{5}$$

$$2\ Bu_2^tSn(NHR')_2 \longrightarrow (Bu_2^tSnNR')_2 + 2\ R'NH_2 \tag{6}$$

The four-membered ring is favoured against the six-membered when bulky substituents are on either the tin or the nitrogen atoms. No stable eight-membered ring has ever been observed.

Besides the synthetic routes mentioned so far, other special preparations of tin–nitrogen rings have been developed. Elimination of a salt and an amine is observed when bis(t-butyl)dichlorstannane is reacted with potassium amide (equation 7):[3]

$$3\ Bu_2^tSnCl_2 + 6\ KNH_2 \longrightarrow (Bu_2^tSnNH)_3 + 3\ NH_3 + 6\ KCl \tag{7}$$

Two quite singular procedures are shown in equations (8)[10] and (9):[11]

$$(Bu_2^tSnNMe)_2 + 2\ Tos\text{---}N{=}S{=}N\text{---}Tos \longrightarrow (Bu_2^tSnN\text{---}Tos)_2$$
$$+ 2\ Tos\text{---}NSN\text{---}Me \tag{8}$$

$$(Me_3Sn)_2NX + (CF_3CO)_2O \longrightarrow \tfrac{1}{3}(Me_2SnNX)_3$$
$$+ Me_3SnOOCCF_3 + CH_3COCF_3 \tag{9}$$

A peculiar type of ring-compound (**3**) is formed, when amines of low-valent tin(II) are synthesized. As the tin atom in these molecules is not electronically saturated, it behaves as a strong Lewis acid and forms dimers by coordination with the Lewis basic amine ligands:[12]

$$2 \ (R_2N)_2Sn: \longrightarrow \quad R_2N-\overset{\ddots}{Sn}\overset{\overset{\displaystyle R \quad R}{\underset{\displaystyle N}{\diagup}}}{\underset{\overset{\displaystyle N}{\diagup}\underset{\displaystyle R \quad R}{}}{}}Sn-NR_2 \qquad (10)$$

(**3**)

This intermolecular Lewis-acid–base interaction is still important, even when the bis(amino)stannylene is coordinated to transition metals.[13] It can be suppressed if bulky substituents are used on the nitrogen atoms.[B 1]

The ring size and composition of cyclostannazanes follow from mass spectra or relative molecular-mass determinations. No direct structural investigations on pure cyclostannazanes by X-ray diffraction have been made.

15.1.2   Reactions of cyclostannazanes

All cyclic tin–nitrogen compounds are sensitive to moisture and air, and must therefore be handled under an N_2 or Ar atmosphere. The reactivity of the cyclodistannadiazane $(Bu_2^tSnN—Me)_2$ has been extensively studied. Besides substitution reactions and ring cleavage, insertions into the Sn—N single bond are observed when this cyclostannazane is allowed to react with C–C or C–N double or triple bonds (R = Bu^t, R' = Me). In the latter reactions either one or two C–C or C–N components may be inserted, leading to six- or eight-membered cyclic compounds (ring-expansion reactions).[10]

Cyclostannazanes react with various organoelement halides, with complete cleavage of the Sn–N ring. Thus with organoboron dihalides RBX_2 (R = Me, X = Cl, Br) the B_2N_2 ring compound is recovered in 75% yield:[1]

$$(Me_2SnNBu^t)_2 + 2R—BX_2 \longrightarrow Bu^t—N \underset{\underset{R}{B}}{\overset{\overset{R}{B}}{\diamond}} N—Bu^t + 2\ Me_2SnX_2 \qquad (11)$$

A similar ring cleavage has been reported for six-membered rings reacting with trimethylchlorosilane:[8]

$$(MeSO_2—NSnMe_2)_3 + 3\ Me_3SiCl \longrightarrow 3\ MeSO_2—N\overset{\diagup SnMe_2Cl}{\diagdown SiMe_3} \qquad (12)$$

^{119}Sn NMR techniques have been used to confirm that in equimolar solutions of $(Me_2SnNEt)_3$ and $(Me_2SnS)_3$ redistribution of the ring components takes place.[14] Besides the starting compounds, the cycles (4) and (5) are observed:

$$\begin{array}{cc}
\text{Et} & \text{Et} \\
| & | \\
\text{N} & \text{N} \\
Me_2Sn \diagup \diagdown SnMe_2 & Me_2Sn \diagup \diagdown SnMe_2 \\
\text{S} \diagdown \diagup \text{S} & \text{Et—N} \diagdown \diagup \text{S} \\
\text{Sn} & \text{Sn} \\
Me_2 & Me_2 \\
(4) & (5)
\end{array}$$

A redistribution of ring components may also be responsible for the formation of (6) in reaction (13).[15] This is a new ring that contains two tin and three nitrogen atoms, and one sulphur atom of oxidation state IV.

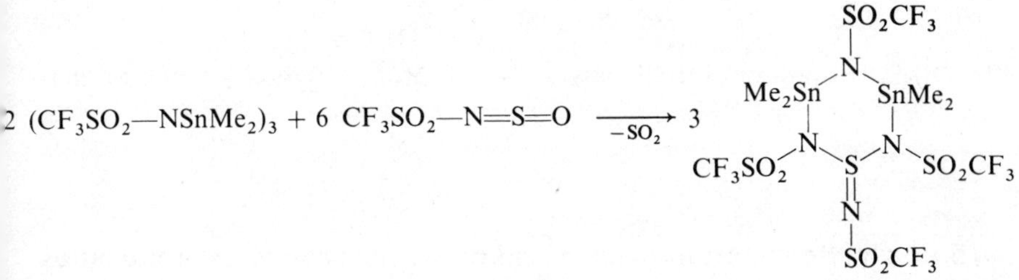

$$2 \ (CF_3SO_2{-}NSnMe_2)_3 \ + \ 6 \ CF_3SO_2{-}N{=}S{=}O \ \xrightarrow[-SO_2]{} \ 3$$

(structure (6))

(6)

(13)

The structure of (6) has been established by X-ray diffraction techniques.[15] The ring is not planar (envelope conformation) with very different Sn–N bond lengths, ranging from 208.4 to 225.1(7) pm.

Cyclic compounds arising from Lewis-acid–base interactions of tin(II) with nitrogen ligands as in (3) are always in competition with noncyclic monomeric species when bases are added:[12]

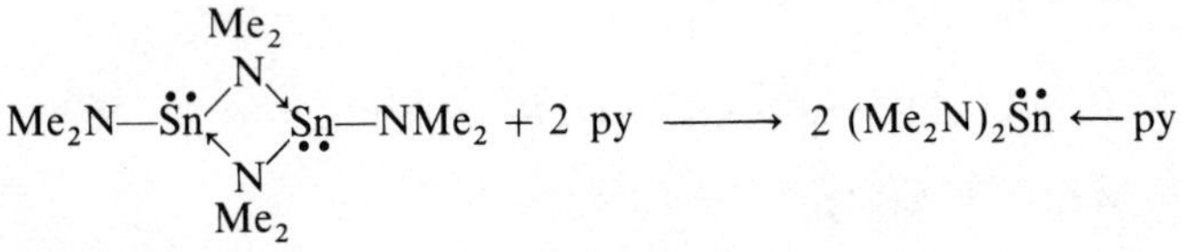

$$Me_2N{-}\overset{\cdot\cdot}{Sn} \ \underset{\underset{Me_2}{N}}{\overset{\overset{Me_2}{N}}{}} \ \overset{\cdot\cdot}{Sn}{-}NMe_2 \ + \ 2 \ py \ \longrightarrow \ 2 \ (Me_2N)_2\overset{\cdot\cdot}{Sn} \leftarrow py \qquad (14)$$

(3)

The same sort of reaction has been observed for the metal-bonded bis(di-methylamino)stannylene dimer (7):[13]

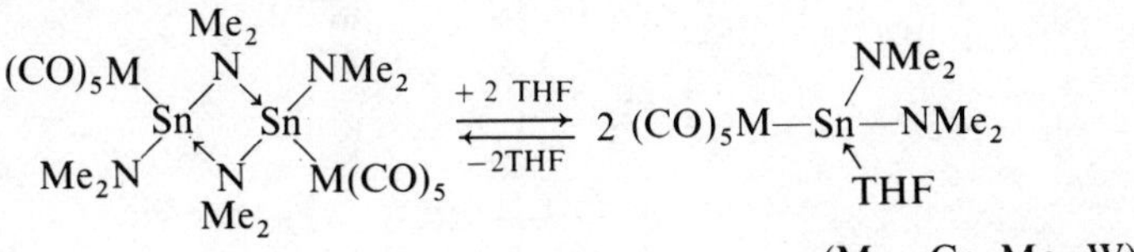

$$(CO)_5M \ \underset{\underset{Me_2}{N}}{\overset{\overset{Me_2}{N}}{}} \ \begin{matrix} NMe_2 \\ M(CO)_5 \end{matrix} \ \underset{-2THF}{\overset{+2\ THF}{\rightleftarrows}} \ 2 \ (CO)_5M{-}Sn \begin{matrix} NMe_2 \\ NMe_2 \\ THF \end{matrix} \qquad (15)$$

(M = Cr, Mo, W)

(7)

In this latter reaction there is equilibrium between the THF-free dimeric compound and the base-stabilized monomer. The tin–nitrogen bond in compounds of type (3) is very sensitive to hydrolysis or alcoholysis.[12] An interesting bicyclic compound is formed as follows (for further discussion of these types of compounds see ref. 16):

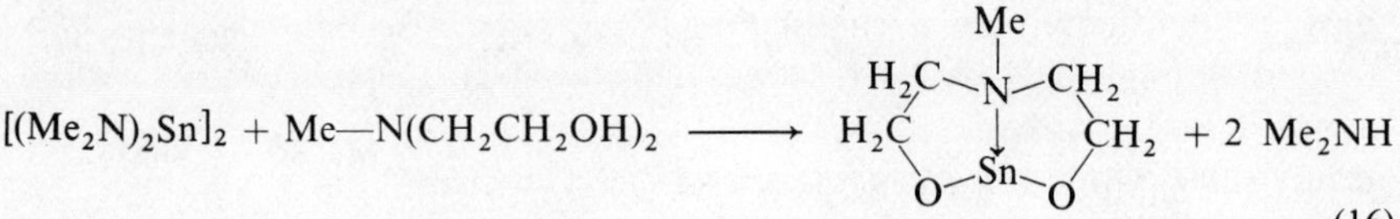

$$[(Me_2N)_2Sn]_2 + Me-N(CH_2CH_2OH)_2 \longrightarrow$$

$$+ 2\ Me_2NH$$

(16)

15.2 Cyclic silastannazanes and other heterocyclostannazanes

A large number of four-membered cyclic compounds with the element combination SnN_2Si or with other elements in the place of silicon have been reported (see also Section 8.5.2). They are synthesized via salt-elimination reactions:[17-20]

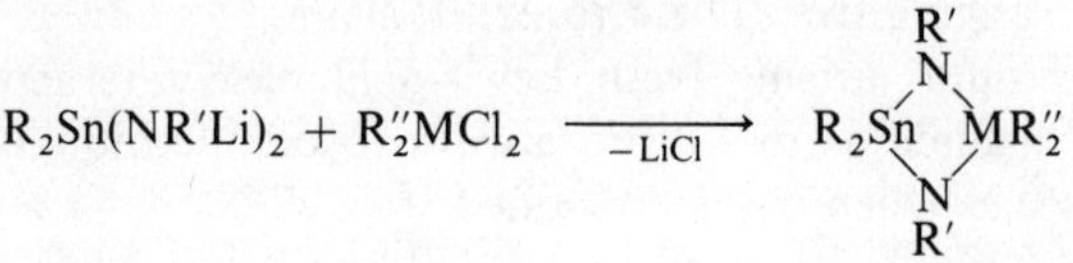

$$R_2Sn(NR'Li)_2 + R_2''MCl_2 \xrightarrow[-LiCl]{}$$

(8)

$M = Si,Ge,Pb;\ R = Me,\ R' = CMe(CF_3)_2,\ R'' = Me$[17]

$M = Ge;\ R = Me,\ R' = SiEt_3,\ R'' = Et$[18]

$M = Ti, Zr;\ R = Me,\ R' = CMe(CF_3)_2,\ R'' = Me$[17]

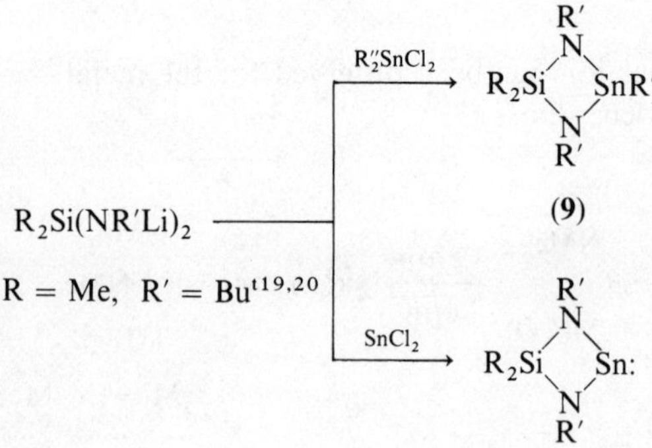

The compounds (10) deserve some comment. In this heterocycle the tin atom is in a formal oxidation state + II, which means that the compound may be considered as a carbene analogue (see also ref 21). This bis-(amino)stannylene is highly reactive: depending on the substituents on the nitrogen atom, it may either be monomeric ($R = Bu^t$) or dimeric ($R =$ iso-propyl) in organic solvents (formulae (10a) and (10b)):[22]

(10a)   (10b)

When a hexane solution of (10a) is allowed to crystallize at $-78\,°C$ monoclinic crystals are formed, which contain equal amounts of the monomer (10a) and the dimer (10b).[22] The tin atom in (10a) can be oxidized very easily,[19,23] it can be coordinated by bases (for example pyridine)[22] or can itself coordinate to transition metals[24] with conservation of the four-membered cycle ($R = Me$, $R' = Bu^t$, $R'' = Pr^i$):

A spirocyclic compound (11) ($R = Me$, $R' = Bu^t$)[19,23,25] is obtained in several reactions:

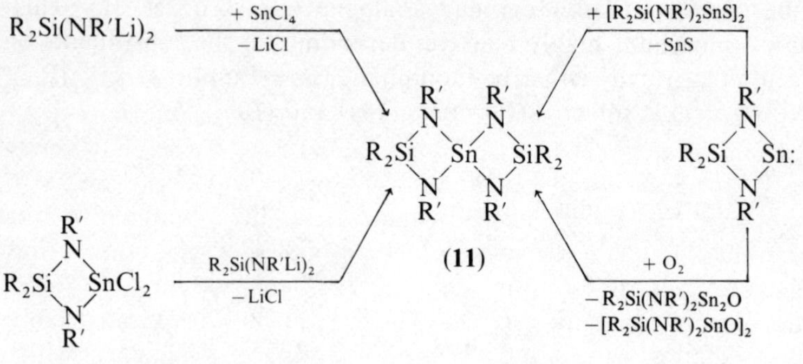

$$R_2Si(NR'Li)_2 \xrightarrow[-LiCl]{+ SnCl_4}$$

$$\xleftarrow[-SnS]{+ [R_2Si(NR')_2SnS]_2}$$

$$R_2Si \underset{\substack{N \\ R'}}{\overset{\substack{R' \\ N}}{\diamond}} Sn \underset{\substack{N \\ R'}}{\overset{\substack{R' \\ N}}{\diamond}} SiR_2$$

(11)

$$R_2Si \underset{\substack{N \\ R'}}{\overset{\substack{R' \\ N}}{\diamond}} SnCl_2 \xrightarrow[-LiCl]{R_2Si(NR'Li)_2}$$

$$R_2Si \underset{\substack{N \\ R'}}{\overset{\substack{R' \\ N}}{\diamond}} Sn:$$

$$\xleftarrow[{-R_2Si(NR')_2Sn_2O \atop -[R_2Si(NR')_2SnO]_2}]{+ O_2}$$

A spirocyclic compound, analogous to (11) but with selenium as spiro-centre, was prepared by reacting $Me_2Sn(NR'Li)_2$ with $SeCl_4$.[17]

The eight-membered cycle (12) has been synthesized in 80% yield by elimination of both tetramethylstannane and trimethylchlorstannane from a bis(trimethylstannyl)amine (R = Me):[26]

$$4\ CF_3SO_2N(SnR_3)_2 + 2\ R_2SiCl_2 \xrightarrow[{-4R_3SnCl \atop -2\ SnR_4}]{}$$

(17)

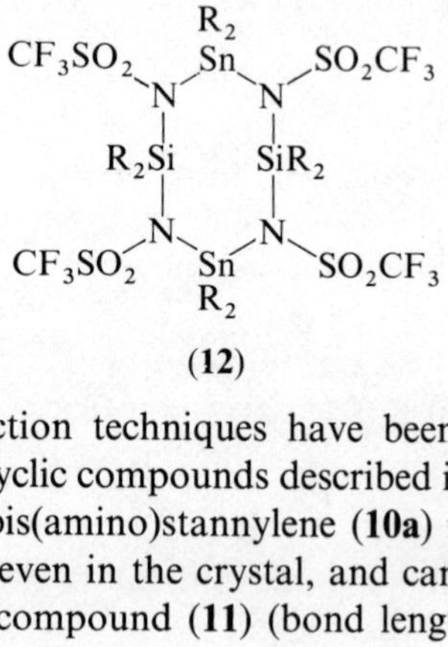

(12)

Structural investigations by X-ray diffraction techniques have been performed on several mono-, spiro- and polycyclic compounds described in this section. The tin atom in the monomeric bis(amino)stannylene (10a) maintains its low coordination number of two even in the crystal, and can thus be readily compared with the spirocyclic compound (11) (bond lengths in pm):

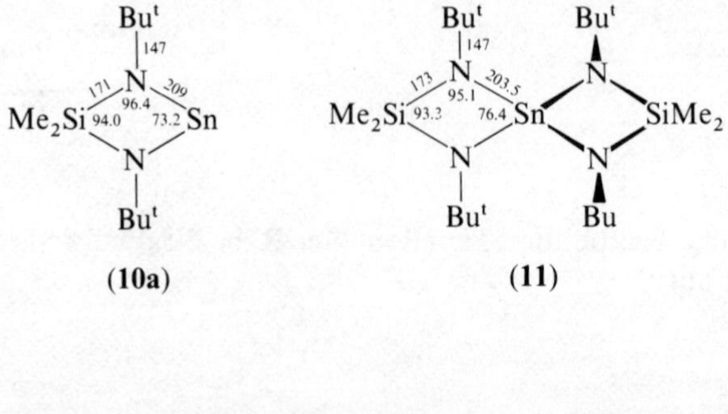

(10a)   (11)

The ring compound (**10a**) has almost C_{2v} symmetry, while D_{2d}-symmetry must be assumed for the free molecule (**11**). The four-membered rings in (**10a**) and (**11**) are strictly planar with nitrogen atoms in a trigonal planar coordination. The Sn–N bond length in the spirocyclic compound is 5.5 pm less than in the stannylene, which must be explained by a smaller covalent radius for the four-valent tin atom.[B 1] The electron density seems to be shifted into the Si–N bonds (small distances); this effect is more pronounced in (**10a**) than in (**11**). As expected, the smallest angle in the rings is connected to the heaviest atom.

In the dimeric analogue of (**10a**), depicted in (**10b**), three four-membered cycles are sharing edges, and are nearly orthogonally aligned with one another (point symmetry: C_i) (bond lengths in pm):[20]

(**10b**)

The chemical application of silastannazanes has been studied in the case of compound (**10a**). The reactions shown on p. 389 give some examples of the reactivity of this cyclic diazasilastannylene. The whole molecule may act as a Lewis acid via the unsaturated tin atom or as a base via the nonbonding electron pairs at the nitrogen and tin atoms. Combining the different electro- and nucleophilic centres, the whole molecule (**10a**) can act as a mono-, di- or tridentate ligand versus other Lewis-acid–base systems. It is beyond the scope of this book to describe all possibilities, and the reader is referred to some review articles.[B 1 – B 3] However one important use will be described in the following section.

15.3   Polycyclic Sn–N compounds and related molecules

When the bis(amino)stannylene (**10a**) is allowed to react with primary amines, three different polycycles are formed besides the dimethylbis-(t-butylamino)silane, depending on conditions:[27,28]

$$Me_2Si(NBu^t)_2Sn \xrightarrow{\ +\,n\ R-NH_2\ } Me_2Si(NBu^tH)_2 + (RN)_4Sn_3H_2 \tag{18}$$

$$+ Me_2Si(NBu^t)_2Sn_2NR + (RN)_4Sn_4$$

If R is t-butyl then the three compounds can be isolated and identified as (**12**) (**13**) and (**14**):[27]

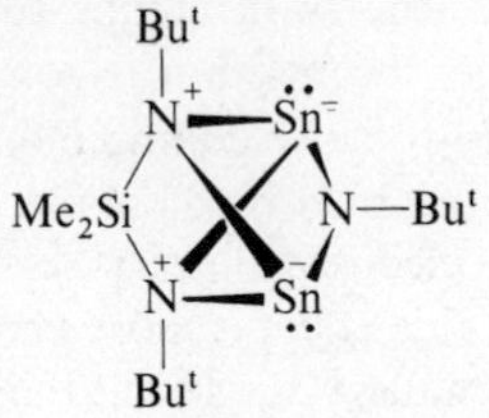

(12)

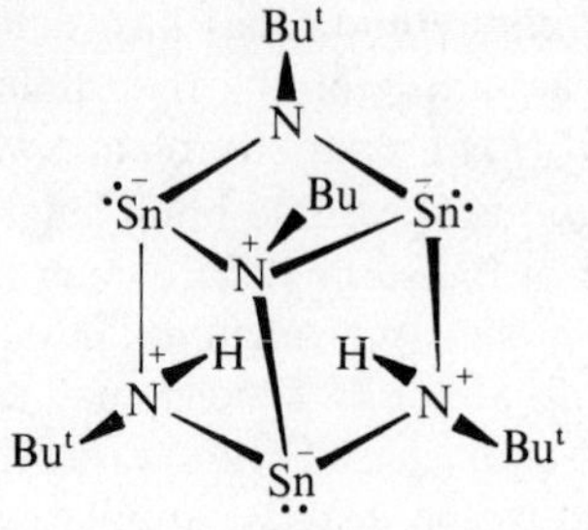

(13)

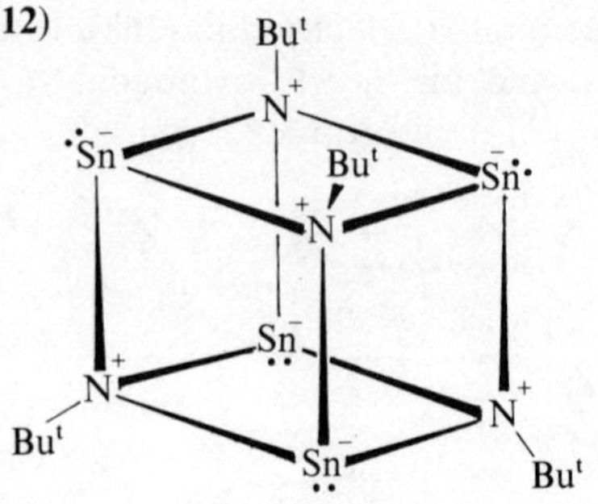

(14)

In (14) the t-butyl group can be replaced by isopropyl, dimethyl and benzyl amino groups. If the substituent on the nitrogen atoms exhibits more reduced steric requirements then the resulting compound becomes polymeric, as in the case of $(SnN-Me)_n$.[29]

The reaction course of equation (18) has been studied very thoroughly. It has been demonstrated that the educts in equation (18) are in equilibrium with a Lewis-acid–base adduct (15), which can be isolated at $-78\,^{\circ}C$.[B 3]

$$Me_2Si(NBu^t)_2Sn + H_2N-Bu^t \rightleftharpoons$$

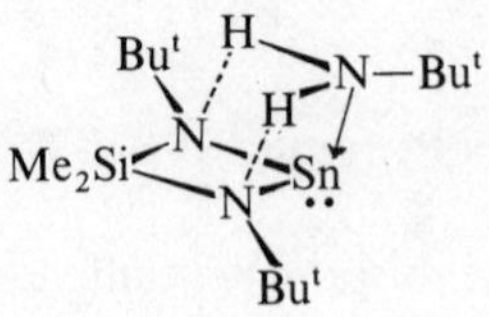

(19)

(15)

An X-ray structural determination of (15) shows that besides the bond between the tin atom and the nitrogen atom of the t-butylamine, $H \cdots N$ contacts to the nitrogen atoms of the ring have to be considered. It can be assumed that the decomposition of the intermediate (15) proceeds via a hydrogen transfer from the t-butylamine to the diazasilagroup, as indicated by the structure. Besides bis(t-butylamino)dimethylsilane, the hypothetical intermediate (16) should be formed:

$$:\!\overset{-}{Sn} \equiv \overset{+}{N}\!\!-\!Bu^t \longleftrightarrow :Sn\!=\!\overset{=}{N} \longleftrightarrow :\overset{+}{Sn}\!-\!\overset{-}{N}\!\!-\!Bu^t$$
$$\underset{Bu^t}{\big\backslash}$$

(16)

Compound (16) may then react with the diazasilastannylene to form (12), with itself and t-butylamine to form (13), or may oligomerize to form (14). Compound (13) can be derived from a seco-norcubane by substituting all carbon by nitrogen or tin atoms,[30] while in (14) the polycycle resembles a cube, the corners of which are alternately occupied by nitrogen or tin atoms. In (13) and (14) the tin atoms are in the top of a pyramid with NSnN angles of about 80°. The four-valent nitrogen atoms have SnNSn angles of about 100°. The average Sn–N distance is 220.2 pm and corresponds well with a bond, which should have $\frac{2}{3}$ single- and $\frac{1}{3}$ dative-bond character.[B 1,B 2]

The basic structure of (12) consists of a bisphenoid of two tin and two nitrogen atoms, two opposite edges being bridged by a t-butylamino or a dimethylsilyl group. The three-valent nitrogen atom in (12) can be replaced by an oxygen atom to form (17), which can be isolated as a $SnCl_2$ adduct (X-ray structure[31]):

$$
\begin{array}{c}
Bu^t \\
| \\
\overset{\cdot\cdot}{^+N}\!-\!Sn^{\,-} \\
Me_2Si \diagup \;\; \diagdown \;\; \diagup O \\
\;\;\;\;\diagdown \;\; \diagup \;\; \diagdown \\
^+N\!-\!Sn^{\,-} \\
| \\
Bu^t
\end{array}
$$

(17)

Other types of polycyclic SnNSi compounds are obtained when Me_2Si-$(NBu^t)_2Sn$ is allowed to react with bis(methylamino)dimethylsilane:[29]

$$3\ Me_2Si(NBu^t)_2Sn + 3\ Me_2Si(NMeH)_2 \longrightarrow 3\ Me_2Si(NBu^tH)_2 +$$
$$0.9\ (Me_2Si)_3(NMe)_5Sn_2 + 0.3\ (Me_2Si)(NMe)_5Sn_4 \tag{20}$$

Bis(t-butylamino)dimethylsilane is recovered quantitatively besides two other products, their structures being established by NMR and X-ray diffraction techniques.[29] The compound $(Me_2Si)_3(NMe)_5Sn_2$ (18) is a tricyclic system with a central Sn_2N_2 ring, while $(Me_2Si)(NMe)_5Sn_4$ (19) resembles a basketane:

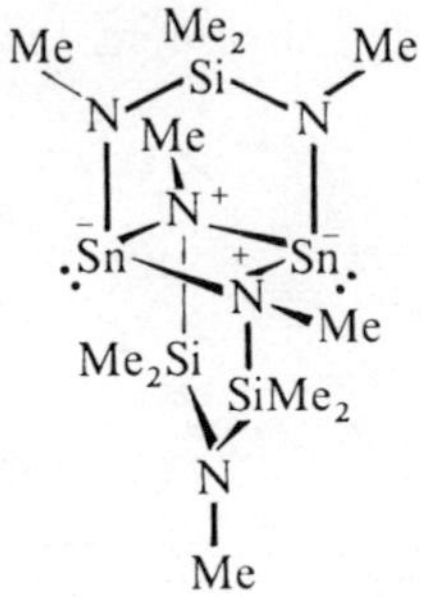

(18) 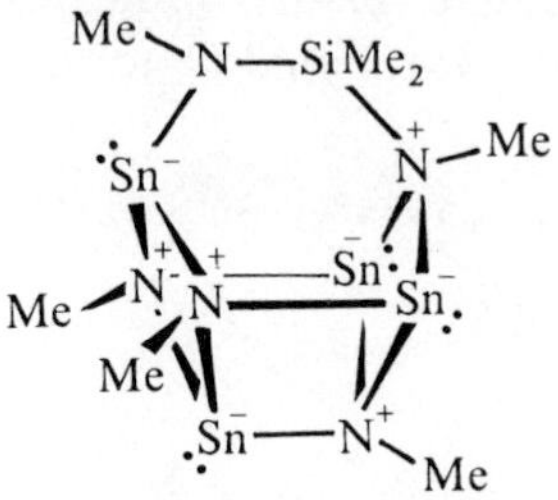   (19)

Compound (19) shows fluxional properties in solution, as established by NMR techniques.[29] It follows from the product (18) that condensation of two molecules bis(methylamino)dimethylsilane must have occurred during the reaction.

A complicated polycyclic system has been found as a product in the reaction of $Me_2Si(NBu^t)_2Sn$ with oxygen (see p. 389).[32] Two molecules, (17) and the dispiro compound (20), interact via oxygen–tin donor bonds to give a polycyclic compound consisting of thirteen four-membered cycles sharing edges or corners (see Fig. 15.1).

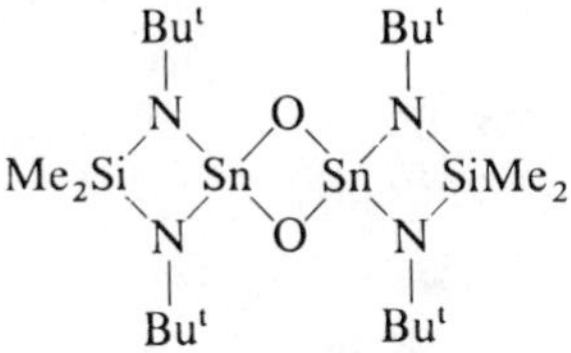

(20)

A derivative of the cage compound (14) with tin(II), nitrogen and oxygen atoms can be synthesized by careful hydrolysis of (13):[33]

$$4\ Sn_3(NBu^t)_4H_2 + 3\ H_2O \longrightarrow 3\ Sn_4(NBu^t)_3O + 7\ Bu^t\!-\!NH_2 \quad (21)$$

Besides t-butylamine, the cubane-like compound (21a) is obtained in 44% yield:[33]

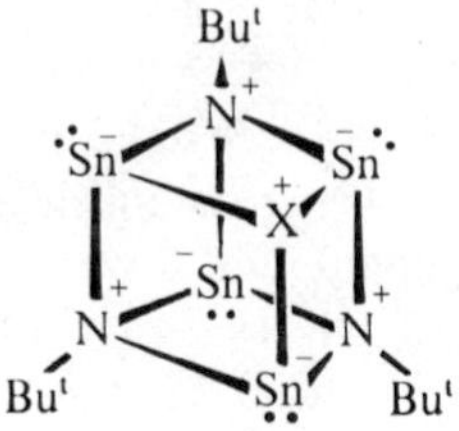

(21a) X = O
(21b) X = O—AlMe₃

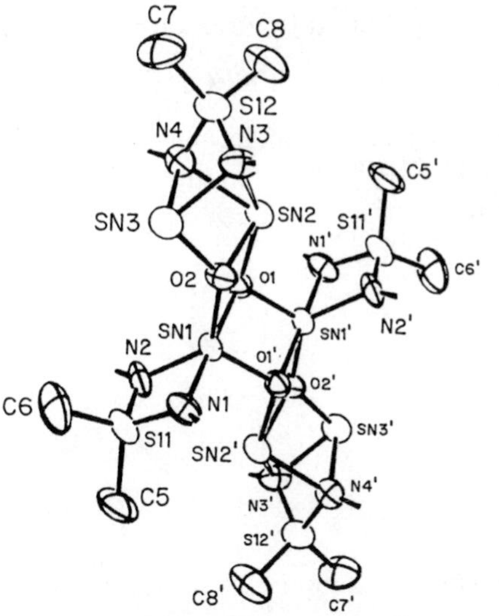

Figure 15.1   The structure of $[Me_2Si(NBu^t)_2]_4Sn_6O_4$ (t-butyl groups have been omitted). (Reproduced with permission from Z. Anorg. Allg. Chem. **459** (1979) 208.)

The cubane (**21a**) can be transformed to (**21b**) (X-ray structural characterization[34]) by a simple addition of trimethylaluminium[33] to the still basic oxygen atom in (**21a**):

$$Sn_4(NBu^t)_3O + \tfrac{1}{2} Al_2Me_6 \longrightarrow Sn_4(NBu^t)_3(O\!-\!AlMe_3) \qquad (22)$$

15.4 Cyclostannaphosphanes

15.4.1 Cyclodistannadiphosphanes and cyclotristannatriphosphanes

In contrast with tin–nitrogen cycles and polycycles, there are only few tin–phosphorus cyclic compounds known at the present time. The only four-membered cycle with tin and phosphorus seems to be present in compound (**22**), which can be rationalized by a Lewis-acid–base interaction of two diphosphinostannylenes:

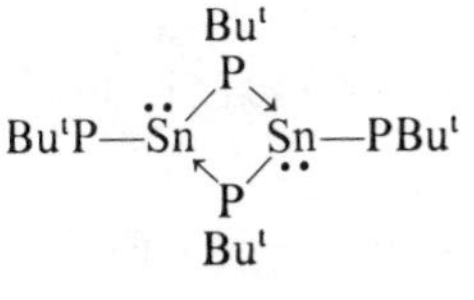

(**22**)

Compound (22) can be prepared by reaction (23) in 45% yield.[35]

$$2 \text{ KPBu}_2^t + \text{Et}_3\text{PSnCl}_2 \longrightarrow 2 \text{ KCl} + \text{Et}_3\text{P} + \tfrac{1}{2} [(\text{Bu}_2^t\text{P})_2\text{Sn}]_2 \qquad (23)$$

The structure of (22) follows from molecular-mass determination and NMR spectroscopic data (^{1}H, ^{31}P, ^{119}Sn). No direct structural determination has been performed so far.

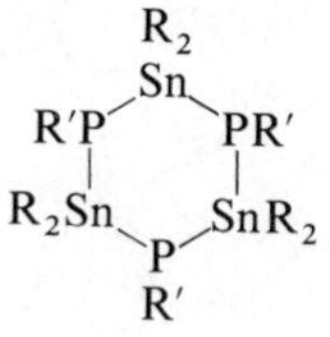

(23)

Some cyclic tin–phosphorus compounds with a six-membered ring (23) (R = Me,[36–38] Bu, Ph[36,37]) are obtained by condensation of dichlorodiorgano-stannanes with phenylphosphine (equation 24) or by a salt elimination between trichlorophosphane and lithium triphenylstannane (equation 25)[39] (for other synthetic routes to Sn–P bonds, see ref. 40):

$$3 \text{ R}_2\text{SnCl}_2 + 3 \text{ Ph—PH}_2 \xrightarrow[-6 \text{ Et}_3\text{N.HCl}]{+ 6 \text{ Et}_3\text{N}} (\text{R}_2\text{SnPPh})_3 \qquad (24)$$

(23)

$$\text{PCl}_3 + \text{Ph}_3\text{SnLi} \longrightarrow \tfrac{1}{3}(\text{Ph}_3\text{Sn—PSnPh}_2)_3 + \text{other products} \qquad (25)$$

Very recently an alternative synthetic procedure has been described, which is believed to proceed via a Me$_2$Sn=P—Ph monomeric intermediate:[38]

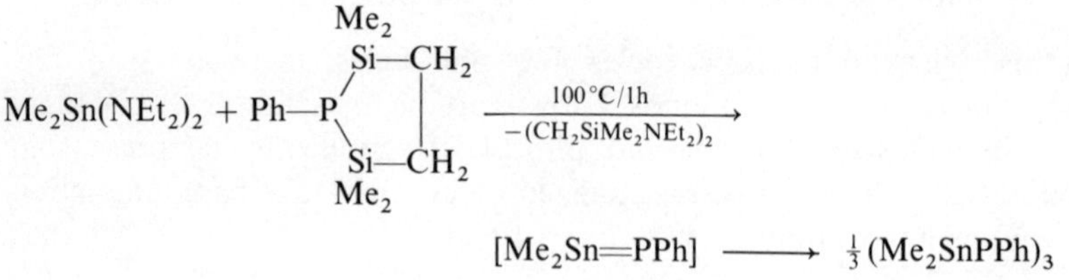

$$[\text{Me}_2\text{Sn=PPh}] \longrightarrow \tfrac{1}{3}(\text{Me}_2\text{SnPPh})_3 \qquad (26)$$

The rings are cleaved by thermolysis, yielding cyclic organophosphanes:[37]

$$(\text{PhSnMe}_2)_3 \xrightarrow{\Delta} (\text{Me}_3\text{Sn})_2\text{P—Ph} + \tfrac{2}{5} (\text{PhP})_5 + \text{Sn} \qquad (27)$$

Tin–phosphorus ring compounds are very sensitive to moisture and air.

15.4.2. Phosphorus-rich and tin-rich phosphorus-tin cycles and polycycles

A number of polycyclic compounds have been claimed to result when organohalogenostannanes are reacted with phosphine:[41,42]

$$4 \text{ PhSnCl}_3 + 4 \text{ PH}_3 + 12 \text{ Et}_3\text{N} \longrightarrow (\text{PSnPh})_4 + 12 \text{ Et}_3\text{N.HCl} \qquad (28)$$

$$6 \text{ Ph}_2\text{SnCl}_2 + 4 \text{ PH}_3 + 12 \text{ Et}_3\text{N} \longrightarrow \text{P}_4(\text{SnPh}_2)_6 + 12 \text{ Et}_3\text{N.HCl} \qquad (29)$$

$$2 \text{ PhSnCl}_3 + 3 \text{ Ph}_2\text{PH} \longrightarrow (\text{PhSn})_2(\text{PPh})_3 + \text{polymer} \qquad (30)$$

The yields of the products in these reactions are very low (5–10%) and the compounds are poorly characterized (for example by IR spectra[42]). The structures (24)–(26) have been proposed; although they are reasonable, they should nevertheless be regarded with some doubt:

(24)   (25)   (26)

Tin-rich polycyclic compounds have been synthesized recently by reaction of white phosphorus with tin hydrides:[43–45]

$$5 \text{ Me}_2\text{SnH}_2 + \text{P}_4 \xrightarrow[\text{Et}_2\text{O}/295\,\text{K}]{\text{DMF}} 2 \text{ H}_2 + 2 \text{ PH}_3 + (\text{Me}_2\text{Sn})_5\text{P}_2 \qquad (31)$$

$$3 \text{ Me}_4\text{Sn}_2\text{H}_2 + \text{P}_4 \xrightarrow[273\,\text{K}]{\text{Et}_2\text{O}} 2 \text{ PH}_3 + (\text{Me}_2\text{Sn})_6\text{P}_2 \qquad (32)$$

While $(\text{SnMe}_2)_5\text{P}_2$ is colourless, decomposing above $400\,^\circ\text{C}$,[43,44] $(\text{Me}_2\text{Sn})_6\text{P}_2$ is pale yellow with a melting point of $220\,^\circ\text{C}$.[45] Both compounds have been characterized by ^1H and ^{31}P NMR techniques, and their structures have been determined by X-ray diffraction methods. The two molecules are depicted with their structural parameters in Fig. 15.2; the molecular symmetry of $(\text{Me}_2\text{Sn})_5\text{P}_2$ is C_{2v}, while $(\text{Me}_2\text{Sn})_6\text{P}_2$ has D_3 symmetry. When irradiated, $(\text{Me}_2\text{Sn})_6\text{P}_2$ decomposes with elimination of dimethylstannylene, which subsequently polymerizes:[45]

$$(\text{Me}_2\text{Sn})_6\text{P}_2 \longrightarrow (\text{Me}_2\text{Sn})_5\text{P}_2 + \text{Me}_2\text{Sn:} \qquad (33)$$

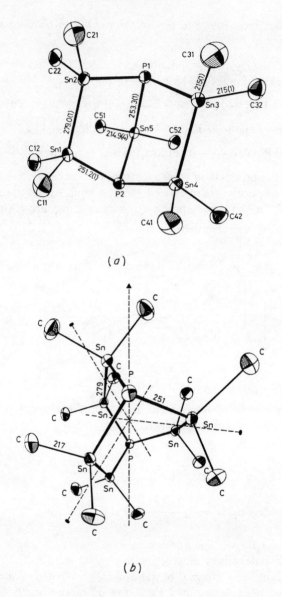

(a)

(b)

Figure 15.2 (a) The structure of $(Me_2Sn)_5P_2$. (b) The structure of $(Me_2Sn)_6P_2$. Bond lengths in pm. (Reproduced with permission from *Angew. Chem. Int. Ed. Engl.* **17** (1978) 767; **20** (1981) 1029.)

Phosphorus-rich five-membered cycles of the P_4Sn type can be synthesized by the following reactions (R' = Et):[46]

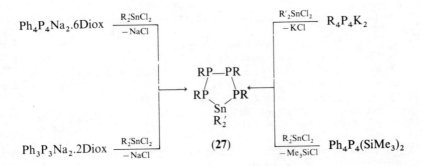

Spectroscopic methods (NMR, mass spectra) have been used to confirm the structure of these five-membered rings (27).[46] The cyclic compounds (27) are decomposed by oxidation with iodine or cleaved by hydrogen chloride:[46]

$$(RP)_4SnR_2' + 9\,I_2 \longrightarrow 4\,RPI_4 + R_2'SnI_2 \qquad (34)$$

$$(RP)_4SnR_2' + 2\,HCl \longrightarrow R_2'SnCl_2 + \left\{ \begin{array}{c} R\quad R\quad R\quad R \\ H\!-\!P\!-\!P\!-\!P\!-\!P\!-\!H \end{array} \right\} \qquad (35)$$

$$\longrightarrow \tfrac{3}{n}\,(RP)_n + RPH_2$$

Bibliography

B 1   M. Veith and O. Recktenwald, *Top. Curr. Chem.* **104** (1982) 1.
B 2   M. Veith, *J. Organomet. Chem. Libr.* **12** (1981) 319.
B 3   M. Veith, *Nachr. Chem. Tech. Lab.* **30** (1982) 940.
B 4   Yu. I. Dergunov, V. F. Gerega, and O. S. Dyachkovskaya, *Russ. Chem. Rev. (Engl. Transl.)* **46** (1977) 1132.
B 5   M. F. Lappert, P. P. Power, A. R. Sanger and R. C. Srivastava, *Metal and Metalloid Amides* (Ellis Horwood, Chichester, 1980), p. 235.

References

1   W. Storch, W. Jackstiess, H. Nöth and G. Winter, *Angew. Chem.* **89** (1977) 494; *Angew. Chem. Int. Ed. Engl.* **16** (1977) 478.
2   A. G. Davies and J. D. Kennedy, *J. Chem. Soc.* C **1970**, 759.
3   D. Haenssgen, J. Kuna and B. Ross, *J. Organomet. Chem.* **92** (1975) C 49.
4   D. Haenssgen, J. Kuna and B. Ross, *Chem. Ber.* **109** (1976) 1797.
5   K. E. Peterman and J. M. Shreeve, *Inorg. Chem.* **15** (1976) 743.

6 D. Haenssgen and I. Pohl, *Angew. Chem.* **86** (1974) 676; *Angew. Chem. Int. Ed. Engl.* **13** (1974) 607.

7 K. Jones and M. F. Lappert, *J. Chem. Soc.* **1965**, 1944.

8 H. W. Roesky and K. Ambrosius, *Z. Anorg. Allg. Chem.* **445** (1978) 211.

9 H. W. Roesky, M. Diehl, B. Krebs and H. Heim, *Z. Naturforsch.* **34b** (1979) 814.

10 D. Haenssgen and I. Pohl, *Chem. Ber.* **112** (1979) 2798.

11 H. W. Roesky and H. Wiezer, *Chem. Ber.* **107** (1974) 1153.

12 P. Foley and M. Zeldin, *Inorg. Chem.* **14** (1975) 2264.

13 W. Petz, *J. Organomet. Chem.* **165** (1979) 199.

14 B. Wrackmeyer, *Z. Naturforsch.* **34b** (1979) 1464.

15 H. Fuess, J. W. Bats, M. Diehl, L. Schoenfelder and H. W. Roesky, *Chem. Ber.* **114** (1981) 2369.

16 A. Tzschach, H. Weichmann and K. Jurkschat, *J. Organomet. Chem. Libr.* **12** (1981) 293.

17 T. Kitazume and J. M. Shreeve, *Inorg. Chem.* **16** (1977) 2040.

18 O. J. Scherer and D. Biller, *Z. Naturforsch.* **22b** (1967) 446.

19 M. Veith, *Angew. Chem.* **87** (1975) 287; *Angew Chem. Int. Ed. Engl.* **14** (1975) 263.

20 M. Veith, *Z. Naturforsch.* **33b** (1978) 7.

21 C. D. Schaeffer and J. J. Zuckerman, *J. Am. Chem. Soc.* **96** (1974) 7160.

22 M. Veith, *Z. Naturforsch* **33b** (1978) 1.

23 M. Veith, O. Recktenwald and E. Humpfer, *Z. Naturforsch.* **33b** (1978) 14.

24 M. Veith, H. Lange, K. Bräuer and R. Bachmann, *J. Organomet. Chem.* **216** (1981) 377.

25 M. Veith, *Z. Anorg. Allg. Chem.* **446** (1978) 227.

26 H. W. Roesky, M. Diehl and M. Banek, *Chem. Ber.* **111** (1978) 1503.

27 M. Veith, M. L. Sommer and D. Jäger, *Chem. Ber.* **112** (1979) 2581.

28 M. Veith and G. Schlemmer, *Chem. Ber.* **115** (1982) 2141.

29 M. Veith, M. Grosser and O. Recktenwald, *J. Organomet. Chem* **216** (1981) 27.

30 M. Veith, *Z. Naturforsch.* **35b** (1980) 20.

31 M. Veith, *Chem. Ber.* **111** (1978) 2536.

32 M. Veith and O. Recktenwald, *Z. Anorg. Allg. Chem.* **459** (1979) 208.

33 M. Veith and H. Lange, *Angew. Chem.* **92** (1980) 408; *Angew Chem. Int. Ed. Engl.* **19** (1980) 401.

34 M. Veith and O. Recktenwald, *Z. Naturforsch.* **36b** (1981) 144.

35 W. W. du Mont, and H. J. Kroth, *Angew. Chem.* **89** (1977) 832; *Angew. Chem. Int. Ed. Engl.* **16** (1977) 792.

36 H. Schumann and H. Benda, *Angew. Chem.* **80** (1968) 845; *Angew. Chem. Int. Ed. Engl.* **7** (1968) 812.

37 H. Schumann and H. Benda, *Chem. Ber.* **104** (1971) 333.

38 C. Couret, J. D. Andriamizaka, J. Escudié and J. Satgé, *J. Organomet. Chem.* **208** (1981) C3.

39 H. Schumann, H. Köpf and M. Schmidt, *Chem. Ber.* **97** (1964) 2395.

40 H. Schumann and M. Schmidt, *Angew. Chem.* **77** (1965) 1039; *Angew. Chem. Int. Ed. Engl.* **4** (1965) 991.

41 H. Schumann, *Angew. Chem.* **81** (1969) 970; *Angew. Chem. Int. Ed. Engl.* **8** (1969) 937.

42 H. Schumann and H. Benda, *Angew. Chem.* **80** (1968) 846; *Angew. Chem. Int. Ed. Engl.* **7** (1968) 813.

43 B. Mathiasch and M. Dräger, *Angew. Chem.* **90** (1978) 814; *Angew. Chem. Int. Ed. Engl.* **17** (1978) 767.

44 B. Mathiasch, *J. Organomet. Chem.* **165** (1979) 295.

45 M. Dräger and B. Mathiasch, *Angew. Chem.* **93** (1981) 1079; *Angew. Chem. Int. Ed. Engl.* **20** (1981) 1048.

46 K. Issleib, F. Krech and E. Lapp, *Synth. React. Inorg. Metalorg. Chem.* **7** (1977) 253.

Tin–Oxygen, Tin–Sulphur, Tin–Selenium and Tin–Tellurium Heterocycles

B. MATHIASCH

University of Mainz, Federal Republic of Germany

16.1 Introduction

This chapter describes inorganic heterocycles containing tin and a chalcogen (oxygen, sulphur, selenium or tellurium).

The monocyclic tin–Group VI heterocycles are illustrated by formulae **(1)**–**(15)**:

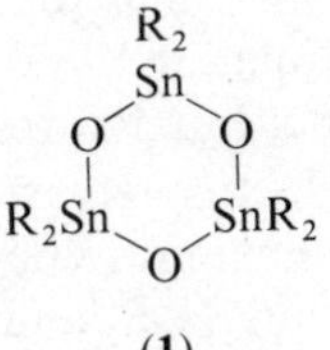

(1)

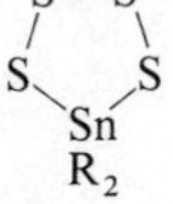

(2) 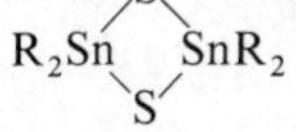　　(3) 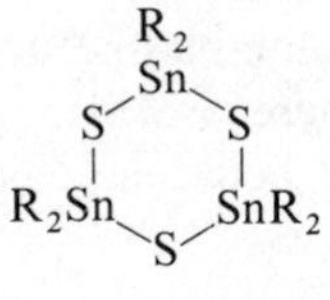　　(4) 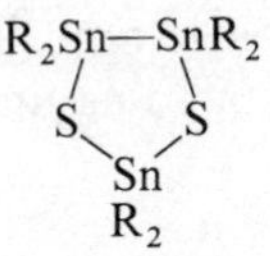　　(5) 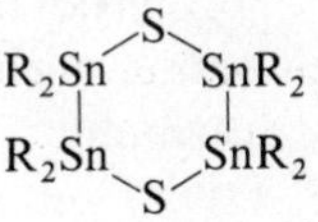　　(6)

THE CHEMISTRY OF INORGANIC HOMO-
AND HETEROCYCLES VOL 1 ISBN 0-12-655775-6

$$R_2Sn\underset{Se}{\overset{Se}{\diagdown}}SnR_2 \qquad R_2Sn\underset{Se}{\overset{Se}{\diagdown}}\underset{R_2}{\overset{Sn}{\diagup}}\underset{Se}{\overset{Se}{\diagdown}}SnR_2 \qquad R_2Sn-SnR_2 \atop Se\diagdown \underset{R_2}{Sn} \diagup Se \qquad R_2Sn\overset{Se}{\diagup}SnR_2 \atop R_2Sn\diagdown Se \diagup SnR_2$$

(7)   (8)   (9)   (10)

$$R_2Sn\underset{Te}{\overset{Te}{\diagdown}}SnR_2 \qquad R_2Sn\underset{Te}{\overset{Te}{\diagdown}}\underset{R_2}{\overset{Sn}{\diagup}}\underset{Te}{\overset{Te}{\diagdown}}SnR_2 \qquad R_2Sn-SnR_2 \atop Te \diagdown \underset{R_2}{Sn} \diagup Te \qquad R_2Sn\overset{Te}{\diagup}SnR_2 \atop R_2Sn\diagdown Te \diagup SnR_2 \qquad R_2Sn-SnR_2 \atop R_2Sn-SnR_2 \diagdown Te$$

(11)   (12)   (13)   (14)   (15)

Tin–oxygen rings (cyclostannoxanes) are little known. The only well-characterized compound of this type is a cyclotristannoxane (1) with $R = Bu^t$; apparently all other diorganotin oxides R_2SnO are high-polymeric stannoxanes. In addition, distannoxanes of the type $(XR_2Sn)_2O$ tend to dimerize or polymerize with formation of four-membered Sn_2O_2 rings (see (16) and (17)) via additional coordination from oxygen to tin.

Tin–sulphur heterocycles are better represented and include sulphur-rich heterocycles like SnS_4 (cyclostannatetrasulphane (2)), rings formed by regular alternation of tin and sulphur, Sn_2S_2 and Sn_3S_3, cyclostannathianes (3) and (4), and tin-rich heterocycles Sn_3S_2 and Sn_4S_2 (5) and (6) containing tin–tin bonds. Similar rings are known with selenium (7)–(10) and tellurium (11)–(15).

Polycyclic species with adamantane-like structures are known, as well as a tin–sulphur–nitrogen ring. There are also mixed species, resulting from redistribution of building units between some of these rings.

16.2   Cyclostannoxanes $(R_2SnO)_n$, $n = 2$ or 3

16.2.1   Preparation and structure

Tin–oxygen rings are usually prepared by hydrolysis of a tin–halogen or tin–nitrogen bond followed by removal of H_2O. The only known Sn–O ring containing tetracoordinated Sn is a trimer, $(Bu_2^tSnO)_3$,[1] (1), $R = Bu^t$, which can be obtained by the following reaction:

$$3 Bu_2^tSnCl_2 + 6 NaOH \xrightarrow[\text{reflux}]{\text{toluene}} 6NaCl + 3H_2O + (Bu_2^tSnO)_3 \qquad (1)$$

The solubility in organic solvents and molecular-weight determinations prove the existence of discrete small molecules. Other Sn–O rings are linked by the interaction of the lone pairs on oxygen with the 5d Sn orbitals, thus giving 5- or 6-coordinate Sn.

Dimethyltin diisothiocyanate and $(Me_2SnO)_x$ in a 1 : 1 molar ratio react to give $[(Me_2SnNCS)_2O]_2$ in refluxing benzene.[2] This compound contains a four-membered Sn_2O_2 ring. Partial hydrolysis of R_2SnCl_2, where $R = Me$ or Et, in the presence of organic bases is used to prepare polymeric dichlorotetramethyl-1,3-distannoxane and its ethyl homologue,[3] a method originally stated for the preparation of 1,2-dichloroditins.[4] These contain Sn_2O_2 rings in their structure, and a comparable ring is assumed for dibutyltin glycolate.[5]

An X-ray structure determination shows that $(Bu_2^tSnO)_3$ (1) contains discrete, planar six-membered Sn–O rings with Sn–O and Sn–C distances of 196 and 219 pm respectively.[6] The OSnO, SnOSn and CSnC angles are 106.9°, 133.1° and 119.9° respectively.

The structure of $[(Me_2SnNCS)_2O]_2$ shown in (16) is of the "ladder-type" with different Sn–O bond lengths (199 and 215 pm) in the central four-membered ring:[2]

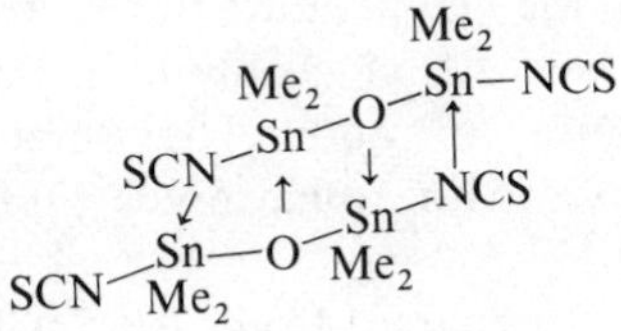

(16)

The endocyclic angles are 75° and 105° at tin and oxygen respectively.

The structure determination of $(Me_4Sn_2Cl_2O)_2$[3] shows an essentially planar Sn_2O_2 ring with slightly elongated bonds. In addition to the ring bonds, the oxygen atoms are each connected to a tin atom of a Me_2SnCl_2 group where, as shown by (17) (bond length in pm), one chlorine acts as a bridging anion for both endo- and exocyclic tin atoms. The geometry of both Sn types lies between octahedral and trigonal bipyramidal co-ordination.[3]

$$
\begin{array}{c}
\text{C} \xrightarrow{209.0} \text{C} \\
\text{Cl} \xrightarrow{243.8} \overset{\displaystyle \text{Sn}}{} \underset{278.8}{\overset{212.0}{\cdots}} \\
202.9 \mid \quad \text{Cl} \\
\text{C} \diagdown \; {}_{211.5} \; \text{O} \\
105° \\
\overset{209.0}{\text{Sn}} \quad 75° \; \text{Sn} \xrightarrow{214.0} \text{C} \\
\text{C} \diagup \; {}_{205.4} \quad \\
\text{Cl} \cdots \text{O} \quad \text{C} \\
\mid \\
\text{Sn} - \text{Cl} \\
\text{C} \diagup \quad \text{C}
\end{array}
$$

(17)

The ethyl homologue incorporates two structural types: one comparable to the methyl compound, the other being more complicated.[3] Other structure determinations of dimeric distannoxanes forming Sn_2O_2 rings have been carried out for $[(CF_3CO_2Vi_2Sn)_2O]_2$, $Vi = vinyl$,[7] and for the Me homologue.[8]

16.2.2 Spectra

In the series $[(R_2SnNCS)_2O]_2$, $R = Me$, Et, Pr^n or Bu^n, the Me species has the highest melting point, the highest value for the CN stretching frequency in the solid, and the lowest solubility in organic solvents. To explain this behaviour, a sulphur bridge has been proposed.[2]

Mössbauer studies in the temperature range from 77 to 175.5 K on $(R_4Sn_2Cl_2O)_2$, $R = Me$ and Et, give only a single somewhat broadened quadrupole split resonance for both. The isomer shifts are 1.28 and $1.50 \, \text{mm s}^{-1}$ respectively, with quadrupole splittings of 3.17 and $3.41 \, \text{mm s}^{-1}$ respectively.[3]

Dibutyltin dialkoxides prepared from Bu_2SnCl_2 and NaOR, where $R = Me$, Et, Pr, Bu, Bu^i, together with $Bu_2Sn(OPh)_2$ are found by ^{119}Sn NMR spectroscopy to be associated, forming four membered Sn–O rings, in contrast with $Bu_2Sn(OBu^t)_2$, $Bu_2Sn(OBu^s)_2$ and $Bu_2Sn(OSiPh_3)_2$. Two distinct ranges of chemical shifts can be recognized; species dimeric in the liquid state have resonances between -150 and -160 ppm (relative to Me_4Sn), whereas the compounds known as monomers are found between -34 and -45 ppm.[9] In the solid state, however, the relatively small quadrupole splitting of $2.4 \, \text{mm s}^{-1}$ is indicative of neither tetra- nor penta-coordination at tin. Thus an octahedral $cis\text{-}R_2SnX_4$ configuration with a polymeric structure has been proposed for the solid.[9]

From ^1H, ^{13}C, ^{15}N and ^{119}Sn NMR data, a six-membered $(SnO)_3$ ring structure (18) is assumed for alkyl- and phenylstannatranes in which the Sn atoms become 6-coordinated by trimerization, $\delta(^{119}Sn) = -530$ to -540 and -350 to -360 ppm respectively:[10]

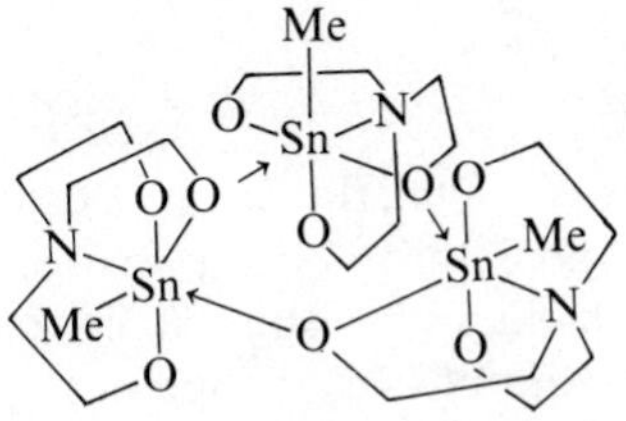

(18)

The reaction of liquid SO_2 with $(R_3Sn)_2O$, where R = Et or Ph, at room temperature gives the following disproportionation:

$$(R_3Sn)_2O + 2SO_2 \longrightarrow R_3SnO_2SR + R_2SnSO_3 \qquad (2)$$

From spectroscopic results, a "foil-type" structural model (19) has been proposed with $(SnOSO)_2$ eight-membered rings crosslinked by Sn–O–S bridges with pentacoordinated tin atoms:[11,12]

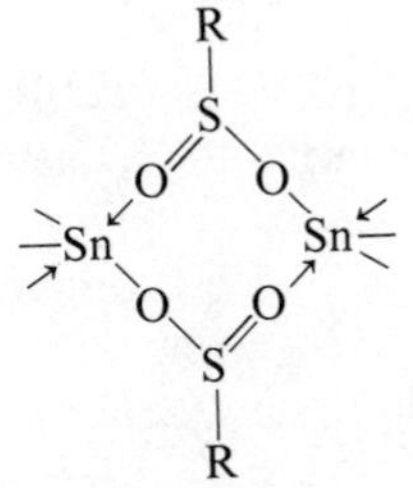

(19)

16.3 Tin–sulphur rings

16.3.1 Cyclostannatetrasulphane R_2SnS_4

The highest sulphur content is found in a transition metal stabilized cyclostannatetrasulphane $(\pi\text{-}CpFe(CO)_2)_2SnS_4$, formed as orange microcrystals by the following condensation reaction:[13]

$$(\pi\text{-}CpFe(CO)_2)_2SnCl_2 + Na_2S_4 \longrightarrow 2NaCl + \quad (3)$$

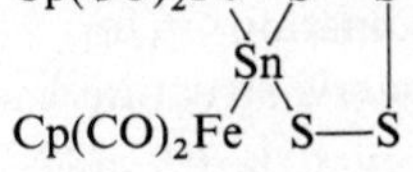

Using K_2S_5, the same five-membered SnS_4 ring is formed along with elemental sulphur, probably a result of a subsequent ring contraction.

16.3.2 Cyclostannathianes $(R_2SnS)_n$, n = 2 or 3

Preparation and structure. Dimeric (3) and trimeric (4) organotin sulphides $(R_2SnS)_n$ ($n = 2$ and 3) have been prepared by earlier workers, for example Pfeiffer,[14] Kozeschkow,[15] Harada;[16] the more recent investigations concern chemical and physical properties.

A crystallographic study of $(Ph_2SnS)_3$ shows it to be monoclinic ($P2_1$), but the molecular structure has not been determined.[17]

Two independent complete structure determinations show the six-membered ring of hexamethylcyclotristannathiane $(Me_2SnS)_3$ to be in a twisted-boat conformation with near C_2 symmetry.[18,19] The C_2SnS_2 units are almost tetrahedral, with Sn–S and Sn–C bond lengths of 241 and 213 pm respectively.[19]

The hitherto unique four-membered organotin–sulphur ring is found in $(Bu_2^tSnS)_2$ (3), and a complete structure determination shows that the four-membered ring is almost planar, with Sn–S distances of 238 and 249 pm and Sn–C distances of 217 and 221 pm. The SSnS and SnSSn angles are 94.3° and 84.0° respectively.[20] Comparable to this is the structure of the hexathiodistannate anion (20) with Sn–S distances of 238 pm and planar $(SnS)_2$ rings.[21]

(20)

Spectra. The ring-stretching frequencies in $(R_2SnS)_3$ compounds, where R = Me, Bu and Ph, are independent of the nature of R.[22]

^{119}Sn NMR measurements on $(R_2SnS)_3$ rings show only a small variation in chemical shifts depending on R, for R = Me, Et, Pr^i and for Bu^t in $(Bu_2^tSnS)_2$, but there are larger changes in the two bond coupling constants.[23]

The effective vibrating mass (EVM) model applied to $(Me_2SnS)_3$ by measuring the temperature dependence of the ^{119M}Sn Mössbauer resonance area and using lattice vibrations from the Raman spectrum gives a value of 552 mass units, thus demonstrating the existence of discrete six-membered rings even in the solid.[25]

Chemical properties. Only a few reactions have been carried out with the compounds $(R_2SnS)_3$. Iron pentacarbonyl and $(Bu_2SnS)_3$ give the Sn_2Fe_2 ring compound in good yield:[26]

$$\tfrac{2}{3}\,(Bu_2SnS)_3 + 2Fe(CO)_5 \longrightarrow 2\,COS + (CO)_4Fe\!\!\underset{\underset{Bu_2}{Sn}}{\overset{\overset{Bu_2}{Sn}}{\diagup\!\!\diagdown}}\!\!Fe(CO)_4 \qquad (4)$$

The reactivity towards SO_2 insertion into the Sn–S bonds of $(R_2SnS)_3$ shows R = Ph to be less reactive. Whereas R = Me or Et give thiosulphite products in 70–75% yields (equation 5) during 20 h at 20 °C, $(Ph_2SnS)_3$ is only converted to sulphites (equation 6) with an excess of SO_2 at 60 °C:[27]

$$\tfrac{1}{3}\,(R_2SnS)_3 + SO_2 \xrightarrow{\;20\,°C\;} R_2SnS_2O_2 \;\; (R = Me,\ Et) \qquad (5)$$

$$\tfrac{2}{3}\,(Ph_2SnS)_3 + 3\,SO_2 \xrightarrow{\;60\,°C\;} 2\,R_2SnSO_3 + 3\,S \qquad (6)$$

The observation of exchange reactions between $(Me_2SnS)_3$ and Me_2SnI_2 to give $(Me_2SnI)_2S$[24] and between $(Me_2SnS)_3$ and $[Me_2Sn(NMe_2)_3]_3$[28] leads to the formation of an intermediate transition complex resulting from nucleophilic attack of Y and/or Z on tin shown schematically below.[24]

$$WXY_2Sn + WXZ_2Sn \;\rightleftharpoons\; WXSnY\!\!\underset{Z}{\overset{Y}{\diagup\!\!\diagdown}}\!\!ZSnXW \;\rightleftharpoons\; 2\,WXYZSn \qquad (7)$$

Similar observations from ^{119}Sn NMR show that on mixing solutions of $(Me_2SnS)_3$ with $(Me_2SnSe)_3$ or $(Me_2SnTe)_3$ all possible six-membered rings result from the redistribution of ring fragments:[23]

(a) R = Me, X = Se or Te   (b) R = Me, X = NEt

As $(Bu^t_2SnS)_2$ and $(Me_2SnS)_3$ redistribute to six-membered products but $(Bu^t_2SnS)_2$ and $(Bu^t_2SnX)_2$, X = Se, Te do not,[29] the model of a pentacoordinated transition complex is strongly supported. The intimate contact in the latter case is prevented by the steric requirements of the Bu^t groups.

In a mixture of $(Me_2SnS)_3$ with $(Me_2SnNEt)_3$ investigated by INDOR NMR,[30] the redistribution products $[(Me_2SnS)_2(Me_2SnNEt)]$ and $[(Me_2SnS)(Me_2SnNEt)_2]$ (see equation 8 b) appear in addition to the parent molecules.

Depending on the method of preparation or on the recrystallization conditions, different modifications of $(Et_2SnS)_n$ are obtained. The degree of oligomerization (n) in $(Et_2SnS)_n$ prepared from $(Et_2SnO)_x$ and CS_2 is 3, m.p. 24 °C, but from $(NH_4)_2S$ or Na_2S, n is much greater, m.p. 147–150 °C.[27] A small polymer fraction of $(Me_2SnS)_n$ is also observed when solutions of the original trimer are allowed to stand.[18]

Similarly, when a solution of $(Pr^i_2SnS)_3$ in dimethylformamide is allowed to stand, a polymer containing zig-zag chain molecules separates. The Sn–S distances are 245 and 240 pm and the chain angles SSnS and SnSSn are 100.7° and 105.6° respectively.[31]

16.3.3 Polycyclic stannathianes $(RSn)_4S_6$

These rings are part of an adamantane-like cage system (21) built from four six-membered SnS rings:

(21)

The preparation methods for these compounds are essentially the same as for $(R_2SnS)_n$ rings,[32,33] where the appropriate organotin halide $RSnX_3$ reacts exothermically with an aqueous solution of Na_2S. The sesquisulphides $(RSnS_{1.5})_4$, which are poorly soluble in all solvents, can thus be prepared in high yields.[34]

A structure determination on $(RSn)_4S_6$ (21), R = Me, shows an adamantane-like molecule, described alternatively as an octahedron of sulphur atoms penetrated by a tetrahedron of MeSn groups.[35] The Sn–S and Sn–C bond lengths are approximately 239 and 214 pm respectively, and lie unequivocally in the expected single-bond range. As the Mössbauer data (isomer shift 1.30–1.42, quadrupole splitting 1.17–1.49 mm s^{-1}) for these compounds are characteristic of tetrahedral $RSnX_3$ units, there is a good reason to exclude higher coordinated Sn species within this series[34] and even for $(ClSn)_4S_6$.[36] Based on the structural data for $(MeSn)_4S_6$,[35] two normal-coordinate analyses have been undertaken to evaluate the force constants.[37,38]

16.3.4   Cyclotri(tetra)stannadithianes $(R_2Sn)_nS_2$, $n = 3, 4$

Five- (5) and six- (6) membered tin–sulphur rings, with an atomic ratio $Sn : S > 1 : 1$, necessarily contain at least one Sn–Sn bond.

The five-membered ring, hexamethylcyclo-1,2,4-tristanna-3,5-dithiane, is prepared in two ways:

(a) by partial oxidation of oligomeric dimethylstannylene $(Me_2Sn)_x$ with sulphur or $(Me_2SnS)_3$:[39]

$$\left.\begin{array}{l} \dfrac{3}{x}(R_2Sn)_x + 2S \\[20pt] \dfrac{1}{x}(R_2Sn)_x + \tfrac{2}{3}(R_2SnS)_3 \end{array}\right\} \longrightarrow \underset{(5)}{R_6Sn_3S_2} \qquad (9)$$

(b) by hydrostannolysis of sulphur, combined with catalytic dehydrogenation of Me_2SnH_2:[40]

$$3Me_2SnH_2 + 4S \xrightarrow{\text{catalyst}} 2H_2S + H_2 + Me_6Sn_3S_2 \qquad (10)$$

In principle the same reaction schemes could produce octamethylcyclo-1,2,4,5-tetrastanna-3,6-dithiane (6) $(Me_8Sn_4S_2)$, a symmetrical ring with two Sn–Sn bonds, but its isolation is prevented by the thermodynamic instability of the product. Therefore only fast reactions can provide considerable amounts of this six-membered ring compound.

Aqueous solutions of tetramethyl-1,2-dichlorodistannane[41] and Na_2S give almost instantaneously the water-insoluble product (6) $(R = Me)$:[42]

$$2R_4Sn_2{}^{2+} + 2S^{2-} \xrightarrow[H_2O]{} \underset{(6)}{R_8Sn_4S_2} \qquad (11)$$

The Sn–Sn stretching frequency in both (5) and (6) occurs at $190\ cm^{-1}$,[39,42] and the two types of Me_2Sn groups in (5) are clearly shown in NMR spectra in C_6D_6 solutions. NMR (^1H, ^{13}C and ^{119}Sn) parameters for compounds $Me_6Sn_3S_2$ (5), $R = Me$, and $Me_8Sn_4S_2$ (6), $R = Me$, are reported in the literature.[23,39,42–44]

Especially in solution, the ring compounds containing the tetramethylditin group are susceptible to reaction with atmospheric oxygen. The six-membered compound (6) disproportionates on standing as shown in equation (12) to the thermodynamically more stable products (4), (5) and $(Me_2Sn)_6$.[42] The final products were identified by mass and NMR spectroscopy.

$$\underset{(6)}{5Me_8Sn_4S_2} \longrightarrow \underset{(5)}{4Me_6Sn_3S_2} + \underset{(4)}{\tfrac{2}{3}(Me_2SnS)_3} + (Me_2Sn)_6 \qquad (12)$$

16.3.5 Cyclostannadithiadiazenes $Me_2SnS_2N_2$

In addition to the redistribution products $[(Me_2SnS)_2(Me_2SnNEt)]$ and $[(Me_2SnS)(Me_2SnNEt)_2]$, mentioned above,[30] a further Sn–N–S ring compound (22) has been identified. The compound, which is monomeric in the gas phase, is obtained by the reaction of disulphur dichloride and tris(trimethylstannyl)amine:[45]

$$2S_2Cl_2 + 2(Me_3Sn)_3N \longrightarrow \underset{\textbf{(22)}}{\overset{\displaystyle Me_2}{\underset{\displaystyle}{\begin{array}{c} Sn \\ N \diagup \quad \diagdown S \\ \| \qquad | \\ S =\!=\!= N \end{array}}}} + Me_4Sn + 4Me_3SnCl + 2S \quad (13)$$

The same product can be obtained from S_4N_4 as shown in equation (14). On heating in the presence of sulphur the intermediate trimethyltin substituted sulphur diimide is converted to Me_4Sn and the cyclic compound.[45]

$$S_4N_4 + 2(Me_3Sn)_3N \longrightarrow 3Me_3Sn-N=S=N-SnMe_3 + S$$
$$\overset{\Delta}{\longrightarrow} Me_4Sn + Me_2SnS_2N_2 \qquad (14)$$

In both the solid state and solution, $Me_2SnS_2N_2$ forms dimers via $N \rightarrow Sn$ coordination; in the former, linking of the almost-planar five-membered rings gives a new four-membered Sn_2N_2 ring (23) (bond lengths in pm):[45]

(23)

The presence of monomers in the gas phase is indicated by an intense M^+ peak in the mass spectrum ($m/z = 242$, intensity 58% relative to $(M-Me)^+$). At room temperature in solution, a single ^1H NMR signal results from the equivalent Me_2Sn groups ($\delta = 0.95$ ppm, $^2J(SnH) = 69.3$ Hz).[46]

16.4 Tin–Selenium rings

16.4.1 Cyclostannaselenanes $(R_2SnSe)_n$, n = 2 or 3

The simplest route to these compounds is by anion exchange, starting from the organotin halide and a solution containing selenide (equation 15). The four-membered ring compound (7), R = Bu^t can be prepared in this way, as an alternative to the oxidation of $(Bu_2^tSn)_4$ with selenium.[20]

$$n\ R_2SnX_2 + n\ Na_2Se \longrightarrow 2\ n\ NaX + (R_2SnSe)_n \qquad (15)$$

Tetra-t-butyl-cyclodistannadiselenane (7) contains a nearly planar four-membered ring as found for the S analogue.[20] However, when the substituents are, for example, Me or Ph, the ring size is six. Table 16.1 gives some structural details for $(Bu_2^tSnSe)_2$[20] and $(Me_2SnSe)_3$.[47]

Table 16.1 Structural data on tin-selenium rings

Compound	Molecular shape	Sn–Se (pm)	SeSnSe	SnSeSn	CSnC	Ref.
$(Bu_2^tSnSe)_2$	Planar	225	97.5°	82.5°	115.1°	20
$(Me_2SnSe)_3$	Twisted-boat	253	109°	101°	118.4°	47

Vibrational data on $(Me_2SnSe)_3$ are used to exclude a chair conformation for the molecule in the solid. The ring Sn–Se stretching modes lie in the range 245–211 cm^{-1},[48] c. 100 cm^{-1} lower than the corresponding Sn–S modes.

^{119}Sn NMR data show that for the four-membered ring compound there is reduced coupling to [77]Se, in agreement with the longer Sn–Se distance (see Table 16.1). In addition tin–tin coupling cannot be resolved, probably as a consequence of the SnSeSn angle being less than 90°.[23,29]

The Sn–Se compounds are generally stable except for a slight tendency towards oxidation and the liberation of red selenium, especially in light.

The Sn–Se bond is susceptible to nucleophilic attack by R_2SnS or R_2SnTe compounds, leading to a mixture of redistribution products, identified by ^{119}Sn NMR spectroscopy.[23] The redistributions are represented by the following equilibria:

$$ \qquad (15) $$

R = Me, X = S or Te

16.4.2 Polycyclic stannaselenane $(MeSn)_4Se_6$

Methyltin tribromide reacts with NaHSe to give the title compound, an adamantane-type molecule (1,3,5,7-tetramethyl-2,4,6,8,9,10-hexaselena-

1,3,5,7-tetrastannaadamantane). The compound is poorly soluble in most solvents, but has a vapour pressure high enough to yield a mass spectrum; the highest peak, m/z 995, corresponds to $(M-Me)^+$.[49]

X-ray structure determination shows a rigid molecule, based on four six-membered Sn-Se rings in the chair conformation (24):

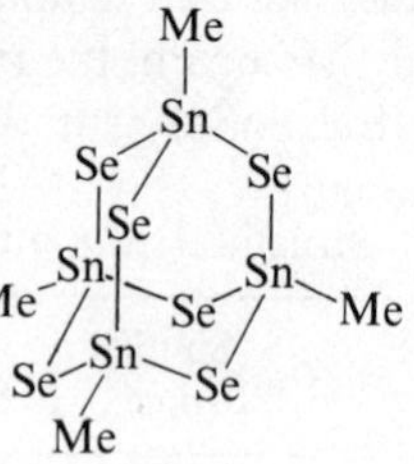

(24)

Some mean values (with standard deviation) are: Sn-Se 252.9(7) pm, Sn-C 214.9(9) pm, SeSnSe 112.5(16)°, SnSeSn 102.9(5)°.[49] Force constants have been calculated from infrared and Raman spectra.[49]

16.4.3   Cyclotri(tetra)stannadiselenanes $(R_2Sn)_nSe_2$

The five-membered ring compound (9) can be prepared by methods analogous to those used for the corresponding sulphur derivatives (see Section 16.3.4).[50] These are

(a) the partial oxidation of $(Me_2Sn)_6$ by selenium; and
(b) combined hydrostannolysis of selenium and catalytic dehydrogenation of Me_2SnH_2.

The product crystallizes from petrol as colourless needles (m.p. 106.5 °C). The six-membered ring (10) results from the following coupling reaction between two R_4Sn_2 units with a mixture of NaHSe and Na_2Se (R = Me):[51]

$$2R_4Sn_2Cl_2 + 2Na_2Se \longrightarrow 4NaCl + \begin{array}{c} Se \\ R_2Sn \diagup \diagdown SnR_2 \\ | \qquad | \\ R_2Sn \diagdown \diagup SnR_2 \\ Se \end{array} \qquad (16)$$

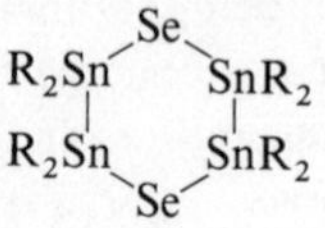

(10)

The five-membered ring compound (9) contains two independent molecules in the asymmetric unit,[52] but each is in the envelope conformation with C_s point symmetry. There are two distinct Sn-Se bond distances of 253 and 257 pm associated with the chemically different tin atoms; the Sn-C distances are 214 pm and do not reflect these differences. Important bond angles are SeSnSe 107°, SnSnSe 104° and SnSeSn 95°. The four-atom sequence Se–Sn–Sn–Se is planar, with no atom deviating by more than 6 pm from the plane; the fifth ring atom on the other hand is some 130 pm from this plane.

Vibrational data, and force-constant calculations, have been obtained,[53] together with full NMR data for $Me_6Sn_3Se_2$ (9), R = Me.[23,43,44]

The Me_4Sn_2 unit in the five-membered ring is comparable to those in octamethylcyclotetrastannadiselenane (10), R = Me, where, from vibrational analysis, the six-membered ring has the boat conformation.[51]

As found for the analogous sulphur derivative, compound (10) disproportionates and rearranges in solution.[51] This type of reaction has been studied in more detail with the tellurium analogue (see Section 16.2).

16.5 Tin–tellurium rings

16.5.1 Cyclostannatelluranes $(R_2SnTe)_n$ n = 2 or 3

The usual preparative methods involve the elimination of NaCl either as an insoluble salt in an organic solvent or as the single soluble species in water. These have been used to prepare $(Bu_2{}^tSnTe)_2$[20] and $(Ph_2SnTe)_3$.[54] The first preparation of $(Me_2SnTe)_3$, however, used the following hydrostannolysis reaction:

$$3Me_2SnH_2 + 6Te \longrightarrow 3H_2Te + (Me_2SnTe)_3 \qquad (17)$$

As $(R_2SnTe)_n$ compounds are insoluble and therefore stable in water, they can be prepared by treating aqueous NaHTe with diorganotin dihalides in a non-miscible organic solvent.[56] The ring compounds are then extracted into the organic phase. For R = Me, however, the preparation can be carried out in aqueous solution, since the $(Me_2SnTe)_3$ product separates as a yellow precipitate.[57]

Only unit-cell data have been obtained for $(Bu_2{}^tSnTe)_2$,[20] but a complete structure determination for $(Me_2SnTe)_3$ shows the presence of a six-membered twisted-boat ring with C_2 symmetry.[57] Mean values for the Sn–Te and Sn–C distances are 275(2) and 220(5) pm respectively, while the SnTeSn angle, 96(1)°, is smaller than in the corresponding Sn–Se and Sn–S ring compounds.

Mass spectrometry shows, from the abundance ratio of M^+ and M—Me^+, increasing stability of the molecular ion in the series $(Me_2SnX)_3$ as X varies from S to Te.[48,55]

Vibrational data on $(Me_2SnTe)_3$ give a range for $\nu(SnC)$ extending from 530 to 490 cm^{-1} and for the ring vibrations from 203 to 173 cm^{-1}.[55]

In the NMR spectra of $(Bu^t_2SnTe)_2$[23] and $(R_2SnTe)_3$ (R = Me and Ph),[58] the t-butyl compound shows exceptionally weak Sn–Te coupling, probably as a consequence of the long bond with small s-overlap.

These compounds are somewhat labile to oxidation of the tellurium, particularly in the light, which leads to darkening of the originally yellow solids. The only reactions investigated are the redistributions mentioned above.[23]

16.5.2   Cyclotri(tetra)stannaditelluranes $(R_2Sn)_nTe_2$ n = 3 or 4

The five membered ring compound (13) is prepared by the combined hydrostannolysis of tellurium combined with catalytic dehydrogenation of Me_2SnH_2.[50] The six-membered ring (14), by analogy with the selenium derivative, results when aqueous $NaHTe^{56}$ reacts with $Me_4Sn_2Cl_2$ solutions.[59] NMR data for the two methyl compounds in C_6D_6 have been reported.[23,43,44,59] Mass spectrometry for (13) shows, from the abundance ratio M^+/M—Me^+, an increase in the stability of the molecular ion M^+ as the Group VI element varies from sulphur to selenium to tellurium.[50]

For the six-membered ring compound (14), vibrational analysis rules out a chair conformation. The Sn–C stretching modes cover the range $519–503\,cm^{-1}$; the ring modes $\nu(SnTe)$, $\nu(SnSn)$ are in the range $203–161\,cm^{-1}$.[59]

The Sn_4Te_2 skeleton is unstable, similarly to the corresponding S and Se analogues, but here distinct light sensitivity can be observed.

The primary step under irradiation with visible light is the excitation of the Te lone pair of electrons. The excited molecule (14) stabilizes itself by the expulsion of dimethylstannylene, and yields the five-membered ring (13) by ring contraction:

$$Me_8Sn_4Te_2 \xrightarrow{h\nu} Me_2Sn: + Me_6Sn_3Te_2 \qquad (18)$$
$$\text{(14)} \qquad\qquad\qquad\qquad \text{(13)}$$

The stannylene attacks excited (14) again, if scavengers such as oxygen or Me_3SnI are absent, to yield the intermediate $\cdot(Me_2SnTe)\cdot$ and a second product (15):

$$Me_2Sn: + Me_8Sn_4Te_2 \longrightarrow \cdot(Me_2SnTe)\cdot + Me_8Sn_4Te \qquad (19)$$
$$\text{(15)}$$

After trimerization of $\cdot(Me_2SnTe)\cdot$ to (12), part of the Me_8Sn_4Te is converted to (13):

$$\tfrac{5}{3}(Me_2SnTe)_3 + Me_8Sn_4Te \longrightarrow 3Me_6Sn_3Te_2 \qquad (20)$$
$$\text{(12)} \qquad\quad \text{(15)} \qquad\qquad\qquad \text{(13)}$$

so that **(12)** disappears from the reaction scheme. The overall reaction is summarized in equation (21). In fact, compound **(12)** was only detected in traces when a sample of **(14)** was irradiated in an inert atmosphere.[59]

$$5Me_8Sn_4Te_2 \xrightarrow{h\nu} 4Me_6Sn_3Te_2 + 2Me_8Sn_4Te \qquad (21)$$

The new compound octamethyl-cyclo-1,2,3,4-tetrastannatellurane **(15)** was not isolated, but was characterized by its ^1H NMR spectrum.

References

1   C. K. Chu and J. D. Murray, *J. Chem. Soc.* A **1971** 360.
2   Y. M. Chou, *Inorg. Chem.* **10** (1973) 673.
3   P. G. Harrison, M. J. Begley and K. C. Molloy, *J. Organomet. Chem.* **186** (1980) 213.
4   O. H. Johnson, H. E. Fritz, D. H. Halvorson and R. Evans, *J. Am. Chem. Soc.* **77** (1950) 5857.
5   S. David and A. Thiéffry, *Tetrahedron Lett.* **22** (1981) 2885.
6   H. Puff, W. Schuh, R. Sievers and R. Zimmer, *Angew. Chem.* **93** (1981) 622.
7   C. D. Garner, B. Hughes and T. J. King, *Inorg. Nucl. Chem. Lett.* **12** (1976) 859.
8   R. Fagliani, J. P. Johnson, I. D. Brown and R. Birchall, *Acta Crystallogr.* **B34** (1978) 3734.
9   P. J. Smith, R. F. M. White and L. Smith, *J. Organomet. Chem.* **40** (1972) 341.
10 K. Jurkschat, C. Mügge, A. Tzschach, A. Zschunke, G. Engelhardt, E. Lippmaa, M. Mägi, M. F. Larin, V. A. Pestunovich and M. G. Voronkov, *J. Organomet. Chem.* **171** (1979) 301.
11 U. Kunze and A. P. Völker, *Chem. Ber.* **107** (1974) 3818.
12 U. Kunze, E. Lindner and J. Koola, *J. Organomet. Chem.* **38** (1972) 51.
13 C. Ungurenasu, G. Stiubianu and E. Streba, *Synth. React. Inorg. Metalorg. Chem.* **3** (1973) 211.
14 P. Pfeiffer and R. Lehnhardt, *Z. Anorg. Allg. Chem.* **68** (1910) 102.
15 K. A. Kozeschkow, *Chem. Ber.* **66** (1933) 1661.
16 T. Harada, *Bull. Chem. Soc. Jpn.* **17** (1942) 281, 283.
17 N. R. Kunchur and S. Borhani, *J. Appl. Crystallogr.* **9** (1976) 508.
18 B. Menzebach and P. Bleckmann, *J. Organomet. Chem.* **91** (1975) 291.
19 H. J. Jacobsen and B. Krebs, *J. Organomet. Chem.* **136** (1977) 333.
20 H. Puff, R. Gattermayer, R. Hundt and R. Zimmer, *Angew. Chem.* **89** (1977) 556.
21 B. Krebs, S. Pohl and W. Schiwy, *Angew. Chem.* **82** (1970) 884.
22 P. G. Harrison and S. R. Stobart, *J. Organomet. Chem.* **47** (1973) 89.
23 A. Blecher, B. Mathiasch and T. N. Mitchell, *J. Organomet. Chem.* **184** (1980) 175.
24 J. D. Kennedy, W. McFarlane, G. S. Pyne, P. L. Clarke and J. L. Wardell, *J. Chem. Soc., Perkin Trans. 2* **1975**, 1234.
25 R. H. Herber and M. F. Leahy, in *Organotin Compounds: New Chemistry and Application*, edited by J. J. Zuckerman (Am. Chem. Soc., Washington, 1976).
26 R. A. Burnham, M. A. Lyle and S. R. Stobart, *J. Organomet. Chem.* **125** (1977) 179.
27 U. Kunze and R. Hengel, *Chem. Ber.* **109** (1976) 2793.
28 J. D. Kennedy and W. McFarlane, *Rev. Silicon, Germanium, Tin, Lead Compounds* **1** (1974) 235.

29 A. Blecher and B. Mathiasch, *3rd Int. Symp. on Inorganic Ring Systems, Graz, 1981.*
30 A. G. Davies, P. G. Harrison, J. D. Kennedy, T. N. Mitchell, K. J. Puddephatt, W. Ramsay and W. McFarlane, *J. Chem. Soc.* C **1969**, 1136.
31 H. Puff, A. Bongartz, R. Sievers and R. Zimmer, *Angew. Chem* **90** (1978) 995.
32 M. Komura and R. Okawara, *Inorg. Nucl. Chem. Lett.* **2** (1966) 93.
33 J. A. Forstner and E. L. Muetterties, *Inorg. Chem.* **5** (1966) 552.
34 A. G. Davies, L. Smith and P. J. Smith, *J. Organomet. Chem.* **39** (1972) 279.
35 D. Kobelt, E. F. Paulus and H. Scherer, *Acta Crystallogr.* **B28** (1972) 2323.
36 C. H. Stapfer, *Inorg. Chem.* **9** (1970) 421.
37 A. Müller, N. Noltes, S. J. Cyvin, B. N. Cyvin and A. J. P. Alix, *Spectrochim. Acta* **34A** (1978) 383.
38 A. Blecher, B. Mathiasch and M. Dräger, *Z. Anorg. Allg. Chem.* **488** (1982) 177.
39 B. Mathiasch, *J. Organomet. Chem.* **122** (1976) 345.
40 B. Mathiasch, *J. Organomet. Chem.* **165** (1979) 295.
41 B. Mathiasch, *Inorg. Nucl. Chem. Lett.* **13** (1977) 13.
42 B. Mathiasch, *Synth. React. Inorg. Metalorg. Chem.* **7** (1977) 227.
43 B. Mathiasch, *Inorg. Nucl. Chem. Lett.* **13** (1977) 271.
44 B. Mathiasch, *Inorg. Nucl. Chem. Lett.* **13** (1977) 5.
45 H. Roesky, *Z. Naturforsch.* **31b** (1976) 680.
46 H. Roesky and H. Wiezer, *Angew. Chem.* **85** (1973) 722.
47 M. Dräger and A. Blecher, *J. Organomet. Chem.* **161** (1978) 319.
48 B. Mathiasch and A. Blecher, *Bull. Soc. Chim. Belg.* **84** (1975) 1045.
49 A. Blecher, M. Dräger and B. Mathiasch, *Z. Naturforsch.* **36b** (1981) 1361.
50 B. Mathiasch, *Z. Anorg. Allg. Chem.* **432** (1977) 269.
51 B. Mathiasch. *J. Organometal. Chem.* **141** (1977) 189.
52 M. Dräger and B. Mathiasch, *Z. Anorg. Allg. Chem.* **470** (1980) 45.
53 M. Dräger, B. Mathiasch and A. Blecher, *Fresenius' Z. Anal. Chem.*, **304** (1980) 268.
54 B. Mathiasch, *Org. Magn. Reson.* **17** (1981) 296.
55 A. Blecher and B. Mathiasch, *Z. Naturforsch.* **33b** (1978) 246.
56 D. L. Klayman and T. S. Griffin, *J. Am. Chem. Soc.* **95** (1973) 197.
57 A. Blecher and M. Dräger, *Angew. Chem.* **91** (1979) 740.
58 B. Mathiasch and M. Dräger, *3rd Int. Symp. on Inorganic Ring Systems, Graz, 1981.*
59 B. Mathiasch, *J. Organomet. Chem.* **194** (1980) 37.

Index

Adamantane–like systems, 110, 160,
    177, 279, 280, 282, 349, 371,
    402, 408, 411, 438, 469, 486,
    693, 708, 720, 725
  cage systems, 408
Alazenes, see Aluminium—Nitrogen
Alcoholysis
  of phosph(III)azanes, 487
  of phosphazenes, 538–544
  thioalcoholysis, 549–550
Alkylamido aluminium derivatives,
    167–171
Aluminium
  in B—N rings, 88–89
  in P(III)–N rings, 474, 480
Aluminium—nitrogen
  Al–N–Al bridges, 184
  association degrees, 168–169
  bond lengths, 180–183
    in B—N rings, 88–89
  cage compounds, 171–178
    bond lengths and bond angles, 181
    list, 172–173
  catalysts, 185–187
  chemical stability, 184
  preparation methods, 168–169
  reactions, 184–185
  spectra, 183–184
  structure, 178–179
  uses, 185–187
Aminoboranes
  with 4–membered rings, 74–77
  with 5–membered rings, 84
  with 6–membered rings, 80–81
Antibacterial activity, boroxanes, 129
Antimony
  heterocycles, 731–733
  homocycles, 729–730
  in P rings, 447–460
  in phosphazine rings, 578
  in S—O rings, 872

Antimony (cont'd)
  in Si—N rings, 255–6
Aminoboroxanes, 124
Arsenic
  homocycles,
    classification, 701
    monocyclic, 702–707
    polycyclic and cage compounds,
      707–709
  in B—N rings, 66–67
  in Ge rings, 362, 364
  in P rings, 447–460
  in P(III)–N rings, 474
  in phosphazine rings, 578
  in P—S cage compounds, 689,
    696–697
  in S—N rings, 815–816, 818
  in Si—N rings, 255–256
  in Si—N—O rings, 267–268
Arsenic—nitrogen
  arsenic (III), 714–716
  arsenic (V), 716–717
Arsenic—oxygen
  arsenic (III), 718
  arsenic (V), 719
  cage compounds, 720
Arsenic—selenium, 724–725
Arsenic–silicon, 725
Arsenic—sulphur
  monocyclic, 721
  cage compounds, 722–724
Arsinoborines, 107–108
Azaborolidines, 55–56, 63–64, 76
Azadiphosphiranes, 449
Azasulphanes, 847–854
Azide radicals, generation, 418

Biological activity, of boroxanes, 129
Bismuth compounds, 733–734

Boracyclohexane/ene, 64–66
  preparation, 71
Boracyclopentanes, 60, 62
Boranes
  allyl-, 79–80
  amino-, 74–77, 80–81, 84
  halogeno-, 86, 159–160
  hydrazino-, 81–83
  imino-, 78–80, 185
  iminoxy-, 85–86
  methylthio- and alkylthio-, 158
  and phosph(III)azanes, 476, 492
  phosphineamino-, 87
  polyboranes, 15–16
  pyrazolyl-, 83
    selenium-containing, 163
  structure, 3
Borates
  metal thioborates, 160
Borates, cyclic
  in aqueous solution, 121–122
  crystal structures, 114–121
  hexaborates, 120
  uses, 128–129
Boratetroxanes, 129–130
Borazines
  chemical reactions, 34–37
  crystal and molecular structure, 45–47
  dipole moment, 49
  fluorescence spectra, 49
  functional derivatives, 21–27
  hydrolysis, 36, 49
  inorganic derivatives, 19–21
  IR spectra, 48
  macrocyclic ring systems, 30–32
  magneto-optic studies, 47
  mass spectra, 44–45
  metal derivatives, 37–38
  miscellaneous derivatives, 27–28
  nuclear magnetic resonance data, 39–41
  nuclear quadrupole resonance, 47
  photochemical reactions, 32–34
  photoelectron spectra, 41–44
  physical properties, 39–41

Borazines (cont'd)
  polycyclic systems, 28–30
  potential commercial uses, 49–50
  Raman spectrum, 48
  silaborazines, 60–62
  spectroscopic properties, 39–49
  trichloroborazine, 76
  unsubstituted, 19
  UV spectra, 47–48
Borazocines, 50–51
Boretanes, 145–146
Boretidines, 51–53
Borinanes, 146–149, 153–156, 161
Borolanes, 149–153, 155–157, 161
Borolidines
  azaborolidines, various, 76
  fused ring systems, with heteroatom, 72–74
  monomeric/dimeric forms, 76
  with selenium, 72, 162
  triazadiborolidines, 56–59, 85
  triazaphosphaborolidines, 63–64
Borolines
  tetrazaboroline, 53–54
Borines
  tetrazadiborines, 55–56
Boron
  chemistry, criteria for inclusion, 3
  disulphide, 145
  homocycles, 15–16
  hydrides, 15–16
  in P rings, 447–460
  in P(III)–N rings, 474, 476
  in phosphazine rings, 577
  in Si rings, 209, 210
  in Si—N—O rings, 267–268
  in Si—S rings, 355–356
Boron—arsenic; 107–108
  boron—nitrogen—arsenic; 66–67
Boron—nitrogen, see also Borazines
  with 3-coordinate boron
    B—N—As, 66–67
    B—N—O, 67–68
    B—N—O—Si, 68
    B—N—P, 62–66
    B—N—S, 68–72

Boron-nitrogen
  with 3-coordinate boron (*cont'd*)
    B—N—Se, 72
    B—N—Si, 60–62
    borazocines, 50
    boretidines, 51–53
    borolidines, 53–54, 56–59
    fused-ring systems, 72–74
  with 4-coordinate boron, 74–89
    4-membered rings, 74–80
    6-membered rings, 80–81
    with heteroatoms, Al, O, P, S, 84–89
    hydrazinoboranes, 81–83
    pyrazaboles, 83–84
  borolidines, 55–56, 63–64, 76
  fused-ring systems with heteroatoms, 72–74
  nomenclature, 17–18
Boron—nitrogen—aluminium, 88–89
Boron—nitrogen—arsenic, 66–67
Boron—nitrogen—oxygen, 67–68, 85–86
Boron—nitrogen—oxygen—silicon, 68–69
Boron—nitrogen—phosphorus, 62–66, 86–87
Boron—nitrogen—selenium, 72
Boron—nitrogen—silicon, 60–62, 68–69
Boron—nitrogen—sulphur, 69–72, 87–88
Boron—oxygen, *see also* Boroxanes
  boron—nitrogen—oxygen, 67–68, 85–86
  cyclic borates, 114–122
  cyclodiboratetroxanes, 129–130
  cyclodiboratrioxanes, 110–111
  cyclodiboroxanes, 110
  cyclotriboroxanes, 111–122
    fluoroborates, 114
    halogen derivatives, 113
    hydride, 112
    organic derivatives, 122–124
    ring types, 109
Boron—phosphorus
  cyclodiphosphinoborines, 104

Boron—phosphorus (*cont'd*)
  cyclotriphosphinoborines, 105–107
  condensed and tetra-, 107
Boron—selenium
  with 3-coordinate boron, 161–162
  with 4-coordinate boron, 163
  boron—nitrogen—selenium, 72
Boron—silicon, 209
  boron—nitrogen—silicon, 60–62, 68–69
  boroxanes, 124
Boron—sulphur
  with 3-coordinate boron, 144–157
    boretanes, 145–146
    borinanes, 146–149, 153–157
    borolanes, 149–153, 155–157
  with 4-coordinate boron, 157–160
    donor–acceptor compounds, 157–158
    metal thioborates, 160
    thioboranes, 158–160
  boron—nitrogen—sulphur, 69–72, 87–88
    donor–acceptor compounds, 157–158
Boroxanes, *see also* Boron—oxygen
  addition reactions, 126–127
  biological activity, 129
  chemical properties, 126–128
  insertion reactions, 127
  organic derivatives, 122–124
  physical properties/structures, 124–126
  ring-cleavage reactions, 127
  types, 110–114
  uses, 128–129
Bridged rings, 4, 5, 7

Cage structures, 7
Catalysts: Al—N compounds, polymerization systems, 186
Catalysts, siloxane–bound metal complexes, 293
Cubane tetramers, 167
Cycloarsanes, *see* Arsenic homocyles
Cycloazasulphanes, 847–854

Cycloboranes, 15–16
Cycloborasilathianes, 355–356
Cyclodiboratrioxanes, 110
Cyclodiboroxanes, 110
Cyclodisilatriazanes, 245–246
Cyclodithiazene, 801–803
Cyclogermanes, 361–363
Cyclophosphanes, *see* Phosphanes
Cyclophosphapolysilanes, 277–278
Cyclophosphates, *see* Phosphorus—
oxygen
Cyclophosphathianes, *see* Phosphorus—sulphur
Cyclophosph(III)azanes, *see* Phos-
ph(III)azanes
Cyclophosphazenes, *see* Phosphazenes
Cyclopolysilanes:
cyclohexasilanes, 202–205
halogenated, 203–205
cyclopentasilanes, 197–202
halogenated, 202
cyclotetrasilanes, 192–197
halogenated, 195–196
cyclotrisilanes, 191–192
heterocyclic, 209–216
with Group III, 209
with Group IV, 209–211
with Group V, 211–213
with Group VI, 213–216
polycyclic systems, 206–208
radicals, 208
Cyclosilaphosphanes, 278–282
cage molecules, 282–283
reactions, 283–285
Cyclosilathianes, 350–354
chemical reactions, 353–354
preparation, 350–352
structure, 352
uses, 354
Cyclosilathioxanes, 356–357
Cyclosilazanes: *see also* Silicon—nitrogen
chemical properties and reactions,
237–244
rearrangements, 239–243
with ring cleavage, 244

Cyclosilazanes: *see also* Silicon—nitrogen (*cont'd*)
with ring preservation, 237–239
cyclodisilazanes, 224–228
cyclotetrasilazanes, 228–230
cyclotrisilazanes, 228–230
physical properties, 236–237
polycyclic, 230–233
fused, 232
spirocyclic, 231–232
spectra, 236
structures, 221–224
and bonding, 233–235
uses 244–245
Cyclosilazoxanes:
cyclodisilazoxanes, 257
cyclohexasiladiazatetroxanes, 261
cyclopentasiladiazatetroxanes, 261
cyclotetrasilazoxanes, 259–261
cyclotrisilazoxanes, 257–259
cyclotrisilazadioxanes, 259
rearrangements, 261–263
with N—N ring units, 263–264
with other hetero atoms, 266–269
with Si—Si units, 264
structures, 223
uses, 264
Cyclosiloxanes:
carbon-functional, 290
chemical properties, 291–293
cyclodisiloxanes, 287–288
cyclotetrasiloxanes, 290, 301–308
cyclotrisiloxanes, 295–300
heterofunctional condensation, 290
heterocyclosiloxanes, 294, 317–319
higher cyclosiloxanes, 309–311
hydrolysis, 289
list of compounds, 295–319
organocyclosiloxanes, 288–289
reactions on substituents, 293
uses, 289
preparation, 289–290
polycyclic siloxanes, 312–316
polymerization, 291–293
depolymerization, 290
silsesquioxanes, 293–294

Cyclosilazoxanes (*cont'd*)
  structures, 287–289
Cyclosilthiazanes, 265–266
Cyclostannadithiadiazenes, 410
Cyclostannanes:
  chemical reactions, 380
  spectra, 380
  structural data, 378–379
  syntheses, 377–378
Cyclostannaphosphanes, 395–399
  polycyclic, 397–399
Cyclostannazanes, 383–388
Cyclostannaselenanes, 410–413
Cyclostannatelluranes, 413–415
Cyclostannatetrasulphane, 405–406
Cyclostannathianes, 406–408
  polycyclic, 408
Cyclostannoxanes, 402–405
Cyclotetrathiazanes, 852–854
Cyclotetrathiazene, *see* Sulphur—nitrogen
Cyclotetrathiazene dioxide, 843–845
Cyclotetrathiazene tetroxide, 845
Cyclothiazenes:
  5–membered, 815
  6–membered, 816–818
  8–membered, 818–820
  metallocyclothiazenes, 820–822
Cyclotriazanes, 417–418
Cyclotriboroxanes, III, *see also* Boroxane and Boron—oxygen
Cyclotrigermanes, 362
Cyclotrisiladithianes, 355
Cyclotrisilazanetrispiro, structure, 235
Cyclotristannadithianes, 409

Diazadiboretidines, 51–53
Diazadiphosphetidines, 620–651
Diboratetroxanes, 129–130
Diboratrioxanes, 110–111
Diboroxanes, 110
Diphosphadithianes, 682–686
Diselenazadiborolidines, 162
Disilazanes, 224–228
Disilatetrazanes, 246–247

Disilathiatriazanes, 265
Disilatriazanes, 245–246
Disilazoxanes, 257
Distannadiphosphanes, 295–296
Disulphur dinitride, 801–803
Dithiadiboretanes, 145–146
Dithiatetraborinanes, 156–157
Dithiatriborolanes, 155–156

Friedel–Crafts reactions, 562, 840
Fused rings, definition, 7

Germanium
  heterocycles, 364
  homocycles, 361–363
  polycycles, 361–363
  in P rings, 447–460
  in Si rings, 209
  in Si—N rings, 249–251
  in Si—N—O rings, 267–268
Germanium—arsenic, 362, 364
Germanium—nitrogen, 368–370
Germanium—oxygen, 362, 364, 371–372
  organocyclogermanes, 371
Germanium—phosphorus, 370–371
Germanium—selenium, 364, 373–374
Germanium—silicon, 209
Germanium—sulphur, 372–373
Germanium—tellurium, 364, 374

Heterocycles, types, 6–8
Heterocyclogermanes, 364
Heterpcyclosilazanes structures, 221–222
Heterocyclosilazoxanes, 266–269
Heterocyclothiazenes, 815–820, 831–833
Hexasiladiazatetroxanes, 261
Homocycles, 6
  transition to heterocycles, 7
HOMO-LUMO, orbital crossing, 16, 825
Hückel sextet, 419

Hydrazinoboranes, 81–83

Iminoboranes, 78–80
Inorganic hetero/homocycles
classification, 5–8
definition, 2, 5
nomenclature, 9–13
Iron-containing cyclopentasilanes, 200

Lead
in P rings, 447–460
in Si—N rings, 252
thioborates, 161

Mercaptoboranes, 77
M.O. calculations, 42–44, 52, 521–523,
852, 855

Nitrogen
dinitrogen dimer, 418
homocycles,
3–membered, 417
4–membered, 418
5–membered, 419
6–membered, 419–420
Hückel sextet, possibility, 419
in P rings, 447–466
in Si rings, 212
in Sn—S rings, 410
tetra–nitrogen
anions, 419
cations, 419
tri-nitrogen cations, 418
Nitrogen—phosphorus heterocycles,
test diffraction studies, 579–
588
Nomenclature, 9–13
additive/replacement, 10
Gmelin, 10
Hantzch–Widman, 9

Nomenclature, (cont'd)
repeating unit, 11
roots, 12

Oxygen
in B—N rings, 67–69, 86
in B—N—Si rings, 68–69
in Ge rings, 364
in P(III)–N rings, 494
in P(V)–S cages, 693
in Si rings, 213–215
in Si—N rings, 257–261
in Si—S rings, 356–357

Pentasiladiazatrioxanes, 261
Pentasilatetrazanes, 249
Pentasulphur hexanitride, 805–806
Pentathiazyl cation, 810–811
Pentazoles, 419
Peroxoborates, 129–130
Phosphanes
monocyclic,
anions, 435–444
chemical properties and reactions,
431–435
miscellaneous formation reactions,
427–429
nomenclature, 423–424
spectra, 429–431
structures, 423, 429–431
syntheses, 425–426
polycyclic
anions, 442–444
chemical properties, 446–447
hydrides, 438
metal polyphosphides, 442–444
organophosphanes thermolysis,
441–442
spectra, structure and bonding,
442–446
syntheses, 439–442
Phosph(III)azanes
alcoholysis, 487
aminocyclophosphazanes, 476

Phosph(III)azanes (*cont'd*)
 aminolysis, 485–487
 borane adducts, 492
 borane elimination reaction, 476
 cage-like compounds, 476
 chemical reactions, 484–495
 cis/trans isomers, 481
 halogen exchange, 485
 NMR spectroscopy, 482–484
 nomenclature, 467
 oxidative addition, 488–492
 phosphazane-related compounds,
    476–478
 preparation, 468–472
 ring and cage molecules, 494
 ring rearrangement, 493
 and silylamines, 472–474
 spectra, 482–484
 structural studies, 478–482
 substitution reactions, 485
 thermal, hydrolytic and oxidative
    stability, 484–485
 transamination, 474–476
Phosph(V)azanes:
 definition, 617
 diazadiphosphetidines
    condensed, 636–651
    monocyclic, 620–629
    spirocyclic, 624, 629–636
 phosphazene alternative, 618–619
 thiadiazaphosphetidines, 651–652
Phosphazenes
 alcoholysis, 538–544
 alkoxyphosphazenes, thermal
    rearrangements, 547–549
 amines, reactions with, 523–535
 ammonolysis, 523–524
 basicity, 515–517
 bicyclic, 550–556
 bond lengths, 507–508
 chromatography, 519–520
 cis and *trans* isomers, 520
 clathrates, 566–567
 conductivity, 520
 diffraction studies, 507–508, 579–588
 dipole moment studies, 520

Phosphazenes (*cont'd*)
 electronic structure, 521–523
 Faraday effect, 519
 Friedel–Crafts reactions, 562–563
 halogenation reactions, 535–538
 heterophosphazenes, 577–578
 hydridophosphazenes, 567–571
 hydrolysis, 544–547
 Lewis acid–base reactions, 563–566
 nomenclature, 501–502
 optical activity, 503
 organometallic reactions, 556–561
 polycyclic, 550–556
 polymers, 575–576
 preparation, 502–507
 reactions at exocyclic positions, 571–
    574
 reagents, organic synthesis, 574
 related compounds, diffraction data,
    579–586
 spectroscopy
    electronic, 513–515
    electron spin resonance, 515
    mass, 517–519
    NMR, 508–510, 527
    nuclear quadrupole resonance,
       512–513
    vibrational, 510–512
 spirocyclic, formation, 505, 550–556
 structures, 579–588
 thermal instability, 469
 thermolysis and ring-opening, 575–
    576
 thioalcoholysis, 549–550
 uses, 576
 vapour-pressure measurement, 520
Phosphine substitution, organocyclo-
    siloxanes, 293
Phosphino-borines, *see* Boron—phos-
    phorus
Phosphiranes, 449
Phosphorus, *see* Phosphanes
 in B—N rings, 62–66, 73, 87
 in Si rings, 213
 in S—N rings, 816–819, 832–834
 in Si—N rings, 253–255

Phosphorus (*cont'd*)
  in Si—N—O rings, 267–268
Phosphorus—arsenic, 447–456
Phosphorus—arsenic—sulphur, 696–
  697
Phosphorus—antimony, 447–452
Phosphorus—boron, 103–107, 447–454
Phosphorus—germanium, 370–371,
  447–452
Phosphorus—nitrogen
  boron—nitrogen—phosphorus, 62–
    66, 86–87
  crystal structure data, 479–481
  cyclophosphane-based, 447–454
  diffraction studies, cyclophospha-
    zenes/related compounds,
    579–586
Phosphorus—nitrogen—boron, 474
Phosphorus—nitrogen—oxygen, 494
Phosphorus—nitrogen—selenium, 488,
  491
Phosphorus—nitrogen—sulphur, 488,
  490, 494
Phosphorus—nitrogen—tellurium, 488
Phosphorus—oxygen
  classification, 659–660
  cyclodecaphosphates, 674
  cyclododecaphosphates, 675
  cyclohexaphosphates, 671–672
  cyclo-octaphosphates, 673–674
  cyclopentaphosphates, 671
  cyclotetraphosphates, 666–671
  cyclotriphosphates, 660–666
  phosphate esters, 676
  phosphate—tellurate, 676
Phosphorus—selenium
  other cage compounds, 695–696
Phosphorus—silicon, 447–454
Phosphorus—sulphur
  cage compounds, 688–695
  classification, 681
  cyclodiphosph(V)adithianes, 682–686
  ring compounds, 686–687
Phosphorus—tellurium, 447–450, 676
Phosphorus—tin, 447–452
Phosphorus—titanium, 449

Phosphorus—zirconium, 449
Polysilanes, 206–208
Pyrazaboles, 83–84
Pyrazolylborates, 83

Selenadiazadiborolidines, 162
Selenium
  homocycles
    cations, 777–781
    neutral rings, 770–776
    thermodynamic properties, 775–
      776
    types, 770–772
  in B—N rings, 72
  in Ge rings, 362
  in P rings, 447–460
  in S rings, 782–789
  in Si rings, 215–216
  in Sn rings, 412
Selenium—boron, 161–163
Selenium—oxygen, 873–874
Selenium—nitrogen, 874–876
Selenium—sulphur
  HPLC, 788
  preparation, 782–785
  reactions, 788–789
  structure and spectra, 785–788
Silaborazines, 60–62
Silathia-azanes, 265–266
Silathioxanes, 356–357
Silanes, *see* Cyclosilanes
Silaphosphanes, 278–285
  cyclophosphanes, 278–282
  cage structures, 282
  chemical behaviour, 283–285
  P—P bonds, 285
  P—Si bonds, 278–283
Silastannazanes, 388–391
Silathianes, 350–354
Silathiazanes, 223, 264–266
  arsenic, 268
  boron, 268
  germanium, 268
  phosphorus, 268
  sulphur, 269

Silathiazanes (*cont'd*)
  tin, 268
Silazanes, *see* Cyclosilazanes and sili-
    con–nitrogen
Silazoxanes, *see also* Cyclosilazoxane
Silicon
  homocycles, *see* Cyclosilanes
  in B—N rings, 60–62, 63, 73
  in B—N—O rings, 68–69
  in B—S rings, 157
  in P rings, 447–460
  in P(III)–N rings, 473–474, 491
  in S—N rings, 818
  in Sn—N rings, 388–395
Silicon—boron, 209
Silicon—germanium, 209
Silicon—nitrogen, *see also* Cyclosila-
    zanes
  classification, 221–224
  germanium in, 211
  with Group IV elements, 249–252
  with Group V, 252–256
  with N—N bonds, 245–247
  with Si—Si bonds, 247–249
  spirocyclic compounds, 247–248
  sulphur-containing, 264–266
Silicon—nitrogen—oxygen:
  cyclosilazoxanes, 257–264
    heterocyclosilazoxanes, 266–269
  with hetero atoms, 267–268
Silicon—nitrogen—phosphorus, 210
Silicon—nitrogen—sulphur, 210, 211,
    223–224, 264–266
Silicon—oxygen, *see* Cyclosiloxanes
Silicon—phosphorus, 278–285
  cage structures, 282
  chemical behaviour, 283–285
  structures, 279
Silicon—selenium, 215–216
Silicon—sulphur
  from cyclosilanes, 215–216
  cycloborasilathianes, 355–356
  cyclosilathianes, 350–354
  cyclosilathioxanes, 356–357
  cyclotrisiladithianes, 355
Silicon—sulphur—boron, 355–356

Silicon—sulphur—oxygen, 356–357
Siloxanes, *see* Cyclosiloxanes
Silsesquioxanes, 293–294
Spirocyclic rings, definition, 7
Stannadithiadiazenes, 410
Stannatetrasulphates, 405
Stannanes, 377–380
Stannaphosphanes, 395–399
Stannaselenanes, 410–413
Stannasulphanes, 405–406
Stannatellurances, 413–415
Stannathianes, 406–410
Stannazanes, 383–388
Stannoxanes, 402–405
Sulphanuric halides, 834
  phosphazene rings, 838–841
  thiazyl rings, 841–842
Sulphides, phosphorus-arsenic, 696–
    697
Sulphur
  cyclosilazoxanes, 269
  fused-ring systems, 74
  homocycles, *see below*
  hydrazides, 854
  imides, 847–854
  in B—N rings, 69–72, 74, 87–88
  in Ge rings, 364
  in P rings, 447–460
  in P(III)–N rings, 471, 494
  in Si rings, 210, 215–216
  in Si—N rings, 264–266
  in Si—N—O rings, 269
  nitrides, 801–803, 805–806
    dinitrides, 801–805
    hexanitrides, 805–806
    pentanitrides, 810, 814–815, 845
    tetranitrides, 794–801
    trinitrides, 812–814
Sulphur homocycles
  cations, 751–756
  chromatography, 748
  classification, 737
  neutral rings, 738–751
    mixtures, 748–750
  oxides, 756–760
  preparation, 738–741

Sulphur homocycles (*cont'd*)
reactions, 750–751
spectra, 745–746
structures, 741–745
sulphur—halide cations, 760–762
thermodynamic properties, 746–748
Sulphur—nitrogen:
classification, 793
saturated, 2-coordinate sulphur
cyclic sulphur hydrazides, 854
cyclic sulphur imides, 847–852
metal complexes, 854–857
tetrasulphur tetraimide, 852–854
unsaturated, 2-coordinate sulphur
cyclopentathiazyl cation, 811–812
cyclotetrathiazyl, 809–810
disulphur dinitride, 801–803
heterocyclothiazenes, 815–820
metallocyclothiazenes, 820–822
pentasulphur hexanitride, 805–806
tetrasulphur dinitride, 803–805
tetrasulphur pentanitride
anion, 814–815
tetrasulphur pentanitride cation,
810
tetrasulphur tetranitride, 794–801
thiodithiazyl cation, 806
thiotrithiazyl cation, 807–809
trisulphur trinitride anion, 812–814
unsaturated, 3-coordinate sulphur
cyclotetrathiazenes, 829–831
cyclotrithiazyls, 825–829
heterocyclothiazenes, 831–834
thiodithiazyls, 822–825
unsaturated, 4-coordinate sulphur
containing NSO$_2$ groups, 842–845
cyclotetrathiazenes, 845
mixed sulphanuric-phosphazene
rings, 838–841
mixed sulphanuric-thiazyl rings,
841–845
sulphanuric halides, 834–838
tetrasulphur pentanitride oxide
anion, 845–846
trioxocyclotrithiazenes, 834–838
Sulphur-oxygen
sulphur trioxides, 871–873

Sulphur—selenium, *see* Selenium—sulphur

Tellurium
cyclophosphates, 676
in Ge rings, 364
in P rings, 447–460
tin heterocycles, 413–415
Tetrasilatetrazanes, 247
Tetrasilatriazanes, 248
Tetrasilazanes, 228–230
Tetrasilazoxanes, 259–261
Tetrasulphur pentanitride,
anion, 814–815
cation, 810–811
Tetrasulphur tetranitride, *see* Sulphur—
nitrogen
Tetrathiadiborinanes, 153
Tetrazaboratrisilacyclooctane, 62
Tetrazaborolines, 53–54
Tetrazadiborines, 55–56
stannylated, 59
Tetrazaphospholenes, 49
Tetrathiazyl cation, 809–810
Thiadiazaphosphetidines, 651–652
Thiadiboretanes, 145–146
Thiadiborolanes, 149–153
Thiasilaborolanes, 157
Thiatetraborinanes, 156
Thiatriborinanes, 146–149, 153–155
Thiatriborolanes, 155–156
Thiazanes, 852–854
Thiazenes
cyclodithiazene, 801–803
cyclotetrathiazenes, 794–801
derivatives, 829–831
cyclotrithiazene, 812–814
heterocyclothiazenes, 815–820, 831–
834
metallocyclothiazenes, 820–822
Thiazyls
cyclopentathiazyl, 811–812
cyclotrithiazyl derivatives, 825–829
halides, 826
sulphanuric-thiazyl rings, 841–845
thiodithiazyl, 806–807, 822–825

Thiazyls (*cont'd*)
  derivatives, 824–825
  oxide, 823–824
  thiotrithiazyl, 806–807, 822–825, *see also* Sulphur nitrides; Thiazenes
Thioboranes, 158
Thioborates, metal, 160
Thioboric acid, 145
Thiodithiazyl cation, 806–807
Thiotrithiazyl cation, 807–809
Tin
  homocycles, 377–380
  -rich cycles and polycycles, 397–399
  in B—N rings, 73
  in P rings, 447–460
  in S—N rings, 320
  in Si rings, 209
  in Si—N rings, 251–252
  in Si—N—O rings, 267–8
  cyclostannazanes, 383–388
  polycyclic Sn—N compounds, 391–395
  polycycles, 391–395
  silastannazanes, 388–391
Tin—nitrogen—oxygen, 394
Tin—nitrogen—silicon, 388–392
Tin—nitrogen—silicon—oxygen, 393
Tin—oxygen, 401–405
  'foil-type' structure, 405

Tin—phosphorus, 395–399
Tin—selenium, 410–413
Tin—sulphur, 405–410
  reaction with atmospheric oxygen, 409
Tin—tellurium, 413–415
Titanium—phosphorus, 449
Triazadiborolidines, 56–59, 84–85
Triazaphosphaborolidines, 63–64
Triazasilaboracyclopentane, 60
Triaziridines, 417–418
Triboroxanes, 111–128
Triselenadiborolanes, 161–162
Triselenatriborinanes, 161
Trisiladiazanes, 247–248
Trisiladiazoxanes, 257–258
Trisiladithianes, 355
Trisilatetrazanes, 247
Trisilazadioxanes, 259
Trisilazanes, 228–230
Tristannadiselenanes, 412–413
Tristannaditelluranes, 414–415
Tristannadithianes, 409
Tristannatriphosphanes, 295–296
Trisulphurtrinitride anion, 812–814
Trithiadiborolanes, 149–153
Trithiatriborinanes, 146–149

Zirconium—phosphorus, 449